REED'S
ADVANCED
ELECTROTECHNOLOGY
FOR ENGINEERS

REED'S
ADVANCED
ELECTROTECHNOLOGY
FOR ENGINEERS

By

EDMUND G. R. KRAAL

D.F.H. (Hons.), C.Eng., A.M.I.E.E., M.I.Mar.E.
*Head of Electrical Engineering and Radio Department
South Shields Marine and Technical College*

PUBLISHED BY THOMAS REED PUBLICATIONS LIMITED
SUNDERLAND, GLASGOW AND LONDON

First Edition - 1970
Second Edition - 1977 (in SI Units)

ISBN 0 900335 41 6

PRINTED BY THOMAS REED AND COMPANY LIMITED
SUNDERLAND, GLASGOW AND LONDON

PREFACE

This book is offered as a companion to Volume 6 (Basic Electrotechnology for Engineers) and covers aspects of theory which are outside the scope of Volume 6. The syllabus followed is close to that of Electrical Engineering for Marine Engineer Cadets (Phase 3) of the Alternative Training Scheme and, as well as covering more fully the requirements of the Department of Trade syllabuses of Electrotechnology for First and Second Class Marine Engineers. It anticipates future extensions of these syllabuses and in this context attention has been given to Brushless A.C. Generator, Excitation Systems for marine alternators and semiconductor theory relating to the diode, transistor and the thyristor. It will also be of value to students who are following a course of study for the Extra First Class Engineers Certificate.

In common with the previous volume in this series, numerous fully-worked problems are included in the text. In addition, test examples and typical examination questions are provided — with solutions, for the student to attempt on his own. As for Volume 6, the subject matter has been treated in the order and in the manner in which it would be taught at a college and this book is thus complementary to lecture notes taken at such a college. It is for this reason that a.c. and d.c. theory is progressed simultaneously and the student is encouraged to advance his "electronics" knowledge at the same time as the individual a.c. machines are studied.

The author again wishes to acknowledge the assistance given by his college colleagues, Mr. J.W. Powell for drawing the diagrams and Mr. T.E. Fox for assisting with the proof reading. Acknowledgement is also made to the Controller of Her Majesty's Stationery Office for permission to reproduce and use the specimen questions from "Examinations of Engineers in the Mercantile Marine" made available by the Department of Trade.

E.G.R. KRAAL

CONTENTS

THE S.I. SYSTEM
PREFIXES, SYMBOLS, MULTIPLES
AND SUB-MULTIPLES

PREFIX	SYMBOL	UNITS MULTIPLYING FACTOR
tetra	T	$\times 10^{12}$
giga	G	$\times 10^{9}$
mega	M	$\times 10^{6}$
kilo	k	$\times 10^{3}$
milli	m	$\times 10^{-3}$
micro	μ	$\times 10^{-6}$
nano	n	$\times 10^{-9}$
pico	p	$\times 10^{-12}$

Examples: 1 megawatt (MW) $= 1 \times 10^{3}$ kilowatts (kW)
1×10^{6} watts (W)
1 kilovolt (kV) $= 1 \times 10^{3}$ volts (V)
1 milliampere (mA) $= 1 \times 10^{-3}$ ampere (A)
1 microfarad (μF) $= 1 \times 10^{-6}$ farad (F)

PHYSICAL QUANTITIES (ELECTRICAL),
SYMBOLS AND UNITS

The table has been compiled from recommendations in B.S. 1991 and the List of Symbols and Abbreviations issued by the I.E.E.

QUANTITY	SYMBOL	UNIT	ABBREVIA–TION OF UNIT AFTER NUMERICAL VALUE
Force	F	newton	N
Work		joule	J
or	W	or	
Energy		newton metre	Nm
Torque	T	newton metre	Nm
Power	P	watt	W
Time	t	second	s
Angular Velocity	ω (omega)	radians per second	rad/s
Speed	N	revolutions per minute	rev/min
	n	revolutions per second	rev/s
Electric charge	Q	coulomb	C
Potential difference (p.d.)	V	volt	V

QUANTITY	SYMBOL	UNIT	ABBREVIATION OF UNIT AFTER NUMERICAL VALUE
Electromotive force (e.m.f)	E	volt	V
Current	I	ampere	A
Resistance	R	ohm	Ω (omega)
Resistivity (specific resistance)	ρ (rho)	ohm metre	Ωm
Conductance	G	siemens	S
Magnetomotive force (m.m.f)	F	ampere-turn	At
Magnetic field strength	H	ampere-turn per metre or ampere per metre	At/m A/m
Magnetic flux	Φ (phi)	weber	Wb
Magnetic flux density	B	tesla	T
Reluctance	S	ampere-turn or amper per weber	At/Wb or A/Wb
Absolute permeability of free space	μ_0 (mu)	henry per metre	H/m
Absolute permeability	μ	henry per metre	H/m
Relative permeability	μ_r	—	—
Self inductance	L	henry	H
Mutual inductance	M	henry	H
Reactance	X	ohm	Ω
Impedance	Z	ohm	Ω
Frequency	f	hertz	Hz
Capacitance	C	farad	F
Absolute permittivity of free space	ϵ_0 (epsilon)	farad per metre	F/m
Absolute permittivity	ϵ	farad per metre	F/m
Relative permittivity (dielectric constant, specific inductive capacity)	ϵ_r	—	—
Electric field strength, electric force		volt per metre	V/m
Electric flux	Ψ (psi)	coulomb	C

QUANTITY	SYMBOL	UNIT	ABBREVIATION OF UNIT AFTER NUMERICAL VALUE
Electric flux density, electric displacement	D	coulomb per square metre	C/m^2
Active power	P	watt	W
Reactive power	Q	volt ampere reactive	VAr
Apparent power	S	volt ampere	VA
Phase difference	ϕ (phi)	degree	°
Power factor (p.f.)	$\cos \phi$	—	—

CHAPTER 1

D.C. MACHINES (i)

Although both the d.c. generator and the d.c. motor were made
the subject of separate chapters in Volume 6, it was frequently
stressed that further work would be necessary to enable the study of
separate but essential points of theory to be completed. The greatest
difficulty encountered by the author, in compiling the volumes on
Electrotechnology, has been in making decisions relating to the
depth and extent of study required for any particular aspects of
theory. Treatment is biased towards the requirements of the marine
engineer and, although this would at first appear to be having a limiting
effect on the scope of an electrical syllabus, it is apparent on further
consideration, that a book devoted to marine electrotechnology must
be related to modern developments and future trends. Within the
last few years marine engineering practice has seen considerable
changes and electrical equipment, which is considered to be both
sophisticated and complex by the electrical specialist, is now being
installed aboard ship as complementary to modern mechanical machinery. Techniques employing static excitation and switching,
methods of effecting and controlling brushless generation, automatic
synchronising, preferential tripping and sequential starting schemes
all require a standard of electrical knowledge which has never before
been required of a marine engineer. The automated engine-room is
now an accepted development for which the engineer must be
trained and conditioned. Although many of the control systems
employed will not be electrically operated, nevertheless complicated
pneumatic/electrical or hydraulic/electrical indicating and alarm systems will be encountered and here again knowledge akin to that of a
specialist may be required. Such knowledge must be based on fundamentals correctly learned, and it is with confidence that the writer
continues with the treatment evolved for the previous book where
essential items are covered under the broad headings of machine
theory, circuit theory or electronics. Treatment, sequence and scope
may differ from that set out in other text books, but this is explained
by the fact that, an attempt is made to achieve uniform progress in
all relevant branches of study thus allowing integration with college
courses and also facilitating resumption of study at a definite starting
point after a period of lapse.

TESTING OF D.C. MACHINES

Before considering testing methods in detail, we examine briefly the necessity for and types of tests usually made by an electrical machine manufacturer.

If the development of a range of small machines is undertaken, this would commence by a prototype being built, which would incorporate all the features considered desirable. This machine would then be subjected to a series of tests in the works by direct loading, in accordance with a programme set out by the design and development team. The results would be used to check the effectiveness of the design since the machine output would be measured and its overall efficiency. The final temperature rise, the behaviour on load, *i.e.* commutation, vibration, noise and response to control of speed and excitation would also be checked. The test results would also be used to confirm or amend the emperical design formulae and would be examined in detail to allow appropriate information to be extracted. Only when the prototype is proved to be efficient and satisfactory for the rating and purpose envisaged will production commence. Again at the early production stages, minor modifications may be introduced to speed manufacture and assist mass production techniques. The first machine offered for sale will again be comprehensively tested in order to provide guaranteed performance figures and results for type tests which are made on specimen sample machines as they come off the production line.

The development procedure described above will apply to sizes of machine which are capable of being mass produced and are not required to perform very special duties. A machine may be considered to be *a special* because of its size, operating requirements, dimensions or enclosure which warrants different constructional methods. There would be no prototype and the designers will need to rely on their experience and skill for building such a machine to a customer's specification. Only limited testing may be possible at the manufacturer's works because of the size of the machine or the specialities of its operating requirements and assessment tests have therefore been devised. Such tests can be made without difficulty and the results used to check the design before the machine is installed. The final testing, certification and acceptance tests would then be made on site under the correct operating conditions.

OUTPUT

The output of a machine can be measured by direct loading. For a motor this can be undertaken by a brake, calibrated generator or electrical dynamometer. For a generator, the load can be a water resistance tank or an appropriately calibrated variable resistor. Measurement of the ouput presents no difficulty, since the correct instruments are readily available, but the waste of energy and its dissipation are factors which are considered when large machines are to be tested.

EFFICIENCY

The efficiency of an electrical machine may be expressed as

$$\text{Efficiency} = \frac{\text{Output}}{\text{Input}} \text{ or } \frac{\text{Input} - \text{Losses}}{\text{Input}} \text{ or } \frac{\text{Output}}{\text{Output} + \text{Losses}}$$

Efficiency (η) is usually expressed as a percentage and for a generator it can be written as

$$\eta_G = \frac{\text{Output } (W \text{ or } kW)}{\text{Output } (W \text{ or } kW) + \text{Losses } (W \text{ or } kW)}$$

Similarly for a motor

$$\eta_M = \frac{\text{Input } (W \text{ or } kW) - \text{Losses } (W \text{ or } kW)}{\text{Input } (W \text{ or } kW)}$$

LOSSES

The losses which occur in an electrical machine can be considered under two main headings (a) *Electrical Losses*. (b) *Mechanical Losses*.

(a) ELECTRICAL LOSSES (P_E). These comprise the *Copper Losses* (P_{Cu}) and the *Iron Losses* (P_{Fe}).

Copper losses occur in the armature conductors and connections due to the passage of current through the ohmic resistance of the material and can readily be determined from the expression $I_a^2 R_a$. A copper loss also occurs in the field windings and is usually some 30 to 50 per cent of the full-load losses. The copper losses are usually kept to a minimum by using the maximum amount of copper on the machine consistent with a commercially economic design. It is quite usual to calculate this loss at the working temperature.

Iron losses occur in the ironwork of the armature, pole-tips and other parts of the magnetic circuit which are subjected to cyclic magnetisation or flux changes. They consist of a *Hysteresis Loss* (P_{Hy})

and an *Eddy-Current Loss* (P_{EC}) which were dealt with in some detail in Volume 6, Chapter 12. The principal points of importance resulting from the study were that (i) Hysteresis Loss is proportional to flux density to the power 1·6 and to speed or frequency. Thus $P_{Hy} \propto B_m^{1·6} N$. (ii) Eddy-Current Loss is proportional to flux density squared and to speed squared or frequency squared. Thus $P_{EC} \propto B_m^2 N^2$.

(b) MECHANICAL LOSSES. (P_M). These comprise the *Windage* and *Friction* (bearing and brush) *Losses.*

Since ventilation is necessary to reduce the temperature rise of electrical machines, the amount of Windage Loss is a matter of compromise. Some machines have a comparatively smooth armature with little fanning effect and most of the machine heat is dissipated through the yoke. This is particularly the case for low-speed machines which rely on radiation for heat loss and are of basically "open" construction. High-speed machines have smaller dimensions and rely on forced ventilation, *i.e.* have an associated fan. The Windage Loss is proportional to the cube of the speed but is comparatively small even when a fan is fitted.

Bearing friction is minimised by using the correct type of bearing for the duty to be undertaken by the machine. Small and medium size high-speed machines are invariably provided with ball or roller bearings. Low-speed machines of these sizes, use sleeve bearings with ample lubrication and for marine work this type of bearing is favoured because of its superior performance in withstanding vibration. For very large machines, special bearing arrangements are devised which frequently involve forced high-pressure lubrication. The rotor weight and speed, type of drive, position and mounting of the machine, are all factors which decide the kind of bearings to be used. Bearing friction loss is roughly proportional to speed.

Brush friction depends on the coefficient of friction, the total pressure and the peripheral speed of the commutator. It should be appreciated that a particular grade of brush is chosen for a machine in accordance with the design requirements and this should not be altered.

Summarising we see that the Mechanical Losses are proportional to speed and attention will be drawn to this relationship whenever it is considered to be necessary.

Example 1. A 220 V shunt motor takes 10·25 A on full load. The armature resistance is 0·8Ω and the field resistance is 880Ω. The losses due to friction, windage and the iron amount to 150 W. Find the output power and the efficiency of the motor on full load.

Motor input on full load $= 220 \times 10{\cdot}25 = 2255$ W

$$\text{Field current} = \frac{220}{880} = 0{\cdot}25 \text{ A}$$

Armature current $= 10{\cdot}25 - 0{\cdot}25 = 10$ A
Armature copper loss on full load $= 10^2 \times 0{\cdot}8 = 80$ W
Field copper loss $= 220 \times 0{\cdot}25 = 55$ W
Friction, windage and iron loss $= 150$ W
Total losses $= 80 + 55 + 150 = 285$ W
Full-load output $=$ input $-$ losses
$= 2255 - 285 = 1970$ W
$= 1{\cdot}97$ kW

$$\text{Full-load efficiency} = \frac{1970}{2255} = 0{\cdot}874 \text{ or } 87{\cdot}4 \text{ per cent.}$$

ROTATIONAL AND CONSTANT LOSSES (P_R and P_C). Since the Mechanical Losses, both friction and windage, are proportional to speed and the Iron Losses are proportional to speed if the flux density is kept constant, the term *Rotational Loss* (P_R) can be used to include both P_M and P_{Fe}. A further description of the losses in a machine are the *Constant Losses* (P_C) and the variable losses. If the losses are examined in this manner, then it is seen that the field copper losses are themselves constant for most working conditions and can be included with the rotational losses for a constant speed to give P_C. The variable losses are seen to be the armature copper losses and an efficiency expression can now be evolved thus

$$\text{Efficiency} = \frac{\text{Output}}{\text{Output} + \text{Losses}} \text{ or, for example, as a generator}$$

$$\eta_G = \frac{\text{Output}}{\text{Output} + \text{Arm.Cu Loss} + \text{Constant Loss}}$$

CONDITION FOR MAXIMUM EFFICIENCY. If the field copper loss is included in the constant loss value and consequently the field current can be considered negligible compared to the armature current on load, then the efficiency expression can be written as set out above. Thus for a generator

$$\eta_G = \frac{VI_a}{VI_a + I_a^2 R_a + P_C}$$

Dividing by VI_a

$$\eta_G = \frac{1}{1 + \dfrac{I_a R_a}{V} + \dfrac{P_C}{VI_a}} \quad \text{or} \quad \frac{1}{\eta_G} = 1 + \frac{I_a R_a}{V} + \frac{P_C}{VI_a}$$

If calculus is used and the proof is taken in a simple manner, then

$$\frac{1}{\eta_G} = 1 + \frac{I_a R_a}{V} + \frac{P_C I_a^{-1}}{V} \quad \text{and differentiating with respect to } I_a$$

Then $\quad \dfrac{d\left(\dfrac{1}{\eta}\right)}{dI_a} = 0 + \dfrac{R_a}{V} - \dfrac{P_C I_a^{-2}}{V}$

The above expression will be zero when $\dfrac{R_a}{V} = \dfrac{P_C I_a^{-2}}{V}$ and the

minimum value of $\dfrac{1}{\eta_G}$ is obtained.

If therefore $\dfrac{R_a}{V} = \dfrac{P_C}{VI_a^2}$ then by simplifying, $P_C = I_a^2 R_a$.

Thus minimum value of $\dfrac{1}{\eta_G}$ or maximum value of η_G occurs when

Copper loss (in the armature) = Constant loss. As $I_a^2 R_a = P_C$

then $\quad I_a = \sqrt{\dfrac{P_C}{R_a}}$ giving the current value for maximum efficiency.

It should be noted that the field copper loss is included in P_C.

The same condition for maximum efficiency can be devised for the motor using a similar method.

Example 2. A 220 V shunt generator is rated to have a full-load current of 200 A. Its armature resistance is 0·06 Ω and its field resistance is 55 Ω. The rotational losses, at correct speed and excitation, are measured to be 3 kW. Find the output power rating of the prime-mover and also the load current for maximum efficiency.

$$\text{Field current} \ = \ \frac{220}{55} \ = \ 4\text{A}$$

$$\begin{aligned}
\text{Armature current} \ &= \ 200 + 4 \ = \ 204 \text{ A} \\
\text{Armature copper loss} \ &= \ 204^2 \times 0 \cdot 06 = 204 \times 12 \cdot 24 = 2497 \text{ W} \\
\text{Field copper loss} \ &= \ 220 \times 4 \ = \ 880 \text{ W} \\
\text{Rotational loss} \ &= \ 3000 \text{ W} \\
\text{Constant loss} \ &= \ \text{Rotational Loss} + \text{Field Cu Loss} \\
\therefore P_C \ &= \ 3000 + 880 \ = \ 3880 \text{ W} \\
\text{Required generator input} \ &= \ \text{Output} + \text{Arm.Cu Loss} + \text{Constant} \\
& \qquad \text{Loss} \\
&= \ 220 \times 200 + 2497 + 3880 = 50\ 377 \text{ W}
\end{aligned}$$

Thus prime-mover rating   =   50·38 kW

$$\begin{aligned}
\text{Current for maximum efficiency} \ &= \ \sqrt{\frac{3880}{0 \cdot 06}} \ = \ 10^2 \sqrt{\frac{38.8}{6}} \\
&= \ 10^2 \sqrt{6 \cdot 466} \ = \ 254 \text{ A}
\end{aligned}$$

This is I_a, load current would be 254 $-$ 4 = 250 A.

STRAY LOSSES. When a machine is operating on load, small losses additional to those already enumerated are known to occur. They are mainly due to an increase of iron loss caused by distortion of the field by the armature reaction effects. Unless mention is made of such losses they can be neglected.

TESTING METHODS

The efficiency and output of an electrical machine can be determined by one of three methods all of which will be considered below in detail. Direct-Loading Methods are well known to the mechanical engineer who uses various types of brake for loading prime-movers such as I.C. Engines and Turbines. Electrical brakes, especially in the dynamometer form, are being used to an ever increasing extent and it is hoped that the brief descriptions given below, will assist in the understanding of any electrical loading arrangement which may have been encountered in the past. Assessment Tests allow an estimation of output and efficiency but the performance on load cannot be checked because the results are obtained from no-load running conditions. Regenerative-Testing Methods, often referred to as "back-to-back" methods are favoured because the full-load performance of large machines can be checked with confidence, whilst testing is made with minimum of power waste.

DIRECT LOADING. A brake is used to measure the output power of a machine and can be applied to a motor which is operated at the correct voltage, speed and load. If the brake measurements are made in newton metres and the speed in rev/min, then the output power can be determined from the well-known expression $\dfrac{2\pi NT}{60}$

N is in revolutions per minute and T in newton metres.

The efficiency is obtained from $\dfrac{2\pi NT}{60\,VI}$.

(a) The Friction Brake. This can be of the rope or band and drum type and may take the forms illustrated by the diagrams of Fig. 1a and 1b.

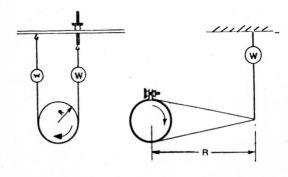

Fig. 1a   Fig. 1b

For type (a) the force, recorded by spring balances and acting on the drum equals $(W - w)$ newtons. If the radius of the drum is R metres then the torque T would be given by $(W - w)\,R$ newton metres.

For type (b) the force, recorded by a spring balance is W newtons and if the radius of the Proney-brake arm is R metres, then the torque T is equal to WR newton metres.

The disadvantages of the friction brake are the difficulties of dissipating the heat generated at the drum and that an excessive load is placed on the bearing nearest the brake which may give a slightly lower efficiency figure. The method is confined to small motors and is fast being superseded by the others described below.

Example 3. In a brake-test on a motor, the effective load on the brake-drum was 222·7N, the effective diameter of the drum 0·5 m and the motor speed 960 rev/min. Under these conditions the input to the motor was 30 A at 230 V. Calculate the output of the motor and its efficiency.

$$\text{Braking torque on motor} = 222 \cdot 7 \times 0 \cdot 25 \text{ newton metres}$$

$$= 55 \cdot 68 \text{ Nm}$$

$$\text{Motor power output} = \frac{2 \times 3 \cdot 14 \times 960 \times 55 \cdot 68}{60}$$

$$= 3 \cdot 14 \times 32 \times 55 \cdot 68$$

$$= 5594 \cdot 7 \text{ W or } 5 \cdot 59 \text{ kW}$$

$$\eta = \frac{5595}{30 \times 230}$$

$$\text{or Efficiency} = 0 \cdot 81 \text{ or } 81 \text{ per cent.}$$

(b) The Calibrated Generator. This is a generator which is precalibrated by careful testing and is retained in the test-room. Its output is fed into a resistive load, where it can be dissipated as heat, or it can be fed into a common busbar system. The diagram of Fig. 2 shows the motor to be tested, direct coupled to the calibrated generator and the electrical arrangement from which the following can be deduced.

Let P be the power transmitted through the coupling.

Then $\eta_M = \dfrac{P}{I_M V_M}$ and $\eta_G = \dfrac{I_G V_G}{P}$ Whence $\eta_M \eta_G = \dfrac{I_G V_G}{I_M V_M}$

or $\eta_M = \dfrac{I_G V_G}{I_M V_M \eta_G}$

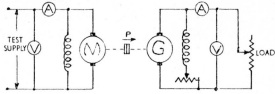

Fig. 2

The efficiency of the generator is known from the calibration tests
and can be determined for any output from the calibration graph
which is shown by Fig. 3. The appropriate value is substituted in the
efficiency expression.

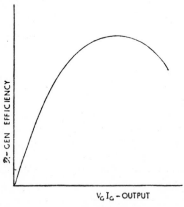

Fig. 3

The advantage of this method is that the energy output can be
put to some useful purpose, *i.e.* I.C. engine-makers frequently use the
generator initially as a motor for the running-in of engines, taking the
supply from other engines which are being loaded and are feeding
into common busbars. Alternatively the energy being supplied during
load tests can be fed back into the factory and is thus used effectively.

(c) The Electrical Dynamometer. This machine is constructed as a
generator or motor and, as shown by the diagrams of Fig. 4, is
mounted on trunnion bearings. Since the stator of the machine will

tend to rotate, when the armature is loaded, due to the magnetic coupling between the field and the armature, then a restraining torque is obviously required to provide the reaction which normally occurs at the feet of a loaded machine. A torque-arm is fitted and the reaction is measured by a spring balance and here $T = WR$ newton metres. The output of the motor or the input to the generator being tested can be measured directly during the test and no calibration curve is required.

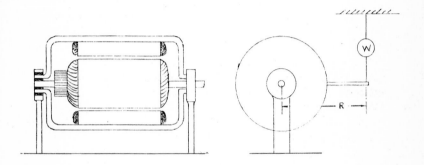

Fig. 4

The dynamometer is constructed as a motor to be retained in the test room for the testing of generators or as a generator for the testing of motors and I.C. engines. Its output as a motor or its input as a generator is checked and used in conjunction with the measurements of the machine to which it is coupled. The electrical energy generated during the testing can be used to useful purpose as for the calibrated generator described above. If the complications necessary to effectively use or feed back this energy into the "mains" are not deemed worthwhile, then the dynamometer can be modified to waste the heat. For such an arrangement, the armature is constructed like the squirrel cage of an induction motor and water is arranged to flow through the machine. The bars of the cage which are heated by the eddy currents, are thus cooled by the water which is flowing continuously and is allowed to run to waste.

The advantages of all the direct-loading methods are that performance of the machine on load can be checked. Not only are output and the efficiency determined but the commutation and temperature rise can be observed and other relevant points of the design specification noted.

ASSESSMENT TESTS. Two methods of making these tests are considered. The first is well-known and is of considerable importance. It is named after Sir James Swinburne, a pioneer in the development of electrical machinery, and is the best example of the technique of assessing efficiency without actually loading a machine. The method of making the test and the theory involved should be fully appreciated. The second method considered is not used to a great extent but is useful for determining the magnitude of the various component losses which occur in a machine. The tests are made with the aid of a small auxiliary motor and the waste of large amounts of power is not involved. The limitations of the assessment methods of testing have already been mentioned and should be remembered when the procedure is undertaken and calculations made.

(a) The Swinburne Method. This is particularly suitable for large machines where loading either as a generator or a motor would present difficulty. To make the test, the machine is run light as a motor irrespective of its final function. The armature and field resistances are measured before the no-load test and the ambient temperature is noted. During the test, the supply voltage and input current are recorded after the speed has been adjusted to the value at which the machine is required to operate on load. The efficiency is then assessed as detailed below. The diagram of Fig. 5 shows the arrangement.

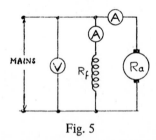

Fig. 5

When the machine is run light no output is taken from the shaft and it follows that only sufficient input power is required to supply the no-load losses of the machine which are copper losses and the rotational losses.

Thus Input Power

= No-load Copper Losses + Iron Losses + Mechanical Losses

= No-load Arm. Cu Loss + Field Cu Loss + Rotational Loss

or $VI_0 = I_{a_0}^2 R_a + I_{f_0}^2 R_f + P_R$

Whence $P_R = VI_0 - I_{a_0}^2 R_a - I_{f_0}^2 R_f$

Once the Rotational Loss has been determined for a particular speed and field excitation condition, it can be used to assess the efficiency at any load, the speed and excitation being assumed constant.

Thus for a generator we can use the expression

$$\text{Efficiency} = \frac{\text{Output}}{\text{Output} + \text{Losses}}$$

$$= \frac{\text{Output}}{\text{Output} + \text{Arm. Cu Loss} + \text{Field Cu. Loss} + \text{Rotational Loss}}$$

or $\eta_G = \dfrac{VI}{VI + I_a^2 R_a + I_f^2 R_f + P_R}$ where I_a and I_f are the load values.

It should be noted that for a shunt-connected machine $I_f = I_{f_0}$ and as a result the calculation can be made easier by determining the Constant Loss. The procedure for making a prediction of efficiency is best illustrated by examples.

Example 4. A 50 kW, 500 V shunt generator is tested by the Swinburne method. When run light as a motor it takes a no-load current of 10·1 A at the correct voltage and speed. $R_a = 0·25\ \Omega$ and $R_f = 500\ \Omega$.

Assess the efficiency at full load as a generator.

No Load.   Supply current 10·1 A $I_{f_0} = \dfrac{500}{500} = 1\ \text{A}$

So $I_{a_0} = 10·1 - 1 = 9·1\ \text{A}$
Input = Cu Losses + Rotational Loss
or 500 × 10·1 = 9·1² × 0·25 + 500 × 1 + P_R
Whence $P_R = 5050 - 20·7 - 500 = 5050 - 520·7 = 4529\ \text{W}$

Full Load.   Line current $= \dfrac{50\ 000}{500} = 100\ \text{A}$

$I_f = \dfrac{500}{500} = 1\ \text{A}$ $I_a = 100 + 1 = 101\ \text{A}$

$$\therefore \ \eta = \frac{50\,000}{50\,000 \ + \ (\text{Cu. Loss} + P_R)}$$

$$= \frac{50\,000}{50\,000 \ + \ (101^2 \times 0.25) \ + \ (1 \times 500) \ + \ 4529}$$

$$= \frac{50}{50 \ + \ 2.55 \ + \ 0.5 \ + \ 4.529} \quad (\text{worked here in } kW)$$

$$= \frac{50}{57.58} \ = \ 0.868 \text{ or Efficiency 86.8 per cent.}$$

This problem can also be solved by finding the Constant Loss Thus:

No Load.   Input = Arm. Cu Loss + Constant Loss
or Constant Loss = $5050 - (9.1^2 \times 0.25) = 5050 - 20.7$
$P_C = 5029.3$ W or 50.3 kW

Full Load. $\eta = \dfrac{\text{Output}}{\text{Output} \ + \ \text{Arm. Cu Loss} \ + \ P_C}$

$$= \frac{50}{50 \ + \ 2.55 \ + \ 5.03}$$

$$= \frac{50}{57.58} \text{ or Efficiency 86.8 per cent.}$$

Example 5. A 500 V shunt motor takes a current of 5 A on no load. The resistances of the armature and field circuits are 0·22 Ω and 250 Ω respectively. Estimate the efficiency when the motor current is 100 A. State the assumption made in estimating the efficiency.

No Load. $I_{f_0} = \dfrac{500}{250} = 2$ A   Supply current = 5 A

$\therefore \ I_{a_0} = 5 - 2 = 3$ A
$P_R =$ Input $-$ Cu Losses
$= 2500 - (3^2 \times 0.22) - (500 \times 2)$
$= 2500 - 1.98 - 1000 = 2500 - 1002$
$= 1498$ W

Full load. $I_f = \dfrac{500}{250} = 2\,A \quad I_a = 100 - 2 = 98\,A$

$\eta = \dfrac{\text{Input} - \text{Losses}}{\text{Input}}$

$= \dfrac{(500 \times 100) - (98^2 \times 0{\cdot}22) - (500 \times 2) - 1498}{500 \times 100}$

$= \dfrac{50 - 2{\cdot}113 - 1{\cdot}0 - 1{\cdot}498}{50}$ (worked here in kW)

$= \dfrac{45{\cdot}389}{50} = 0{\cdot}9078$ or Efficiency 90·78 per cent.

For this problem here again P_C could have been determined making for an easier solution. It should be noted that for the Swinburne test, a closer approximation to the actual efficiency is obtained if the R_a and R_f values are raised to a "hot" value, say a 50°C rise. This should be done if indications of the required adjustments are given in the problem. An example to this effect is given amongst the Practice Examples at the end of the chapter.

In answer to the query regarding the assumption made in solving the above problems, no speed and flux variation between the no-load and full-load tests is allowed. Thus constant iron and mechanical losses are assumed.

(b) Summation of Losses Method. (by Auxiliary Motor). The diagram of Fig. 6 shows the arrangement with two machines coupled together by a belt. In practice if any doubts, as to belt-slip or losses due to belt friction are to be avoided, the machines can be fixed in line and direct coupled. The smaller machine is the auxiliary motor and the tests are made in stages as shown.

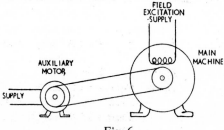

Fig. 6

Let P_0 be the input power to the auxiliary motor when it is running light and at the speed required to drive the large machine at its operational speed of N rev/min.

Let P_1 be the input power to the auxiliary motor when coupled to the large main machine and driving its armature in an unexcited field at its operational speed of N rev/min.
Then $P_1 - P_0 = P_M$ the Mechanical losses (windage and friction) of the large machine.

Let P_2 be the input power to the auxiliary motor, coupled to the main machine and driving its armature in an excited field at a speed of N rev/min.
Then $P_2 - P_1 =$ Iron Losses (P_{Fe}) of the main machine under the correct testing conditions of speed and flux.
Note that since P_M and P_{Fe} constitute P_R then the Rotational Loss (P_R) can be determined from $P_2 - P_0$.

If R_a and R_f are measured then the efficiency of the main machine can be assessed for any loading in a manner similar to that done for the Swinburne Test. This method of testing, using an auxiliary motor, has further interests for the designer since the effects of small alterations can be determined. Thus input power readings could be taken with the brushes lifted off and again with them at the correct pressure. The loss due to brush friction could then be obtained. The tests could be repeated with alterations of bearings, ventilation openings, fans, *etc.*, and data obtained for improving performance. The rig-up is also useful for determining the constituent parts of the Iron Losses and procedure would be as follows.

Separation of Iron Losses. From Volume 6, Chapter 12, it was seen that iron losses were made up of Hysteresis Loss and Eddy-Current Loss or $P_{Fe} = P_{Hy} + P_{EC}$. It was also shown that $P_{Hy} = K_H B_m^{1 \cdot 6}$ and $P_E = K_E B_m^2 f^2$. If the flux-density B_m was to be kept constant during the testing, the iron losses could be written as $P_{Fe} = Hf + Ef^2$ or for a machine as $P_{Fe} = HN + EN^2$ where HN represents the hysteresis loss and EN^2 the corresponding eddy-current loss. Examination of the equation $P_{Fe} = HN + EN^2$ shows that a solution is possible if values of P_{Fe} can be obtained at two speeds. The tests are therefore made by keeping the flux density constant and varying the speed at which the armature is driven. If the auxiliary motor method, as described above, is used the iron loss can be measured at normal speed and excitation. With the excitation kept constant, the testing is repeated at half or double speed and the solution of the equation is then possible, as is illustrated by the following example.

Example 6. A test is made on a 230 V, shunt motor in order to determine the constituent parts of the iron loss. The shunt field is separately excited at 230 V and its excitation is kept constant. Measurements were made of the iron losses at full speed and half speed. Thus at 1500 rev/min the iron losses were 1100 W and at 750 rev/min these were 450 W. Find the magnitude of the hysteresis loss and the eddy current loss at normal speed, *i.e.* 1500 rev/min.

Substituting in the equation $P_{Fe} = HN + EN^2$ we have

$$1100 = HN + EN^2 \quad \dots \quad \dots \quad \dots \quad \dots \quad \dots \text{ (a)}$$

$$450 = \frac{HN}{2} + E\left(\frac{N}{2}\right)^2 \quad \dots \quad \dots \quad \dots \quad \dots \text{ (b)}$$

or

$$1100 = HN + EN^2 \quad \dots \quad \dots \quad \dots \quad \dots \quad \dots \text{ (a)}$$

and

$$1800 = 2HN + EN^2 \quad \dots \quad \dots \quad \dots \quad \dots \quad \dots \text{ (c)}$$

Subtracting (a) from (c) we have

$$700 = HN \quad \text{or the Hysteresis Loss is 700 W}$$

The Eddy-Current Loss is therefore $1100 - 700 = 400$ W

At half speed these losses would be $\dfrac{700}{2} = 350$ W for Hysteresis Loss

and $\dfrac{400}{4} = 100$ W for Eddy-Current Loss.

REGENERATIVE TESTS. The advantages of this method of testing have already been mentioned and the technique is now considered in detail. The direct loading of a machine is obviously the best method of checking performance because its efficiency, commutation and temperature rise can be noted. The difficulties of loading are overcome by operating the machine to be tested in conjunction with a similar machine which either provides the electrical power or absorbs it in accordance with the arrangement. Basically the two machines operate "back-to-back" and only sufficient energy is taken from the mains to supply the losses of both machines. The method, readily adapted to shunt machines and attributed to John Hopkinson, a Professor of Electrical Engineering at London University, is considered first. Appropriate circuit modifications allow the technique to be adapted for compound machines. The testing of series machines presents complications but here again this can be achieved, provided appropriate circuit modifications and protective arrangements are made.

(a) The Hopkinson Method. If the machines are identical, the test is most readily performed but it can be made conveniently with machines which are basically similar and of approximately the same rating. The machines are mechanically and electrically connected as is shown by the diagram of Fig. 7.

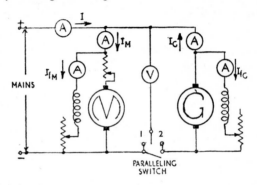

Fig. 7

The machine which is to operate as a motor is started by cutting out the starting resistance and the speed is adjusted by the variation of the field regulator to control I_{f_M}. The generator will self excite and its voltage can be controlled by the regulator varying the current I_{f_G}.

The change-over switch is frequently operated from position 1 to position 2 and the generated voltage E_G is thus adjusted and compared with the mains voltage V. When $E_G = V$, the paralleling switch is closed since no current can circulate between G and M for this condition. If E_G is now increased by raising I_{f_G}, the generator can be loaded, i.e., made to supply current. The motor will consequently be loaded and will require an increased current which can be obtained from the generator. By manipulating the field regulators of both machines the condition can be achieved when either the motor or generator is fully loaded. The power taken from the supply will be a minimum and will then supply the losses of both machines. Under this condition

$$VI = P_{R_M} + P_{R_G} + I^2{}_{a_M}R_{a_M} + I_{f_M}V + I^2{}_{a_G}R_{a_G} + I_{f_G}V$$
where $I_{a_G} = I_G + I_{f_G}$ and $I_{a_M} = I_M - I_{f_M}$

From the above $P_{R_M} + P_{R_G}$ can be found and if they are assumed to be similar then the rotational loss of each machine can be found. Resistance R_a and R_f are usually measured hot after the test and the efficiency can be calculated for any assigned load under normal working conditions either as a generator or motor. The method is illustrated by Example No. 7.

It should be noted that the approximate efficiency can be found on the test-bed for checking purposes by the following deduction.

Let P be the power transmitted through the coupling between motor and generator.

Then $\eta_M = \dfrac{P}{VI_M}$ and $\eta_G = \dfrac{VI_G}{P}$

$\therefore \; \eta_G \, \eta_M = \dfrac{VI_G}{P} \times \dfrac{P}{VI_M} = \dfrac{I_G}{I_M}$ If the efficiency of each ma-

chine is assumed equal then $\eta^2 = \dfrac{I_G}{I_M}$ or $\eta = \sqrt{\dfrac{I_G}{I_M}}$.

Example 7. Two shunt motors on a Hopkinson Test take a current of 15 A from 200 V mains. The current to the motoring machine is 100 A and the field currents are 3 A and 2·5 A. If the resistance of each armature is measured to be 0·05 Ω, estimate the efficiency of each machine under the particular loading conditions of the test.

Total Input $= 15 \times 200 = 3000$ W. Since the generated e.m.f. of the generator must be higher than the back e.m.f. of the motor and both armatures are rotating at the same speed, then the field current of 3 A must refer to the generator and the 2·5 A to the motor.

Since $I_M = 100$ A Then $I_{a_M} = 100 - 2\cdot5 = 97\cdot5$ A

and $I_{a_G} = (100 - 15) + 3 = 88$ A

$\therefore \quad 3000 = 2\,P_R + $ Motor Arm. Cu loss + Motor Field Cu loss + Generator Arm. Cu loss + Generator Field Cu loss

or $3000 = 2\,P_R + (97\cdot5^2 \times 0\cdot05) + (2\cdot5 \times 200)$
$\qquad\qquad + (88^2 \times 0\cdot05) + (3 \times 200)$

$\qquad\quad = 2\,P_R + 475\cdot3 + 500 + 387 + 600$

So $2P_R = 3000 - 1962 \cdot 3 = 1037 \cdot 7 \, \text{W}$

or $P_R = 518 \cdot 8 \, \text{W}$

Then $\eta_M = \dfrac{\text{Input} - \text{Losses}}{\text{Input}}$

$= \dfrac{(100 \times 200) - (518 \cdot 8 + \text{Arm.Cu Loss} + \text{Field Cu Loss})}{20\,000}$

$= \dfrac{20\,000 - (518 \cdot 8 + 475 \cdot 3 + 500)}{20\,000} = \dfrac{20\,000 - 1494 \cdot 1}{20\,000}$

$= \dfrac{18\,505 \cdot 9}{20\,000} \times 100$ or Efficiency $= 92 \cdot 4$ per cent.

and $\eta_G = \dfrac{\text{Output}}{\text{Output} + \text{Losses}}$

$= \dfrac{85 \times 200}{(85 \times 200) + (518 \cdot 8 + \text{Arm.Cu Loss} + \text{Field Cu Loss})}$

$= \dfrac{85 \times 200}{(85 \times 200) + (518 \cdot 8 + 387 + 600)} = \dfrac{17\,000}{17\,000 + 1505 \cdot 8}$

$= \dfrac{17\,000}{18\,505 \cdot 8} \times 100$ or Efficiency $= 91 \cdot 7$ per cent.

Check.

Approximate efficiency of each machine $= \sqrt{\dfrac{85}{100}} = \sqrt{0 \cdot 85}$

or $\eta = 0 \cdot 922$ or $92 \cdot 2$ per cent.

The Hopkinson Test can be made for compound machines if attention is given to the correct connection of the series fields. Thus if two compound generators are to be tested then since one machine is to "motor", the series field must be reverse connected because the direction of armature current is reversed and differential compounding for the motor will result unless this modification is made. The motor, if differentially connected, could not be loaded effectively and correct testing would not be possible. Assuming a satisfactory load test to have been achieved then the rotational loss and efficiency can be determined as described for the shunt machines, due regard being paid to the additional series field copper losses.

(b) Field's Method. Due to the tendency for a series motor to race on no-load and the difficulty of operating a series generator in parallel with a constant voltage supply, the regenerative method of testing is not readily possible without taking elaborate precautions and giving the test-rig continual attention. In its essential form Field's test is not strictly a regenerative method but is invariably made in the form described. Two similar series motors are mechanically coupled, one machine acting as a motor and the other as a separately excited generator. To achieve this both series fields are connected in series with the motor armature and the generator output is absorbed in a resistance load with no switches in this circuit to eliminate the possibility of the load being detached. The method of calculating the efficiency is illustrated by the example.

Example 8. The following results were obtained from a Field's test on two series motors. Motor current 20 A. Generator output current 17·5 A. Generator voltage 160 V. Supply voltage 220 V. Resistance of each armature 0·6 Ω. Resistance of each series field 0·4 Ω. Estimate the full-load efficiency of a machine as a motor.

Input to test rig $= 220 \times 20 = 4400$ W

Output of generator $= 160 \times 17\cdot5 = 2800$ W

Total losses $P = 4400 - 2800 = 1600$ W

Motor armature copper loss $= 20^2 \times 0\cdot6 = 240$ W

Motor and generator field copper losses $= 20^2 \, (2 \times 0\cdot4)$
$$= 400 \times 0\cdot8 = 320 \text{ W}$$

Generator armature copper loss $= 17\cdot5^2 \times 0\cdot6 = 183\cdot75$ W

Rotational loss of 2 machines $= 1600 -$ Total Copper Losses

or $2 P_R = 1600 - (240 + 320 + 183\cdot75)$

$= 1600 - 743\cdot75 = 856\cdot25$ W

and $P_R = 428\cdot125$ W   Then $\eta_M = \dfrac{\text{Input} - \text{Losses}}{\text{Input}}$

Now voltage drop across the generator series field = $20 \times 0.4 = 8$ V

$\therefore$ Voltage applied to motor $= 220 - 8 = 212$ V

Input $= 212 \times 20 = 4240$ W

$$
\begin{aligned}
\text{Losses} &= P_R + \text{Motor Arm.Cu Loss} + \text{Field Cu Loss} \\
&= 428{\cdot}125 + 20^2(0{\cdot}6 + 0{\cdot}4) \\
&= 428{\cdot}125 + 400 = 828{\cdot}125 \text{ W}
\end{aligned}
$$

$$
\text{or } \eta_M = \frac{4240 - 828{\cdot}125}{4240} = \frac{3411{\cdot}875}{4240} = 0{\cdot}8047
$$

or Efficiency = 80·47 per cent.

D.C. GENERATORS IN PARALLEL

The need for and advantages of operating generators in parallel are well known to serving marine engineers and here we consider the theory for satisfactory running and load sharing. The technique of actually paralleling two d.c. generators involves only basic conditions and safeguards are built into the switchboard to ensure a correct sequence of operations. Thus "to put a machine on the bars", we know that the generator must be run up to its correct speed and its voltage adjusted until it equals that of the busbars. The circuit-breaker of the "incoming" machine can then be closed, when it can be said to "float on the bars". If the excitation is next increased by operating the field regulator, an ouput current will be noted which indicates that the generator is now taking load. The busbar voltage will also be seen to increase and adjustment of the field regulators of all the machines in parallel will be necessary before the correct busbar voltage and sharing of load between the generators can be achieved.

Essential conditions for the correct paralleling of d.c. generators are assumed in the above statements. These infer that the polarity of the incoming machine is correct *i.e.* it corresponds to that of the busbars and this is ensured by the use of moving-coil voltmeters. Thus if the polarity had reversed, because of some fault condition, then the "machine" voltmeter would indicate this or no apparent build-up of voltage. The circuit-breakers would not therefore be closed and the condition would then be investigated further. Again as compound generators are almost invariably concerned then the use of an *equalising* connection will be involved. The need for such a connection must be made before the incoming machine circuit-breaker is closed. In earlier days such a connection took the form of a separate single-pole switch and the drill was to close this first. Modern equipment ensures a mechanical interlock between the switch and the circuit-breaker and the most usual arrangement is a three-pole circuit-breaker. The equalising connection is usually the centre pole and this

is given a "lead" *i.e.* its contacts "make" first, before those of the
+ ve and − ve poles. The condition of making the equalising connection first is thus automatically assured.

PARALLEL OPERATION

The possibility of d.c. generators operating effectively in parallel
is best investigated by considering their load characteristics. The
reader is reminded that for the shunt-connected generator, terminal
voltage falls with increasing load current or the characteristic is a
falling one. For the series machine terminal voltage rises with load
current until saturation of the magnetic system occurs. Its characteristic may therefore be described as a rising one. For compound generators, the shape of the characteristics is decided by the strength of
the series field, the existence of which requires special attention when
effecting parallel operation.

(a) Shunt Generators. Such machines are readily paralleled and operate correctly and with stability for the following reasons. If the prime-
mover of No. 1, as shown in the diagram of Fig. 8, was to slow-down
momentarily, the generated e.m.f. would fall and the output current,

I_1, given by the expression $\dfrac{E_1 - V}{R_{a_1}}$ would diminish.

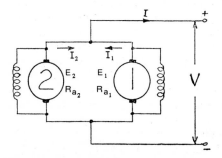

Fig. 8

Stability is assured by the fact that a reduced input means a reduced
output. E_1 might eventually be reduced below V when the current I_1
would reverse and the machine would "motor". Sharing of load with
a similar machine can also be considered thus. Assume two similar
machines to have been paralleled and operating correctly and there is
next an increase of load *i.e.* a demand for extra current. The voltage

of each machine will fall with the resulting effect of each attempting to reject the load. Since both machines tend to "sit down", it follows that eventually the new load must be taken jointly by both machines, the final proportion on each being decided by the slope of the load characteristic which in turn is influenced by the governor character-istics of the prime-mover, by the magnitude of the armature resistance value and to some extent, by the amount of excitation. Stable work-ing conditions are thus a feature of shunt generators in parallel or a shunt generator in parallel with a battery.

(b) Series Generator. Since a series generator will not self-excite until the load circuit is completed and its resistance made less than the critical value, it is evident that such a machine cannot be paralleled onto "live busbars". To investigate parallel operation however, assume two such similar series generators to have been connected in parallel and started up together from rest with the load circuit com-pleted. If the latter is adjusted until self-excitation results, the stabil-ity of operation can be examined in conjunction with the diagram of Fig. 9.

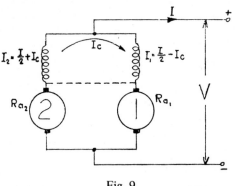

Fig. 9

Let E_1 and E_2 be the e.m.f. of each generator and R_1 and R_2 the resistance, (armature and field). The generators will share load so long as E_1 and E_2 are equal. If due to a momentary slow-down E_1 be-comes less than E_2, then a current will circulate whose value will be

$$I_c \; = \; \frac{E_2 - E_1}{R_1 + R_2} \; .$$

Since the load current is I, then generator No. 2 is now supplying a

current of $\dfrac{I}{2} + I_c$ and No. 1 is supplying a current of $\dfrac{I}{2} - I_c$.

The effect of these new current conditions will be to strengthen the field of No. 2 and weaken the field of No. 1. The generated e.m.f.'s will follow in response, also the machine currents and the effect will be cumulative until No. 2 takes all the load and the current in No. 1 will reverse, once E_1 falls in value below the busbar voltage V. No. 2 will now tend to "motor" No. 1, which once the polarity of the field is reversed will tend to be braked in order to reverse. Generator No. 2 would under this condition supply an exceedingly large current which could burn out both machines. Failing this occurring and that the prime-mover of No. 1 had picked up speed and was sufficiently powerful to drive the generator in the original direction, then the polarity of the machine will reverse and in effect this would result in two generators being in series with each other on a dead short-circuit — again a disastrous condition.

The foregoing shows that series machines would operate in parallel in a highly unstable state and to correct this, the current in the field of No. 1 must be made independent of the e.m.f. The obvious method is to provide a connection between the armatures. This connection, called an "equalising connection", is shown dotted in Fig. 9 and should be of negligible resistance. If the fault condition already considered should now occur, the circulating current I_c would not pass through the field winding thus avoiding reversal of polarity. The fields are now in parallel and their currents will be of substantially equal value, even though the armature currents are different. Stability has now been introduced since with fixed polarity, if the armature current of No. 1 should reverse the machine would motor and assist the prime-mover to pick up speed.

(c) Compound Generators. Such generators have a series field and almost invariably rising characteristics. When operating in parallel, if due to momentary slowing down of a generator its e.m.f. falls, then an unbalance of currents will occur with consequent variation of field strengths as caused by the series fields. Once the e.m.f. of No. 1 falls below V, the current in the series field will reverse and differential compounding will result with even more severe faulty conditions which are cumulative and may even result in reversal of polarity. Thus the machines will be prone to the type of instability already considered for the series generators.

The diagram of Fig. 10 shows how an equalising connection can be introduced, shown dotted. Stability is now assured because the directions of current in the series fields are independent of the reversed armature currents in No. 1 and reversal of polarity will be avoided. Note the point at which this equalising connection is made,

i.e. between the armature and series field. If commutating or interpoles are fitted these are considered to be part of the armature circuit.
Marine regulations require the series fields to be fitted on the − ve
side of the system, to ensure uniformity, and the ammeters to be in
the opposite pole. With this arrangement, the armature current conditions would be revealed.

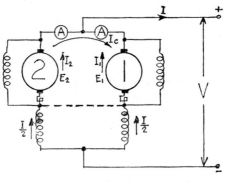

Fig. 10

LOAD SHARING

For d.c. generators as for batteries, no difficulty in sharing load is
encountered provided stable working in parallel can be assured. As a
basic example consider a battery of e.m.f. 4 V connected correctly in
parallel with another of e.m.f. 4 V. No flow of current is observed
and this is explained by the fact that no e.m.f. acts round the complete circuit formed by the two batteries. The 4 V of No. 1 is
opposed by the 4 V of No. 2, which results in no effective e.m.f. for
causing a current. If however, a resistor is connected between the
common + ve line and the common − ve line then a current will
flow, part of which is contributed by battery No. 1.and the rest by
battery No. 2. The proportion of this shared current is decided by
the internal resistance of the batteries and is determined by the application of the basic circuit laws. Methods employing these laws have
been considered in Chapter 14 of Volume 6 but for problems concerned with load sharing the usual procedure is to derive equations
which can be solved simultaneously.

For d.c. generators similar reasoning can be applied but since the
machine characteristics are not always straight lines then load sharing
problems may be better solved by graphical rather than mathematical
methods. Both methods are illustrated by the examples which follow.
These are treated in detail and should be studied with due attention.

Example 9. Two shunt generators A and B, each with straight line load characteristics are operating in parallel. The characteristics are

Machine	Open Circuit Voltage	Terminal voltage at 50 amperes
A	460 V	420 V
B	440 V	410 V

Determine how the machines would share a load of 100 A and determine the common busbar voltage.

Mathematical Solution. For a generator since $E = V + I_a R_a$ then for

Machine A $460 = 420 + (50 \times R_{a_A})$ so $R_{a_A} = \dfrac{460 - 420}{50}$

and $R_{a_A} = \dfrac{40}{50} = 0.8\ \Omega$ 　　　Thus $R_{a_A} = 0.8\ \Omega$

For Machine B $440 = 410 + (50 \times R_{a_B})$ so $R_{a_B} = \dfrac{440 - 410}{50}$

and $R_{a_B} = \dfrac{30}{50} = 0.6\ \Omega$ 　　　Thus $R_{a_B} = 0.6\ \Omega$

When in parallel let I_A = current of machine A and I_B = current of machine B.

Let $I_A + I_B = I$ (the total load current).

Let V = the common busbar voltage.

Substituting in equations $E_A = V + I_A R_{a_A}$

$$\text{or} \quad 460 = V + 0.8\,I_A \quad \dots \quad \dots \quad (a)$$

$$\text{and} \quad E_B = V + I_B R_{a_B}$$

$$440 = V + 0.6\,I_B \quad \dots \quad \dots \quad (b)$$

Using equation (b) then $440 = V + 0.6\,(I - I_A)$

$$= V + 0.6\,(100 - I_A)$$

$$\text{or} \quad 440 - 60 = V - 0.6\,I_A$$

$$\text{giving} \quad 380 = V - 0.6\,I_A \quad \dots \quad \dots \quad (c)$$

Using (a) and (c). Then $460 = V + 0.8\,I_A$

$$380 = V - 0.6\,I_A$$

$$\text{Subtracting} \quad 80 = 1.4\,I_A$$

$$\text{or} \quad I_A = \frac{80}{1.4} = 57.14\ \text{A}$$

Also $I_B = 100 - 57.14 = 42.86$ A

And $V = 460 - (0.8 \times 57.14) = 460 - 45.712 = 414.2$ V.

Graphical Solution. Two methods are possible and both are applied to this problem.

Method 1. The two characteristics are plotted as shown in Fig. 11a with the voltage ordinates spaced 100 A apart. The common terminal or busbar voltage is given by the intersection of the two lines. The required answers can be read off the graph.

This method is suitable for a problem where the load sharing for one particular current value is required. If the problem requires a solution for more than one load current value then Method 2 would be more suitable.

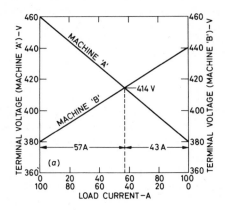

Fig. 11a

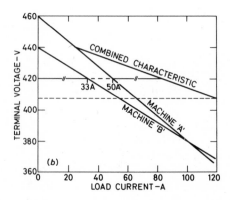

Fig. 11b

Method 2. The procedure here is to draw both characteristics to a common voltage ordinate. This is shown in Fig. 11b. The "combined characteristic" is next drawn by assuming the generators are in parallel and supplying currents to satisfy the condition of a common busbar voltage. Thus assume the busbar voltage is 420 V. Then machine A would supply 50 A and machine B 33 A. The total current supplied would be 83 A. Plot this point and repeat the procedure. Thus for a busbar voltage of 410 V, machine A supplies 63 A and machine B supplies 50 A. The total current is 113 A. Plot this and proceed to obtain the combined characteristic. This latter can be then used to determine the busbar voltage and currents supplied by each machine for any particular value of load current. Thus a load current of 120 A consists of 66 A given by machine A and 54 A given by machine B.

The busbar voltage would then be 408 V.

Example 10. Two shunt generators A and B operating in parallel share equally a total load of 300 A at 220 V. Their load characteristics for the excitations at which they are operating are as follows:

A. Terminal Voltage (V) 230   227·5   224   220   215·5   209·5
    Load Current (A)   0   50   100   150   200   250
B. Terminal Voltage (V)   240   235   228·5   220   209·5
    Load Current (A)   0   50   100   150   200

If the total load changes to (a) 200 A, (b) 400 A, how will the new loads be shared?

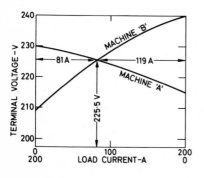

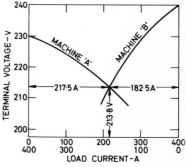

Fig. 12a

Graphical solutions are appropriate since the characteristics are not straight lines. Solutions can be obtained by employing either Methods 1 or 2 as is shown by Fig. 12 a and Fig. 12b.

Either method shows that for:

200 A total load. A delivers 81 A and B delivers 119 A
     The common terminal busbar voltage is 225·5 V
400 A total load. A delivers 217·5 and B delivers 182·5 A
     The common terminal busbar voltage is 213·8 V.

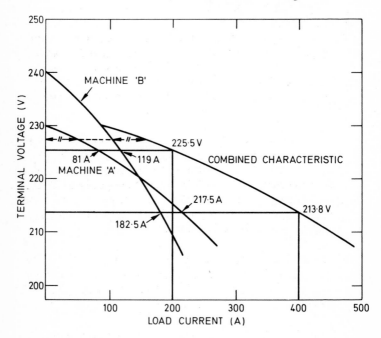

Fig. 12b

CHAPTER 1

PRACTICE EXAMPLES

1.　　A d.c. shunt motor has an armature resistance of 0·9 Ω and takes an armature current of 18 A from 230 V mains. Calculate the power output and overall efficiency of the motor, if the rotational losses are measured to be 112 W and the shunt-field resistance is 300 Ω.

2.　　A ship's 110 V, d.c. "lubricating oil" pump shunt motor is tested by being run light at its correct voltage and at rated speed. It takes 4·5 A. Estimate the efficiency of the motor when loaded to take its rated line current of 40 A. The resistance of the armature including the brushes is 0·12 Ω and the resistance of the field 100 Ω.

3.　　A 4 kW, 105 V, 1200 rev/min shunt motor when run light at normal speed takes an armature current of 3 A at 102 V, normal voltage being applied to the field winding. The field and armature resistances are 95 Ω and 0·1 Ω respectively. Calculate the probable output power rating and efficiency of the motor when operating at 105 V and taking a line current of 40 A. Allow 2 volts drop at the brushes.

4.　　Calculate the light-load current taken by a 100 kW, 460 V shunt motor assuming the armature and field resistances to remain constant and to equal 0·03 Ω and 46 Ω respectively. The efficiency at full load is 88 per cent.

5.　　A 500 V shunt motor takes 4·2 A on no load in a surrounding air temperature of 16°C. The armature plus brush resistance is 0·2 Ω and the field current is 1·2 A, both measurements being made at the same temperature. Estimate the output and the efficiency when the input current is 80 A, the average machine temperature having risen by 22°C. Find the motor current which would give maximum efficiency conditions.

6.　　A 200 V, 15 kW motor when tested by the Swinburne method gave the following results. Running light, the armature current was 6·5 and the field current 2·2 A. With the armature locked the current was 70 A, when a potential difference of 3 V was applied to the brushes. Calculate the efficiency on full load.

7.   The excitation of a d.c. generator is kept constant and the iron loss is measured to be 800 W at 750 rev/min and 2400 W at 1500 rev/min. Deduce its value at 375 rev/min.

8.   A Hopkinson "back-to-back" test made on two shunt generators rated at 150 kW, 220 V, required a current of 140 A to be taken from the mains. The field currents of the two machines were 5 A and 7 A and the armature resistance was measured to be 0·02 Ω. Estimate the efficiency of each machine when loaded to 150 kW.

9.   Two d.c. compound generators operate in parallel with an equalising connection of negligible resistance. Generator A is excited to 235 V and has resistance values of 0·025 Ω for the armature and 0·015 Ω for the series field. Generator B is excited to 237 V and has resistance values of 0·02 Ω for the armature and 0·01 Ω for the series field. If a total load current of 300 A is being supplied, find the busbar voltage and the currents in the equalising connection.

10.   Two d.c. shunt generators having the following load characteristics are operating in parallel to supply a common set of busbars. Find how the machines will share loads of (a) 160 A and (b) 200 A. Find the values of the busbar voltage for each value of load current.

Machine A

Terminal Voltage (V)	265	261	254	243	230	210	175
Load Current (A)	0	20	40	60	80	100	120

Machine B

Terminal Voltage (V)	255	251	246	237	227	210	190
Load Current (A)	0	20	40	60	80	100	120

THE TRANSFORMER (i)

A first introduction to the static transformer principle has already been made in Volume 6, Chapter 6, page 113; but it is necessary here to stress to the student, the importance of understanding its operating theory and applications. The transformer is a basic and essential item of equipment for the utilisation of electrical energy when provided in the alternating-current form, and it is frequently classed as a machine because, although it has no rotating parts, it is capable of converting energy. However, unlike the generator, it receives the energy in the electrical and not the mechanical form, but conversion does take place, in that, the electrical energy is given out at a higher or lower voltage than that at which it is received. Since its operation does not involve the rotation of any armature, field system or commutator, rotational and windage losses do not occur and its efficiency benefits accordingly. For electrical "power" purposes *i.e.* transformers operating at 50 or 60 Hz, an iron core is essential and *iron losses* are to be expected. In addition the *copper losses* of the windings are inherent when current is being supplied but, in spite of these losses, the transformer is the most efficient of the electrical machines and a full-load efficiency figure of 95·5 per cent can be expected for units of 5 kVA and 97·5 per cent for units up to 1 MVA.

PRINCIPLE OF OPERATION

Since the transformer functions by using the principle of mutual induction, it would be of advantage to revise briefly some of the work done on this aspect of theory. When an alternating current is passed through a solenoid, an alternating flux is produced which is in time-phase with the current *i.e.* if the former is varied sinusoidally then the flux will vary sinusoidally. This relationship applies specifically if the solenoid is air-cored, but in order to use the magnetising ampere-turns to maximum advantage, an iron core is necessary. Provided the iron is not worked at too high a flux-density, *i.e.* does not saturate, the conditions of working are similar to those for air. Returning to basic theory, it is seen that a condition of flux-linkages exists, and as the current is varying in a sinusoidal manner then the flux-linkages will vary similarly and an e.m.f. will be induced through the medium of self-induction. The magnitude and direction of this e.m.f. will be in accordance with the laws of electro-magnetic induction, which will be used later in the chapter to develop a formula applicable to transformer operation.

Consider next a second coil, isolated from the first but wound on the same iron core. The first or *primary* coil of insulated wire, when carrying an alternating current will produce the accompanying flux which will now link with the *secondary* or second coil of insulated wire. Since the flux is varying, not only the flux-linkages associated with the primary, but also those associated with the secondary will vary and here again an e.m.f. is induced through the medium of mutual induction. As for the primary, the magnitude and direction of the e.m.f. can be deduced from first principles. This was shown by Example 45, page 117, Chapter 6, Volume 6. Thus the transformer can be considered to be an arrangement consisting of two or more coils, electrically separated but linked by a common magnetic circuit. If an alternating voltage is applied to the primary so as to cause an alternating current then, by mutual induction, a voltage is generated across the secondary. If the secondary circuit is completed, a current will flow and thus energy will be transferred from the primary to the secondary.

The diagram (Fig. 13) illustrates the basic arrangement of the primary and secondary windings and the iron core. Although this form of diagram will continue to be used for explanatory purposes, it should be remembered that this is not a practical arrangement. For efficient working, primary and secondary are never placed on separate limbs of the core.

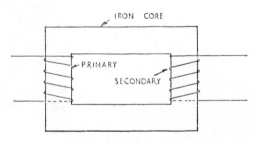

IRON   CORE

PRIMARY

SECONDARY

Fig. 13

Although the practical points of design and construction are best dealt with in a book specialising on the subject, it is convenient here to give an illustration (Fig. 14) of a single-phase, air-cooled, marine-type transformer. The main parts have been labelled.

In accordance with the operating principles already described, the conditions now considered are those which occur when the primary

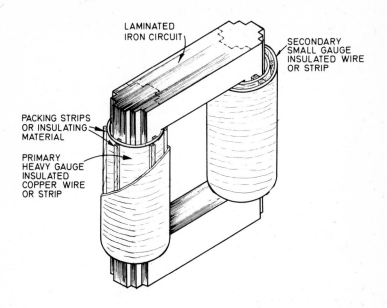

LAMINATED
IRON CIRCUIT

SECONDARY
SMALL GAUGE
INSULATED WIRE
OR STRIP

PACKING STRIPS
OR INSULATING
MATERIAL

PRIMARY
HEAVY GAUGE
INSULATED
COPPER WIRE
OR STRIP

Fig. 14

winding is energised by being connected to the supply voltage as
represented by V_1. The secondary winding can be neglected for the
present. Let I be the resulting current which sets up a flux Φ, the
maximum value of which is Φ_m. The diagram (Fig. 15) illustrates the
waveform relationships for the conditions being considered. I and Φ
are shown in phase as was discussed earlier and an induced e.m.f. E_1
lags the flux by $90°$. This is in accordance with Faraday's law which
states that the value of e.m.f. induced is proportional to the rate of
change of flux-linkages. Since the maximum rate of change of flux
occurs when the wave is passing through its zero value, it follows
that the maximum value of the induced e.m.f. occurs at this instant.
According to Lenz's law this induced e.m.f. E_1 must at all times be
equal and opposite to the applied voltage V_1 which is producing the
magnetising current I and the resulting flux Φ. The phase of the voltage waveforms will be as shown and approximate phasors have also
been drawn to produce a phasor diagram with the flux used as reference. This diagram shows I and Φ to be in phase, with E_1, lagging Φ
by $90°$. V_1 is $180°$ out of phase with E_1 or $90°$ ahead of Φ.

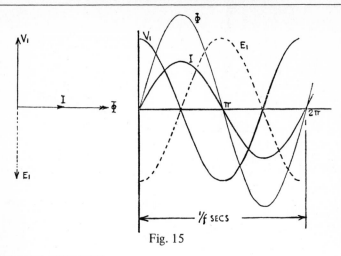

Fig. 15

THE E.M.F. EQUATION

Let the frequency of the applied voltage be f hertz and its r.m.s. value V_1 volts. Let N_1 be the number of turns of the primary winding and N_2 the number of turns of the secondary winding. Let Φ be the flux linking these windings, the maximum value of which is Φ_m.

In a quarter of a cyle or $\dfrac{1}{4f}$ seconds, the flux falls from a value of Φ_m webers to zero.

From Faraday's law, the average value of the e.m.f. induced per turn

$$= \text{ rate of change of flux-linkages } = \frac{(\Phi_m - 0)}{\dfrac{1}{4f}} = 4f\,\Phi_m \text{ volts.}$$

Similarly the average value for N_1 turns is $E_{av} = 4\Phi_m fN_1$ volts. Since sine-wave conditions are to be considered and it is known that the r.m.s. value $= 1 \cdot 11$ times the average value, then for the e.m.f. induced in N_1 turns we have $E_1 = 4 \cdot 44\Phi_m\,fN_1$ volts. Here E_1 is the r.m.s. value of the sine-wave induced voltage.

Again since the induced e.m.f. must be equal and opposite to the supply voltage then:

$$E_1 = V_1 = 4 \cdot 44\,\Phi_m\,fN_1 \text{ volts.}$$

Since the same flux is associated with the secondary winding then by a similar deduction, it can be shown that the value of the e.m.f.

induced in this winding is given by:

$E_2 = 4\cdot44\,\Phi_\mathrm{m}\,fN_2$ volts.

An approach to some proofs by calculus has already been made in Volume 6 of the series. Experience has shown that some students prefer this method of deduction and the above transformer equation can therefore be evolved in the following alternative manner.

If the flux variation is assumed to be sinusoidal then this can be shown as $\phi = \Phi_\mathrm{m} \sin \omega t$.

From Faraday's law, it is known that $e = N_1 \dfrac{\mathrm{d}\phi}{\mathrm{d}t}$. (Reference – see Volume 6, Chapter 9, page 195.) Then the e.m.f. induced in the primary winding can be expressed as $e_1 = -N_1 \dfrac{\mathrm{d}(\Phi_\mathrm{m} \sin \omega t)}{\mathrm{d}t}$ volts.

(The −ve sign indicates that the e.m.f. of self-induction opposes the change of current and flux.)

Carrying out the necessary differentiation, the result is
$e_1 = -N_1\,\omega\,\Phi_\mathrm{m} \cos \omega t$ volts or $e_1 = N_1\,\omega\,\Phi_\mathrm{m} (\sin \omega t - 90°)$.
Thus the induced e.m.f. takes the form of a sine wave lagging behind the flux wave by $90°$.

Now the maximum value of e occurs when $\cos \omega t = 1$ and this condition can be written as $E_{\mathrm{m}_1} = 2\pi fN_1 \Phi_\mathrm{m}$

Remember that $\omega = 2\pi f$

Since the r.m.s. value of a sine wave is $0\cdot707 E_\mathrm{m}$ or $\dfrac{E_\mathrm{m}}{\sqrt{2}}$, then:

$E_1 = \dfrac{2\pi fN_1 \Phi_\mathrm{m}}{\sqrt{2}} = 4\cdot44\,\Phi_\mathrm{m}\,fN_1$ volts – as already deduced.

Example 11. The core of a single-phase, 6600/440 V, 50 Hz transformer is of square cross-section, each side being 160 mm. If the maximum flux density in the core iron is not to exceed $1\cdot4$ T, find the number of turns required for each winding.

Since Flux = Flux density × Area.

Then Flux = $1\cdot4 \times 160 \times 10^{-3} \times 160 \times 10^{-3}$

$= 1\cdot4 \times 25600 \times 10^{-6}$

$= 1\cdot4 \times 2\cdot56 \times 10^{-2} = 3\cdot584 \times 10^{-2}$ webers.

To avoid saturation of the iron, the value of Φ_m in the formula, must not exceed $0\cdot03584$ Wb or $35\cdot85$ mWb.

Thus for the primary winding

$$6600 = 4\cdot44 \times 3\cdot584 \times 10^{-2} \times 50 \times N_1$$

or $N_1 = \dfrac{6600}{2\cdot22 \times 3\cdot584} = \dfrac{6600}{7\cdot956} = 829\cdot6$ (Say 830 turns)

For the secondary winding

$$N_2 = \dfrac{440}{2\cdot22 \times 3\cdot584} = 55\cdot33 \qquad \text{(Say 55 turns)}$$

As a check; it can be deduced that:

$$\dfrac{N_2}{N_1} = \dfrac{V_2}{V_1} \quad \text{or} \quad N_2 = \dfrac{830 \times 440}{6600} = 83 \times \dfrac{2}{3} = 55\cdot33$$

(Say 55 turns)

THE TRANSFORMER ON NO LOAD

The conditions of working on no load, *i.e.* with the secondary circuit "open", can be investigated if suitable instruments are connected into the circuit. Since a temperature rise of the iron core would be noted soon after energising the primary, it is evident that some electrical energy is taken from the supply and is converted into heat. A wattmeter, capable of giving an accurate measurement at a low power factor, should therefore be included in the test circuit together with voltmeters and an ammeter. The diagram (Fig. 16) shows the arrangement.

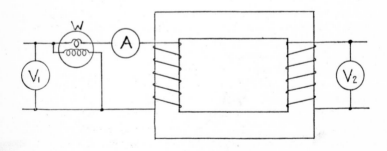

Fig. 16

PRIMARY PHASOR DIAGRAM

The transformer phasor diagram has already been introduced and the "readings" of the test instruments can now be used to explain the development of the diagram. The no-load current would be the ammeter reading I_0 and the wattmeter reading would indicate the power input $P_0 = V_1 I_0 \cos \phi_0$. From these readings it can be deduced that the no-load current I_0 can be resolved into a power component $I_0 \cos \phi_0$, which can be looked upon as a working component, *i.e.* that responsible for the unwanted heat which is here regarded as a loss. $I_0 \cos \phi_0$ can be denoted by I_w, the power component in the phasor diagram, since it is in phase with the voltage. The no-load phasor diagram is illustrated (Fig. 17a and b) and is considered separately for primary and secondary.

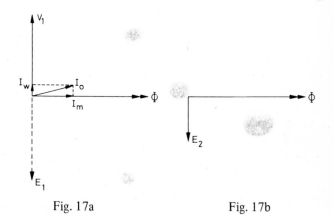

Fig. 17a   Fig. 17b

The primary phasor diagram also shows the other component of the no-load current I_0. This component would be $I_0 \sin \phi_0$, *i.e.* a reactive or wattless component, which, being at right angles to the voltage or in phase with the flux Φ, can be regarded as producing the latter. Thus $I_0 \sin \phi_0$ can be denoted by I_m and is called the magnetising component. The following can also be deduced:

$I_0 = \sqrt{I_w^2 + I_m^2}$ where $I_w = I_0 \cos \phi_0$ and $I_m = I_0 \sin \phi_0$.

We also have $V_1 I_0 \cos \phi_0 = V_1 I_w$ = heat loss due to the iron loss in the core + a small copper loss in the primary.

It should be noted that as the no-load current is so small and the resistance of the winding so low, the copper loss can be neglected and the input power taken as a measure of the iron loss which remains constant over the load range of the transformer. This last statement assumes a constant V_1 and f at any load, which in turn ensures a constant B value of flux density.

SECONDARY PHASOR DIAGRAM

The illustrations of Figure 17 show two phasor diagrams. The first is for the primary winding where all the phasors shown have been explained. That for E_1 is drawn dotted because as will be explained later it is customary to omit this for further development. The second diagram of Fig. 17, is for the secondary winding and it will be seen that the induced e.m.f. E_2 is drawn at right angles to the common core flux Φ. Unlike the primary since no supply voltage is applied, no anti-phase voltage phasor is drawn *i.e.* there is no phasor corresponding to V_1. When the secondary circuit is completed, the induced e.m.f. becomes an active voltage which is responsible for the current through the connected load and it can also be expressed as V_2 (the secondary terminal voltage). Thus on no-load $V_2 = E_2$.

Example 12. For the no-load test on a transformer, the ammeter was found to read 0·18 A and the wattmeter 12 W. The reading on the primary voltmeter was 400 V and that on the secondary voltmeter was 240 V. Calculate, the magnetising component of the no-load current, the iron loss component and the transformation ratio.

$$\text{Iron-loss component} \ = \ I_w \ = \ \frac{P}{V_1} \ = \ \frac{12}{400} \ = \ 0\cdot03 \text{ A}$$

$$
\begin{aligned}
\text{Magnetising component } I_m \ &= \ \sqrt{I_0{}^2 - I_w{}^2} \ = \ \sqrt{0\cdot18^2 - 0\cdot03^2} \\
&= \ 10^{-1} \sqrt{1\cdot8^2 - 0\cdot3^2} \\
&= \ 10^{-1} \sqrt{3\cdot24 - 0\cdot09} \\
&= \ 10^{-1} \sqrt{3\cdot15} \ = \ 10^{-1} \times 1\cdot775 \\
&= \ 0\cdot178 \text{ A}
\end{aligned}
$$

$$\text{Transformation Ratio} \ = \ \frac{400}{240} \ = \ \frac{5}{3} \ \text{ or } \ 1\cdot66:1$$

COMBINED PHASOR DIAGRAM

From the secondary diagram it will be seen that there is a $180°$ phase displacement between the primary supply voltage V_1, and the secondary induced e.m.f. E_2 or the secondary terminal voltage V_2. The phasor E_1, through explaining the action of the transformer, is not essential to the phasor diagram and if omitted, its place can be taken by E_2. The primary and secondary diagrams are thus combined with flux, Φ being used as the reference phasor to show the primary and secondary phase relationships. This arrangement is more satisfactory, since it will be seen later that any effects in one winding will be reflected in the other. The illustration (Fig. 18), shows the combined phasor diagram for a transformer with a turns ratio of $2:1$. The phasor for voltage V_2 is thus drawn half the length of the phasor for the voltage V_1.

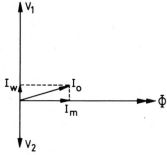

Fig. 18

VOLTAGE AND TURNS RELATIONSHIP. Since the induced e.m.f.'s in the primary and secondary windings are in accordance with the e.m.f. equation then

$$E_1 = 4{\cdot}44\ \Phi_m\ fN_1 \text{ and } E_2 = 4{\cdot}44\ \Phi_m\ fN_2$$

Also since $E_1 = V_1$ and E_2 can be written as V_2 then:

$$\frac{V_1}{V_2} = \frac{4{\cdot}44\ \Phi_m\ fN_1}{4{\cdot}44\ \Phi_m\ fN_2} \text{ or } \frac{V_1}{V_2} = \frac{N_1}{N_2}$$

Attention has been drawn to this relationship earlier but it has been repeated here because of its similarity to a current relationship which is shortly to be deduced.

Example 13. A 440 V transformer has 3000 turns on the primary. If a tapping is to be available for 400 V, find its position on the winding. If one secondary winding suitable for 200, 220 and 240 V is also to be provided, find the necessary number of turns and the position of the tappings.

Since voltage is proportional to turns then:

For the Primary $\dfrac{440}{400} = \dfrac{3000}{N}$ or $N = \dfrac{3000 \times 400}{440} = \dfrac{30\ 000}{11}$

$$= 2727{\cdot}2$$

The tapping should be 3000 − 2727 or 273 turns from the end which is remote from that made common for 440 and 400 volts. Fig. 19 shows the arrangement.

For the Secondary $\dfrac{V_2}{V_1} = \dfrac{N_2}{N_1}$ or $N_2 = \dfrac{220 \times 3000}{440} = 1500$ turns

Thus 1500 turns will give a voltage of 220 V

For 200 volts, the turns required would be half the number required for 400 volts *i.e.* $\dfrac{2727}{2} = 1364$

Thus $1500 − 1364 = 136$ turns would make the difference for 20 V
The turns required for 240 V would be $1500 + 136 = 1636$
The secondary would consist of 1636 turns with tappings at 136 and 272 turns from the end which is remote from that made common for 240, 220 and 200 volts.

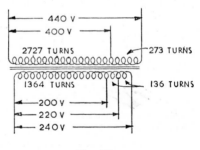

Fig. 19

THE TRANSFORMER ON LOAD

The condition now to be considered is that which occurs when the secondary circuit of the transformer is completed and current flows. Since the secondary will now function as a generator, the magnitude of the secondary current would be decided by the impedance of the load being supplied. Assume a current of I_2 to be produced by the terminal voltage V_2. The passage of I_2 through the turns of the secondary winding, constitutes demagnetising ampere-turns which result in a flux, shown in the diagram (Fig. 20) as Φ_{L_2}. The use of the suffix L_2 will be explained in due course.

This secondary flux also links with the primary and by Lenz's law, tends to produce a demagnetising effect which results in the core flux Φ being reduced. The reduction of flux through the primary results in a small reduction of the primary induced e.m.f. E_1. The effect of the supply voltage V_1 is then such as to produce a larger primary current I_1, most of which counters the demagnetising effect of the secondary current. It should be noted that the increase in I_1 due to I_2 flowing, caused by the greater difference between V_1 and E_1, is made up of an extra primary-current component called I_1' which will be additional to the no-load current I_0 which continues to flow to produce the components I_m and I_w. The new primary current I_1 thus consists of I_1' and I_0, but a phasor summation is involved as is to be shown.

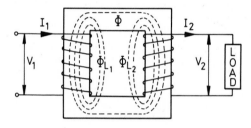

Fig. 20

Returning to the demagnetising effect of the secondary flux Φ_{L_2}, it will be remembered that the value of the core flux Φ is constant, being decided by the expression $V_1 = 4{\cdot}44\,\Phi_m\,fN_1$. If V_1, f and N_1 are constant, Φ_m and thus Φ must be constant, so any demagnetising effect of I_2N_2 must be offset by the magnetising effect of the extra primary current. The resulting primary magnetising ampere-turns $I_1'N_1$ restores the main flux to its original value by producing the additional primary flux Φ_{L_1}. It follows that since Φ_{L_1} must equal Φ_{L_2} then $I_1'N_1$ must equal I_2N_2 or $I_1'N_1 = I_2N_2$.

The full representation of the flux conditions is shown by the diagram (Fig. 20), but some explanation of the phases of these fluxes is still necessary in order to show the complete action.

Fluxes Φ_{L_1} and Φ_{L_2} are equal in magnitude and since they nullify the effects of each other on the non-related windings, it follows that they are in antiphase and oppose each other in the main iron core. They are accordingly forced out into the air, as shown, for the major path of their magnetic circuits and are called *"Leakage Fluxes"*. − hence the suffix L_1 and L_2 as used for Φ_{L_1} and Φ_{L_2}. Also since the reluctance of the magnetic circuits are due mainly to air paths, the fluxes are proportional to the energising currents and are in phase with them. This is shown on the on-load phasor diagram — to be developed.

ON-LOAD PHASOR DIAGRAM

To follow the development of the diagram it is essential to understand fully the theory of transformer working. To this end it would be well to summarise transformer action by considering its similarity to that of a d.c. motor-generator set. If the generator is loaded the motor will immediately take more current from the supply. This is explained by the fact that as load is applied the speed falls. The back e.m.f. of the motor falls accordingly, and the supply voltage ensures more current is supplied until the condition is attained when, the power put in is sufficient to provide the output power plus the power required to supply the losses. For the transformer, it is seen that, the output current causes a demagnetising effect on the core which lowers the back e.m.f. of the primary. Here again the supply voltage ensures that additional current is supplied until the balance condition is attained, when input power equals the output power plus the power required to overcome the losses.

The phasor diagram (Fig. 21) is drawn to represent the conditions occurring when an inductive load is supplied. An inductive load is the general condition and represents most cases. Assume a secondary current of I_2 for a load operating at a power factor of $\cos \phi_2$. On the diagram for the secondary side, I_2 is drawn lagging V_2 by an angle ϕ_2 and, as before, a transformation ratio of $2:1$ is assumed. On the primary side, the original no-load current conditions are shown but in addition we have I_1', which being antiphase with I_2, lags V_1 by an angle equal to ϕ_2. It is drawn half the size of I_2 because of the transformation ratio. It will be seen that I_1 is the phasor sum of I_0 and I_1' but it is stressed that I_0 is drawn much larger than it would be for an actual practical transformer. It is also necessary to point out here that although $I_1'N = I_2N_2$, for purposes of approximation, we are justified in writing $I_1N_1 = I_2N_2$ because the no-load current I_0 is small enough to be neglected when the transformer is on load. The assumption of I_1 being equal to I_1' is acceptable for all practical purposes and I_0 will be omitted from most phasor diagrams from now on. The current/turns relationship to be deduced shortly is worked on this basis.

The phasor diagrams shows Φ_{L_2} drawn in phase with I_2 and Φ_{L_1} to be in phase with I_1. It is pointed out that both Φ_{L_2} and Φ_{L_1} are small in comparison to Φ because, although they are produced by currents of load magnitudes compared to the very small value of I_m, it must be remembered that the magnetic circuits are through air, whilst that for Φ is through iron. The fact that the leakage fluxes can be drawn in phase with the energising currents has been mentioned already, but in the case of the primary since I_1' and I_1 are assumed identical, I_0 being neglected, then no error is introduced if Φ_{L_1} is assumed to be caused by I_1 and is drawn in phase with it.

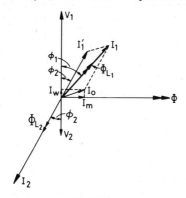

Fig. 21

THE CURRENT AND TURNS RELATIONSHIP. Theory of operation has shown that the demagnetising ampere-turns of the secondary are equalled and nullified by magnetising ampere-turns produced by the primary. Thus $I_2 N_2 = I_1' N_1$. Also since I_1 is assumed equal to I_1' — the effect of I_0 being neglected, the more generally accepted

expression is $I_1 N_1 = I_2 N_2$ or $\dfrac{I_2}{I_1} = \dfrac{N_1}{N_2}$

The inversion of the proportionality should be noted *i.e.* unlike that

for the voltage where $\dfrac{V_1}{V_2} = \dfrac{N_1}{N_2}$

Example 14. A single-phase transformer with a ratio of 440/200 V takes a no-load current of 8 A at a power factor of 0·25 (lagging). If the secondary supplies a current of 220 A at a power factor of 0·8 (lagging), estimate the current taken by the primary from the supply.

Load component of primary current $I_1' = \dfrac{I_2 \times N_2}{N_1}$

$$\text{or } I_1' = \frac{I_2 \times V_2}{V_1} = \frac{220 \times 200}{440} = 100 \text{ A}$$

Resolving I_1' and I_0 into horizontal and vertical component the resultant can be obtained in accordance with the accepted method of phasor summation:

Sum of vertical components $I_V = I_1' \cos \phi_2 + I_0 \cos \phi$

$$= 100 \times 0 \cdot 8 + 8 \times 0 \cdot 25 = 80 + 2$$

$$= 82 \text{ A}$$

Sum of horizontal components $I_H = I_1' \sin \phi_2 + I_0 \sin \phi_0$

$$= 100 \times 0 \cdot 6 + 8 \times 0 \cdot 97$$

$$= 60 + 7 \cdot 76 = 67 \cdot 76 \text{ A}$$

So $I = \sqrt{I_H{}^2 + I_V{}^2} = \sqrt{82^2 + 67 \cdot 76^2} = 10\sqrt{8 \cdot 2^2 + 6 \cdot 78^2}$

$$= 10\sqrt{67 \cdot 24 + 45 \cdot 97} = 10\sqrt{113 \cdot 21} = 10 \times 10 \cdot 64$$

$$= 106 \cdot 4 \text{ A}.$$

THE kVA RELATIONSHIP. The deductions from the voltage, current and turns relationships are combined here to produce the following:

From the voltage/turns relationship $\dfrac{V_1}{V_2} = \dfrac{N_1}{N_2}$

From the current/turns relationship $\dfrac{I_2}{I_1} = \dfrac{N_1}{N_2}$

By equating $\dfrac{V_1}{V_2} = \dfrac{N_1}{N_2} = \dfrac{I_2}{I_1}$ or $\dfrac{V_1}{V_2} = \dfrac{I_2}{I_1}$ or $V_1 I_1 = V_2 I_2$

This can be written as

Primary volt-amperes = Secondary volt-amperes

Dividing by 1000 $\quad \dfrac{\text{Primary } VA}{1000} = \dfrac{\text{Secondary } VA}{1000}$

or $\qquad\qquad$ Primary kVA = Secondary kVA

The above shows that there is only one kVA rating for a transformer and this is applicable to either the primary or secondary.

Example 15. If the iron core of the transformer of Example 11 is considered to be suitable for a unit of 50 kVA, estimate the area of the wire to be used for the primary and secondary windings. The wire should not be worked at a current density of more than 2·33 amperes per square millimetre, to ensure a satisfactory working temperature.

Data already known about this transformer from Example 11.

Primary voltage 6600 V   Turns 830

Secondary voltage 440 V Turns 55

Estimated primary current $\quad I_1 \;=\; \dfrac{50\,000}{6600} \;=\; \dfrac{250}{33} \;=\; 7\text{·}576\ \text{A}$

Estimated secondary current $\;I_2 \;=\; \dfrac{50\,000}{440} \;=\; \dfrac{1250}{11} \;=\; 113\text{·}636\ \text{A}$

or secondary current from $I_1 V_1 \;=\; I_2 V_2$ gives $\;I_2 \;=\; \dfrac{I_1 V_1}{V_2}$

whence $I_2 \;=\; 7\text{·}576 \times \dfrac{6600}{440} \;=\; 7\text{·}576 \times 15 \;=\; 113\text{·}64\ \text{A}$

Cross-section of primary $\;=\; \dfrac{7\text{·}576}{2\text{·}33} \;=\; 3\text{·}25\ \text{mm}^2$

Cross-section of secondary $\;=\; \dfrac{113\text{·}64}{2\text{·}33} \;=\; 48\text{·}77\ \text{mm}^2$

ON-LOAD VOLTAGE DIAGRAM (Continued)

Since the diagram involves voltage phasors it cannot be considered to be complete until all causes of e.m.f. or voltage drop have been taken into account. Any voltage drops would be internal since their effect only becomes evident when the transformer is on load. They would be due to the normal causes of current limitation in the a.c. circuit namely, resistance and reactance. Their existence and effects on transformer operation are now examined.

INTERNAL VOLTAGE DROPS (RESISTANCE). The ohmic resistances of the primary and secondary windings are quite independent of each other being determined by the length, area and specific resistance of the wire. Since the currents passing through these windings are also dissimilar in value, it follows that the primary and secondary voltage drops should be considered separately at first and then co-related if a suitable relationship can be devised.

The transformer can be represented as an ideal unit with all resistance of windings contained in external units R_1 and R_2 as shown. The illustrations (Fig. 22 and 23) show the arrangement and the phasor diagram is discussed below.

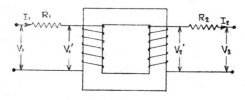

Fig. 22

The passage of load current will cause voltage drops $I_1 R_1$ and $I_2 R_2$ which, being due to resistance, will be in phase with the current. These voltage drops, if taken into account when considering the primary and secondary terminal voltages, can be likened to the armature voltage drops in the motor-generator set already introduced to understand transformer working. Thus for the generator or secondary of the transformer, since the terminal voltage is less than the generated voltage by the internal voltage drop, it follows that the latter must be subtracted from the generated voltage. For the transformer a construction is involved or $\bar{V}_2 = \bar{V}_2' - \bar{I_2 R_2}$, the generated or induced e.m.f. being called V_2' and the voltage-drop phasor $I_2 R_2$ being drawn in phase with the current. Note the slight phase displacement of the terminal-voltage phasor V_2 from the original no-load position V_2'.

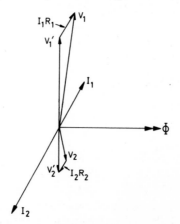

Fig. 23

Similarly for the primary, it will be seen that as for a motor, the supply voltage V_1 must overcome the induced e.m.f. $V_1{}'$ and the internal resistance of the armature or, in this case, the primary winding. Here we have $\overline{V}_1 = \overline{V}_1{}' + I_1\overline{R}_1$ and due allowance must be made for a phasor construction, I_1R_1 being drawn in phase with I_1.

INTERNAL VOLTAGE DROPS (REACTANCE). When the secondary circuit is completed and current flows, action results whereby the demagnetising secondary ampere-turns I_2N_2 are countered by magnetising primary ampere-turns I_1N_1. This has already been described earlier, and is stressed here by saying that the two ampere-turn effects or m.m.f.'s set up fluxes which oppose each other and take the path shown by Fig. 20. These fluxes are termed "leakage fluxes" and being mainly through air, they are in phase with the currents which produce them. The fluxes cut the coils which produce them and, being alternating, they result in e.m.f.'s of self-induction which are in quadrature with the fluxes or the energising currents. From first principles, we know the term "reactance" was introduced for convenience. In fact, an a.c. current, through the medium of the flux, produces a back e.m.f. which is proportional to the current. This e.m.f. would be written as $E_b \propto I$ or $E_b = IX$. Thus no error is introduced in crediting the winding with a current limiting factor which is additional to the resistance and is itself responsible for a voltage drop. This factor or quantity is termed *leakage reactance* and, provided the phase of its voltage drop with respect to the producing current is allowed for, then the terms: − reactance, resistance and impedance can be used for the primary and secondary.

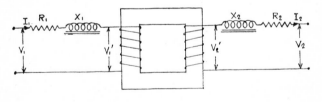

Fig. 24

The on-load diagrams (Figs 24 and 25) are drawn to take into account both the resistance and the reactance voltage drops. Points to note are

(i)   The secondary can be looked upon as a generator and the internal voltage drop should be treated by phasors to give the terminal voltage on load. Thus $\overline{V}_2 = \overline{V}_2{}' - I_1\overline{R}_2 - I_2\overline{X}_2$ where $V_2{}'$ is the

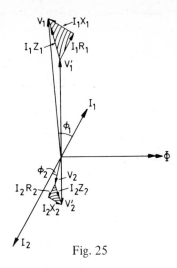

Fig. 25

open-circuit induced voltage, $I_2 R_2$ is the resistance voltage drop in phase with I_2, $I_2 X_2$ is the leakage reactance voltage drop in quadrature with I_2 and V_2 is the resulting terminal voltage. The shaded triangle is called the secondary impedance triangle and $V_2 V_2'$ is the secondary voltage drop given by $I_2 Z_2$. Note that, as can be expected, $Z_2 = \sqrt{R_2{}^2 + X_2{}^2}$.

(ii) The primary can be looked upon as a motor and the supply voltage V_1 must overcome all the internal voltage drops which are added as phasors. Thus $\bar{V}_1 = \bar{V}_1{}' + \bar{I}_1 R_1 + \bar{I}_1 X_1$. On no-load V_1 will coincide with $V_1{}'$ but on load there are the internal voltage drops due to the resistance $I_1 R_1$ in phase with I_1 and leakage reactance $I_1 X_1$ in quadrature with I_1. These must be added as phasors to the primary induced voltage $V_1{}'$ as produced by the main flux Φ. The shaded triangle is the primary impedance triangle and $V_1 V_1{}'$ is the impedance voltage drop given by $I_1 Z_1$, As can be surmised $Z_1 = \sqrt{R_1{}^2 + X_1{}^2}$.

REFERRED VALUES OF RESISTANCE, REACTANCE AND IMPEDANCE

Since the voltages at the primary and secondary are related by the ratio of the turns, it follows that any secondary voltage drop could be considered in terms of the primary side if the turns ratio is applied. The same reasoning could be applied to a primary voltage drop if referred to the secondary and, although much more work is to be done in this direction, it is convenient at this stage to deduce the *equivalent resistance* and *equivalent reactance* for one winding when considered in terms of the other winding.

Consider the resistance voltage drop in the primary which is given by $I_1 R_1$. Then if the turns ratio is applied this would give the equivalent voltage drop in the secondary namely, $I_1 R_1 \left(\dfrac{N_2}{N_1} \right)$.

But $I_1 = I_2 \left(\dfrac{N_2}{N_1} \right)$ so the primary voltage drop in secondary terms can be written as $I_2 R_1 \left(\dfrac{N_2}{N_1} \right)^2$ There is also the secondary voltage drop due to the secondary current and resistance given by $I_2 R_2$. Thus the total resistance voltage drop referred to the secondary is

$$I_2 R_2 + I_2 R_1 \left(\frac{N_2}{N_1} \right)^2 \quad \text{or} \ = \ I_2 \left\{ R_2 + R_1 \left(\frac{N_2}{N_1} \right)^2 \right\}.$$

Thus the primary can be regarded as having no resistance and the secondary as having a resistance of $R_2 + R_1 \left(\dfrac{N_2}{N_1} \right)^2$. This total resistance value is called the *Equivalent Resistance* written as $\bar{R}_2$. Thus the Equivalent Resistance of the transformer, referred to the secondary side is $\bar{R}_2 = R_2 + R_1 \left(\dfrac{N_2}{N_1} \right)^2$.

Similarly the Equivalent Resistance referred to the primary side is

$$\bar{R}_1 = R_1 + R_2 \left(\frac{N_1}{N_2} \right)^2.$$

By considering the reactance voltage drops, a similar deduction for *Equivalent Reactance* can be produced. Thus for the primary, reactance voltage drop is $I_1 X_1$, which is equivalent to a voltage drop on the secondary side given by $I_1 X_1 \left(\dfrac{N_2}{N_1} \right)$. But $I_1 = I_2 \left(\dfrac{N_2}{N_1} \right)$ therefore the primary voltage drop becomes $I_2 X_1 \left(\dfrac{N_2}{N_1} \right)^2$, if referred to the secondary. However, there already exists the secondary reactance voltage drop $I_2 X_2$ so the total reactance voltage drop, referred to the secondary is $I_2 X_2 + I_2 X_1 \left(\dfrac{N_2}{N_1} \right)^2 \quad \text{or} \ = \ I_2 \left\{ X_2 + X_1 \left(\dfrac{N_2}{N_1} \right)^2 \right\}.$

As for resistance, the transformer primary can now be regarded as having no reactance and the secondary to have a value given by:

$X_2 + X_1 \left(\dfrac{N_2}{N_1}\right)^2$. This expression gives $\bar{X}_2$, the *Equivalent*

Reactance value referred to the secondary and a similar expression can be derived for the primary side. Thus:

Equivalent Reactance of the transformer referred to the secondary

side is $$\bar{X}_2 = X_2 + X_1 \left(\dfrac{N_2}{N_1}\right)^2$$ and

Equivalent Reactance referred to the primary side is

$$\bar{X}_1 = X_1 + X_2 \left(\dfrac{N_1}{N_2}\right)^2$$

The expression for the *Equivalent Impedance* is self evident in that $Z_1 = \sqrt{R_1{}^2 + \bar{X}_1{}^2}$ and $\bar{Z}_2 = \sqrt{R_2{}^2 + \bar{X}_2{}^2}$.

Example 16. A 6600/440 V, single-phase transformer has a primary resistance of 140 Ω and a secondary resistance of 0·25 Ω. Calculate its equivalent resistances, referred to the secondary winding and primary winding respectively.

For the expression $\bar{R}_2 = R_2 + R_1 \left(\dfrac{N_2}{N_1}\right)^2$ the turn ratio is

$\dfrac{N_1}{N_2} = \dfrac{6600}{440} = \dfrac{15}{1}$ or $\bar{R}_2 = 0\cdot25 + 140 \left(\dfrac{1}{15}\right)^2 = 0\cdot25 + \dfrac{140}{225}$

$$= 0\cdot25 + 0\cdot622 = 0\cdot872 \ \Omega$$

Also
$$\bar{R}_1 = R_1 + R_2 \left(\dfrac{N_1}{N_2}\right)^2$$
or

$\bar{R}_1 = 140 + (0\cdot25 \times 15^2) = 140 + (0\cdot25 \times 225)$

$$= 140 + 56\cdot25 = 196\cdot25 \ \Omega$$

Equivalent resistance of windings referred to the secondary side is 0·872 Ω

Equivalent resistance of windings referred to the primary side is 196·25 Ω.

Example 17. A marine, dry-type, 17·5 kVA, 460/115 V, single-phase, 50/60 Hz transformer has primary and secondary resistances of 0·36 Ω and 0·02 Ω respectively and the leakage reactances of these windings are 0·82 Ω and 0·06 Ω respectively. Determine the voltage to be applied to the primary to obtain full-load current with the secondary winding short-circuited. Neglect the magnetising current.

$$\text{Full load primary current } I_1 \;=\; \frac{17\,500}{460} \;=\; 38 \text{ A}$$

$$\text{Transformation ratio} \;=\; \frac{460}{115} \;=\; \frac{4}{1} \text{ or } \frac{N_1}{N_2} \;=\; \frac{4}{1}$$

$$\text{Then } \bar{R}_1 \;=\; R_1 + R_2\left(\frac{N_1}{N_2}\right)^2 \;=\; 0\cdot36 + 0\cdot02 \times \left(\frac{4}{1}\right)^2$$

$$= 0\cdot36 + (0\cdot02 \times 16)$$

$$= 0\cdot36 + 0\cdot32 = 0\cdot68 \ \Omega$$

$$\bar{X}_1 \;=\; X_1 + X_2\left(\frac{N_1}{N_2}\right)^2 \;=\; 0\cdot82 + 0\cdot06 \times \left(\frac{4}{1}\right)^2$$

$$= 0\cdot82 + (0\cdot06 \times 16)$$

$$= 0\cdot82 + 0\cdot96 = 1\cdot78 \ \Omega$$

$$\bar{Z}_1 \;=\; \sqrt{0\cdot68^2 + 1\cdot78^2} \;=\; \sqrt{0\cdot4624 + 3\cdot1684} \;=\; \sqrt{3\cdot6308} \;=\; 1\cdot905 \ \Omega$$

$$\text{Voltage to be applied to the primary} \;=\; I_1\bar{Z}_1$$
$$= 38 \times 1\cdot905 = 72\cdot4 \text{ V.}$$

EFFICIENCY OF A TRANSFORMER

Only basic work is done at this stage in connection with losses and efficiency. Further aspects will be considered in Chapter 4 as more theory on transformer working is developed. Since the transformer is a static piece of apparatus, the only losses which occur are (i) Iron Losses, (ii) Copper Losses.

IRON LOSSES (P_{Fe}). These consist of the *hysteresis* and *eddy current* losses which occur in the core. These are affected by flux density and frequency as was shown in Volume 6, Chapter 12, Page 293. For a constant applied voltage and frequency, core flux is unaltered *i.e.* it is unaltered by load current. The iron losses P_{Fe} can therefore be assumed constant and independent of load.

COPPER LOSSES (P_{Cu}). From first principles it is known that these are proportional to current squared and resistance.

Then Primary copper loss $= I_1{}^2 R_1$ Secondary copper loss $= I_2{}^2 R_2$

or the total copper loss $P_{Cu} = I_1{}^2 R_1 + I_2{}^2 R_2$ But $I_1 = I_2\left(\dfrac{N_2}{N_1}\right)$

so the total copper loss $= I_2{}^2\left(\dfrac{N_2}{N_1}\right)^2 R_1 + I_2{}^2 R_2$

$$= I_2{}^2\left\{ R_2 + R_1\left(\frac{N_2}{N_1}\right)^2 \right\}.$$

It is already known that $\bar{R}_2 = R_2 + R_1\left(\dfrac{N_2}{N_1}\right)^2$

so total copper loss $= I_2{}^2 \bar{R}_2$.

In a similar manner the total copper loss can be shown to be given by $I_1{}^2 \bar{R}_1$. Thus $P_{Cu} = I_2{}^2 \bar{R}_2$ or $I_1{}^2 \bar{R}_1$.

EFFICIENCY.   The expression for efficiency can be developed from first principles thus

$$\text{Efficiency} = \frac{\text{Output}}{\text{Input}} = \frac{\text{Output}}{\text{Output} + \text{Losses}}$$

$$= \frac{\text{Output}}{\text{Output} + \text{Iron Loss} + \text{Copper Loss}}$$

Introducing the appropriate units the above can be expressed as

$$\eta = \frac{kW}{kW + (P_{Fe} + P_{Cu})}$$

where P_{Fe} is the iron loss in kW and P_{Cu} is the copper loss in kW or

$$\eta = \frac{kVA \cos \phi}{kVA \cos \phi + P_{Fe} + P_{Cu}}$$

$$= \frac{kVA \cos \phi}{kVA \cos \phi + \left(P_{Fe} + \dfrac{I_2{}^2 \bar{R}_2}{1000}\right)}$$

or $\eta = \dfrac{kVA \cos \phi}{kVA \cos \phi + \left(P_{\text{Fe}} + \dfrac{I_1^2 \bar{R}_1}{1000}\right)}$

It will be shown later that the maximum efficiency of a transformer will occur when the copper loss is equal to the iron loss.

Example 18. A 20 kVA, 2000/220 V, single-phase transformer has a primary resistance of $2 \cdot 1 \Omega$ and a secondary resistance of $0 \cdot 026 \Omega$. If the total iron loss equals 200 W, find the efficiency on (a) full load and at a power factor of $0 \cdot 5$ (lagging); (b) half load and a power factor of $0 \cdot 8$ (leading).

(a) Iron loss $= 200$ W   Full-load primary current $= \dfrac{20\,000}{2000} = 10$ A

Turns ratio $= \dfrac{2000}{220} = \dfrac{100}{11} = \dfrac{9 \cdot 09}{1}$

Full-load secondary current $= 9 \cdot 09 \times 10 = 90 \cdot 9$ A

Since $\bar{R}_2 = R_2 + R_1 \left(\dfrac{N_2}{N_1}\right)^2$ or $\bar{R}_2 = 0 \cdot 026 + 2 \cdot 1 \left(\dfrac{1}{9 \cdot 09}\right)^2$

$= 0 \cdot 026 + 0 \cdot 0255 = 0 \cdot 0515 \Omega$

$\eta_{\text{FL}} = \dfrac{20 \times 0 \cdot 5}{20 \times 0 \cdot 5 + \left(\dfrac{200}{1000} + \dfrac{90 \cdot 9^2 \times 0 \cdot 0515}{1000}\right)}$

$= \dfrac{10}{10 + (0 \cdot 2 + 0 \cdot 426)} = 0 \cdot 941$ or Full-load Efficiency

$= 94 \cdot 1$ per cent

(b) Half-load primary current $= \dfrac{10\,000}{2000} = 5$ A

Half-load secondary current $= 9 \cdot 09 \times 5 = 45 \cdot 45$ A

As before $\bar{R}_2 = 0 \cdot 0515 \Omega$

$$\eta_{HL} = \frac{10 \times 0.8}{10 \times 0.8 + \left(\frac{200}{1000} + \frac{45.45^2 \times 0.0515}{1000} \right)} = \frac{8}{8.306}$$

Thus: $\eta_{HL} = 0.963$ or Half-load Efficiency 96·3 per cent.

Note that as the load current is halved the copper loss, being proportional to current squared is quartered *i.e.* $P_{Cu} = \frac{0.426}{4} = 0.106$ kW

This approach makes for easier working. Thus:

$$\eta_{HL} = \frac{8}{8 + 0.2 + \frac{0.426}{4}} = \frac{8}{8 + 0.2 + 0.106}$$

or $\eta_{HL} = 0.963$ or Half-load Efficiency 96·3 per cent.

Example 19. The primary resistance of a 440/110 V, single-phase transformer is 0·28 Ω and the secondary resistance is 0·018 Ω. If the iron loss is measured to be 160 W when the correct primary voltage is applied, find the *kW* loading to give maximum efficiency at unity power factor.

$$\text{Turns ratio} = \frac{440}{110} = \frac{4}{1}$$

$$\text{so } \bar{R}_2 = 0.018 + 0.28 \left(\frac{1}{4} \right)^2$$
$$= 0.018 + 0.0175 = 0.0355 \ \Omega$$

For maximum efficiency, copper loss equals the iron loss so $P_{Cu} = 160$ W or $I_2^2 \bar{R}_2 = 160$

Thus: $I_2^2 = \frac{160}{0.0355}$ giving $I_2 = \frac{12.64}{0.188} = 67.4$ A

$$\text{Load rating} = \frac{110 \times 67.4}{1000} = 7.43 \text{ kVA}$$

or since $\cos \phi = 1$ and $kW = kVA$

then "loading" for maximum efficiency $= 7.43$ kW.

Example 20. The core of a single-phase trasnformer has a cross-sectional area of 15 000 mm^2 and the windings are chosen to operate the iron at a maximum flux density of 1.1 T from a 50 Hz supply. If the secondary winding consists of 66 turns, estimate the kVA output if the winding is connected to a load of 4 Ω impedance value.

Since area $= 15\ 000 \times 10^{-6}$ and $B_m = 1.1$ T

$$\therefore \Phi_m = 150 \times 10^{-4} \times 1.1 = 165 \times 10^{-4} \text{ Wb}$$

Substituting in the e.m.f. equation.

$$\begin{aligned} \text{Then} \quad V_2 &= 4.44\ \Phi_m\ fN_2 = 4.44 \times 165 \times 10^{-4} \times 50 \times 66 \\ &= 2.22 \times 165 \times 10^{-2} \times 66 = 1.11 \times 33 \times 66 \times 10^{-1} \\ &= 241.758 \text{ V} \end{aligned}$$

Estimated current $= \dfrac{241.758}{4} = 60.44$ A

Output rating $= \dfrac{241.8 \times 60.44}{1000} = 2.42 \times 6.04 = 14.62$ kVA.

CHAPTER 2

PRACTICE EXAMPLES

1.　　A single-phase transformer is designed to operate at 2 V per turn and a turns ratio of 3:1. If the secondary winding is to supply a load of 8 kVA at 80 V, determine: (a) the primary supply voltage; (b) the number of turns on each winding; (c) the current in each winding.

2.　　A 25 kVA, 440/110 V, 50 Hz, single-phase, marine-type, step-down, "engine room" transformer is designed to work with 1·5 V per turn with a flux density not exceeding 1·35 T. Calculate: (a) the required number of turns on the primary and secondary windings respectively; (b) the cross-sectional area of the iron core; (c) the secondary current.

3.　　A single-phase, marine-type, step-down transformer has the following particulars: Turns ratio 4:1. No-load current 5·0 A at 0·3 at a power factor of 0·3 (lagging). Secondary voltage 110 V. Secondary load 10 kVA at 0·8 power factor (lagging). Calculate: (a) the primary voltage, neglecting the internal voltage drop; (b) the secondary current on load; (c) the primary current; (d) the primary power factor.

4.　　A 6·6 kV, 50 Hz, single-phase transformer with a transformation ratio of 1:0·06 takes a no-load current of 0·7 A and a full-load current of 7·827 A when the secondary is loaded to 120 A at a power factor of 0·8 (lagging). What is the no-load power factor?

5.　　A 1 kVA, single-phase transformer has an iron loss of 15 W and a full-load copper loss of 30 W. Calculate its efficiency on full-load output at a power factor of 0·8 (lagging).

6.　　A single-phase power transformer supplies a load of 20 kVA at a power-factor of 0·81 (lagging). The iron loss of the transformer is 200 W and the copper loss at this load is 180 W. Calculate: (a) the efficiency; (b) if the load is now changed to 30 kVA at a power-factor of 0·91 (lagging), calculate the new efficiency.

7.   A 20 kVA, 2000/220 V, single-phase transformer has a
primary resistance of 2·5 Ω and a secondary resistance of
0·028 Ω, the corresponding leakage reactance being 2·8 Ω and
0·032 Ω. If the secondary terminals were accidentally short-
circuited, estimate the current which would flow in the primary
circuit.

8.   The resistance of the primary and secondary windings of a
27·5 kVA, 450/112 V, single-phase, marine-type transformer
are 0·055 Ω and 0·00325 Ω respectively. At the rated supply
voltage the iron loss is 170 W. Calculate for this transformer:
(a) the full-load efficiency at a power factor of 0·8 (lagging);
(b) the kVA output at which efficiency is a maximum at a
power factor of 0·8 (lagging); (c) the value of maximum
efficiency at a power factor of 0·8 (lagging).

9.   A three-phase, marine, dry-type transformer is used to step
down the voltage of a three-phase, star-connected alternator to
provide the supply for 120 V lighting. The transformer has a
4:1 phase turns ratio and is delta connected on the primary
side and star connected on the secondary side. If the lighting is
supplied at the line voltage of the transformer, what must be
the phase voltage of the alternator.

10.   A 200 kVA, 6600/415 V, three-phase transformer connected
in delta/star supplies a 120 kW, 415 V, 50 Hz three-phase motor
whose power factor and efficiency are 0·8 (lagging) and 83 per
cent respectively. Neglecting the transformer losses, calculate
the current in each transformer winding.

D.C. MACHINES (ii)

Before considering the fuller treatment of Commutation and Armature Reaction for the d.c. machine, the reader is advised to revise the work already done in connection with basics, as introduced in Volume 6, Chapter 6, pages 127 to 135 and Chapter 8, pages 158 and 167. The development of the modern d.c. armature has resulted in refinements such as *equalising rings, duplex windings, etc.* but, in its simplest form, the pattern followed by the connecting up of the coil elements to each other and to the commutator segments is called a winding. The coil elements can be preformed and can consist of several turns in series or even of similar multi-turn coils connected in parallel. These coil elements are connected up to form a closed circuit and the method used results in the winding being classed as either a "lap" or "wave".

The diagram of Fig. 88 (Vol. 6), has been repeated here as Fig. 26 to show a simple lap winding, having 16 conductors arranged in eight slots for use in a four-pole field system. Each coil element consists of two conductors in slots spaced almost 90° apart *i.e.* one pole-pitch, and the ends of the coils are connected to adjacent segments on the commutator. The equivalent ring winding has also been drawn out to show that there are four paths in parallel between the + ve and —ve terminals of the machine. Note that for a lap winding, it is a rule that the number of parallel paths is equal to the number of poles. For a large machine due to unequal air-gaps, the e.m.f.s generated across the various parallel paths may not be equal, an assumption which has already been made for a simple lap winding. Such dissimilarity of e.m.f.s tend to make the current-sharing unequal in the parallel paths and to rectify this, use is made of "equalising rings". Such rings are made of copper conductors, insulated from the armature but mounted on it at the end remote from the commutator. They are connected to points on the winding which should have the same potential and provide a path for the circulating currents which would result from the unbalanced e.m.f.s. They prevent such currents passing through the brushes and giving rise to undesirable heating and commutation effects. The number of equalising rings and connections depends on the design and size of the machine.

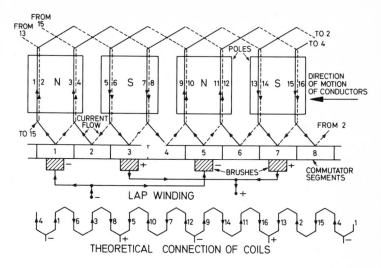

Fig. 26

The wave winding of Fig. 89, (Vol. 6) is also shown here as Fig. 27. It is seen to be for 18 conductors arranged in nine slots and is for use in a four-pole machine. Due to the method of making the winding, the ends of each coil element are connected to segments which are spaced some two pole-pitches apart *i.e.* 180°. In passing from one commutator segment to the next adjacent segments we pass through two coil elements in series or, in this case, we pass round the armature by way of two coil elements in series.

The ring winding, as drawn out, shows that there are only two paths in parallel, namely the generated e.m.f.s converge at only one point *i.e.* at conductors Nos. 2 and 13. Conductors 5 and 10 are in the interpolar-gap and have no e.m.f.s induced in them, Brushes can therefore be placed at either end of the coil element so formed or, as in this case, at both ends. Since the wave winding only has two parallel paths, it follows that in general there are more conductors in series than for the corresponding conductor arrangement of a lap winding. This statement only holds for a machine with more than two poles. The wave winding would thus appear to be more suitable for machines where a given *kW* rating is required at a comparatively high voltage and correspondingly lower current. For a machine of the same rating, but lower voltage and higher current, the lap winding would appear suitable. This is a general rule but is not always followed in practice. The choice of winding is decided by experience and for large machines, many variations are used, one principal object being to reduce the voltage between commutator segments. Thus for the "duplex winding" arrangement, two completely independent windings may be used which are connected to alternate segments. The windings are put in parallel by the brushes which could span more than one segment.

Another expedient is to choose a winding pitch which ensures that the winding traverses the armature more than once before it closes on itself. This arrangement makes for a voltage reduction between segments and allows better commutation conditions.

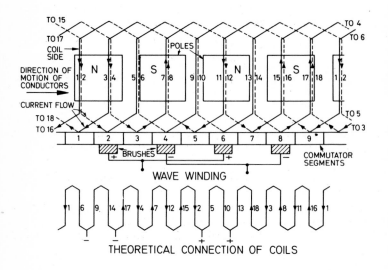

Fig. 27

Example 21. The armature of a six-pole generator has a lap winding with 120 single-turn coils. Calculate the e.m.f. at no load when running at a speed of 500 rev/min, when the flux per pole is 0·055 Wb. If the resistance of the armature winding, measured between diametrically opposite segments on the commutator, is 0·1 Ω, calculate the terminal voltage for a load current of 400 A, the speed and field current remaining unchanged. Allow 0·5 V for the drop at a brush and 10 A for shunt excitation. Neglect the effects of armature recreation.

E.m.f. as given from the voltage equation for a generator.

$$= \frac{240 \times 0 \cdot 055 \times 500}{60} \times \frac{6}{6} = 4 \times 5 \cdot 5 \times 5 = 110 \text{ V}$$

Note. The point of this problem is the determination of R_a. Obviously the armature resistance has not been measured with the six brushes making contact. If the test has been made as stated, then there were only two paths in parallel. The diagram of Fig. 28 shows the reasoning.

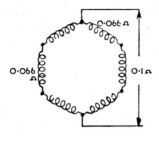

Fig. 28

As the joint resistance was 0.1 Ω then the resistance of one parallel path during test was 0.2 Ω. One such parallel path would consist of three of the armature parallel paths $= \dfrac{0.2}{3} = 0.066$ Ω. By an armature parallel path, is meant the actual circuits along which currents flow when the machine is running on load. Under these conditions the six paths would be in parallel or the armature resistance R_a would be

$$\frac{0.066}{6} = 0.011 \ \Omega$$

The on-load terminal voltage is given by:

$V = E - I_a R_a - $ brush-volt drop. Here $I_a = 400 + 10 = 410$ A
$= 110 - (410 \times 0.011) - (2 \times 0.5) = 110 - 4.51 - 1.0$
$= 110 - 5.51 = 104.49$ V.

COMMUTATION

For both lap and wave windings it is seen that, as a brush passes from one segment to the next, a coil element, consisting of a single coil or a series of coils, is being short-circuited. This is the period when commutation is occurring. It should also be remembered that the brushes are placed at the points on the commutator where the currents converge from and diverge to the parallel paths of the armature. At these points the brushes contact adjacent segments which are connected to coil elements lying in the inter-polar gaps and in the optimum position for being short-circuited. Summarising, we see that the current in the armature coil elements, on either side of the brushes, are in opposite directions and as the commutator turns and a complete short-circuited element passes under a brush, from one side to the other, the current in this element must be reversed. If the arrangement is illustrated by the diagrams of Fig. 29, we see that as the adjacent segments pass under a brush, the current in the connected coil element X must be stopped and restarted in the opposite direction. To assist the following reasoning, a width of brush equal to that of a commutator segment has been assumed.

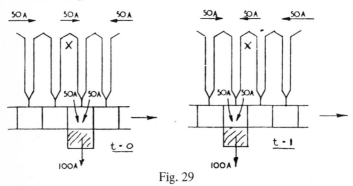

Fig. 29

The diagram of Fig. 30 shows the current/time curve for the coil element X which is being commutated. The time marked as "t" represents the period during which the current direction is being reversed *i.e.* the time taken for the brush to pass completely from one segment to the next, which may be of the order of 1/1000th second or less. Note that the $+$ ve and $-$ ve signs of the current ordinate are purely arbitrary and are merely used to indicate the reversal of current.

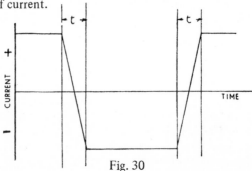

Fig. 30

The graph illustrates "straight-line commutation" where each section of the brush contact area takes its correct proportion of current. This is further illustrated by the diagrams of Fig. 31.

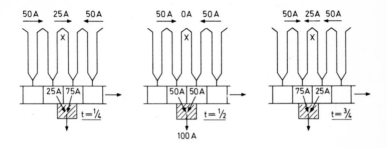

Fig. 31

The commutation condition considered so far has been the ideal case and cannot be attained in practice. For the coil element being commutated, because of the self and mutual inductance effects due to the rapid collapse of the flux associated with the changing current, an induced e.m.f. is set up which is called the "reactance voltage". According to Lenz's law, this e.m.f. opposes and tends to slow down the rate of current change. Thus the current in the short-circuited coil element tends to change more slowly and this condition is shown by the new current/time graph of Fig. 32.

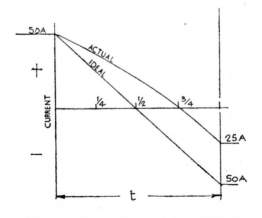

Fig. 32

It is seen from the graph of the actual working condition that although the current in the coil has reversed, it has not reached its full value when commutation should be completed. The effect of this "under-commutation" is further illustrated by Fig. 33, from which it will be seen that, for the instant being considered, *i.e.* when $t = \frac{3}{4}$, the last $\frac{1}{4}$-section of the brush is carrying 50 A instead of the 25 A for ideal straight-line commutation. The effect of this crowding of current into the trailing edge of the brush results in sparking, burning and overheating. Furthermore, since at the instant $t = 1$, when the segment is due to break contact with the brush, the current in the coil element being commutated is only 25 A, then the remaining 25 A has to pass between the segment and brush as an arc, which is drawn out to persist till the current in coil X rises to its correct value. This arcing and the heating, already mentioned, results in damage to the commutator. The condition of arcing is shown by Fig. 34.

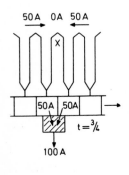

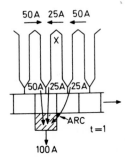

Fig. 33   Fig. 34

Since the inductance of the coil element X and the resulting reactance voltage is the principal cause of poor commutation, it is apparent that to improve the latter, some outside assistance must be provided. It will be appreciated that, if an e.m.f. can be generated in the coil element X which will produce a current in the same direction as will occur when the element enters the armature-current path after it has been commutated, then the effect of under-commutation can be nullified. This therefore, is the principal method of improving commutation but, before considering it in detail, attention is drawn to the fact that commutation can benefit by increasing the resistance of the brushes. "Resistance commutation" is invariably used for small machines and the associated theory will also be considered amongst the methods employed for improving commutation.

Example 22. The following data relates to a machine for which the conditions of commutation are being considered. If the current collected by a brush is 100 A and the speed of the commutator is 1200 rev/min, calculate the reactance voltage. There are 50 comutator segments and the inductance of the coil element connected to adjacent segments is 10 μH.

Speed of commutator $= \dfrac{1200}{60} = 20$ revolutions per second

Thus 1 revolution takes $\dfrac{1}{20}$ second

The time taken for a segment to pass under a brush $= \dfrac{1}{20 \times 50}$

$= 1 \times 10^{-3}$ second

Current in a coil element being commutated $= \dfrac{100}{2}$ or 50 A

Rate of change of current in such a coil $= \dfrac{50 - (-50)}{1 \times 10^{-3}}$

$= 100 \times 10^{3} = 1 \times 10^{5}$ amperes per second

So reactance voltage $= L \dfrac{di}{dt} = 10 \times 10^{-6} \times 10^{5} = 1$ V.

METHODS OF IMPROVING COMMUTATION. These are considered under three distinct headings but a combination of two methods is frequently used.

1. Resistance Commutation. The material most commonly used for the brushes of d.c. machines is carbon, the resistivity of which is much higher than that of copper. The effect of a high resistance brush is now considered in conjunction with the diagram of Fig. 35.

As segment A contacts and moves past the brush, the current from coil element Y can pass along two alternative paths. Thus it can flow directly down the section of brush marked **a** or can continue to pass through the coil element X and into the other section of the brush, shown as **b**. As a result of the movement of commutator under brush, section **b** gets smaller and its resistance rises whereas section **a** gets larger and its resistance falls.

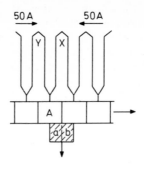

Fig. 35

The overall effect of brush resistance on commutation is now seen. At the first instant of contact of segment A with a high resistance brush, little current passes and the correct unchanged value of current continues to flow through coil element X. As more of the brush is contacted, the resistance of brush path a falls inversely in proportion to the movement, and current is diverted to take this falling resistance path. The current value in coil element X thus falls in accordance with the movement of the commutator segments and more closely follows the required condition of straight-line commutation. Similarly when the current in X has reversed and is beginning to increase, correct commutation is assisted by the rising resistance of section **b** and the falling resistance of section **a**. The rising reversed current is thus forced through the coil in accordance with the ideal requirements.

2. E.M.F. Commutation. Mention has been made already of a method of improving commutation by generating a reversing e.m.f. in the short-circuited coil element, Such a reverse e.m.f. would oppose the reactance voltage and be achieved, (a) by giving the brushes a "lead" or, (b) by using Commutating Poles — these are frequently referred to as Interpoles.

(a) Displacement of Brushes. It has been seen that, for a drum winding to operate correctly, the brushes are placed in a neutral position to contact segments which are connected to coil elements which are in the interpolar gaps. For this condition, no e.m.f. is induced in the coils which are being short-circuited by the brushes but if, for a generator, the brushes are moved forward — in the direction of rotation, to a position slightly in advance of the neutral position, then commutation takes place in a reversing field. Since the coil element, being short-circuited, is now no longer in the interpolar gap but is in the field of the main pole ahead, an e.m.f. is induced because of the

armature rotation and its direction is opposite to that which resulted
from the coil element passing under the poles prior to commutation.
The new e.m.f. thus opposes the reactance voltage *i.e.* it tends to
stop the original current and accelerates its build-up in the reverse direction to satisfy the condition of ideal commutation. The method
needs a different "angle of lead" of the brush-rocker for each value
of load current, but for small machines a satisfactory compromise
can be achieved by fixing the brushes at the "lead" required for about
three-quarter load. This position, together with resistance commutation, can give a reasonably sparkless performance over the working
load range. For a motor, the displacement of the brushes is in the
opposite direction *i.e.* it is against the direction of rotation and is an
angle of "lag". This is because the motoring current in the armature
would be in a direction opposite to that for generating and the reversing e.m.f. must be opposite *i.e.* obtained from the pole behind.

(b) Use of Commutating Poles. It is apparent that, instead of
moving forward the coil element being commutated to be under the
influence of the pole ahead, the same effect could be achieved by
moving the magnetic field backwards. This can be effected by introducing small extra electromagnetically-energised poles midway between the main poles, whose polarity and flux density can be decided
by the direction and magnitude of the current passing through the
exciting coils. These commutating poles (compoles) are run at low
flux densities to avoid their being saturated, and are energised by
coils which consist of a few turns of wire or strip conductor,
connected in series with the armature. The energising ampere-turns
will thus vary the commutating flux density in accordance with the
armature current and the commutating e.m.f. generated in the short-
circuited coil, will thus vary in relation to the reactance voltage which,
in turn, varies with the load *i.e.* the magnitude of the current in the
coil being commutated.

Although armature reaction has yet to be studied in detail, the
diagram (Fig. 36a) shows that the revolving armature itself behaves
as a stationary electromagnet when the conductors are carrying current. If small poles are placed as shown then flux caused by armature
ampere-turns will pass as indicated. Note that flux passes downwards
through the armature and round the yoke as indicated. The polarity
shown is that allocated to the compoles *i.e.* the flux passes from a N
pole through the armature to the S pole.

Since the flux due to the armature reaction must be neutralised
by the magnetic field of the commutating poles, if the latter are to
achieve their purpose, then the ampere-turns of the compoles should
be sufficient to achieve this in addition to producing the induced e.m.f.

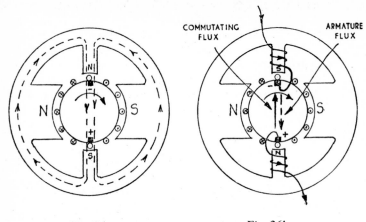

Fig. 36a   Fig. 36b

of commutation. The arrangement (Fig. 36b), means that the brushes need not be moved from the neutral position and the ampere-turns per pole of the commutating poles are usually made to exceed the armature ampere-turns by roughly 25 per cent. For a generator, the interpoles have the same polarity as the main poles ahead of them and for a motor, the same polarity as the main poles behind them. The direction of rotation is taken as the reference.

Example 23. A 150 kW, 250 V, six-pole d.c. generator has a lap-wound armature with 432 conductors. Calculate the number of turns per pole required for the commutating poles, assuming the brushes to be placed at the neutral point and the compole ampere-turns per pole to be about 1·3 times the armature turns per pole.

This being a multi-pole machine, a solution is more readily effected if one pole pair or two-pole pitches of the armature are considered.

Full-load current $= \dfrac{150\,000}{250} = 600$ A

Current per conductor $= \dfrac{600}{6} = 100$ A

Effective conductors in two pole-pitches $= \dfrac{432}{3} = 144$

Effective armature turns $= 72$

Thus armature ampere-turns $= 7200$

Commutating ampere-turns $= 7200 \times 1 \cdot 3 = 9360$ At

Turns for one pole pair $= \dfrac{9360}{600} = 15 \cdot 6$ say 16

Turns per pole $= 8$.

Note. The total armature current passes through the interpole winding.

ARMATURE REACTION

An introduction to the effect of the armature conductors producing flux, when the machine is supplying or taking current, has already been made. This is now investigated to a greater extent and explanation is made of the causes of the armature reaction effect, as was mentioned, when considering the characteristics of the various types of d.c. generator and motor. In this context armature reaction can be defined as the effect on the main-field flux caused by the armature flux. This armature flux, due to the armature current, distorts and weakens the main-field flux.

BRUSHES ON NEUTRAL AXIS. Consider the simple two-pole machine as already introduced in Fig. 36a, but without compoles. To conform with basic theory, the brushes should be on the neutral axis which is midway between the main-poles *i.e.* on the axis referred to as the Magnetic Neutral Axis. Here the M.N.A. coincides with the G.N.A. — the Geometric Neutral Axis. If the armature is loaded as a generator and carries current then, if no interpole projections are provided, the associated armature flux will take the path of lowest reluctance and will cross the air-gap as shown by the illustration of Fig. 37.

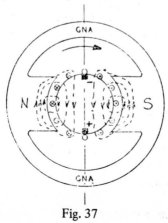

Fig. 37

The armature flux is at right angles to the main flux and is termed a "cross-magnetising flux". It crosses each air-gap twice *i.e.* if the *N* pole-piece is considered, it is seen that the armature flux is opposite to the main flux at the *leading* pole-tip and in the same direction at the *trailing* pole tip. The terms leading and trailing are commonly used in this connection and, here again, the direction of rotation of the armature is taken as the reference for making the distinction. The net effect of the cross-magnetising flux can be illustrated by the diagrams (Fig. 38a and 38 b) where it is seen that the total flux appears to be twisted and the M.N.A. has therefore moved through an angle of lead.

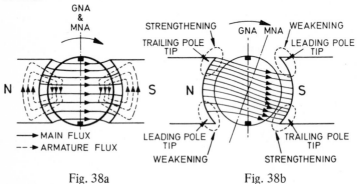

Fig. 38a   Fig. 38b

If the flux effects are illustrated by an m.m.f. vector diagram, then Fig. 39 shows the effect of armature reaction as developed up to this point of theory. F_m is the main field m.m.f. and F_a is the armature or cross-magnetising m.m.f.

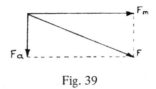

Fig. 39

It now becomes evident that, since the magnetic axis moves forward, then the brushes must be on this axis in order to function satisfactorily and must also be moved forward onto the new M.N.A. The effect of such movement will be considered next but it is pointed out that, since the angle of lead will change as the armature current changes so the brush position should be changed. This makes for an unsatisfactory arrangement and indicates that any attempt to nullify the effect of the cross-magnetising flux would be advantageous. Mention of this has already been made when considering commutating poles.

BRUSHES GIVEN AN ANGLE OF LEAD.   The arrangement is shown by Fig. 40, with the brushes in optimum position for a generator *i.e.* contacting the commutator at the appropriate tapping points and at the same time allowing correct commutation of the coil element being short-circuited.

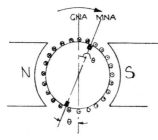

Fig. 40

With the passage of current, the axis of the armature electro-magnet is seen to have been turned through an angle θ° and the m.m.f. vector F_a is now as shown (Fig. 41).

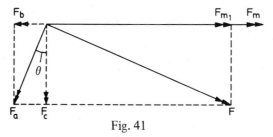

Fig. 41

Since the armature flux combines with the main flux to give the resultant flux, the arrangement can be represented by the new m.m.f. vector diagram. The resultant F is obtained by resolving F_a into its quadrature components F_c and F_b. F_b is a demagnetising component which weakens F_m to give the new value F_{m_1}. F_c is the cross-magnetising component which acts with F_{m_1} to give the new result-ant F.

It is now apparent that, in order to ensure satisfactory commuta-tion, the new armature reaction condition results in the two effects mentioned earlier, namely (a) the main field is weakened and (b) the main-field flux density in the air-gap is distorted. The results of such distortion are yet to be considered, but it should be noted that a forward movement of brushes or angle of lead is only required for the generator. For the machine operating as a motor, the direction of armature current would be reversed for the same direction of rotation and polarity of the main field. As a result the armature flux would be reversed and the brushes would need to be given an angle of lag for satisfactory operation. The main flux would be distorted and weakened as before, and if the brushes are moved when a motor is running, it will be noted that the speed can be varied. This is not a recommended method of speed variation since optimum commutation conditions will be desired, but the cause of the speed variation is explained in that, the flux density is being varied.

CROSS-MAGNETISING AND DEMAGNETISING AMPERE-TURNS. The basic diagram of Fig. 40 has been redrawn for Fig. 42, to show that the armature ampere-turns can be considered to consist of two distinct portions. The band of conductors, between the lines XX and YY, constitute an energising electromagnet whose effect can be represented by the m.m.f. F_b of the diagram (Fig. 41). These conductors when carrying current, form ampere-turns which oppose F_m, the main-field ampere-turns and are described as demagnetising ampere-turns producing the final value of main field shown as F_{m_1}. It will be seen that the demagnetising ampere-turns are proportional to the angle of lead or lag and this corresponds to the deductions which can be made from the vector diagram *i.e.* $F_b = F_a \sin \theta$.

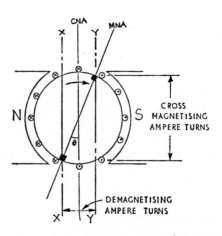

Fig. 42

The conductors outside the lines XX and YY constitute an ampere-turn band or energising solenoid when carrying current, which results in the cross-magnetising m.m.f., F_c as already described. It should be noted that this cross-magnetising m.m.f. results from the total armature ampere-turns, if the brushes are not moved off the geometrical neutral axis. This condition has already been covered when the commutating poles were being considered.

It should also be remembered that since armature reaction reduces the main field, then the armature demagnetising ampere-turns, as set up by the movement of the brushes, should not be too great when related to the ampere-turns required for the main field under full-load conditions. The ratio of the latter to the former is always made greater than 1, in order to ensure that the effects of armature reaction are not too pronounced and that the machine characteristics are not unduly altered.

Example 24. A 150 kW, 250 V, six-pole d.c. generator has a lap-wound armature with 432 conductors. If the brushes are given an angle of lead of 5° from the geometrical neutral axis, estimate the additional ampere-turns per pole required to neutralise the demagnetising effect of the armature when the machine is on full load. Neglect the shunt field current.

As for Example 23 the solution is effected by considering two pole-pitches.

Full-load current $= \dfrac{150\ 000}{250} = 600$ A

Current per conductor $= \dfrac{600}{6} = 100$ A

Effective conductors in two pole-pitches $= \dfrac{432}{3} = 144$

Effective armature turns $= 72$

Thus armature ampere-turns $= 7200$

Since this is a six-pole machine then $\dfrac{360}{6}$ or 60 mechanical degrees

are equivalent to 180 electrical degrees or

$1°$ (mechanical) $= 3°$ (electrical)

If θ is the brush shift in electrical degrees then here $\theta = 15°$(elec). Again the proportion of armature demagnetising ampere-turns to the total armature ampere-turns is given by $\dfrac{2\theta}{180}$ (Ref. Fig. 42).

Thus demagnetising ampere-turns per pole pair

$$= \dfrac{2 \times 15}{180} \times 7200 = 1200$$

or demagnetising ampere-turns/pole $= 600$.

Note that although the cross-magnetising ampere-turns have not been asked for, these can be determined thus:

Total armature ampere-turns $= 7200$ Demagnetising ampere-turns

Total armature ampere-turns $= 7200$

Demagnetising ampere-turns $= 1200$

$\therefore$ Cross-magnetising ampere-turns $= 7200 - 1200 = 6000$

Cross-magnetising ampere-turns/pole $= 3000$.

FLUX-DENSITY DISTRIBUTION AROUND THE ARMATURE. The armature and field poles have been drawn (Fig. 43) in a developed form.

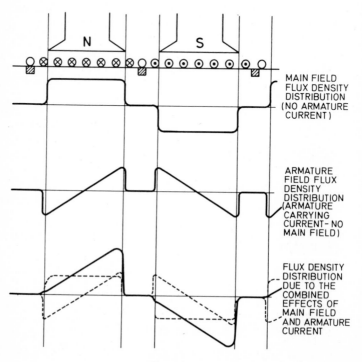

Fig. 43

The flux-density distribution under the main poles and in the interpolar gap, as caused by the main-field ampere-turns and the armature
ampere-turns, are shown both separately and combined. It is seen
that the armature-current m.m.f. acts with the main-field m.m.f. to
produce a resultant m.m.f. and flux-density distribution pattern which
is uneven over the pole arc. It is also evident that the e.m.f.s generated
in the armature-coil elements are not uniform as these pass across the
pole face and the voltages between commutator segments may, for
some positions, be appreciably higher than the mean value. If the field
distortion becomes very pronounced there is a danger of a "flash-
over" across the commutator. The effect of the armature m.m.f. can
be countered to some degree by fitting a *compensating winding.*
Such a winding will be described, but it is stressed that this is an
expensive solution and is only used for large machines which are subjected to heavy and sudden current peaks. Examples of such applications include, the motors for steel-rolling mills and special generators
such as those used for some types of electric locomotives.

The next diagram (Fig. 44) shows in more detail the m.m.f. and
resulting flux-density distribution under a pole with the armature
carrying current. The effect of moving the brushes and the introduction of a compole is also shown. The illustration is for a generator.

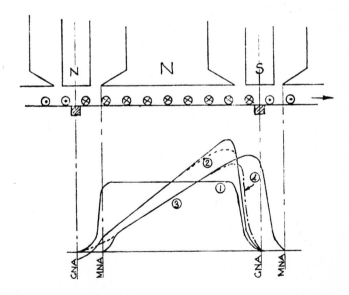

Fig. 44

If the brushes are on the G.N.A. and assuming no saturation of the iron it is seen that the area under the flux-density distribution curve No. 2, which is for the armature carrying current, is the same as that for the No. 1 curve (no-load). However since saturation of the iron does occur on load, especially at the armature teeth, the curve would be more like that shown dotted. This indicates an over-all reduction of flux. Again if the brushes are given a lead, as for a generator then, due to the demagnetising effect of the armature ampere-turns, the area will be further decreased and the distribution would be as indicated by curve No. 3. If interpoles are provided then the brushes are not moved and the curve is as No. 4. The overall effect of armature reaction is illustrated by the final curves which confirm the field weakening and distorting effect of the armature current.

Compensating Windings. These will not be dealt with in any detail since their usage is very restricted. The windings are fitted into the main pole faces as illustrated by Fig. 45. The coils are connected in series with the armature and compoles. Since the winding is close to the armature conductors and current in the opposite sense to the latter is carried, then the armature flux is nullified and the distortion of the main field becomes negligible. The provision of a compensating winding also ensures that the compoles can achieve their purpose since they do not now have to overcome the armature cross-magnetising field and can readily create the correct commutating

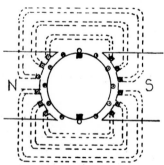

Fig. 45

field over a wide range of load current. It should be remembered that although compoles can be arranged to control the flux due to the armature in the interpolar gap, they do not assist with the distortion of the field in the air-gaps under the poles. Compensating windings however, do just this.

SPECIAL D.C. MACHINES

Types of d.c. machine are now considered which are required to
perform rather special duties and because of this may have special
constructional features or operating characteristics. Many of these
machines, especially in the smaller sizes have applications in the
marine field and will therefore be of interest to the engineer.

THE ROTARY TRANSFORMER

This machine is a d.c. generator and d.c. motor combined and
consists of an armature having separate windings on the same core
and operating in a common field system. Each winding is connected
to its own commutator and its physical dimensions are decided by
the factors already considered when considering d.c. machine construction. Thus a popular model of ship's recording echometer has
been standardised to operate off a 24 V d.c. supply. The ship's supply
is assumed to be 220 V d.c., so a static transformer cannot be used
and for this application a small rotary transformer would be suitable.
If the echometer required 36 W then one widing would be wound for

$$\frac{36}{24} = 1\cdot5 \text{ A and would serve the generator section. Assuming an}$$

overall efficiency of some 60 per cent then the input would be some

$$\frac{100}{60} \times 36 = 60 \text{ W. The other winding would be rated at}$$

$$\frac{60}{220} = \frac{3}{11} = 0\cdot273 \text{ A or 273 mA and would function as a motoring}$$

winding. The field flux can be provided by a permanent magnet or
by field coils which are usually shunt-connected and energised from
the motor supply. Commutation requirements for these machines
are relatively straightforward since the armature reaction effect is
much reduced by the fluxes of the motoring and generating currents
tending to cancel each other out. There is no provision in the field
for voltage control of the generating side, since this would alter the
common flux and thus vary the speed of the motor. The machine is
most commonly installed for a specific duty and operates at pre-
determined speed and conversion conditions. It is started like a d.c.
motor.

THE ROTARY CONVERTER

As its name implies this machine is capable of converting a.c. to d.c. or vice versa, but with the development of the semi-conductor types of rectifier, its usage for the former purpose is being discontinued. It can be run "inverted" *i.e.* it will supply a.c. from a d.c. source, but because there is a fixed relation between the a.c. and d.c. voltage and the fact that varying the field will alter the speed and frequency, the arrangement of the more usual motor-alternator set is usually preferred for ship work.

In construction the rotary converter is similar to a d.c. generator except that slip-rings are mounted on the shaft on the side of the armature opposite to the commutator. The slip-rings are connected to suitable points on the common armature winding and single-phase or polyphase supply can thus be obtained. The field is excited and connected like a d.c. machine and the machine is started like a normal d.c. motor if a suitable supply is available. Starting from the a.c. side presents difficulty and special arrangements have to be made. This is another disadvantage of the machine, if required to operate on an a.c. ship. The fixed ratio between the voltage of the a.c. supply to the d.c. output is explained by the fact that the same armature conductors are used which rotate in the same field system. The ratio depends on the way the winding is tapped on the a.c. side. Thus for a single-phase to d.c. converter, the ratio of a.c. to d.c. voltage is 0·707:1 whereas, for a three-phase to d.c. machine the ratio is 0·612:1. Since the ratios are not such as to give a standard d.c. voltage from a standard a.c. voltage, it is apparent that a transformer must be used to obtain a suitable input a.c. voltage for the desired d.c. output voltage.

Voltage and Current Relations. A fundamental difference exists between the slip-ring method and the commutator method of tapping the e.m.f. generated in the armature winding. The voltage appearing between two diametrically opposite points of the winding is made up from all the e.m.f.s in all the conductors around half the armature. On the a.c. side the armature winding presents itself as a coil which is rotating through two pole-pitches to generate one cycle of e.m.f. For the position of cutting maximum flux, the voltage generated at the tapping points is the maximum value of the a.c. waveform. If E_m is the maximum value of the waveform then the r.m.s. value $= 0·707\,E_m$ and the voltage between the slip-rings $E_{ph} = 0·707\,E_m$ or $E_{ph} = 0·707 \times$ the maximum value of the total voltage of all conductors over one half of the armature.

On the d.c. side as a coil element is being commutated, it passes from one half of the armature winding into the other half with the consequence that the winding functions as though it is fixed in space and is always of a maximum value. The voltage across the brushes is unvarying and is equal in value to the sum of the e.m.f.s round half the armature. As before it is equal to E_m so for single-phase working the a.c. voltage $E_{ph} = 0.707 E_m$ and d.c. voltage $E = E_m$.

Thus a.c. voltage: d.c. voltage = 0.707:1 or
$$E_{ph}: E = 0.707:1 \text{ for single-phase working.}$$
The current values on the a.c. and d.c. sides of the rotary converter are not the same and can be deduced for single-phase working as a ratio of a.c. current to d.c. current. Thus neglecting the losses:

a.c. output $=$ d.c. output

or $E_{ph}I_{ph} \cos \phi = EI$

If $\cos \phi = 1$ then

$$I_{ph} = \frac{EI}{E_{ph} \cos \phi} \text{ or } \frac{E_m I}{0.707 E_m} = 1.414 I$$

Thus $I_{ph} : I = 1.414:1$

The following example illustrates the full significance of the voltage and current ratios.

Example 25. The d.c. load supplied by a single-phase rotary converter is 20 A at 24 V as used for battery charging. Find the ratio and rating of a suitable transformer to enable the machine to be energised from 230 V a.c. mains.

A.C. voltage on input side to converter $\begin{aligned} &= 24 \times 0.707 \\ &= 16.97 \text{ V} \end{aligned}$

A.C. current on input side to converter $\begin{aligned} &= 20 \times 1.414 \\ &= 28.28 \text{ A} \end{aligned}$

Rating of transformer $= 16.97 \times 28.28 = 479.9$ say 480 VA

Ratio of transformer $= \dfrac{230}{16.97} \doteqdot \dfrac{13.55}{1}$ or 13.55:1

Primary current $\dfrac{28.28}{13.55} = 2.087$ A.

THE ROTATING AMPLIFIER
This is also known as a *cross-field generator* (C.F.G.) and appears in forms developed for specific applications. As such it is given a

proprietory name such as the Amplidyne, a machine manufactured by The B.T.H. Co., or the Metadyne, manufactured by The Metropolitan Vickers Co. These two firms now constitute the main companies of the G.E.C./A.E.I. combine, but the names of the machines are retained. Similarly we have the Magnicon, a cross-field machine developed by The Macfarlane Engineering Co. Ltd. and the Magnavolt of The English Electric Co. Ltd. Since the basic principles of operation are substantially the same, only a general description of the "amplifying dynamo" will be considered.

From knowledge of basic theory, it will be appreciated that, any d.c. generator is in fact a rotating amplifier, since a small input field current can be used to produce and control a much larger output current. By modern standards the ratio of amplification between output and input is small and the rate of response is poor because of the high inductance of the field windings. The cross-field machine has been developed to overcome these disadvantages and has an unusually high *power amplification ratio.* This ratio is usually expressed as $\dfrac{\text{output watts}}{\text{control watts}}$ and a figure of 5000 or more is possible, but it should be noted that, as this value increases the speed of response decreases. The high amplification is achieved by making the C.F.G. operate as a main generator and exciter combined into one machine. By controlling the field of the exciter, which supplies the field of the main generator, a small value of control power controls a very much larger output power. The control watts are independent of the machine output and the speed of response is improved by designing the field coils with a low *"time constant" i.e.* a high resistance to inductance ratio.

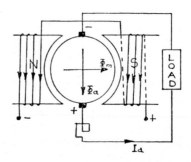

Fig. 46

Theory of Operation. This is readily followed if the work done on armature reaction has been understood. For this machine the armature reaction effect is encouraged and the illustration (Fig. 46) shows a two-pole generator where Φ_m is the main flux and Φ_a is the armature flux as produced by the currents in the armature conductors when this latter circuit is completed. Note that Φ_a is in quadrature with the main flux Φ_m and is a cross-magnetising flux. Fig. 46 is in accordance with the diagrams of Figs. 36 and 37.

If the load-resistance value was to be reduced so that eventually a short-circuit was applied across the brushes, the armature current would be large and in order to reduce it to the normal value, the field current would have to be reduced in order to reduce Φ_m. Thus for the same armature current I_a and armature flux Φ_a a very much smaller value of field current I_f is required. If consideration is now given to using Φ_a as the main flux then the possibility of having a very small control current I_f will be realised. In practice this is achieved by placing a second pair of brushes CD midway between the short-circuited brushes AB as is shown by Fig. 47.

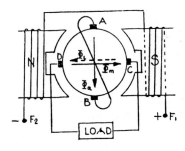

Fig. 47

Action is now as follows. A voltage is applied to the control field $F_1 F_2$ to cause a small current I_f to flow. This current creates a flux Φ_m to generate a low voltage across brushes AB, which being short-circuited, allows a large current to flow and produces a high value of armature flux Φ_a. The armature conductors which cut this latter flux will have generated in them a relatively large e.m.f. which, if brushes are placed at the appropriate tapping points i.e. mid-way between the short-circuited brushes, will allow a load circuit to be energised. Note that a region of minimum flux is introduced by shaping the main poles, as shown, in order to assist commutation for the brushes CD. In the final arrangement the magnetic circuit is so modified to allow the use of compoles.

For the cross-field generator, due to the position of the load circuit brushes CD in relation to the main field, when the machine is supplying load current the associated armature-reaction m.m.f. is on the same axis and in opposition to the main field m.m.f. The condition thus arises where armature flux, Φ_L, is almost equal to Φ_m and the effect is to give a load characteristic similar to that illustrated on Fig. 48. Any increase of "called for" load current will result in an increase of Φ_L. Since Φ_m is fixed, the difference will diminish resulting in a reduction of current in AB, a reduction in Φ_a, a reduction in generated voltage and a fall off of output current. The shape of the graph indicates that for a fixed field current I_f the output current is constant and this constant current characteristic can now be used to advantage for certain motor-control systems. Fig. 48 also shows the characteristic of a shunt generator for comparison purposes.

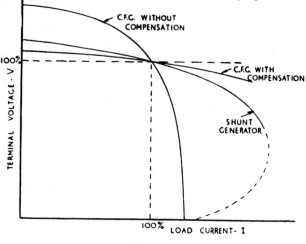

Fig. 48

THE COMPENSATED C.F.G. The constant-current feature of the cross-field generator is an undesirable feature for the rotary amplifier since in order to increase output current, the field current must be increased appreciably. The obvious method of overcoming the armature reaction counter m.m.f. is by the use of a compensating winding. This winding is suitably disposed in relation to the armature conductors, located in the main pole faces and connected in series with the load circuit. The opposing effect of the armature-reaction flux Φ_L is neutralised by the additional flux Φ_C, produced by the compensating winding, and the circuit arrangement is illustrated by Fig. 49. F is the control field, I is the compoles and C the compensating winding.

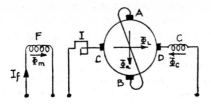

Fig. 49

We can now summarise by saying that the opposition and weakening effect of the armature-reaction flux Φ_L on the main flux Φ_m, produced by winding F is annulled by the compensating winding C which provides a flux to counter the armature-reaction flux. The third graph of Fig. 48 shows a machine with full compensation *i.e.* m.m.f. of winding C is made equal to the armature-reaction m.m.f. For practical purposes the control field current I_f is small (in the region of 50 to 250 milliamperes), the control ampere-turns being only 1 to 2 per cent of those required for a normal d.c. generator and it follows that the inductance is very small and rapid changes of I_f are possible. This allows the high speed of response required. The power input is usually under 1 W and can be obtained from electronic control equipment. The winding occupies so little space that several such windings can be provided, each of which can be made to control a definite feature of an operation so allowing an automatic sequence.

Applications. The machine is used to supply power where rapid control of voltage or current is needed, the property of a small I_f making it suitable for control systems. It can be used to reverse the output current and will accept regenerated power. A disadvantage in operation, is the heavy current circulating but this can be minimised by fitting a "quadrature series field winding", which is connected in series with the short-circuited brushes to increase the armature flux Φ_a and allow a reduction of the armature current and size of conductors. For practical applications, the C.F.G. can itself be used as a generator for ratings up to about 25 kW but its main application is as an exciter for a main generator. For current or voltage control of a generator, the "closed loop" system is used where the quantity being controlled is compared with a fixed reference quantity and any difference is used to adjust the C.F.G. excitation to reduce the discrepancy. Simple examples are shown below but specific marine applications have not been detailed. The rotary amplifier has been used for the control of main electric a.c. propulsion and for constant current d.c. propulsion systems, for automatic voltage regulation of auxiliary generating sets and for constant tension winches.

Voltage Control. A simple arrangement (Fig. 50a) is shown, where the main generator output voltage is compared with a fixed constant reference voltage, the value of which is determined from the potentiometer setting. The difference between these two voltages produces the control-field current and if load is applied to the generator causing its voltage to fall then this difference will rise and an increased excitation current will be applied to the control field to raise and maintain constant the generator output voltage.

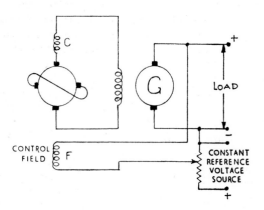

Fig. 50a

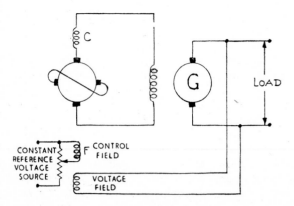

Fig. 50 b

An alternative arrangement (Fig. 50b), uses an additional field winding on the C.F.G. connected across the terminals. If the field ampere-turns are arranged to oppose or "buck" each other, it is apparent that the difference in ampere-turns, produces the necessary excitation flux and this is dependent on the load voltage of the main generator.

Current Control. Fig. 51 shows how this can be arranged. The load current value can be represented by a small voltage value by passing it through a shunt. The field is connected as shown and it is apparent that if the load current increases, the voltage across the shunt varies and so the current in the control field is varied to alter the excitation in sympathy and reduce the output voltage and load current to its acceptable value. Similarly if load current falls, the reference voltage exceeds the shunt voltage drop and the excitation is raised to increase the generated voltage and increase the current. The arrangement can be used for constant tension winch control.

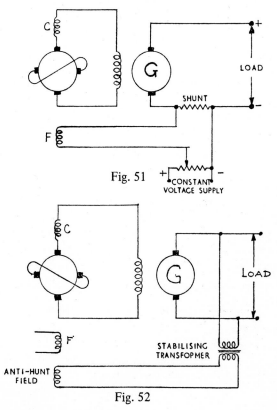

Fig. 51

Fig. 52

STABILISATION. An anti-hunting or stabilising feature is often fitted as shown, (Fig. 52). Fluctuations of the output voltage induce an e.m.f. into the transformer which, in turn, causes the anti-hunt field to be excited and to damp down the changes of output voltage. If no voltage fluctuation exists no e.m.f. is induced in the transformer secondary and the anti-hunt field is not energised.

For interest a simplified diagram of connections (Fig. 53) is given to illustrate the use of the amplidyne in the control of the propulsion machinery for the well known T2 Tankers. Although these ships are now obsolete, the use of the cross-field machine continues for special marine applications and the reader would be well advised to maintain interest in this type of electrical control. For the diagram the basic system of control can be readily followed but, by way of explanation, it should be noted that T is the stabilising transformer.

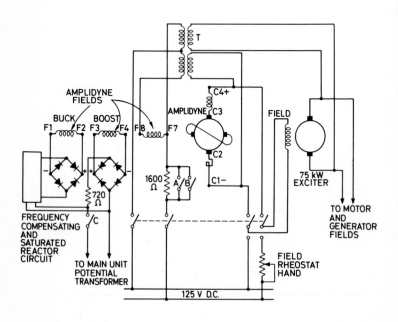

Fig. 53

Since the bulk of this chapter has been devoted to the descriptive treatment of d.c. machine variations and their specialised applications, calculations relating to points of theory have been avoided. The remaining examples of this chapter are therefore devoted to general exercises on d.c. machines and should be treated by the reader as very necessary revision illustrations.

Example 26. The field resistance of a d.c. shunt generator is 45 Ω and when driven at 500, 750 and 1000 rev/min the machine builds up to 55 V, 102·5 V and 150 V respectively. Determine the voltage to which the machine will self-excite, if driven at 750 rev/min with an additional 15 Ω included in its field circuit.

For the first condition, field current would be $\dfrac{55}{45}$ = 1·22 A and

if this current was maintained and the machine driven at 750 rev/min,

the O.C.C. voltage would be $55 \times \dfrac{750}{500}$ = 82·5 V

Similarly for the second condition. Field current $= \dfrac{102·5}{45}$

$$= 2·28 \text{ A}$$

If driven at 750 rev/min the O.C.C. voltage would be $150 \times \dfrac{750}{1000}$

$$= 112·5 \text{ V}$$

The O.C.C. curve at 750 rev/min with field resistance of 45 Ω can be drawn from the following table. The results are illustrated by Fig. 54.

I_f (amperes)	1·22	2·28	3·33
E (volts)	82·5	102·5	112·5

With an additional 15 Ω included, the field resistance becomes 60 Ω. Assuming a field current of 2 A the field voltage drop = 2 × 60 = 120 V. From this data the field voltage-drop line can be drawn and the voltage to which the machine excites is seen to be 88 V.

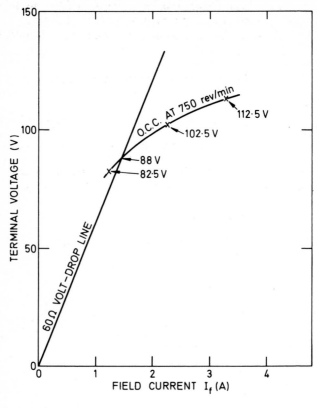

Fig. 54

Example 27. The relationship between the field current and the generated voltage of a d.c. generator driven at constant speed is as follows:

E (volts)	50	120	138	145	149	151	152
I_f (amperes)	0·5	1·0	1·5	2·0	2·5	3·0	3·5

The armature resistance is 0·1 Ω. The machine is connected as a shunt generator with the field circuit resistance adjusted to 53 Ω and is run at the same speed as before, compute and find (a) the O.C. voltage; (b) the load current when the terminal voltage is 140 V; (c) the terminal voltage when the load resistance is 0·4 Ω.

(a) One graph of Fig. 55 shows the O.C.C. plotted from the test result given. The field voltage-drop line is drawn for the point shown. This point is obtained by assuming a field current of 2·0 A *i.e.* $2 \times 53 = 106$ V. From the point of intersection, the O.C. voltage = 150 V.

(b) This is solved by deducing the load characteristic of the machine which is done thus:

Assume the load voltage to fall to 140 V then the field current will be $\dfrac{140}{53} = 2\cdot64$ A. Now from the O.C.C. a field current of 2·64 A will result in an e.m.f. of 149·V. Therefore the difference between the e.m.f. of 149·5 V and a terminal voltage of 140 V must be due to

the armature voltage drop $I_a R_a$ so $I_a R_a = 9\cdot5$ V and

$$I_a = \frac{9\cdot5}{0\cdot1} = 95 \text{ A}$$

∴ Load Current $I_L = 95 - 2\cdot64 = 92\cdot36$ A.

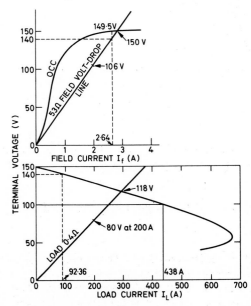

Fig. 55

Thus when the terminal voltage is 140 V the load current is 92·36 A. Simultaneously other points can be found and the load characteristic plotted as shown by the graph of Fig. 55. The values as deduced are tabulated thus

V volts	I_L amperes	V volts	I_L amperes
150	0	80	580
140	92·4	70	635
130	185	60	675
120	280	50	660
100	438	40	560

Note. As field current is very small compared with the large values of armature current, the graph V/I_a approximates very closely to V/I_L and the former can be used for approximate purposes.

(c) This is solved by drawing the load-resistance line. The point of intersection gives the voltage for this condition as 118 V. See Fig. 55.

Example 28. A 220 V shunt motor, when running light, takes 7 A and runs at 1250 rev/min. The armature resistance is 0·1 Ω and the shunt-field resistance 110 Ω. A series winding of resistance 0·05 Ω is added, long-shunt and cumulatively connected. This winding increases the flux per pole by 20 per cent when the motor is taking its full load current of 62 A. Assuming the increase in flux proportional to the armature current and neglecting the effect of armature reaction, find the speed (a) when running light; (b) when taking 62 A.

(a) No load.

$$\text{Shunt-field current} = \frac{220}{110} = 2 \text{ A}$$

$$I_{a_0} = 7 - 2 = 5 \text{ A} \quad E_{b_0} = 220 - (5 \times 0·1) = 219·5 \text{ V}$$

When the series winding is added and the motor takes its full load current of 62 A then $I_{a_1} = 62 - 2 = 60$ A. This is the value of current which, when flowing in the series winding, gives 20 per cent increase of no-load flux ∴ 5 A in the series winding gives

$$20 \times \frac{5}{60} = 1·66 \text{ per cent increase of flux or}$$

$$\Phi_{0_1} = \Phi_0 + \frac{1\cdot66}{100}\,\Phi_0 = 1\cdot0166\,\Phi_0$$

So when compounded
$$\begin{aligned} E_{b0_1} &= 220 - (5 \times 0\cdot1) - (5 \times 0\cdot05) \\ &= 220 - 0\cdot5 - 0\cdot25 \\ &= 220 - 0\cdot75 = 219\cdot25 \text{ V} \end{aligned}$$

$$\therefore \frac{N_{0_1}}{N_0} = \frac{E_{b0_1}}{E_{b0}} \times \frac{\Phi_0}{\Phi_{0_1}}$$

or $N_{0_1} = 1250 \times \dfrac{219\cdot25}{219\cdot5} \times \dfrac{\Phi_0}{1\cdot0166\,\Phi_0} = 1228$ rev/min

(b) Full load.

$$I_{a_1} = 62 - 2 = 60 \text{ A} \quad \Phi_1 = \Phi_0 + \frac{20}{100}\,\Phi_0 = 1\cdot2\,\Phi_0$$

$$E_{b_1} = 220 - (60 \times 0\cdot1) - (60 \times 0\cdot05) = 220 - 6 - 3 = 220 - 9$$

$$= 211 \text{ V}$$

$$\therefore N_1 = 1250 \times \frac{211}{219\cdot5} \times \frac{\Phi_0}{1\cdot2\,\Phi_0} = 1000 \text{ rev/min.}$$

Example 29. A marine shunt motor driving a "circulating water" pump is rated to take a current of 26·5 A at 220 V when running on full load. The speed under this condition is measured to be 725 rev/min., the armature resistance is 0·2 Ω and the field resistance 146·66 Ω. Estimate the new speed if a resistor unit of value 16·24 Ω is inserted into the field circuit by means of the speed regulator. Assume the load torque to remain unchanged and the flux is proportional to field current.

Condition 1. Original speed. Field current $I_{f_1} = \dfrac{220}{146\cdot66} = 1\cdot5$ A

Armature current $I_{a_1} = 26\cdot5 - 1\cdot5 = 25$ A

$$E_{b_1} = 220 - (25 \times 0\cdot2) = 220 - 5 = 215 \text{ V}$$

Condition 2. New Speed. The following deduction is necessary.
Since load torque remains unchanged $T_2 = T_1$
and since torque is proportional to flux and armature current and flux in turn is proportional to field current, then

$$T = k I_f I_a \quad \text{or} \quad \frac{T_2}{T_1} = \frac{k I_{f_2} I_{a_2}}{k I_{f_1} I_{a_1}} \quad \text{but} \quad T_2 = T_1$$

$$\therefore \; I_{f_2} I_{a_2} = I_{f_1} I_{a_1} \quad \text{and} \quad I_{a_2} = \frac{1 \cdot 5 \times 25}{1 \cdot 35} = 27 \cdot 78 \text{ A}$$

Note. $\; I_{f_2} = \dfrac{220}{146 \cdot 66 + 16 \cdot 24} = \dfrac{220}{162 \cdot 9} = 1 \cdot 35 \text{ A}$

Now back e.m.f. $E_{b_2} = 220 - (27 \cdot 78 \times 0 \cdot 2) = 220 - 5 \cdot 556$
$$= 214 \cdot 44 \text{ V}$$

Now $E_b \propto \Phi$ and N or $E_b \propto I_f$ and N
Since $E_b = k I_f N$

so then $\dfrac{E_{b_2}}{E_{b_1}} = \dfrac{I_{f_2} N_2}{I_{f_1} N_1}$ or $N_2 = \dfrac{E_{b_2} N_1 I_{f_1}}{E_{b_1} I_{f_2}}$

$$\therefore \quad N_2 = \frac{214 \cdot 44 \times 725 \times 1 \cdot 5}{215 \times 1 \cdot 35} = 802 \text{ rev/min.}$$

Example 30. A d.c. shunt machine generates 220 V at 800 rev/min on open circuit. The armature resistance including brushes is 0·4 Ω and the field resistance is 160 Ω. The machine takes 5 A running as a motor on no-load at 220 V. Calculate the speed and the efficiency of the machine taking 45 A at 220 V. Assume armature reaction to weaken the field by 3 per cent.

Total machine losses $= 5 \times 220 = 1100 \text{ W}$
Field current $= \dfrac{220}{160} = 1 \cdot 375 \text{ A}$

Field copper loss $= 1 \cdot 375^2 \times 160 = 302 \text{ W}$
Armature current $= 5 - 1 \cdot 375 = 3 \cdot 625 \text{ A}$
Armature copper loss $= 3 \cdot 625^2 \times 0 \cdot 4 = 5 \cdot 25 \text{ W}$
Rotational loss $= 1100 - 302 - 5 \cdot 25 = 792 \cdot 75 \text{ W}$
Back e.m.f. on no-load $= E_{b_0} = V - I_{a_0} R_a = 220 - (3 \cdot 625 \times 0 \cdot 4)$
$$= 220 - 1 \cdot 45 = 218 \cdot 55 \text{ V}$$

On load.   Field copper loss = 302 W

Rotational loss = 792·75 W

Armature current = $45 - 1·375$ = 43·625 A

Armature copper loss = $43·63^2 \times 0·4$ = 765 W

Total losses = $302 + 792·75 + 765$ = 1860 W

Motor input = 220×45 = 9900 W

$$\text{Efficiency} = \frac{\text{Input} - \text{Losses}}{\text{Input}} = \frac{9900 - 1860}{9900} = \begin{array}{l} 0·812 \text{ or} \\ 81·2 \text{ per cent} \end{array}$$

Back e.m.f. on load = $220 - (43·625 \times 0·4)$ = $220 - 17·45$

or E_{b_1} = 202·55 V

Now $N \propto \dfrac{E_b}{\Phi}$ or $\dfrac{\text{No-load speed}}{\text{Full-load speed}} = \dfrac{E_{b_0}}{E_{b_1}}$ (with constant Φ)

Thus speed with constant flux = $\dfrac{800}{N}$ = $\dfrac{218·55}{202·55}$

or $N = 800 \times \dfrac{202·55}{218·55}$ = 741 rev/min

Since flux is weakened then speed increases in accordance with

$$N_1 = 741 \times \frac{100}{97} = 765 \text{ rev/min.}$$

CHAPTER 3

PRACTICE EXAMPLES

1.   Estimate the turns per pole for the compoles of a 150 kW, 250 V, four-pole, d.c. generator. The armature is lap wound with 342 conductors. Assume a flux density of 0·32 T in the interpolar air-gap and a gap length of 12 mm. The reluctance of the iron circuit and leakage can be neglected.

2.   If the brushes of a six-pole generator are rocked forward through a distance equal to three commutator segments from the neutral position, estimate the number of demagnetising ampere-turns per pole set up by a current of 600 A passing through the armature. The armature winding is made up from 128 two-turn coils and is wave wound.

The remaining practice examples for this chapter should be undertaken as revision exercises. They have been chosen as important examples of fundamental theory for the d.c. machine.

3.   The results of a test made to determine the open-circuit characteristic of a ship's shunt-connected generator are set out below. The test was made with the machine being separately excited and driven at 700 rev/min.

Terminal
Voltage (V) 10 20 40   80 120 160 200 240 260
Field
Current (A) 0   0·1 0·24 0·5 0·77 1·2   1·92 3·43 5·2

(a)   If the generator is to be coupled to a 850 rev/min engine and is to be made self-exciting, find the range of the field rheostat required to vary the O.C. voltage between the limits of 200 V and 250 V. The resistance of the shunt field was measured to be 52 Ω.

(b)   If the armature resistance was 0·04 Ω, estimate the O.C. voltage and rheostat setting if the armature is to deliver 200 A at a terminal voltage of 220 V. Neglect the effects of armature reaction and brush-contact voltage drop.

4.   A d.c. shunt motor running at 1200 rev/min has an armature resistance of 0·15 Ω. The current taken by the armature is 60 A when the applied voltage is 220 V. If the load is increased by 30 per cent find the variation in the speed.

5.   A ventilating fan is driven by a 220 V, 10 kW series motor and runs at 800 rev/min at full load. The total resistance of the armature circuit is 0·6 Ω. Calculate the speed and percentage change in torque if the current taken by the motor is reduced by 50 per cent of the full load value. The efficiency of the motor is 82 per cent. Assume the flux to be proportional to the field current.

6.   A d.c. shunt generator has the following O.C.C.

Generated e.m.f. E (volts)

| 50 | 100 | 250 | 300 | 350 | 400 | 450 | 500 | 550 |

Field Current I_f (amperes)

| 0·077 | 0·231 | 0·387 | 0·477 | 0·555 | 0·663 | 0·805 | 1·00 | 1·29 |

If the field resistance is 480 Ω and the armature resistance is 0·1 Ω, deduce the load characteristic of the generator if the effects of armature reaction are neglected.

7.   An engine-room ventilator fan motor is a series machine with a total resistance of 0·5 Ω and runs from a 110 V supply at 1000 rev/min when the current is 28 A. What resistance in series with the motor will reduce the speed to 750 rev/min. The load torque is proportional to the square of the speed (*e.g.* for a fan) and the field strength can be assumed to be proportional to the current.

8.   A 230 V, 10 kW shunt motor with a stated full-load efficency of 85 per cent runs at a speed of 1000 rev/min. At what speed should the motor be driven, if it is used as a generator to supply an emergency lighting load at 220 V? The armature resistance is 0·2 Ω and the field resistance is 115 Ω. Find the *kW* "rating" of the machine under this condition.

9.   A ship's d.c. shunt generator when on test and driven at 900 rev/min gave the following O.C.C.

Exciting current I_f (amperes)

| 0·5 | 1·0 | 1·5 | 2·0 | 2·5 | 3:0 | 3·5 | 4·0 |

Generated e.m.f. E (volts)

| 61 | 111 | 148·5 | 175·5 | 194·5 | 207 | 214 | 221 |

If the governor setting of the prime-mover is set to give a speed of 1000 rev/min find the voltage to which the shunt generator will self-excite, if the field-circuit resistance is adjusted to 80 Ω.

If the resistance of the armature is 0·14 Ω, deduce the curve relating terminal voltage and load current. The effect of armature reaction and brush-contact voltage drop can be neglected.

If the load circuit resistance should be adjusted to 0·95 Ω, estimate the load current and generator terminal voltage.

10.   A series d.c. motor is run on a 220 V circuit with a regulating resistance of R ohms for speed adjustment. The armature and field coils have a total resistance of 0·3 Ω. On a certain load, with R zero, the current is 20 A and the speed is 1200 rev/min. With another load and R set at 3 Ω, the current is 15 A. Find the new speed and also the ratio of the two values of the power output of the motor. Assume the field strength at 15 A to be 80 per cent of that at 20 A.

CHAPTER 4

THE TRANSFORMER (ii)

The operating principle of the transformer, both on no load and on load, has been introduced in Chapter 2 and its study developed until the stage was reached where the causes of internal voltage drop were considered. The associated phasor diagrams were also shown and attention was given to the fact that secondary effects can be seen to be reflected in the primary side, provided they are equated, by using the turns ratio. Thus consider a step-down transformer having a ratio of 2:1. Then a voltage drop in the secondary of 3 volts would be equivalent to a voltage drop of 6 V in the primary. It would thus appear possible to convert the secondary phasor diagram to indicate corresponding primary conditions provided the turns ratio relationship is used. This technique will now be pursued.

THE COMBINED PHASOR DIAGRAM

Since the primary and secondary currents are drawn anti-phase, the following procedure is possible if the primary current I_1 is used as the reference phasor and is drawn horizontally. The no-load current I_0 is neglected. The diagrams (Fig. 56a and Fig. 56b) show the stages of the operation. The secondary diagram is rotated through $180°$, and the current I_2 is then seen to line up with the reference phasor I_1. The secondary voltage-drop triangle is now seen to stand upright and the overall secondary diagram (Fig. 56b) is seen to be

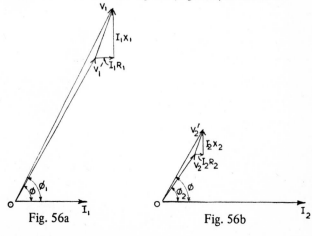

Fig. 56a   Fig. 56b

similar to the primary diagram (Fig. 56a). It could moreover, be matched to the latter if the secondary voltages are adjusted to primary conditions by multiplying by the turns ratio. OV_2' is made

equal to coincide with OV_1' if multiplied by $\left(\dfrac{N_1}{N_2}\right)$.   The other

secondary voltages and voltage drops will also be required to be multiplied by the turns ratio, and the adjustment can be carried

further by using the current-turn relationship, since $I_2 = I_1\left(\dfrac{N_1}{N_2}\right)$.

The diagrams (Fig. 57a and b) now take the form shown. From these it will be seen that the secondary resistance voltage drop, originally

I_2R_2, becomes $I_2R_2\left(\dfrac{N_1}{N_2}\right)$ if referred to the primary, and further

becomes $I_1R_2\left(\dfrac{N_1}{N_2}\right)^2$ if expressed completely in primary current and

voltage terms. Similarly the secondary reactance voltage drop I_2X_2

becomes $I_1X_2\left(\dfrac{N_1}{N_2}\right)^2$ when this is referred to the primary. The next

logical step would be to draw the combined diagram, with the in-phase voltage components being added, to give the total resistance and reactance voltage drops.

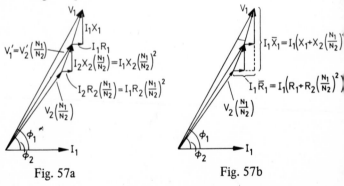

Fig. 57a   Fig. 57b

If the total resistance voltage drop is examined, it is seen to be

$$I_1 \left\{ R_1 + R_2 \left(\frac{N_1}{N_2} \right)^2 \right\}$$ and this expression has already been deduced

in connection with the Equivalent Resistance $\overline{R}_1$. Thus:

$$\overline{R}_1 = R_1 + R_2 \left(\frac{N_1}{N_2} \right)^2$$ and is the equivalent resistance referred to the

primary side *i.e.* the total resistance of the windings is considered to be in the primary only.

The diagrams show the reactance voltage drops to be given by

$$I_1 \left\{ X_1 + X_2 \left(\frac{N_1}{N_2} \right)^2 \right\}.$$ This expression has already been deduced in

Chapter 2 for the Equivalent Reactance $\overline{X}_1$.

Thus $\overline{X}_1 = X_1 + X_2 \left(\dfrac{N_1}{N_2} \right)^2$ and is the equivalent reactance

referred to the primary side *i.e.* the total reactance of the windings is considered to be in the primary only.

The equivalent impedance $\overline{Z}_1$ follows the accepted relation thus:

$$\overline{Z}_1 = \sqrt{\overline{R}_1^2 + \overline{X}_1^2}$$

The simplified combined phasor diagram is shown in Fig. 58. *Note* $V_2' = V_2$, since all voltage drops are now accounted for on the primary side.

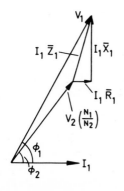

Fig. 58

The above deductions could have been made by referring the primary diagram to the secondary side. In this case it could have been shown that

$$\overline{R}_2 \;=\; R_2 + R_1 \left(\frac{N_2}{N_1}\right)^2 \quad \text{and} \quad \overline{X}_2 \;=\; X_2 + X_1 \left(\frac{N_2}{N_1}\right)^2$$

also $\overline{Z}_2 \;=\; \sqrt{\overline{R}_2{}^2 + \overline{X}_2{}^2}$

The following example is set out to revise work already done in Chapter 2.

Example 31. A 660/220 V, single-phase transformer has a primary resistance of 0·29 Ω and a secondary resistance of 0·025 Ω. The corresponding reactance values are 0·44 Ω and 0·04 Ω. Estimate the primary current which would flow if a short-circuit was to occur across the secondary terminals.

$$\text{Turns ratio} \quad \frac{V_1}{V_2} \;=\; \frac{N_1}{N_2} \;=\; \frac{660}{220} \;=\; \frac{3}{1}$$

$$\text{Also} \quad \overline{R}_1 \;=\; R_1 + R_2 \left(\frac{N_1}{N_2}\right)^2 \;=\; 0{\cdot}29 + (0{\cdot}025 \times 9)$$
$$=\; 0{\cdot}29 + 0{\cdot}225 \;=\; 0{\cdot}515 \;\Omega$$

$$\text{Similarly} \quad \overline{X}_1 \;=\; X_1 + X_2 \left(\frac{N_1}{N_2}\right)^2 \;=\; 0{\cdot}44 + (0{\cdot}04 \times 9)$$
$$=\; 0{\cdot}44 + 0{\cdot}36 \;=\; 0{\cdot}8 \;\Omega$$

$$\text{and} \quad \overline{Z}_1 \;=\; \sqrt{\overline{R}_1{}^2 + \overline{X}_1{}^2} \;=\; \sqrt{0{\cdot}515^2 + 0{\cdot}8^2} \;=\; 10^{-1}\sqrt{5{\cdot}15^2 + 8^2}$$
$$=\; 10^{-1}\sqrt{26{\cdot}52 + 64} \;=\; 10^{-1}\sqrt{90{\cdot}52}$$
$$=\; 10^{-1} \times 9{\cdot}514 \;=\; 0{\cdot}95 \;\Omega$$

The short-circuit current will be $\dfrac{660}{0.95} \;=\; 694{\cdot}7$ A.

THE EQUIVALENT CIRCUIT

In Chapter 2, the phasor diagram of the transformer was developed with the aid of circuit diagrams which illustrated the conditions being considered. The diagram of Fig. 24, should be referred to when considering the illustrations now showing the development of the equivalent circuit. The first illustration of Fig. 59 shows the primary and secondary resistances and reactance values as separate components outside the windings themselves. The next illustration shows how these values can be transferred from one winding to the other, in line with the work done in connection with referred values.

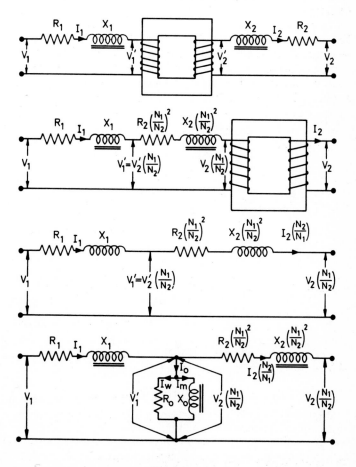

Fig. 59

Since the turns ratio has been taken into account when referring the voltages and values of resistance and reactance from one winding to the other, the transformer can be shown with a 1:1 ratio and can even be omitted from the diagram. This is shown in the third illustration of the equivalent circuit.

The final illustration shows how the equivalent circuit can be modified to allow for the loading resulting from the iron loss and magnetising magneto-motive force.

It has been seen earlier that the no-load current I_o flows when the primary winding is energised, and that its value is constant irrespective of the load on the secondary. It can thus be likened to the current in a parallel circuit. Furthermore since I_o is made up from I_m a magnetising component and I_w a power component, which are respectively in quadrature and in phase with the applied voltage, the parallel circuit can be made up from a resistor and a pure inductor.

The equivalent circuit, as drawn in the final illustration, shows all aspects of transformer working, and is used for the more involved type of investigations and problems. However for the more practical considerations, since the no-load current is omitted from the phasor diagram, the parallel section of the equivalent circuit is also omitted. The circuit is often used in its simplest form as illustrated (Fig. 60), and is that for the phasor diagram of Fig. 58.

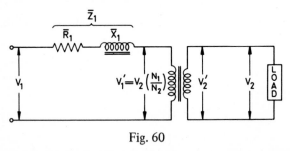

Fig. 60

The example following, shows how the values of the components of the "no-load" parallel circuit can be found from data given for the no-load condition. The example uses the figures given for the second problem of Chapter 2.

Example 32. For the no-load test on a transformer, the ammeter was found to read 0·18 A and the wattmeter 12 W. The reading on the primary voltmeter was 400 V and that on the secondary voltmeter was 240 V. Calculate, the magnetising component of the no-load current, the iron-loss component and the transformation ratio. Find the equivalent resistance and reactance of the "no-load" circuit.

From Chapter 2, Example 12, the magnetising component of no-load current $I_m = 0.178$ A

Then the equivalent inductor value $X_o = \dfrac{400}{0.178} = 2247\ \Omega$

Iron-loss component of no-load current $I_w = \dfrac{12}{400} = 0.03$ A

The equivalent resistor value $R_o = \dfrac{400}{0.03} = 13\,333\ \Omega$

Transformation ratio $= \dfrac{400}{240}$ or $1.66{:}1$.

Figure 58 has been drawn to show the transformer phasor diagram as associated with the simplified equivalent circuit. It must be stressed that, although treatment up to now has shown all resistance and reactance to be referred to the primary side, it can just as readily be deduced for referring the resistance and reactance to the secondary side. The simplified equivalent circuit and phasor diagram would then be as shown (Fig. 61). Here $V_1' = V_1$ since all voltage drops are accounted for on the secondary side.

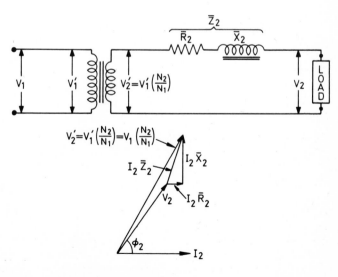

Fig. 61

It has already been stated that $\bar{R}_2 = R_2 + R_1 \left(\dfrac{N_2}{N_1}\right)^2$ and $\bar{X}_2 = X_2 + X_1 \left(\dfrac{N_2}{N_1}\right)^2$

These can be deduced by procedure already introduced, and it follows that $\bar{Z}_2 = \sqrt{\bar{R}_2{}^2 + \bar{X}_2{}^2}$.

VOLTAGE REGULATION

This is the first time that this expression has been encountered and it refers to the "sitting down" of terminal voltage on load. It can be expressed as a percentage or a per-unit value and is the ratio of change in secondary terminal voltage between no-load and full-load to the no-load terminal voltage. This would be for an assumed constant input voltage at the primary.

Thus:

$$\text{Voltage regulation (per unit)} = \frac{\text{No load} - \text{Full-load voltage}}{\text{No-load voltage}}$$

or in terms of the secondary (for example) as a percentage

$$= \frac{\text{Secondary voltage at no load} - \text{Secondary voltage on load}}{\text{Secondary voltage at no load}} \times 100 \text{ per c}$$

It should be noted that for the above a numerical difference and not a phasor difference is involved. This is illustrated by the examples which follow.

Since the definition for regulation has been given in terms of the secondary, then reference to the phasor diagram of Fig. 61 shows that

$$\text{Percentage regulation} = \frac{V_2' - V_2}{V_2'} \times 100 \text{ which would also be}$$

$$= \frac{V_1'\left(\dfrac{N_2}{N_1}\right) - V_2}{V_1'\left(\dfrac{N_2}{N_1}\right)} \times 100 \quad \text{or} \quad = \frac{V_1\left(\dfrac{N_2}{N_1}\right) - V_2}{V_1\left(\dfrac{N_2}{N_1}\right)} \times 100$$

From the phasor diagram it is seen that

$$V_1\left(\frac{N_2}{N_1}\right) = \sqrt{(V_2 \cos \phi_2 + I_2\overline{R}_2)^2 + (V_2 \sin \phi_2 + I_2\overline{X}_2)^2}$$

If $\overline{R}_2$ and $\overline{X}_2$ are known, then the regulation for any load current can be estimated if the transformer open-circuit terminal voltage is known.

Example 33. Data already used for examples is repeated. A 660/220 V, single-phase transformer has a primary resistance of 0·29 Ω and a secondary resistance of 0·025 Ω. The corresponding reactance values are 0·44 Ω and 0·04 Ω. Estimate the percentage regulation for a secondary load current of 50 A at a power factor of 0·8 (lagging).

$$\text{Turns ratio} = \frac{3}{1} \text{ and } \overline{R}_2 = R_2 + R_1\left(\frac{N_2}{N_1}\right)^2$$

$$= 0·025 + (0·29 \times \frac{1}{9}) = 0·025 + 0·0322 = 0·057 \ \Omega$$

also $\quad \overline{X}_2 = X_2 + X_1\left(\frac{N_2}{N_1}\right)^2 = 0·04 + \quad 0·44 \times \frac{1}{9}$

$$= 0·04 + 0·049 = 0·089 \ \Omega$$

Substituting we have

$$220 = \sqrt{\{(V_2 \times 0·8) + (50 \times 0·057)\}^2 + \{(V_2 \times 0·6) + (50 \times 0·089)\}^2}$$

or $220^2 = (0·8 \, V_2 + 2·85)^2 + (0·6 \, V_2 + 4·45)^2$

$$= \{0·64 \, V_2{}^2 + (2 \times 0·8 \, V_2 \times 2·85) + 8·12\} + \{0·36 \, V_2{}^2$$
$$+ (2 \times 0·6 \, V_2 \times 4·45) + 19·8\}$$

$$= 0·64 \, V_2{}^2 + 4·56 \, V_2 + 8·12 + 0·36 \, V_2{}^2 + 5·34 \, V_2 + 19·8$$

$$= V_2{}^2 + 9·9 \, V_2 + 27·92$$

Thus $V_2{}^2 + 9·9 \, V_2 - 48\,400 + 27·92 = 0$

giving the quadratic equation $V_2{}^2 + 9·9 \, V_2 - 48\,372 = 0$

Solving $V_2 = \dfrac{-9 \cdot 9 \pm \sqrt{9 \cdot 9^2 + (4 \times 48\ 372)}}{2}$

$= \dfrac{-9 \cdot 9 \pm \sqrt{98 \cdot 01 + 193\ 488}}{2}$

$= \dfrac{-9 \cdot 9 \pm \sqrt{193\ 586}}{2}$

$= \dfrac{-9 \cdot 9 \pm 10^2 \sqrt{19 \cdot 36}}{2} = \dfrac{-9 \cdot 9 + (10^2 \times 4 \cdot 4)}{2}$

$= \dfrac{-9 \cdot 9 + 440}{2} = \dfrac{430 \cdot 1}{2} = 215 \cdot 05 \text{ V}$

Voltage regulation $= \dfrac{220 - 215 \cdot 05}{220} \times 100$

$= \dfrac{4 \cdot 95}{2 \cdot 2} = 2 \cdot 25 \text{ per cent.}$

It will be seen from the above example that, although the method employed for estimating the regulation is quite straightforward, it nevertheless needs a mathematical solution which can be tedious. To assist in this direction, a well known approximation formula for evaluating the internal voltage drop has been evolved. This gives a result so close to that obtained by the method illustrated above, that it is invariably used. The explanation, as to how the formula referred to is deduced, will follow and here it is introduced to show its accuracy.

Since the difference between the no-load voltage and the load voltage is due to the causes of internal voltage drops in both windings, it follows that, if this voltage drop can be estimated then

Voltage regulation $= \dfrac{\text{Internal voltage drop}}{\text{No-load voltage}} \times 100 \text{ per cent}$

The approximation formula has been deduced for this internal voltage drop and is given by $I_2\,(\overline{R}_2 \cos \phi_2 \pm \overline{X}_2 \sin \phi_2)$ and

Voltage regulation $= \dfrac{I_2\,(\overline{R}_2 \cos \phi_2 \pm \overline{X}_2 \sin \phi_2)}{V_2} \times 100 \text{ per cent}$

Substituting with the values obtained for the previous example

$$\text{Voltage regulation} = \frac{50(0 \cdot 057 \times 0 \cdot 8 \pm 0 \cdot 089 \times 0 \cdot 6)}{220} \times 100$$

$$= \frac{50(0 \cdot 0456 \pm 0 \cdot 0534)}{220} \times 100 = \frac{50 \times 0 \cdot 099 \times 100}{220}$$

$$= \frac{5 \times 99}{220} = \frac{45}{20} = 2 \cdot 25 \text{ per cent.}$$

It will be noted that the formula uses a plus or minus sign and that with using the latter, a negative regulation can be obtained. This is explained by the fact that with a capacitive load, the secondary voltage rises with increased loading. This would result in a −ve regulation value or regulation "UP", − as is frequently written. Sitting down of terminal voltage means a + ve regulation figure or regulation "DOWN", although this latter definition may not always be stated.

Loading conditions, as influenced by the power factor, are shown by the diagrams of Figure 62.

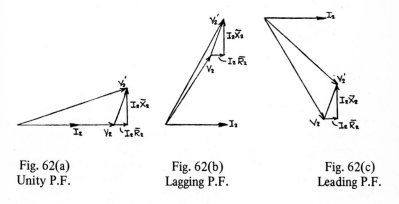

Fig. 62(a)   Fig. 62(b)   Fig. 62(c)
Unity P.F.   Lagging P.F.   Leading P.F.

From condition (c), it will be seen that the on-load terminal voltage phasor V_2 is longer than that for the induced e.m.f. or O.C. voltage condition V_2'. Thus a rise of terminal voltage with load is evident and a − ve or Regulation "UP" figure is possible.

Transformer internal voltage drop and regulation can be determined from the primary side, as can be expected and with the aid of the phasor diagram of Fig. 58, the treatment for obtaining an expression for regulation can be repeated. Thus:

$$\text{Voltage regulation} \quad = \quad \frac{V_1 - V_2'\left(\dfrac{N_1}{N_2}\right)}{V_1} \times 100$$

$$\text{or better} \quad = \quad \frac{V_1 - V_2\left(\dfrac{N_1}{N_2}\right)}{V_1} \times 100 \text{ per cent}$$

since $V_2 = V_2'$ all causes of voltage drop being considered on the primary side.

To calculate $V_2\left(\dfrac{N_1}{N_2}\right)$ proceed as before using the expression

$$V_1 = \sqrt{\left[V_2\left(\frac{N_1}{N_2}\right)\cos\phi_2 + I_1\bar{R}_1\right]^2 + \left[V_2\left(\frac{N_1}{N_2}\right)\sin\phi_2 + I_2\bar{X}_1\right]^2}.$$

It may often be more convenient to use the regulation expressions in this form *i.e.* referred to the primary, since data is frequently given as $\bar{R}_1$ and $\bar{X}_1$. Such data may be obtained from the open-circuit and short-circuit tests, as are to be shortly described, and these can be made to give referred primary results. An expression for the internal voltage drop (approximated value) can also be deduced for the primary side and will give the desired result. Its application is illustrated by the following example, which uses the figures for the transformer as treated originally from the secondary side. The formula used for the approximate voltage drop is yet to be deduced, but in "primary form" is written as

$$\text{Voltage drop (referred to primary)} = I_1(\bar{R}_1 \cos\phi_1 \pm \bar{X}_1 \sin\phi_1).$$

Example 34. A 660/220 V, single-phase transformer has a primary resistance of 0·29 Ω and a secondary resistance of 0·025 Ω. The corresponding reactance values are 0·44 Ω and 0·04 Ω. Estimate the percentage regulation for a secondary load current of 50 A at a power-factor of 0·8 (lagging).

Turns ratio $= \dfrac{3}{1}$

and $\overline{R}_1 = R_1 + R_2 \left(\dfrac{N_1}{N_2}\right)^2 = 0\cdot29 + (0\cdot025 \times 9)$

$\qquad = 0\cdot29 + 0\cdot225 = 0\cdot515 \ \Omega$

also $\overline{X}_1 = X_1 + X_2 \left(\dfrac{N_1}{N_2}\right)^2 = 0\cdot44 + (0\cdot04 \times 9)$

$\qquad = 0\cdot44 + 0\cdot36 = 0\cdot8 \ \Omega$

Primary current would be $\dfrac{50}{3} = 16\cdot67$ A $\cos \phi = 0\cdot8$ (lagging)

Voltage regulation $= \dfrac{16\cdot67 \left\{(0\cdot515 \times 0\cdot8) + (0\cdot8 \times 0\cdot6)\right\}}{660} \times 100$

$\qquad = \dfrac{16\cdot67 \ (0\cdot412 + 0\cdot48)}{660} \times 100$

$\qquad = \dfrac{16\cdot67 \times 0\cdot892}{6\cdot6} = \dfrac{14\cdot87}{6\cdot6} = 2\cdot25$ per cent.

This is the same value as that obtained by applying the formula to the secondary side.

INTERNAL VOLTAGE-DROP FORMULA (For approximation)

The reasoning for this expression is based on the phasor diagram as shown in Fig. 58. Since this was not drawn to scale, it can be assumed that the transformer internal voltage-drop triangle is small, and so far away from the point O that phasors V_1 and $V_2\left(\dfrac{N_1}{N_2}\right)$ are parallel.

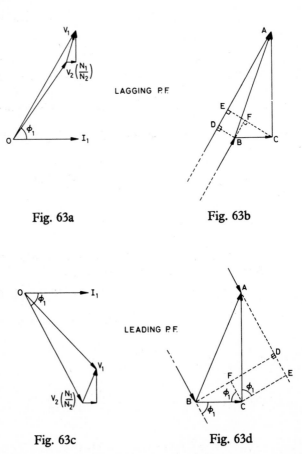

Fig. 63a   Fig. 63b

Fig. 63c   Fig. 63d

The diagrams (Fig. 63a and b), show this triangle and the phasors referred to, which are drawn parallel and designated by OA and OB. If the additional constructions, as shown by the dotted lines, are made then we can proceed with the deduction.

Since the numerical difference between the no-load terminal voltage and the on-load terminal voltage is the measure of the internal voltage drop, then on the phasor diagram, it is the difference in the lengths of V_1 and $V_2 \left(\dfrac{N_1}{N_2}\right)$ or the lines OA and OB. This difference is given by the length AD.

$$\text{Now DA} = \text{DE} + \text{EA}$$
$$= \text{BF} + \text{EA} \quad \text{or} \quad \text{BC} \cos \phi_1 + \text{AC} \sin \phi_1$$

Thus voltage drop $= I_1 \bar{R}_1 \cos \phi_1 + I_1 \bar{X}_1 \sin \phi_1$

For a capacitive load (Fig. 63c and d), the voltage drop is the difference between V_1 and $V_2 \left(\dfrac{N_1}{N_2}\right)$ or the length of the lines OA and OB. Since OA is shorter, then this difference is given by the length $- \text{AD}$.

$$\text{Now} - \text{AD} = - (\text{AE} - \text{DE}) = \text{DE} - \text{AE}$$
$$= \text{FC} - \text{AE} \quad \text{or} \quad \text{BC} \cos \phi_1 - \text{AC} \sin \phi_1$$
$$\text{Thus voltage drop} = I_1 \bar{R}_1 \cos \phi_1 - I_1 \bar{X}_1 \sin \phi_1$$

Example 35. A 20 kVA, 2000/220 V, single-phase transformer has a primary resistance of 2·1 Ω and a secondary resistance of 0·026 Ω. The corresponding leakage reactances are 2·5 Ω and 0·03 Ω. Estimate the regulation at full load under power-factor conditions of (a) unity (b) 0·5 (lagging) and (c) 0·5 (leading).

$$\text{Turns ratio} = \frac{2000}{220} = \frac{9 \cdot 09}{1}$$

$$\text{Full-load secondary current } I_2 = \frac{20\,000}{220} = 90 \cdot 9 \text{ A}$$

$$\bar{R}_2 = R_2 + R_1 \left(\frac{N_2}{N_1}\right)^2 \quad \text{or} \quad \bar{R}_2 = 0 \cdot 026 + 2 \cdot 1 \left(\frac{1}{9 \cdot 09}\right)^2$$
$$= 0 \cdot 026 + 0 \cdot 0255 = 0 \cdot 0515 \ \Omega$$

$$\bar{X}_2 = X_2 + X_1 \left(\frac{N_2}{N_1}\right)^2 \quad \text{or} \quad \bar{X}_2 = 0 \cdot 03 + 2 \cdot 5 \left(\frac{1}{9 \cdot 09}\right)^2$$
$$= 0 \cdot 03 + 0 \cdot 0302 = 0 \cdot 0602 \ \Omega$$

(a)   Voltage regulation at unity power factor

$$= \frac{I_2(\overline{R}_2 \cos \phi_2 + \overline{X}_2 \sin \phi_2)}{V_2} \times 100$$

$$= \frac{90 \cdot 9 \left\{ (0 \cdot 0515 \times 1) + (0 \cdot 0602 \times 0) \right\}}{220} \times 100$$

$$= \frac{90 \cdot 9 \times 0 \cdot 0515}{220} \times 100 \ = \ 2 \cdot 12 \text{ per cent (down).}$$

(b)   Regulation at a power factor of 0·5 (lagging)

$$= \frac{90 \cdot 9 \left\{ (0 \cdot 0515 \times 0 \cdot 5) + (0 \cdot 0602 \times 0 \cdot 866) \right\}}{220} \times 100$$

$$= \frac{90 \cdot 9 (0 \cdot 02575 + 0 \cdot 0523)}{220} \times 100$$

$$= \frac{90 \cdot 9 \times 0 \cdot 078}{220} \times 100 \ = \ 3 \cdot 23 \ \text{ per cent (down).}$$

(c)   Regulation at a power factor of 0·5 (leading)

$$= \frac{90 \cdot 9 \left\{ (0 \cdot 0515 \times 0 \cdot 5) - (0 \cdot 0602 \times 0 \cdot 866) \right\}}{220} \times 100$$

$$= \frac{90 \cdot 9 (0 \cdot 02575 - 0 \cdot 0523)}{220} \times 100$$

$$= \frac{90 \cdot 9 \times - 0 \cdot 02655}{220} \times 100 \ = \ \frac{- 241 \cdot 34}{220}$$

$$= - 1 \cdot 09 \text{ per cent or } 1 \cdot 09 \text{ per cent (Up).}$$

TRANSFORMER TESTING

As for all electrical machines, testing of a transformer is necessary to check its overall performance. Since direct loading involves the wastage of energy then methods, which enable the design specification to be checked by assessment, are used whenever possible. Such methods are used to determine the magnitude of the losses, overall efficiency and regulation and, for the transformer, they involve two complementary tests. These are (a) the Open-circuit Test, (b) the Short-circuit Test. In basic form, they provide the appropriate necessary information and their importance is stressed. The principle of the technique is to measure the Iron Losses and the Copper Losses separately and the results are applied to appropriate formulae. Similar testing procedure is used for other a.c. machines such as the alternator and induction motor. Variations of the basic O.C. and S.C. tests may be made to obtain additional information and will be mentioned whenever they are considered appropriate.

THE OPEN-CIRCUIT TEST

Mention of this test has already been made in Chapter 2 when the conditions of working on no load were being investigated. One winding, whichever is most convenient, is connected to a supply of normal voltage value and frequency whilst the other is left open-circuited. The diagram (Fig 64) shows the circuit which incorporates appropriate instruments. The input power P_o (in watts) is measured by a special wattmeter designed for use on low power-factor circuits. A low-range ammeter, of suitable accuracy, measures the no-load current I_o and suitable voltmeters measure the primary and secondary voltages. As was mentioned in Chapter 2, since a temperature rise of the iron core is noted when the winding is energised, it is apparent that some energy is taken from the supply and converted into heat. This unwanted heat is obviously a "loss" and must be due to the cyclic magnetisation of the iron core and to the copper loss in the energised winding. Since the resistance of the winding is very small and the no-load current is usually less than a tenth of the full-load value, the copper loss, under this condition, is less than a hundredth of full-load value and is negligible. Thus the input power is assumed to be due to P_{Fe} — the iron losses only. Furthermore, since the applied voltage and frequency can be assumed to be constant under all conditions of loading then, the flux-density value remains fixed and the iron losses, as measured by the O.C. test, are considered constant for all loads.

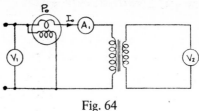

Fig. 64

In the introduction to the efficiency of a transformer, mention was made of the component losses of the iron loss. These are the hysteresis loss (P_{Hy}) and the eddy current loss (P_{EC}). Although, for ordinary purposes, it is not usual to require to know their relative magnitudes nevertheless, mention should be made of the technique used for determining these when required.

SEPARATION OF IRON LOSSES.   The usual method is in line with that described in Chapter 1 for d.c. machines. It will be recalled that, such a test was made by keeping the flux density constant and varying the speed at which the armature was driven. For the transformer, since $V = 4.44\,\Phi_m\,fN$ then it follows that, if the applied voltage is halved and the frequency is halved, the flux density would be kept constant. From Volume 6, Chapter 12, it is seen that iron loss, $P_{Fe} = P_H + P_{EC}$ or $P_{Fe} = K_H B_m^{1.6} f + K_E B_m^2 f^2$. If the flux density B_m is kept constant during the testing, then the expression can be written as $P_{Fe} = HN + EN^2$ for a machine or $P_{Fe} = Hf + Ef^2$ for a transformer. The quantity Hf is the hysteresis loss and Ef^2 the eddy-current loss at that frequency. If the test is made at a particular frequency and then repeated at another frequency — preferably a multiple or sub-multiple, then enough information will be obtained to allow determination of the constituent losses at these particular frequencies. The following example shows the application of this technique.

Example 36.   When an O.C. test is made on the primary of a transformer at 440 V and 50 Hz, the iron loss is 2·5 kW. When the test is repeated at 220 V and 25 Hz, the corresponding loss is measured to be 850 W. Determine the hysteresis and eddy-current loss at normal voltage and frequency.

At 50 Hz we can write $2500 = Hf + Ef^2$ ...   ...   (a)

At 25 Hz we can write $850 = H\dfrac{f}{2} + E\left(\dfrac{f}{2}\right)^2$ ...   ...   (b)

Solving by multiplying (b) by 4, we have:

$$2500 = Hf + Ef^2 \quad \ldots \quad \ldots \text{(a)}$$

$$\text{and } 3400 = 2Hf + Ef^2 \quad \ldots \quad \ldots \text{(c)}$$

Subtracting (c) from (a) $-900 = -Hf$

Thus the Hysteresis Loss $= 900$ W at 50 Hz

and the Eddy-Current Loss $= 2500 - 900$
    $= 1600$ W at 50 Hz.

THE SHORT-CIRCUIT TEST

The diagram (Fig. 65) shows the test circuit. One winding, whichever is convenient, is short-circuited through an ammeter, of suitable range, and a reduced voltage is applied to the other winding. This applied voltage should be just sufficient to circulate full-load current through the short-circuited winding, and, it is stressed that, since the only current-limiting factor is the impedance of the transformer, the resistance of the instruments being negligible, then the applied voltage will be very low. Under this condition of testing, the core flux, being proportional to the applied voltage, is also small and the iron losses are therefore assumed negligible. The wattmeter reading (P_{SC}) therefore indicates the copper losses. An ammeter and voltmeter, of appropriate ranges, are also included in the energising circuit and all readings are used to obtain the values required for estimating efficiency and regulation.

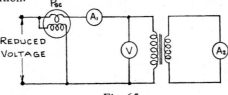

Fig. 65

The procedure of using the O.C. and S.C. test results for calculation purposes will be illustrated by an example set out below, but the reasoning can be assisted by reference to the phasor diagram.

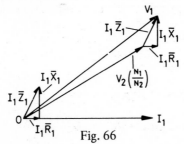

Fig. 66

Consider the diagram (Fig. 66). Since the terminal voltage of the secondary is zero due to the short circuit, the applied voltage is used to drive current through the equivalent impedance of the transformer as referred to the primary. Thus point B of Fig. 63 coincides with point O and

$$\overline{Z}_1 = \frac{V_{SC}}{I_{SC}} \quad \overline{R}_1 = \frac{P_{SC}}{I^2_{SC}} \quad \text{and} \quad \overline{X}_1 = \sqrt{\overline{Z}_1{}^2 - \overline{R}_1{}^2}$$

The results of these expressions can then be used in the appropriate efficiency and regulation formulae.

Example 37. The following results were obtained in tests on a 50 kVA, single-phase, 3300/400 V transformer.

O.C. Test   Primary voltage, 3300 V. Second voltage, 400 V.
Input power 430 W.

S.C. Test   Reduced voltage on primary, to give full secondary current, was 124 V. Primary current, 15·3 A.
Input power = 525 W.

Calculate   (a)   The efficiency at full load and ½ full load – both at 0·707 power factor (lagging).

                 (b)   The regulation at full load for a power factor of 0·707 (lagging and leading).

                 (c)   Full-load terminal voltage under the condition of 0·707 power factor (lagging).

From O.C. Test.   Iron Losses   =   430 W
From S.C. Test.   Copper Loss   =   525 W

Also $\overline{Z}_1 = \dfrac{124}{15\cdot3} = 8\cdot12\ \Omega$

Also $\overline{R}_1 = \dfrac{525}{15\cdot3^2} = 2\cdot25\ \Omega \quad \text{and} \quad \overline{X}_1 = \sqrt{8\cdot12^2 - 2\cdot25^2}$

$$= 7\cdot78\ \Omega$$

Therefore:

(a)   Efficiency at full load, 0·707 (lagging) power factor

$$\text{or} \quad \eta_{FL} = \frac{kVA \cos\phi}{kVA \cos\phi + P_{Fe} + P_{Cu}}$$

$$= \frac{50 \times 0.707}{(50 \times 0.707) + 0.43 + 0.525} = 97.34 \text{ per cent}$$

and

$$\eta_{\frac{1}{2}\text{FL}} = \frac{25 \times 0.707}{(25 \times 0.707) + 0.43 + 0.131}$$

Note that on ½ full load, current is halved and copper loss $\propto I^2$, is quartered. $\therefore \eta_{\frac{1}{2}\text{FL}} = $ Efficiency (half full load) $= 96.9$ per cent.

(b)   Voltage regulation, 0.707 (lagging) power factor

$$= \frac{I_1 \overline{R}_1 \cos \phi + I_1 \overline{X}_1 \sin \phi}{V_1} \times 100$$

$$= \frac{15.3 (2.25 \times 0.707) + (7.78 \times 0.707)}{3300} \times 100$$

$$= 3.3 \text{ per cent (down)}.$$

Voltage regulation, 0.707 (leading) power factor

$$= \frac{15.3 (2.25 \times 0.707) - (7.78 \times 0.707)}{3300} \times 100$$

$$= 1.81 \text{ per cent (up)}.$$

(c)   Terminal voltage $=$ O.C. voltage $-$ voltage drop
$$= 400 - (15.3 \times 2.25 \times 0.707) +$$
$$(15.3 \times 7.78 \times 0.707) \times \frac{400}{3300}$$

$$= 400 - 13.2 = 386.8 \text{ V}$$

or using the percentage regulation expression: $400 - 400 \times \dfrac{3.3}{100}$

$$= 400 - 13.2 = 386.8 \text{ V}.$$

DIRECT-LOADING TEST (Sumpner's Test)

The disadvantages of the O.C. and the S.C. tests lie in the fact that the iron losses and copper losses do not occur at the same time and thus the temperature rise of a transformer cannot be checked by these tests alone. If more than one unit is available, to avoid waste of power by direct loading, a back-to-back test similar, to the Hopkinson test for d.c. machines, has been devised.

Sumpner's test is performed with two transformers which should be identical but it could be made with units which are approximately similarly rated, provided the turns ratios are the same. The simple diagram of connections is shown by Figure 67. The secondaries are connected in opposition but the H.T. windings carry no current since the voltages are equal and in antiphase in the closed loop.

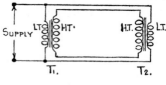

Fig. 67

To upset the balance the voltage applied to the L.T. of one transformer must be greater than that applied to the other. There are several methods of achieving this in practice and one is illustrated by the circuit diagram (Fig. 68).

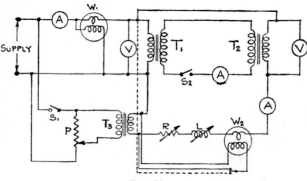

Fig. 68

T_3 is a small auxiliary transformer whose secondary is capable of carrying the full-load current of the units T_1 and T_2. The voltage applied to the primary of T_3 is controlled by varying the position of the sliding contact on the potentiometer P. The variable resistor R and reactor L are included to adjust the power factor.

With switches S_1 and S_2 open, both T_1 and T_2 operate under no-load conditions and being identical transformers, then the reading of W_1 gives the combined loss

or iron loss of one unit $= \dfrac{P_{Fe}}{2}$, where P_{Fe} is the reading of watt-meter W_1.

With switches S_1 and S_2 closed and P adjusted to increase the supply voltage on the primary of T_3, the secondary voltage rises and adds to the mains voltage to increase the applied voltage across the primary of T_2. The secondary voltage of T_2 now rises above that of T_1 and a circulating current I_2 passes as is indicated by the ammeter. Since T_1 is now being excited from the secondary side its primary voltage rises above that of the mains and current is forced round the circuit from T_1 to T_2. Thus the energy output from T_2 into T_1 is now being used to feed T_2. The circulating current flows through wattmeter W_2 and, if its pressure coil is connected across the secondary of the auxiliary transformer, it records the copper losses of T_1 and T_2. This is because the voltages on the L.T. sides of T_1 and T_2 balance out except for the impedance voltage drop $I_1 \times 2\overline{Z}_1$. The wattmeter reads the product of current I_1 and the in-phase component of $I_1 \times 2\overline{Z}_1$ namely $I_1 \times 2\overline{R}_1$ or W_2 reads $I_1{}^2 \times 2\overline{R}_1$ which is the total copper loss in both transformers.

Thus copper loss of one unit $= \dfrac{P_{Cu}}{2}$ where P_{Cu} is the reading of wattmeter W_2.

It will be noted that W_2 has been connected through a change-over switch and that if the latter is moved into the dotted circuit position then the total output of transformer T_1 would be measured.

The efficiency of a transformer can be calculated from the expression

$$\eta = \frac{kVA \cos \phi}{kVA \cos \phi + P_{Fe}(kW) + P_{Cu}(kW)} \times 100$$

Since both transformers are loaded the temperature rise can also be found with the minimum of power wastage.

PERCENTAGE RESISTANCE, REACTANCE AND IMPEDANCE

It is sometimes convenient to express these values in the form of a percentage voltage drop of the open-circuit voltage. Here the treatment has been made in terms of the primary. Thus:

$$\text{Percentage resistance} = \frac{\text{Resistance voltage drop}}{\text{O.C. voltage}} \times 100$$

$$= \frac{I_1 \overline{R}_1}{V_1} \times 100$$

Similarly

$$\text{Percentage reactance} = \frac{\text{Reactance voltage drop}}{\text{O.C. voltage}} \times 100$$

$$= \frac{I_1 \overline{X}_1}{V_1} \times 100$$

and

$$\text{Percentage impedance} = \frac{\text{Impedance voltage drop}}{\text{O.C. voltage}} \times 100$$

$$= \frac{I_1 \overline{Z}_1}{V_1} \times 100$$

A further useful relationship much used by transformer manufacturers is derived from

$$\text{Percentage voltage regulation} = \frac{\text{Voltage drop}}{\text{O.C. voltage}} \times 100$$

$$= \frac{I_1 (\overline{R}_1 \cos \phi_1 \pm \overline{X}_1 \sin \phi_1)}{V_1} \times 100$$

$$= \frac{I_1 \overline{R}_1 \cos \phi_1}{V_1} \times 100 \pm \frac{I_1 X_1 \sin \phi_1}{V_1} \times 100$$

or
Percentage voltage regulation = percentage resistance $\times \cos \phi \pm$ percentage resistance $\times \sin \phi$

Note also that since percentage resistance $= \dfrac{I_1 \overline{R}_1}{V_1} \times 100$

$$= \dfrac{I_1^2 \overline{R}_1}{V_1 I_1} \times 100 = \dfrac{\text{Cu. Loss}}{\text{Full-load } VA} \times 100$$

or Percentage resistance = Percentage copper loss of full-load.

Example 38. A 50 kVA, 3·3 kV/230 V, single-phase transformer has an impedance of 4·2 per cent and a copper loss of 1·8 per cent at full load. Calculate the percentage reactance and ohmic value of resistance, reactance and impedance referred to the primary side. Estimate the primary short-circuit current, assuming the supply voltage to be maintained.

$$\text{Primary full-load current} = \frac{50\ 000}{3300} = 15 \cdot 15 \text{ A}$$

Since

$$\text{Percentage impedance} = \frac{I_1 \overline{Z}_1}{V_1} \times 100$$

$$\text{Then } 4 \cdot 2 = \frac{15 \cdot 15 \times \overline{Z}_1}{3300} \times 100$$

$$\text{or } \overline{Z}_1 = \frac{4 \cdot 2 \times 33}{15 \cdot 15} = 9 \cdot 15 \ \Omega$$

Again

Percentage resistance = Percentage copper loss of full-load
= 1·8 per cent

$$\therefore \text{ Percentage resistance} = 1 \cdot 8 = \frac{I_1 \overline{R}_1}{V_1} \times 100$$

$$\text{or } \overline{R}_1 = \frac{1 \cdot 8 \times 33}{15 \cdot 15} = 3 \cdot 92 \ \Omega$$

The percentage reactance can be obtained in two ways:

$$\text{either} \quad \overline{X} \quad = \quad \sqrt{\overline{Z}^2\% - \overline{R}^2\%} \quad = \quad \sqrt{4{\cdot}2^2 - 1{\cdot}8^2}$$

$$= \quad \sqrt{17{\cdot}64 - 3{\cdot}24} \quad = \quad \sqrt{14{\cdot}4} \quad = \quad 3{\cdot}8 \text{ per cent}$$

$$\text{or since} \quad \overline{X}_1 \quad = \quad \sqrt{\overline{Z}_1{}^2 - \overline{R}_1{}^2} \quad = \quad \sqrt{9{\cdot}15^2 - 3{\cdot}91}$$

$$= \quad \sqrt{83{\cdot}72 - 15{\cdot}37} = \quad \sqrt{68{\cdot}35}$$

$$= \quad 8{\cdot}27 \ \Omega$$

Then percentage reactance $= \dfrac{8{\cdot}27 \times 15{\cdot}15}{3300} \times 100 = 3{\cdot}8$ per cent

Under short-circuit conditions, the impedance of the transformer will limit the current. Therefore:

$$I_{SC} \ = \ \frac{V_1}{\overline{Z}_1} \ = \ \frac{3300}{9{\cdot}15} \ = \ 361 \text{ A.}$$

EFFICIENCY

Introduction to the efficiency expression for a transformer has been made already in Chapter 2 and here we consider further aspects of importance. The shape of the efficiency curve, as shown (Fig. 69a), is as expected and the effect of power factor should be noted.

The statement that maximum efficiency occurs when the copper loss is equal to the iron loss can now be proved. (Fig. 69b).

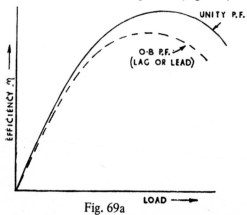

Fig. 69a

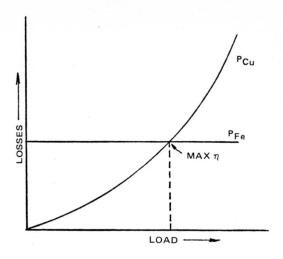

Fig. 69b

CONDITIONS FOR MAXIMUM EFFICIENCY. Maximum efficiency occurs

when $\dfrac{1}{\eta}$ is a minimum. Now $\quad \eta \;=\; \dfrac{kVA \cos \phi}{kVA \cos \phi \;+\; P_{Fe} \;+\; P_{Cu}}$

$$\text{Thus}\quad \frac{1}{\eta} \;=\; \frac{kVA \cos \phi \;+\; P_{Fe} \;+\; P_{Cu}}{kVA \cos \phi}$$

Substituting the current and voltage symbols and referring to the secondary side (for example) then

$$\frac{1}{\eta} \;=\; \frac{V_2 I_2 \cos \phi_2 \;+\; P_{Fe} \;+\; I_2{}^2 \overline{R}_2}{V_2 I_2 \cos \phi_2}$$

$$=\; \frac{V_2 I_2 \cos \phi_2}{V_2 I_2 \cos \phi_2} \;+\; \frac{P_{Fe}}{V_2 I_2 \cos \phi_2} \;+\; \frac{I_2{}^2 \overline{R}_2}{V_2 I_2 \cos \phi_2}$$

$$\text{or}\;\; \frac{1}{\eta} \;=\; 1 \;+\; \frac{P_{Fe}}{\overline{V}_2 I_2 \cos \phi_2} \;+\; \frac{\overline{R}_2 I_2{}^2}{V_2 I_2 \cos \phi_2}$$

$$= 1 + \frac{P_{Fe}I_2^{-1}}{V_2 \cos \phi_2} + \frac{\overline{R}_2 I_2}{V_2 \cos \phi_2}$$

Now differentiating with respect to I_2 the load current, then

$$\frac{d\left(\dfrac{1}{\eta}\right)}{dI_2} = 0 - \frac{P_{Fe}I_2^{-2}}{V_2 \cos \phi_2} + \frac{\overline{R}_2}{V_2 \cos \phi_2}$$

The above expression is zero when $\dfrac{P_{Fe}}{V_2 \cos \phi_2 \, I_2^2} = \dfrac{\overline{R}_2}{V_2 \cos \phi_2}$

and the minimum value of $\dfrac{1}{\eta}$ is obtained.

If therefore $\dfrac{P_{Fe}}{V_2 I_2^2 \cos \phi_2} = \dfrac{\overline{R}_2}{V_2 \cos \phi_2}$ by simplification

$$\frac{P_{Fe}}{I_2^2} = \overline{R}_2 \quad \text{or} \quad P_{Fe} = I_2^2 \overline{R}_2$$

Thus minimum value of $\dfrac{1}{\eta}$ or maximum value of η occurs when

the Copper Loss equals the Iron Losses.

It should be noted also that for a given output in kW at constant

terminal voltage, since $I_2 = \dfrac{kW}{V_2 \cos \phi_2}$ then $I_2 \propto \dfrac{1}{\cos \phi_2}$

Now Copper Loss $= I_2^2 \overline{R}_2$ so Copper Loss $\propto I_2^2$ or $\dfrac{1}{\cos^2 \phi_2}$.

Thus as the power factor falls, copper loss rises and since η is adversely affected by losses, it follows that, efficiency falls as the load power factor falls below unity. This is shown by the graph (a) of Fig. 69. Graph (b) illustrates the condition of maximum efficiency from curves of the losses plotted to a base of load.

ALL-DAY EFFICIENCY.   It is usual for the primary of a transformer to be connected permanently to the supply and for the switching of load to be carried out in the secondary circuit. Since the copper loss varies with load but iron loss is constant and continuous, then it is apparent that the efficiency figure, depending on loading and losses, varies throughout the day. For units which are continuously excited but supply loads only intermittently, a low iron loss is particularly desirable but a low copper loss is specially important where the load factor is high. Again, for a transformer working on full load for the greater part of the day, maximum efficiency should be arranged to occur somewhere around the full-load value but for a transformer whose full-load value may be supplied for only ¼ of the day and the unit is only lightly loaded for the rest of the time, it would be desirable to arrange maximum efficiency to occur at about ½ full-load value.

Consideration of factors such as those described, indicate that the efficiency of a transformer is better estimated on an energy rather than a power ratio and thus we have the term "all-day efficiency" which can be defined as

$$\text{All-day Efficiency} \; = \; \frac{\text{Output in kWh for 24 hours}}{\text{Input in kWh for 24 hours}}$$

Example 39. A lighting transformer rated at 10 kVA has a full-load loss of 0·3 kW, which is made up equally from the iron loss and the copper loss. The duty cycle consists of full load for 3 hours, half full-load for 4 hours and no-load for the remainder of a 24 hour period. If the load operates at unity power factor, calculate the all-day efficiency.

$$
\begin{aligned}
\text{Energy output} \; &= \; 10 \text{ kW for 3 hours} \; = \; 30 \text{ kWh} \\
&\quad\;\; 5 \text{ kW for 4 hours} \; = \; 20 \text{ kWh} \\
\text{Total energy output} \; &= \; 30 + 20 \; = \; 50 \text{ kWh} \\
\text{Energy input} \; = \; \text{Output} + \text{Losses} \; &= \; 50 \text{ kWh} + \text{Energy wasted} \\
&\qquad\qquad\qquad\qquad\;\; \text{in losses}
\end{aligned}
$$

and Energy wasted in losses = Iron Losses + Copper Losses

Iron loss energy $= 0.15$ kW for 24 hours $= 3.6$ kWh

Copper loss energy $= 0.15$ kW for 3 hours $+ \dfrac{0.15}{4}$ kilowatts

for 4 hours

$= 0.45 + 0.15 = 0.6$ kWh

Energy wasted in losses $= 3.6 + 0.6 = 4.2$ kWh

All-day efficiency $= \dfrac{50}{50 + 4.2} = \dfrac{50}{54.2} =$

$= 0.9225$ or 92.25 per cent

Note that for the solution, since copper loss is proportional to current squared, then for ½ load, current would be halved and copper loss quartered; hence the figure $\dfrac{0.15}{4}$ kW for 4 hours.

Example 40. A 11/3·3 kV, 1 MVA, three-phase transformer has a star-connected primary and a delta-connected secondary. The resistance values per phase of these windings are: primary $0.375\ \Omega$, secondary $0.095\ \Omega$. The leakage reactance values per phase are primary $9.5\ \Omega$, secondary $2\ \Omega$. Calculate the voltage to be applied, at normal frequency, to the primary terminals in order to obtain full-load current in the secondary windings when the terminals are short-circuited. Calculate also the power required to be supplied under these conditions.

Primary phase voltage $= \dfrac{11}{\sqrt{3}}$ kilovolts. Secondary phase voltage $= 3.3$ k

Turns ratio $= \dfrac{11}{\sqrt{3}} \times \dfrac{1}{3.3} = 1.925{:}1$

Resistance of secondary per phase referred to primary

$= 0.095 \times 1.925^2 = 0.352\ \Omega$

$\bar{R}_1$/phase $= 0.375 + 0.352 = 0.727\ \Omega$

Similarly

Reactance of secondary per phase referred to primary $= 2 \times 1 \cdot 925^2$

$$= 7 \cdot 411 \ \Omega$$

$$\bar{X}_1/\text{phase} = 9 \cdot 5 + 7 \cdot 411$$

$$= 16 \cdot 91 \ \Omega$$

$$\therefore \ \bar{Z}_1/\text{phase} = \sqrt{0 \cdot 727^2 + 16 \cdot 91^2} = 16 \cdot 92 \ \Omega$$

Normal full-load primary current $I_1 = \dfrac{10^6}{\sqrt{3} \times 11 \times 10^3}$

$$= \dfrac{1 \cdot 732 \times 11}{10^3} = 52 \cdot 5 \ \text{A}$$

This is phase current, therefore voltage per phase to be applied on S.C. test to circulate this current $= 52 \cdot 5 \times 16 \cdot 92$ volts

For a star-connected primary, voltage to be applied to transformer terminals $= \sqrt{3} \times 52 \cdot 5 \times 16 \cdot 92 = 1530$ V or $1 \cdot 53$ kV

Power required per phase $= 52 \cdot 5^2 \times 0 \cdot 727$

Power required for three phases $= 3 \times 52 \cdot 5^2 \times 0 \cdot 727 = 6000$ W

$$= 6 \ \text{kW}.$$

CHAPTER 4

PRACTICE EXAMPLES

1. On open-circuit, a single-phase, marine, dry-type transformer gives 115 V at its secondary terminals when the primary winding is supplied with 460 V. The resistance and leakage reactance of the primary windings are 0·36 Ω and 0·83 Ω respectively while those of the secondary windings are 0·02 Ω and 0·06 Ω respectively. If the secondary terminals are accidently short-circuited while the transformer is connected to the supply, what would be the value of current flowing in the primary winding? Assume that the supply voltage is maintained at 460 V and the magnetising current value can be neglected.

2. For a 25 kVA, 450/121 V, single-phase transformer, the iron and full-load copper losses are respectively 165 W and 280 W. Calculate (a) the efficiency at full load, unity power factor and at half full load 0·8 power factor (lagging); (b) the load at which the efficiency is a maximum.

3. A marine, dry-type, 17·5 kVA, 450/121 V, 50/60 Hz, single-phase transformer gave the following data on test:

    O.C. Test   450 V,   1·5 A,   115 W   at   50 Hz
    S.C. Test   15·75 V,   38·9 A,   312 W   at   50 Hz

    Estimate the voltage of the secondary terminals and the efficiency of the transformer when supplying full-load current, at a power factor of 0·8 (lagging), from the secondary side. Assume the input voltage to be maintained at 450 V, 50 Hz.

4. A single unit of the three units making up a three-phase, marine, dry-type transformer, was subjected to a short-circuited test and the following results were obtained

    Voltage applied to the H.V. side: 14·3 V at 60 Hz
    Current supplied: 55·6 A
    Power taken: 316 W
    The unit is rated at 25 kVA, 450/121 V, 60 Hz.

    Determine the approximate value of the secondary terminal voltage when the transformer is operating at full-load power factors of 0·8 (lagging) and 0·8 (leading).

5.   The resistances of the primary and secondary windings of a
27·5 kVA, 450/121 V, single-phase, marine, dry-type transformer are 0·055 Ω and 0·00325 Ω respectively. The iron loss is
170 W. (a) At a power factor of 0·8 (lagging), calculate the
full-load efficiency. (b) At a power-factor of 0·8 (lagging), calculate the *kVA* output at which efficiency is a maximum and
find the maximum efficiency value.

6.   A 50 kVA, 440/110 V, single-phase transformer has an iron
loss of 250 W. With the secondary windings short-circuited,
full-load currents flow in the windings when 25 V is applied to
the primary; the power input being 500 W. For this transformer
determine (a) the percentage voltage regulation at full-load,
0·8 power factor (lagging). (b) the fraction of full load at which
the efficiency is a maximum.

7.   A 10 kVA, 440/110 V, single-phase, marine, dry-type transformer was tested and gave the following results:

O.C. Test Primary applied voltage 440 V. Power input 75 W.
S.C. Test. Primary applied voltage   30 V for full-load current.
Power input 135 W.

    Draw the equivalent primary circuit indicating the values of
the circuit constants, but neglecting no-load conditions, and
then calculate, (a) the secondary terminal voltage when the
transformer is operating at full load, power factor 0·8 (lagging)
with the rated voltage applied to the primary, (b) the efficiency
of the transformer for the latter condition.

8.   A 50 kVA, 440/230 V, marine, lighting-transformer has
primary resistance and reactance of 0·09 Ω and 0·19 Ω respectively and secondary resistance and reactance of 0·015 Ω and
0·042 Ω respectively. Calculate the secondary terminal voltage
when the transformer is supplying full-load current at a power
factor of 0·8 (lagging). If the secondary terminals were accidently
short-circuited, what would be the current taken by the primary
winding, assuming the primary supply voltage remained constant at 440 V?

9.   A 17·5 kVA, 460/115 V, 60 Hz, single-phase transformer
has primary and secondary resistances of 0·125 Ω and 0·008 Ω
respectively and primary and secondary leakage reactances of
0·39 Ω and 0·025 Ω respectively. The iron loss, when normal
voltage is applied to the primary winding, is 300 W. Draw the

equivalent circuit and calculate, for the full-load, power factor 0·8 (lagging) condition, the voltage at the secondary terminals. What will be the efficiency of the transformer? Neglect the no-load current.

10.   A marine, dry-type, 50 kVA, 440/110 V, single-phase transformer gave the following data on test:

O.C. Test   Primary applied voltage 440 V. Secondary voltage 110 V. Power input to primary 250 W.

S.C. Test   Primary applied voltage 25 V. Primary current 113·6 A. Power input at primary 500 W.

Calculate   (a) the efficiency at half full-load, power factor 0·7 (lagging) and (b) the voltage regulation at full-load, power factor 0·8 (lagging).

CHAPTER 5

D.C. THEORY (Continued): RECTIFICATION

To allow further study in the fields of more advanced electrotechnology some revision of work already covered will always be required. The phenomena of electromagnetic induction were described in Volume 6, Chapters 6 and 9, and the reader may find it advantageous to make a complete reappraisal of points of theory by reading back, especially if it is some time since he left the subject. For the more confident student however, sufficient reference to basics is made here to enable study to proceed. The subject of static rectification will be new, in that, no detailed reference has been made to it earlier but a knowledge of terms, such as, r.m.s. and average values together with their relationships for a sinusoidal waveform is assumed.

ELECTROMAGNETIC INDUCTION

In Volume 6, an introduction was made to Faraday's and Lenz's laws. These stated the magnitude and direction of the e.m.f. which is self-induced in an inductor when the associated flux-linkages are changing. A basic formula was devised to show that the

$$\text{average induced e.m.f.} \quad = \quad \frac{\text{Flux-linkages}}{\text{Time of change}} \quad \text{or} \quad E_{av} \quad = \quad \frac{N\Phi}{t}$$

Here E_{av} gives the value of the induced e.m.f., when N is the number of turns on the coil, Φ is the flux value (in webers) associated with the coil and t the time taken for the flux to change. The formula in this form is useful, if a steady rate of change is occurring but if the change occurs in a very short space of time or is varying from instant

to instant then it is possible to show the condition as $e = \dfrac{Nd\Phi}{dt}$.

Note that here it is the flux that is changing and not the number of turns. The calculus method of stating the rate of change of flux has already been introduced and is particularly suitable here since, instantaneous values are involved. Here e gives the instantaneous magnitude of the induced e.m.f. resulting from a rate of change of flux. This rate of change can also differ from instant to instant.

INDUCTANCE. A changing current in an inductor results in a changing flux which is responsible for an induced e.m.f. The e.m.f. is

proportional to the rate of change of flux-linkages or $e = N\dfrac{d\Phi}{dt}$.

Now any value of flux is given by (the flux set up by 1 ampere) ×
(the value of current in amperes).
Thus Φ = (Flux set up by 1 ampere) × I and
Rate of change of flux = (Flux set up by 1 ampere) × (rate of change
of current).

or $\dfrac{d\Phi}{dt}$ = (Flux set up by 1 ampere) $\dfrac{di}{dt}$

Thus $e = N\dfrac{d\Phi}{dt}$ = (N × Flux set up by 1 ampere) $\dfrac{di}{dt}$

If the value within the bracket is called L, the coefficient of self-
induction or more simply the Inductance of the circuit then the
expression becomes

$$e = L\frac{di}{dt}.$$

It is in basic form and has already been introduced in Volume 6,
Chapter 9.
The assumption $L = N$ × Flux set up by 1 ampere, can also be

developed to $L = \dfrac{N\Phi}{I}$ or L = Flux-linkages per ampere.

The unit of inductance is the Henry which can be defined in several
ways but that given earlier is the easiest to remember and is repeated
here.
A circuit has an inductance of 1 Henry, if an e.m.f. of 1 volt is
induced in it. when the current changes at the rate of 1 ampere per
second.

A further development from L = the flux linkages per ampere is obtained thus:

Since Flux $=$ $\dfrac{\text{MMF}}{\text{Reluctance}}$ then Φ $=$ $\dfrac{F}{S}$ $=$ $\dfrac{NI}{l/\mu A}$ $=$ $\dfrac{NI\mu A}{l}$.

Note. l is used here for the length of the magnetic circuit to avoid confusion with L, the inductance.

Since Φ $=$ $\dfrac{\mu ANI}{l}$ then the flux set up by 1 ampere $=$ $\dfrac{\mu AN}{l}$

It should be remembered that μ $=$ $\mu_r \mu_o$.

Also, since L $=$ $\dfrac{N\Phi}{I}$ $=$ $N \times$ Flux set up by 1 ampere

$\therefore$ L $=$ $N \times \dfrac{\mu AN}{l}$ or L $=$ $\dfrac{\mu AN^2}{l}$.

The formula in this form can be used to determine the inductance of a coil, if the area and length of the magnetic circuit and its working permeability figure is known. It is also worth noting that for a given inductor working at a constant permeability, the inductance is proportional to the coil turns squared.

THE DIRECT CURRENT LR CIRCUIT

Consider the diagram (Fig. 70). A coil of resistance R_L and inductance L is carried on an iron core and is connected in parallel with a non-inductive resistor R. The arrangement can be energised from a d.c. supply. When the switch is closed, current I_2 — through the resistor R, rises to its maximum value instantaneously as is shown by the graph of Fig. 71. Current I_1 — through the inductor L, takes time to grow to its maximum value. The maximum final values of each current, in both cases, is given by Ohm's Law i.e. I_2 $=$ $\dfrac{V}{R}$ and

$I_1 = \dfrac{V}{R_L}$ but the coil circuit appears to have electrical inertia which,

as we now know, is due to its inductance.

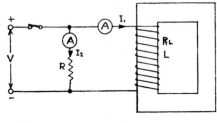

Fig. 70

GROWTH OF CURRENT

Briefly the explanation for the shape of the I_1 curve (graph b) is as follows. An increase in current is accompanied by an increase of flux which, because it is changing, induces an e.m.f. in the coil. By Lenz's law the direction of this e.m.f. is such as to try to oppose the change which takes place *i.e.* it tends to counter the applied voltage and slows up the growth of current. The e.m.f. can be described as a back e.m.f. and its value at any instant would be as already stated,

namely $e = L \dfrac{di}{dt}$.

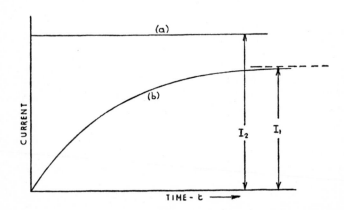

Fig. 71

From earlier work on induction and inductance, the point was made that, the effects of induction are ever present and were to be considered even under the steady running conditions of the circuit. For our study here, we are now concerned with *transient* conditions, *i.e.* the effects which occur when an inductive circuit is switched "on" or "off". Our main interest is in d.c. circuit conditions but a mention will be made of corresponding a.c. circuit conditions.

SHAPE OF CURVE. This is considered for a condition of inductance without and with resistance. The first condition is hypothetical but allows understanding of the second.

(a) CIRCUIT WITH INDUCTANCE (L) AND NEGLIGIBLE RESISTANCE. In accordance with theory developed so far, we know that the induced back e.m.f. which is created when the circuit is switched on, will counter the applied voltage V. There is no voltage drop due to resistance and e must be equal to V.

Thus $V = e$ but $e = L\dfrac{di}{dt}$ $\therefore V = L\dfrac{di}{dt}$ or $\dfrac{di}{dt} = \dfrac{V}{L}$

Thus as V volts are applied, current will increase at the steady rate of $\dfrac{V}{L}$ amperes per second. This current will not reach a steady value since there is no resistance. It will thus continue to grow and the graph will be as shown in Fig. 72a.

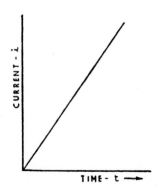

Fig. 72a

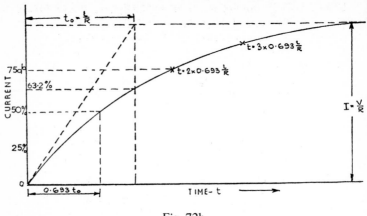

Fig. 72b

(b) CIRCUIT WITH INDUCTANCE (L) AND RESISTANCE (R). At any instant after the circuit switch is closed, the current value would be i amperes. The voltage condition would then be given by

Applied voltage = Resistance voltage drop + Induced back e.m.f.

$$\therefore V = iR + e \quad \text{or} \quad iR + L \frac{di}{dt}$$

Rewriting, it is seen that the voltage, producing the change of current at the instant being considered, would be $\dfrac{V - iR}{L}$

Thus $\dfrac{di}{dt} = \dfrac{V - iR}{L}$ and this, being a differential equation, can be solved to $i = I(1 - e^{-\frac{Rt}{L}})$.

•Note here e is not the symbol used for induced e.m.f. but is the base of Naperian Logarithms.

Thus $e = 2{\cdot}718$ and $\log_{10} e = 0{\cdot}4343$.

Differential equations of this form will be derived at appropriate points in this chapter and the method of solution will be of interest to students. This is set out below for the above equation.

Since $V = iR + L \dfrac{di}{dt}$ then $V - iR - L \dfrac{di}{dt} = 0$

or $\dfrac{V}{R} - i - \dfrac{L}{R}\dfrac{di}{dt} = 0$ and $I - i = \dfrac{L}{R}\dfrac{di}{dt}$

Rearranging $\dfrac{di}{I - i} = \dfrac{R}{L}\,dt$ and by integration

$$-\log_e(I - i) = \dfrac{R}{L}t + K$$

K is a constant of integration and its value is obtained thus:

When $t = 0$ $i = 0$ and $K = -\log_e I$

Using this value for K $\log_e(I - i) = -\dfrac{Rt}{L} + \log_e I$

or $\log_e(I - i) - \log_e I = -\dfrac{Rt}{L}$

and $\log_e \dfrac{I - i}{I} = -\dfrac{Rt}{L}$

Thus $\dfrac{I - i}{I} = e^{-\frac{Rt}{L}}$

or $i = I(1 - e^{-\frac{Rt}{L}})$.

This is frequently known as an Helmholtz Equation and shows that the growth curve is of exponential form. The further investigation of the mathematics leads to the term *time-constant* and its meaning.

This follows after the example.

Example 41. A 20 H inductor with a resistance of 20 Ω is connected to a 200 V d.c. supply. Calculate the value to which the current will have risen 0·1 seconds after switching-on and how long it takes to grow up to a value of 5 A.

Here $I = \dfrac{200}{20} = 10$ A and $i = 10(1 - e^{-\frac{20 \times 0 \cdot 1}{20}})$

or $i = 10(1 - e^{-0 \cdot 1}) = 10 - 10e^{-0 \cdot 1}$

let $x = 10e^{-0 \cdot 1}$ then $\log x = \log 10 - 0 \cdot 1 \log e$

or $\log x = 1 - 0 \cdot 1 \times 0 \cdot 4343$
$\qquad\quad = 1 - 0 \cdot 04343$
$\qquad\quad = 0 \cdot 95657$
giving $x = 9 \cdot 048$

Thus $i = 10 - 9 \cdot 048 = 0 \cdot 952$ A

Similarly substituting in $i = I(1 - e^{-\frac{Rt}{L}})$

$\qquad 5 = 10(1 - e^{-t})$ $\qquad$ Here $\dfrac{R}{L} = \dfrac{20}{20} = 1$

Thus $\dfrac{1}{2} = 1 - e^{-t}$ or $\dfrac{1}{2} = e^{-t}$ whence

$\qquad\qquad\qquad 2 = e^{t}$ giving $t \log e = \log 2$

$\qquad\qquad\qquad$ or $0 \cdot 4343t = 0 \cdot 301$

Thus $t = \dfrac{0 \cdot 301}{0 \cdot 4343} = 0 \cdot 693$ seconds.

THE TIME CONSTANT. If the initial rate of growth of current was to be maintained, the time taken for the current to attain its full value would be given by $\dfrac{\text{Current value}}{\text{Initial rate of change of current}}$

Let this time be t_o then $t_o = \dfrac{\dfrac{V}{R}}{\dfrac{V}{L}} = \dfrac{L}{R}$

The time t_o is called the "time constant" of the circuit. Actually, in this time, the current rises to a value given by

$$i = I(1 - e^{-\frac{R}{L}\frac{L}{R}}) = I(1 - e^{-1}) = I\left(1 - \frac{1}{e}\right)$$

Thus $i = I\left(\dfrac{2 \cdot 718 - 1}{2 \cdot 718}\right)$ or $i = \dfrac{1 \cdot 718}{2 \cdot 718}I = 0 \cdot 632I$

The value of i would be 63·2 per cent of I. This is also shown on the graph of Figure 72b.

The time constant can now be defined as the ratio $\dfrac{L}{R}$ and is equal to the time taken for the current to attain 63·2 per cent of the full value.

The following deduction is also of interest.

When the current increases to half its value in a time t

Then $\dfrac{I}{2} = I - I e^{-\frac{Rt}{L}}$ or $\dfrac{I}{2} = I e^{-\frac{Rt}{L}}$

Thus $e^{-\frac{Rt}{L}} = \dfrac{1}{2}$ or $2 = e^{\frac{Rt}{L}}$ So $\log 2 = \dfrac{Rt}{L} \log e$

and $\dfrac{Rt}{L} = \dfrac{\log 2}{\log e} = \dfrac{0 \cdot 301}{0 \cdot 4343} = 0 \cdot 693$ and $t = 0 \cdot 693 \dfrac{L}{R}$

The current reaches half value in $0 \cdot 693 \dfrac{L}{R}$ seconds

or $0 \cdot 693 t_o$ seconds.

Similarly the current reaches three-quarter value in a further

$0 \cdot 693 \dfrac{L}{R}$ seconds.

Example 42. A lighting circuit is operated by a relay of which the coil has a resistance of 5 Ω and an inductance of 0·5 H. The relay coil is supplied from a 6 V, d.c. source through a push-button switch. If the relay operates when the current in the relay coil attains a value of 50 mA, find the time interval between the pressing of the push-button and the closing of the lighting circuit. Also determine the time constant value of the circuit.

$$\text{Here } I = \frac{6}{5} = 1·2 \text{ A} \quad \text{and} \quad i = \frac{50}{1000} = 0·05 \text{ A}$$

$$\text{Then } 0·05 = 1·2(1 - e^{-\frac{5t}{0·5}}) \quad \text{or} \quad 0·5 = 12(1 - e^{-10t})$$

$$\text{giving } 0·5 = 12 - 12e^{-10t} \quad \text{or} \quad 12e^{-10t} = 11·5$$

$$\text{whence } e^{10t} = \frac{24}{23} = 1·043$$

$$\text{and } 10t \log e = \log 1·043 \quad \text{or} \quad t = \frac{0·0183}{10 \times 0·4343} = \frac{0·0183}{4·343}$$

$$\text{giving } t = 0·00421 \quad \text{or} \quad 4·21 \text{ milliseconds}$$

$$\text{Also} \qquad t_o = \frac{L}{R} = \frac{0·5}{5} = 0·1 \text{ second.}$$

DECAY OF CURRENT

Assume for simplicity that the inductor, after being connected to the supply, is short-circuited at the same instant as the supply is switched off. As the field collapses the energy put into the magnetic field must be released through the agency of a current and the heat generated in the resistance of the circuit. If the arrangement is as shown in the diagram (Fig. 73) then the current decays in a circuit having the same constants as before.

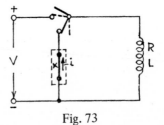

Fig. 73

For the condition being considered $V = 0$ and the induced e.m.f. at any instant tends to oppose the decay of current. The voltage condition at any instant would be given by

$0 = iR + L \dfrac{di}{dt}$ This is a differential equation which can be solved to

$$i = I e^{-\frac{Rt}{L}}$$

The solution would be $\dfrac{di}{dt} = -\dfrac{iR}{L}$ or $\dfrac{di}{i} = -\dfrac{R}{L} dt$

Integrating both sides $\log_e i = -\dfrac{Rt}{L} + K$

K is the constant of integration and its value is obtained thus:

When $t = 0$ $i = I$ $\therefore K = \log_e I$

$$\text{giving } \log_e i = -\dfrac{Rt}{L} + \log_e I$$

$$\text{or } \log_e i - \log_e I = -\dfrac{Rt}{L}$$

which can be written as $\dfrac{i}{I} = e^{-\frac{Rt}{L}}$

$$\text{or } i = I e^{-\frac{Rt}{L}}$$

SHAPE OF CURVE.   Here again the decay of current curve is of exponential form and is a reflection of the growth curve. It is shown by Fig. 74 and if t_0 is the time constant, then the current falls to a

value equal to $i = I e^{-\frac{R}{L}\frac{L}{R}} = I e^{-1} = \dfrac{I}{e} = \dfrac{I}{2 \cdot 718}$

or $\quad i \ = \ 0{\cdot}368I$

As before it can be shown that the current will decay to half value in a time given by $0{\cdot}693 \dfrac{L}{R}$ seconds.

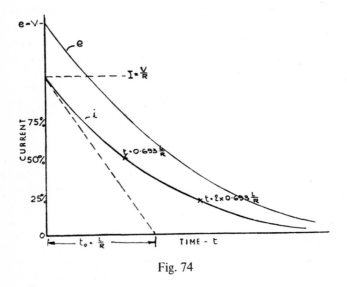

Fig. 74

The e.m.f. or voltage-drop curve is also shown, since e at any instant is responsible for the discharge current and equals the circuit voltage drop v where $v \ = \ iR$.

THE FIELD SWITCH AND DISCHARGE RESISTOR

If the field circuit of a large machine such as an alternator, synchronous motor or d.c. generator is opened quickly, a very high and dangerous e.m.f. could be induced. Since $e \ = \ L \dfrac{di}{dt}$ and, in theory, the current is required to fall to zero instantaneously, it follows that e would be infinitely high. In practice this does not happen. As the current begins to fall an e.m.f. is induced which tends to maintain the current at its original value. The direction of the e.m.f. will assist the mains voltage and the current flow is prolonged and made evident by an arc across the switch contacts. Additional resistance is introduced into the circuit by the resistance of the arc and a condition of balance exists during the circuit-breaking period, when the value of the applied voltage plus that of the induced e.m.f. is equal to the voltage drop across the field resistance plus that across the resistance of the arc. The arrangement is unsatisfactory due to the fact that the voltage developed across the field depends on the speed of opening the field circuit. Thus quite contrary conditions could arise in that it may, for safety purposes, be desirable to "kill" the field quickly by providing a fast opening switch. This would, however, mean a large voltage induced across the field which would result in a break-down of insulation and damage to the machine. The alternative would be to make the switch open comparatively slowly and tolerate a long arc which would persist and burn the contacts. The field would die slowly but the induced voltage would be kept at a safe value.

A compromise arrangement is to provide an extra contact on the field switch which connects-in a non-inductive "discharge resistance" R_D in series with the main field coils, at the instant when they are being disconnected from the supply. The circuit would be as shown in Fig. 73 except that instead of the direct link at X, the resistor R_D would be included. The arrangement allows the current to decay in a closed circuit and the magnetic energy stored in the field to be discharged as heat and the e.m.f. and arcing at the main contacts to be minimised.

If the field was short-circuited there would be no rise in voltage, the e.m.f. induced being only sufficient to momentarily maintain the same value of current, i.e. $e = iR$. Since no sudden alteration of current value is desirable in order to maintain the e.m.f. condition, then i, at start of discharge, should be made equal to I, the full field current. To avoid a short-circuit on the supply when the switch is operating and, the main arc is conducting, the discharge resistance R_D is needed and must be of considerable value. The maximum value of induced e.m.f. will then be given by $i(R_D + R)$ i.e. $I(R_D + R)$. If the field has to be killed rapidly, the time constant must be small. L cannot be reduced, so R_D is increased. A working value is $R_D = R$. The time constant is thus a half of that for the field circuit alone and the induced voltage is, at most, double the working voltage.

Example 43. The six field coils of a d.c. generator, each containing 1500 turns, are connected in series to a 500 V d.c. supply. The field current is 5 A and the flux per pole 0·05 Wb. Calculate (a) the inductance of the field circuit, (b) the value of the induced e.m.f. if the circuit is broken in 0·1 second, (c) the value of non-inductive discharge resistor to be connected across the field terminals, if the induced e.m.f. on opening the switch is not to exceed 750 volts, (d) the time taken for the field current to decay to 10 per cent of its steady value under these conditions.

(a) $L = N$ (Flux set up by one ampere)

$$= 1500 \times \frac{0 \cdot 05}{5} = 15 \text{ H per coil}$$

$$= 6 \times 15 = 90 \text{ H for the complete field}$$

(b) $e = L \times$ Rate of change of current

$$= 90 \times \frac{5}{0 \cdot 1} = 4500 \text{ volts or } 4 \cdot 5 \text{ kV}$$

(c) Let R_T be the total resistance of the circuit. Then $5R_T = 750$

But $R_T = R + R_D$ As $R = \dfrac{500}{5} = 100 \ \Omega$

and since $R_T = 150 \ \Omega$

So $R_D = 150 - 100 = 50 \ \Omega$

(d) Here $i = \dfrac{10}{100} \times 5 = 0 \cdot 5 \text{ A}$ So $0 \cdot 5 = 5 \, e^{-\frac{150t}{90}}$

$$\therefore \ 0 \cdot 1 = e^{-\frac{5t}{3}} \quad \text{or} \quad 10 = e^{\frac{5t}{3}}$$

giving $\log 10 = \dfrac{5}{3} t \times \log e$

or $1 = 1 \cdot 666t \times 0 \cdot 4343$ and $t = \dfrac{1}{0 \cdot 7235} = 1 \cdot 38 \text{ seconds}$

THE DIRECT CURRENT *CR* CIRCUIT

Because the behaviour of the *LR* circuit has been considered in some detail, it is appropriate to give considerable attention to that of the capacitor when connected to a d.c. supply. In Chapter 13 of Vol. 6 introduction was made to the capacitor and its ability to store electricity. The charging action was described, in that, a current would pass on closing the switch to the supply, and the magnitude of this current would gradually fall to zero as the potential difference across the plates rose to equal eventually the value of the applied e.m.f. of the supply. The discharging condition was also described and attention is now given to a more complete investigation of the charging and discharging action with acceptance of the fact that the quantity of electricity which can be stored by a capacitor is proportional to the charging voltage.

CAPACITANCE. Before deducing the mathematical expressions for the current and voltage growth, we revise that C, the capacitance of any capacitor depends on the number and dimensions of the plates, upon their spacing and upon the nature of the dielectric. The unit is the Farad and although other definitions are possible, we repeat that already given, in terms of unit quantity and unit voltage. Thus a capacitor is said to have a capacitance of 1 Farad if 1 Coulomb of electricity is stored when 1 Volt is applied across the plates. Note also the expression already deduced for series and parallel capacitor arrangements. Thus if C is the equivalent capacitance then for a

Series connection $\dfrac{1}{C} = \dfrac{1}{C_1} + \dfrac{1}{C_2} + \dfrac{1}{C_3} +$ etc.

Parallel connection $C = C_1 + C_2 + C_3 +$ etc.

GROWTH OF CURRENT

Charging is considered and since, as stated above, $Q \propto V$, we can write $Q = CV$ where C is a constant, termed the capacitance.

Also since $Q = It$ it follows that $It = CV$ or $I = C\dfrac{V}{t}$. It is thus evident that the current, at any instant, is dependent on $\dfrac{V}{t}$ or on the rate of change of voltage. As for the inductor, since we are more interested here in instantaneous values, we can write the relation as $i = C\dfrac{dv}{dt}$. The expression shows that, if the rate of change of voltage is uniform for a period of time, then a constant current flows. This condition has already been considered in Vol. 6, Chapter 13 when the absence of resistance was also assumed. Resistance is, however, present and the effect of a comparatively high value is now considered. If the values of current and p.d. across the plates are noted at various instants after closing the switch and these are plotted to a time base, then the graphs as shown (Fig. 75) would be seen to be of exponential form. The reason for this shape is accounted for in the mathematics set out below.

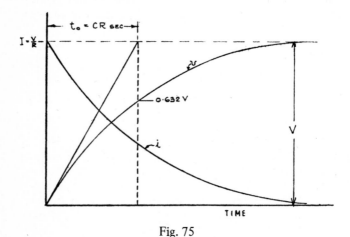

Fig. 75

SHAPE OF CURVE. Consider a circuit of C farads and R ohms, connected as shown in the diagram (Fig. 76). On switching on, there is no potential difference across the plates and maximum current value will flow, given by $I = \dfrac{V}{R}$.

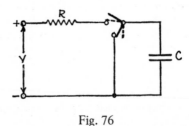

Fig. 76

Let $v = $ the capacitor p.d. after t seconds. Then $i = \dfrac{V - v}{R}$

Also i is the rate of charge build up $\therefore\ i\ =\ \dfrac{Cdv}{dt}$

$\therefore\ \dfrac{Cdv}{dt}\ =\ \dfrac{V-v}{R}$ or $\dfrac{dv}{V-v}\ =\ \dfrac{dt}{CR}$. This is a differential

equation which can be solved to $v\ =\ V(1-e^{-\frac{t}{CR}})$.

The solution would be by integration

$-\log_e(V-v)\ +\ K\ =\ \dfrac{t}{CR}$. Where $K\ =$ a constant of integration.

When $t\ =\ 0\quad v\ =\ 0$ and $K\ =\ \log_e V$

So $\log_e\left(\dfrac{V}{V-v}\right)\ =\ \dfrac{t}{CR}$ whence $\dfrac{V}{V-v}\ =\ e^{\frac{t}{CR}}$

or $\dfrac{V-v}{V}\ =\ e^{-\frac{t}{CR}}$ giving $V-v\ =\ Ve^{-\frac{t}{CR}}$

and $v\ =\ V-Ve^{-\frac{t}{CR}}$ or $v\ =\ V(1-e^{-\frac{t}{CR}})$

THE TIME CONSTANT. If the initial rate of current build up *i.e.*

$I\ =\ \dfrac{V}{R}$ was to be maintained and the time taken to charge the

capacitor to V volts was t_o seconds, then the quantity of electricity

stored is $Q\ =\ It_o\ =\ \dfrac{V}{R}\,t_o\ \therefore\ t_o\ =\ \dfrac{QR}{V}\ =\ \dfrac{CVR}{V}\ =\ CR$ seconds.

This therefore, would be the time constant of the circuit or $t_o\ =\ CR.$

It can be shown that, in this time, the voltage builds up to 63·2 per cent of its maximum value. As before, the voltage reaches half value in 0·693 CR seconds and so on.

During charging, the current is shown to decay exponentially and its value at any instant can be found thus:

Let i = the current value at any instant $\dfrac{V - v}{R}$

or $i = \dfrac{V - V(1 - e^{-\frac{t}{CR}})}{R}$ $\therefore i = \dfrac{V}{R} e^{-\frac{t}{CR}}$

giving $i = I e^{-\frac{t}{CR}}$

Example 44. A 1 MΩ resistor is connected in series with a 20 μF capacitor. The arrangement is connected across a 2·5 kV, d.c. supply. Find the value of the charging current ten seconds after closing the circuit switch. Find also the time constant of the circuit.

Time constant $t_o = CR = 20 \times 10^{-6} \times 1 \times 10^6 = 20$ seconds

The maximum value of charging current $= \dfrac{V}{R} = \dfrac{2·5 \times 10^3}{1 \times 10^6}$

$$= 2·5 \times 10^{-3} = 2·5 \text{ mA}$$

then $i = 2·5 e^{-\frac{t}{CR}}$ Here $-\dfrac{t}{CR} = -\dfrac{10}{20} = -0·5$

$$\text{so } i = 2·5 e^{-0·5}$$

or $\log i = \log 2·5 - 0·5 \log 2·718$

$= \log 2·5 - 0·5 \times 0·4343$

$= 0·3979 - 0·2172 = 0·1807$

giving $i = 1·52$ mA.

DISCHARGE CONDITIONS

If the circuit switch of Fig. 76 is operated so that the capacitor is disconnected from the supply and allowed to discharge through a short-circuit, since no resistance is present, the discharge current will be very large indeed and may well damage the capacitor. It is usual to include resistance in the circuit for this condition and this could readily be achieved by moving R to the hinged side of the switch. The time constant of the circuit would thus be the same as for the charging condition, or it may be altered by including an additional discharge resistance R_D in the short-circuiting link.

DECAY OF CURRENT. The current and voltage curves are shown by the diagram (Fig. 77). The current graph is drawn below the horizontal to remind the reader that it is a reversed current.

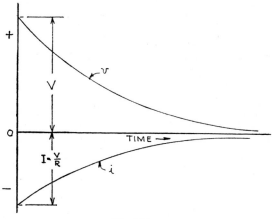

Fig. 77

Let the capacitor be charged to a voltage of V volts and discharged through a resistance of R ohms. The maximum value of current is

$$I = \frac{V}{R}$$

The voltage across the resistance after a time t seconds is v volts

and the current at that instant $i = \dfrac{v}{R}$. The current is also caused by

the voltage change or $i = -C\,\dfrac{dv}{dt}$.

The negative sign is used to show the reversal of current due to a decrease of voltage of value dv. Thus $-i = \dfrac{Cdv}{dt}$

$$\therefore \; -\frac{v}{R} = \frac{Cdv}{dt} \quad \text{so} \quad \frac{dv}{v} = -\frac{1}{CR} \, dt$$

This is a differential equation which is solved to $v = Ve^{-\frac{t}{CR}}$. As before the time constant is CR and the current follows the same exponential law since $\dfrac{v}{R} = \dfrac{V}{R} e^{-\frac{t}{CR}}$ or $i = Ie^{-\frac{t}{CR}}$

Solution for the differential equation is similar to those already detailed earlier and is not repeated.

Example 45. A voltmeter having a resistance of 30 MΩ is connected across a capacitor which is charged by a d.c. supply to a p.d. of 500 V. If after three minutes the reading is noted to indicate 100 V, calculate the value of the capacitor.

$$\text{Since } v = Ve^{-\frac{t}{CR}} \quad \text{then} \quad \frac{v}{V} = e^{-\frac{t}{CR}} \quad \text{or} \quad \frac{V}{v} = e^{\frac{t}{CR}}$$

$$\text{giving } \log V - \log v = \frac{t}{CR} \log e$$

$$\text{and } \log 500 - \log 100 = \frac{180}{C \times 30 \times 10^6} \log e$$

$$\text{or } 2{\cdot}699 - 2{\cdot}0 = \frac{6 \times 10^{-6}}{C} \times 0{\cdot}4343$$

$$\text{Thus } 0{\cdot}699 = \frac{2{\cdot}6058}{C} \times 10^{-6}$$

and $C = \dfrac{2 \cdot 6058}{0 \cdot 699} \times 10^{-6} = 3 \cdot 73 \times 10^{-6}$ farads

or $C = 3 \cdot 73 \, \mu$F.

Example 46. A 2 μF capacitor is connected across the terminals of an electrostatic voltmeter having negligible capacitance and a resistor of 50 MΩ is placed in series with the parallel arrangement. Determine what should be the reading on the voltmeter one minute after connection to a 500 V, d.c. supply.

The voltmeter of negligible capacitance is assumed to have no effect on the circuit.

The charging current 1 minute after switching on is given by

$$i = \frac{V}{R} e^{-\frac{t}{CR}} = \frac{500}{50 \times 10^{6}} e^{-\frac{60}{2 \times 10^{-6} \times 50 \times 10^{6}}}$$

$$\text{or } i = \frac{10}{10^{6}} e^{-\frac{6}{10}}$$

$$\text{and } i = 10e^{-0 \cdot 6} \text{ microamperes}$$

$$\text{So } \log i = \log 10 - 0 \cdot 6 \log e$$

$$= 1 - 0 \cdot 6 \times 0 \cdot 4343 = 1 - 0 \cdot 26058$$

$$= 0 \cdot 7394$$

giving $i == 5 \cdot 488 \, \mu$A. For this value of charging current, the voltage drop across the resistor will be

$$5 \cdot 488 \times 10^{-6} \times 50 \times 10^{6} = 274 \cdot 4 \text{ V}$$

∴ Reading on voltmeter would be $500 - 274 \cdot 4 = 225 \cdot 6$ V.

THE ALTERNATING CURRENT *LR* AND *CR* CIRCUIT

It is not intended to give here a detailed treatment of the conditions which occur at the instant of switching on or off an inductive or capacitive a.c. circuit, but rather to draw attention to their importance in relation to the switching duty of circuit-breakers and fuses. The circuit conditions and the terms used for the non-resistive a.c. circuit during transient periods have not yet been considered since only steady-state operating theory has received attention. This introduction is to the basic fundamentals and applies to both *LR* and *CR* circuits. For convenience only the former will be considered since corresponding relationships can be deduced for the capacitive-resistive circuit.

CURRENT ASYMMETRY.

Consider an a.c. circuit in which the ohmic values of inductive reactance and resistance are comparable *i.e.* the operating power factor would be in the region of 0·7 (lagging). When connected to sinusoidal supply voltage, the current must start from zero no matter at what point in the voltage cycle the switch is closed. The diagram of Fig. 78 illustrates the resulting current condition. After an interval of time the current settles down to the steady or permanent condition given by $I = \dfrac{V}{Z}$. This is the r.m.s. value or the instantaneous value is $i_1 = \dfrac{V_m}{Z} \sin(\omega t - \phi)$ where ϕ is the circuit phase angle and the applied voltage can be expressed as $v = V_m \sin \omega t$.

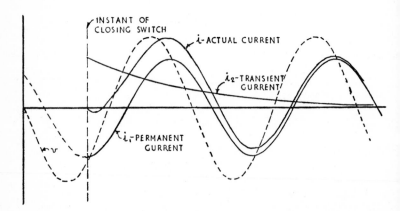

Fig. 78

Let the actual current after closing the switch at any instant, be $i = i_1 + i_2$. The simplest condition to understand is that shown. This assumes that the supply voltage is going through its normal value when the switch is closed. Since the current for the normal closed circuit condition, would have had a definite value and phase angle, but is actually zero when the switch is closed, then, it follows that, a component of direct current must be induced which decays exponentially to zero. This is shown by the diagram above and i_2 is assumed to be the instantaneous value of the transient current which may or may not be present, according to the circuit conditions and the instant of switching.

The expression for the value of the transient current can be obtained from the following deduction where i_1 is the permanent current.

Instantaneous applied voltage = Instantaneous voltage drop + Induced back e.m.f.

or $\quad v = Ri + L\dfrac{di}{dt}$

Since $i = i_1 + i_2$ Then $v = R(i_1 + i_2) + L\dfrac{d(i_1 + i_2)}{dt}$

or $\quad v = Ri_1 + L\dfrac{di_1}{dt} + Ri_2 + L\dfrac{di_2}{dt}$ ... ... (a)

Under steady conditions $i_2 = 0$ and $v = Ri_1 + L\dfrac{di_1}{dt}$ ... (b)

Subtracting (b) from (a)

$$0 = Ri_2 + L\dfrac{di_2}{dt} \quad \text{so} \quad \dfrac{di_2}{i_2} = -\dfrac{R}{L}\,dt.$$

By integration $i_2 = I_2 e^{-\frac{Rt}{L}}$ where I_2 is the initial maximum value of the d.c. transient current at the instant of closing the switch.

We now see that the current flowing in the circuit at any instant after closing the switch, is the result of the sum of the a.c. and d.c. components. Points to note about an asymmetrical current are, (1) the fact that the alternate "loops" of current are dissimilar in magnitude and are in fact major and minor loops, (2) the time intervals between the loops are unequal, unlike those for the permanent current.

The condition of current asymmetry is of particular interest when considering a circuit under fault conditions. Thus if a short-circuit developed on an alternator, it is apparent that at the instant of fault, the short-circuit current would rise to a high value, being limited only by the resistance and reactance of the alternator which are of comparatively low value. Furthermore, if the reactance value was substantially greater than the resistance then the fault current, due to the low power factor of the circuit, would lag the voltage by nearly 90°. Again, if the fault occurred at the condition of zero voltage then complete asymmetry of current would result, and the initial maximum value of fault current during the first loop may be as much as twenty times the final steady value, and this could take a time interval of some three seconds to fall to the final permanent value. If a fuse or circuit-breaker was made sufficiently fast acting, it would operate immediately the fault occurred and would have to break a current substantially greater than that which is switched normally.

The author hopes that this brief introduction to a.c. transients will provide sufficient knowledge for the student to appreciate the importance of equipment being designed to function under fault conditions. The testing and certification of circuit-breakers and fuses for particular duties is a requirement of both British and foreign specifications, compiled to ensure safe operational characteristics. If the reader is interested in the aspect of such devices operating under fault conditions, he can obtain more factual information from the study of relevant British Standard Specifications such as Nos. 116, 936 and 82.

RECTIFICATION

The word "rectification" is used to describe the conversion of an alternating voltage or current waveform into a constant or direct value. The basic principles of rectification were discussed when the action of the commutator, as fitted to the d.c. generator, was introduced, but the more accepted meaning of the term is in connection with the operation of static devices such as the thermionic diode, the semiconductor diode or metal rectifier. Mention has already been made of the uni-lateral conducting property of the thermionic and diode semiconductor (Vol. 6, Chapter 15) and although other types are named in this chapter, it is not proposed, at this point, to describe the construction or action of any specific type of rectifier, but to point out that rectification can be achieved by any one of these devices irrespective of how it operates, and that the general terms and relationships between output (d.c.) and input (a.c.) values hold good for any specific device as used to achieve rectification.

TERMS. (a) Peak Inverse or Reverse Voltage (P.I.V.). This is the maximum value of the reversed voltage which the rectifier unit has to withstand during the non-conducting period. (b). When a rectifier is used with a "reservoir" capacitor smoothing system, it can pass pulses of current which are greater than the load current. The greatest value of current, which can be safely passed, is called the Peak Current Rating of the device. (c) Ripple. This is the variation of output voltage which produces a current in the load consisting of a steady value and a superimposed sinusoidal fundamental and harmonic values. The ripple is usually expressed as the ratio of r.m.s. ripple current to the mean current or r.m.s. ripple voltage to the mean voltage.

RECTIFIER ARRANGEMENTS

The property of a diode to conduct when the anode voltage is made positive with respect to the cathode and to isolate when the voltage is negative, is made use of in rectifier circuits. The most common arrangements are considered.

HALF-WAVE (1 phase). The simplest usage is the half-wave connection as is illustrated by the diagram (Fig. 79a). A transformer may not be necessary. Note that, for all the rectifier arrangements, a straight line characteristic is assumed *i.e.* the "forward resistance" − in the conducting direction is constant and thus the current will, at all times, be directly proportional to the applied voltage.

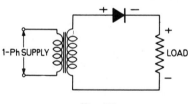

Fig. 79a

The applied voltage and the rectified voltage have the waveforms shown (Fig. 79b) and the latter is half sinusoidal as is the current if the load is resistive. If the load is inductive or capacitive the shape of the current waveform is altered and by suitable choice of *L* and *C* values, smoothing can be introduced.

To estimate the effect of the current obtained from half-wave rectification it is essential to consider the process for which the current is being utilised. Thus for electrolysis or battery-charging, the current is unidirectional and is being used in half-wave pulses, as is depicted by the diagram. The effect of the current will be equivalent to that of a smaller direct current flowing at a constant value over the time of a complete cycle. Now the average value of a sine wave is $\frac{2}{\pi}$ or 0·6365 times the maximum value but here, since one half of the wave is suppressed, the average of the output current I_D must be taken over a complete cycle and is therefore $\frac{1}{\pi}$ or 0·318 times I_m (the peak value).

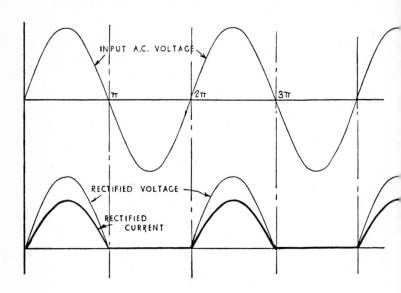

Fig. 79b

Again if the current is being used for its heating or electro-magnetic effect then the r.m.s. value over a full cycle must be considered.

The r.m.s. value for a sinusoidal waveform was deduced mathematically from $\sqrt{\dfrac{I_m^2}{2}} = \dfrac{I_m}{\sqrt{2}} = 0\cdot707\,I_m.$

For a full cycle, half of which is non-conducting, the r.m.s. value is $\sqrt{\dfrac{I_m^2}{2 \times 2}} = \dfrac{I_m}{2} = 0\cdot5\,I_m.$

The relation of r.m.s. and average values to maximum or peak value for a sinusoidal half wave has not been deduced completely from first principles, but would form a useful exercise and could be attempted by the student for either a mathematical or graphical solution.

The example illustrates the significance of these relationships.

Example 47. A half-wave rectifier is connected in series with a 18·8 Ω resistor and copper-sulphate voltmeter of negligible resistance. The rectifier may be assumed to have a constant resistance of 1·2 Ω in the forward direction. Determine the equivalent values of the direct currents which would deposit the same amount of copper and produce the same heat as the rectified current. The supply voltage has an r.m.s. value of 40 V. (Neglect the effects of polarisation of the voltmeter.)

$$\text{Peak value of applied voltage } = \frac{40}{0 \cdot 707} = 56 \cdot 56 \text{ V}$$

$$\text{Average value of rectified current } = \frac{1}{\pi} \times \frac{56 \cdot 56}{(18 \cdot 8 + 1 \cdot 2)}$$

$$= \frac{1}{\pi} \times 2 \cdot 828 = 0 \cdot 9 \text{ A}$$

Thus the same amount of copper would be deposited by a direct current of 0·9 A flowing for the same time as the rectified current.
Similarly

$$\text{R.M.S. value of rectified current } = \frac{1}{2} \times 2 \cdot 828 = 1 \cdot 414 \text{ A}$$

Thus the same amount of heat would be produced in the resistor by a direct current of 1·414 A flowing for the same time as the rectified current.

During the non-conducting period the P.I.V. applied to the rectifier is equal to the maximum voltage V_m of the supply. To compare different rectifier arrangements, it is usual to evaluate the P.I.V. in terms of the average output voltage. Thus for a half-wave arrangement

$$\text{the ratio is } \frac{\text{Peak Inverse Voltage}}{\text{Average Ouput Voltage}} = \frac{V_m}{\dfrac{V_m}{\pi}} = 3 \cdot 14.$$

FULL-WAVE. (1 PHASE. Bi-phase connection.) The diagram of Fig. 80a shows the arrangement. Here two diodes conduct on alternate half-cycles of the supply voltage and the shape of the output voltage and current waveforms are illustrated by Fig. 80b—a resistive load is assumed.

The average value (I_D and V_D) for the current and voltage waves is the same as that for a full sine wave or is $\dfrac{2}{\pi} = 0.6365$ times the maximum value. Similarly for the r.m.s. value (I and V), which is 0.707 times the maximum value. Thus:

$V_D = 0.6365 \, V_m$ and $I_D = 0.6365 \, I_m$, also $V = 0.707 \, V_m$ and $I = 0.707 \, I_m$. From these can be deduced:

$V_D = 0.9 \, V$ and $I_D = 0.9 \, I$ where V and I are r.m.s. values.

It should be noted that as V_m is the peak value of the forward voltage applied to a diode, it is the voltage of each half of the transformer secondary winding and the maximum peak inverse voltage which a diode has to withstand is $2 \, V_m$. This is applied to two rectifiers in series, but since one is conducting then the other has to withstand the full voltage or

$$\frac{\text{Peak Inverse Voltage}}{\text{Average Output Voltage}} = \frac{2 \, V_m}{\dfrac{2 \, V_m}{\pi}} = 3.14 \quad i.e. \text{ the same as for}$$

the half-wave arrangement.

The arrangement, which employs a centre-tap transformer, is used for thermionic valve diodes especially when the two anodes and a single cathode are accommodated in the one glass envelope. It is rarely used with metal rectifiers except where low regulation and high efficiency are required. A transformer is essential, although its usage is uneconomical, since each half of the secondary is used only half the time, while insulation for twice the voltage of a half-winding is necessary. The d.c. side can however be completely isolated from the a.c. mains by means of such a transformer.

Example 48. A half-wave rectifier is connected in series with a moving-iron ammeter and a permanent magnet moving-coil ammeter. The supply is sinusoidal. The reading on the moving-iron instrument is 10 A, find the reading on the other ammeter. Estimate the readings on both meters if full-wave rectification was used.

The moving-iron meter will indicate the r.m.s. value of current whilst the moving-coil meter indicates the average value of the rectified current.

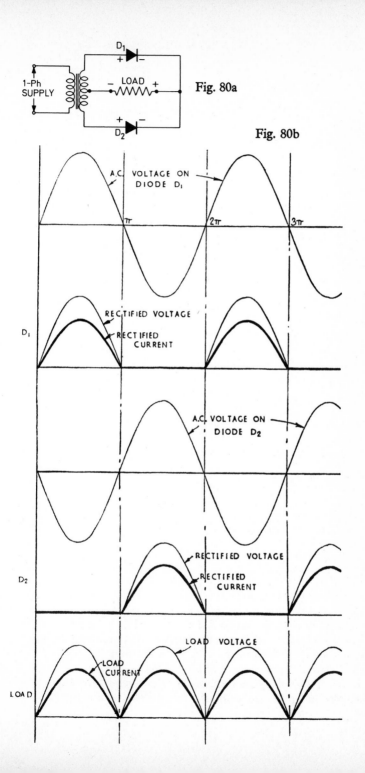

D₁

D₂

LOAD

Fig. 80a

Fig. 80b

1-Ph SUPPLY

LOAD

A.C. VOLTAGE ON DIODE D₁

π   2π   3π

RECTIFIED VOLTAGE
RECTIFIED CURRENT

A.C. VOLTAGE ON DIODE D₂

RECTIFIED VOLTAGE
RECTIFIED CURRENT

LOAD VOLTAGE
LOAD CURRENT

For a full wave,   R.M.S. Value $= \dfrac{\text{Max. Value}}{\sqrt{2}}$

$= 0.707$ Max. Value

Average Value $=$ Max. Value $\times \dfrac{2}{\pi}$

$= 0.6365$ Max. Value

For a half wave,   R.M.S. Value $= \dfrac{\text{Max. Value}}{2}$

$= 0.5$ Max. Value

Average Value $=$ Max. Value $\times \dfrac{1}{\pi}$

$= 0.318$ Max. Value

On half wave,   Maximum Value $= 2 \times 10 = 20$ A

Average Value $= 20 \times 0.318 = 6.36$ A

Thus the moving-coil ammeter would register 6.36 A

On full wave, since Maximum Value $= 20$ A

Then r.m.s. value $= 20 \times 0.707 = 14.14$ A

The moving-iron meter would indicate 14.14 A

Also, average value $= 20 \times 0.6365 = 12.73$ A

The moving-coil meter would indicate 12.73 A.

FULL-WAVE. (1 PHASE Bridge connection.) The arrangement is illustrated by the diagram (Fig. 81a).

Fig. 81b shows this to be a full-wave system when operating as shown and thus the average and r.m.s. values of voltage and current are the same as for full-wave rectification. Thus for example $V_D = 0.6365\,V_m$ or $0.9\,V$ where V_D is the d.c. output voltage. V_m the peak and V the r.m.s. value of the applied a.c. voltage.

The P.I.V. is the full transformer voltage since in relation to the applied potential two rectifiers are in series but one is conducting.

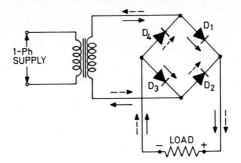

Fig. 81 a

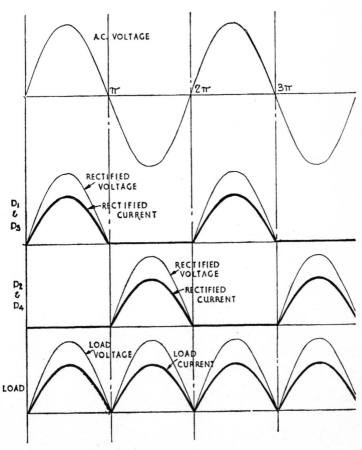

Fig. 81b

Then $\dfrac{\text{Peak Inverse Voltage}}{\text{Average Output Voltage}} = \dfrac{V_m}{\dfrac{2}{\pi} V_m} = \dfrac{3\cdot14}{2} = 1\cdot57$

A transformer is generally used to provide the desired output voltage and also provides isolation from the mains.

Example 49. A permanent magnet moving-coil voltmeter which requires 10 mA for full-scale deflection is to be used on a 110 V a.c. system. A suitable bridge-connected instrument rectifier with a forward resistance of $2\cdot5\ \Omega$ per unit is available. Find the value of the necessary series resistor.

For a full-wave sinusoidal supply, since the ratio of r.m.s. to average values is the form-factor, then this value can be used here to advantage. Thus $\dfrac{\text{R.M.S. value}}{\text{Average value}} = 1\cdot11$

So average value of rectified direct voltage $= \dfrac{110}{1\cdot11} = 99\cdot1$ V

Let R be the value of the required series resistor. Note that there are two rectifier units in series when conduction occurs during a half cycle.

Thus average value of direct current required by instrument for full-scale deflection $i.e.$ to read 110 V (a.c.) is 10×10^{-3} A.

So $10 \times 10^{-3} = \dfrac{99\cdot1}{R + 2\cdot5 + 2\cdot5}$ or $R + 5 = \dfrac{99\cdot1}{10 \times 10^{-3}}$

$$= 9910$$

Required series-resistor value $= 9910 - 5 = 9905\ \Omega$.

HALF-WAVE (3 PHASE). By using a three-phase supply and three diodes (Fig. 82a), the pulsating nature of the unidirectional current is decreased — a nearer approach to direct current. Current conduction occurs at the diode whose anode is at a + ve potential with respect to the cathode. Reference to the diagram of Fig. 82b shows that any one diode is seen to conduct for $\frac{1}{3}$ of a cycle. Assume diode D_1 as conducting. As the potential across it falls and that across D_2 rises, the conduction must transfer when the load potential, taken as the common datum value, equals that across D_1. As the red phase voltage decreases, the potential across D_1 falls further and the load potential now acts as a back voltage, so D_1 ceases to conduct.

The load is now supplied by the yellow phase via D_2 and the current is passed from one diode to the next and its value never falls to zero.

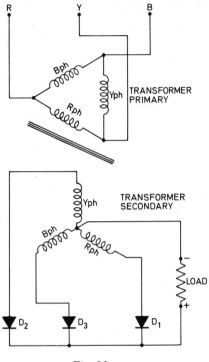

Fig. 82a

For a resistive load the waveform of output current and voltage is similar and it is frequently represented by rectangular blocks. The circuit is used extensively for plating equipment since the arrangement has a higher efficeincy and a lower voltage regulation than a three-phase bridge rectifier.

If this and subsequent polyphase rectifier arrangements are regarded more in the nature of a commutating switch, which connects each diode, into circuit at the appropriate instant then, it can be shown mathematically that there is a definite relation between the d.c. output voltage and the a.c. applied voltage. This relation can be

written as $V_D = V_m \dfrac{N}{\pi} \sin \dfrac{\pi}{N}$ where N is the number of diodes

or rectifying circuits. V_D is the average value of the output direct voltage and V_m is the maximum or peak value of the applied forward voltage — here it is a phase voltage.

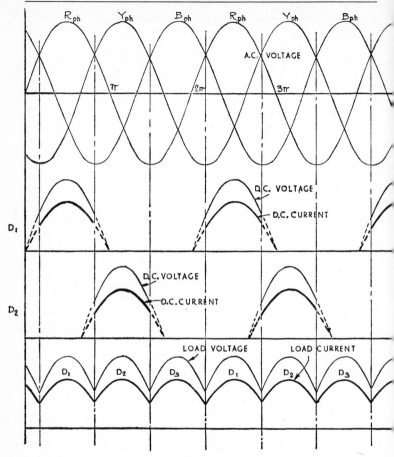

Fig. 82b

Thus for the above three-phase arrangement there are three recti-fying circuits and

$$V_D = V_m \frac{3}{3 \cdot 14} \sin \frac{180}{3} = V_m \frac{3}{3 \cdot 14} \sin 60 = \frac{3}{3 \cdot 14} \times \frac{\sqrt{3}}{2} V_m$$

Thus $V_D = 0 \cdot 827 V_m$

or $V_D = \dfrac{0 \cdot 827}{0 \cdot 707} V = 1 \cdot 17 V$ where V is the r.m.s. value of the applied voltage.

Here the P.I.V. is the full peak line voltage which is applied across a rectifier unit when it is not conducting. Here the ratio is

$$\frac{\text{Peak Inverse Voltage}}{\text{Average Output Voltage}} = \frac{\sqrt{3}\ V_m}{0.827\ V_m} = \frac{1.732}{0.827} = 2.1$$

FULL-WAVE (3 PHASE Bridge connection). For this arrangement six diodes are used, being connected into the three-phase lines, as shown by the diagram of Fig. 83(a). For an explanation of circuit action consider the voltage between the $R-Y$ lines (Fig. 83b).with Red + ve with respect to Yellow. Disregarding the effect of the other line voltages, it will be seen that diodes D_1 and D_5 will conduct until the voltage falls to zero and reverses with the Yellow line becoming + ve with respect to Red. Diodes D_2 and D_6 will now conduct and full-wave rectification will occur. The d.c. load voltage and current conditions for one cycle of supply voltage are shown by the second illustration of the diagram. Here as stated diodes 1, 2, 5 and 6 are involved.

If now the conditions for the $Y-B$ and $B-R$ lines are considered the appropriate diodes will be conducting and the further illustrations of Fig. 83 can be studied. If all three voltage conditions between the lines are considered, it will be evident that a datum of output voltage is attained and that, as the voltage across any conducting diode circuit falls below this value, the conduction ceases but d.c. continuity is maintained through the pair of diodes in a parallel circuit across which the voltage is rising. The idea of a commutating switch, as already introduced, can be used here and each diode circuit is seen to conduct for a sixth of a cycle or the switch has six positions. In the general formula already introduced, the number of rectifying diode circuits is six — two for each voltage cycle, one for forward and one for reverse. Thus $N = 6$ and if V_m is the maximum value of applied line voltage then

$$V_D = V_m \frac{6}{3.14} \sin \frac{180}{6} = V_m \frac{6}{3.14} \sin 30 = V_m \frac{6}{3.14 \times 2}$$

$$= \frac{3}{3.14}\ V_m = 0.954\ V_m$$

or $V_D = \dfrac{0.954}{0.707}\ V = 1.35\ V$ where V is the line r.m.s. value.

It should be noted that, as for the single-phase bridge arrangement, there are two diodes in series but as one is conducting, then here the

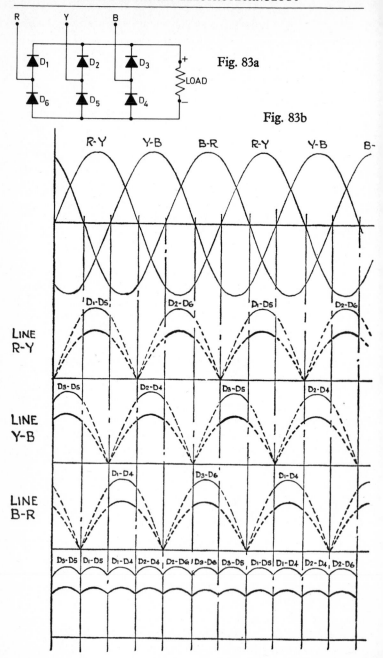

Fig. 83a

Fig. 83b

ratio of $\dfrac{\text{Peak Inverse Voltage}}{\text{Average Output Voltage}} = \dfrac{V_m}{0.954\,V_m} = 1.05$

The expression for V_D has been developed in terms of the line voltage *i.e.* no transformer has been considered. In practice such a transformer is usual since it provides the method for obtaining the required d.c. voltage value from a fixed a.c. supply. Such a transformer may well be delta/star connected and V_D could be developed in terms of the transformer phase voltage — as was done for the three-phase, half-wave condition.

If $V_m = \sqrt{3}\,V_{ph_m}$ then $V_D = \sqrt{3} \times 0.954\,V_{ph_m} = 1.654\,V_{ph_m}$

or $V_D = \sqrt{3}\,V_{ph} \times \dfrac{0.954}{0.707} = 2.34\,V_{ph}$

Here V_{ph_m} is the maximum value of a phase voltage and V_{ph} is the phase r.m.s. value.

Example 50. For a three-phase, half-wave rectifier express (a) the output voltage variation as a percentage of the mean value. (b) If the latter is 2000 V calculate the P.I.V. (c) If the supply is at 400 V and the transformer primary is in delta with the secondary in star, find the step-up values for each phase.

(a) For three-phase, half-wave rectification, any one diode conducts for a third of a cycle. The conducting period is for $120°$, spread $60°$ on either side of the maximum value. The change-over occurs when

the voltage is at the $30°$ value or $v = V_m \sin 30° = V_m \times \dfrac{1}{2} = 0.5\,V_m$

Thus the output voltage varies from V_m to $0.5\,V_m = 0.5\,V_m$ and mean value of d.c. output $= 0.827\,V_m$

Voltage variation $= \dfrac{0.5\,V_m}{0.827\,V_m} \times 100 = 60.5$ per cent.

(b) If $V_D = 2000$ Then P.I.V. $= 2000 \times 2.1 = 4200$ V

Note. The ratio of $\dfrac{\text{Peak Inverse Voltage}}{\text{Average Output Voltage}} = 2.1$ as already

developed.

(c) Also $V_D = 1 \cdot 17 \, V$ where V is the r.m.s. value of a phase

applied voltage. So $V = \dfrac{2000}{1 \cdot 17}$. This is a secondary phase voltage

but the primary voltage per phase $= 400 \, V.$

So step-up ratio $= \; 400 : \dfrac{2000}{1 \cdot 17} \; = \; 1 : \dfrac{5}{1 \cdot 17} \;$ or $1 : 4 \cdot 28.$

BATTERY CHARGING BY RECTIFIER

When a rectifier is operating in the forward direction, the supply voltage is countered by the battery e.m.f. which acts to reduce the voltage effective in producing the charging current. The action is best illustrated by a half-wave arrangement as in Fig. 84a.

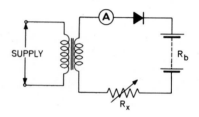

Fig. 84a

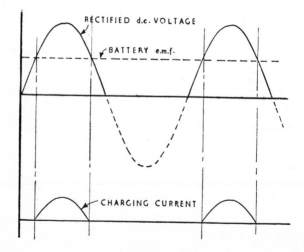

Fig. 84b

When the output voltage of the rectifier circuit exceeds that of the battery, current will flow into the battery to charge it. During the complementary half-cycles, the rectifier offers no conduction and a reverse current does not flow. Thus charging is possible and the actual effective current value is best determined by making a graphical solution in accordance with the diagrams of Fig. 84b.

The rectified voltage wave is plotted, the battery e.m.f. set off and the resultant voltage waveform drawn. Instantaneous values of current can be deduced from $i = \dfrac{v}{R_b + R_x}$ where R_b is the battery internal resistance and R_x the variable control resistance value. The average for the complete cycle is then determined in accordance with procedure already established.

Thus $i = \dfrac{i_1 + i_2 + i_3 \cdots i_n}{n}$ where n is the number of mid-ordinates for a complete cycle. The deduced value of current is that which would be indicated by the ammeter.

CHAPTER 5

PRACTICE EXAMPLES

1.   A d.c. supply of 100 V is applied to a coil of resistance 10 Ω and inductance 10 H. Find the current 0·1 s after switching on and the time for the current to reach 5 A.

2.   A resistor is connected across the terminals of a 20 μF capacitor which has previously been charged from a d.c. supply to 500 V. If the p.d. falls to 300 V in 0·5 min, determine the value of the resistor.

3.   The field winding of a separately excited d.c. generator has an inductance of 10 H and a resistance of 50 Ω and there is a discharge resistance of 50 Ω in parallel with the coil. The coil is energised by a d.c. supply at 200 V which is suddenly switched off. Find the value of the field current 0·04 s after the instant the supply is switched off.

4.   A resistor of 100 kΩ is connected in series with a 50 μF capacitor to a d.c. supply of 200 V. Calculate the voltage across the capacitor and the current at a time of 0·2 s after switching on.

5.   The time constant of a coil is known to be 2·0 and the inductance is 15 H. Determine the value of the current 0·2 s after connecting the coil to 300 V, d.c. mains. Find also the time taken for the current to reach one-half of its maximum steady value.

6.   A 1 μF capacitor is charged from a d.c. supply to 50 V, then discharged through a 5 MΩ resistor. After 5 s another 5 MΩ resistor is connected in parallel with the first. Determine the voltage after a further 5 s have elapsed and find the capacitor current.

7.   A relay coil has a 1 kΩ non-inductive resistor connected across it and the parallel arrangement is connected to a 50 V d.c. supply. The relay data is: Resistance 2 kΩ. Inductance 100 H. Operating current 10 mA. Release current 1 mA. Find the time taken by the relay to: (a) operate when the supply is applied; (b) release when the supply is removed. Assume instantaneous operation when the current reaches the operate and release values.

8.   A circuit made up from a 2 μF capacitor in series with a 100 kΩ resistor is used to control another device in such a manner that, when the p.d. across the capacitor reaches 63·2 V its instant discharge is effected and the voltage, as applied to the circuit, is allowed to recharge the capacitor as before. If the applied voltage is 100 V, find: (a) the initial current and, (b) the charging current at the instant before the capacitor is discharged. Sketch a current/time curve for the current in the resistor throughout the charging and discharging sequence and, (c) determine the frequency of the sequence.

9.   A metal rectifier, a moving-coil ammeter and a thermo-junction ammeter are connected in series across a 2 V supply. The rectifier has a forward resistance of 20 Ω, an infinite reverse resistance and the resistances of the ammeters can be neglected. When a d.c. supply is applied, the ammeters give identical readings. With an a.c. supply at 50 Hz and 2 V (r.m.s.) the readings are different. Calculate the actual meter readings.

10.   Given that a three-phase, 50 Hz supply is at 450 V between lines and feeds a delta-star transformer with primary to secondary turns ratio 1:5, determine the direct output voltage on no-load from a suitably connected full-wave rectifier. Determine the magnitude and frequency of the ripple voltage.

CHAPTER 6

THE TRANSFORMER (iii)
THE TRANSDUCTOR

This chapter is devoted to the various types of transformer and reactor which have been developed for special applications. Mention is made of instrument transformers, auto-transformers, both of fixed and variable ratio, saturable reactors or transductors and magnetic amplifiers. Since the treatment is mostly of a descriptive nature, the opportunity is taken to introduce miscellaneous examples of a.c. theory for revision and further instruction. It will be appreciated that, in this volume, attention is often given to matters which are deemed by the author to be essential. Here he may be at variance with the examining authority in that no relevant questions are asked in the examinations for marine engineers. It must, however, be conceded that, if the reader has a desire to increase his electrical engineering knowledge then, he must be introduced to items of equipment which will be encountered to an ever increasing extent as the size and complexity of the ship-board electrical system develops.

INSTRUMENT TRANSFORMERS

This heading covers current transformers and voltage transformers: the latter are sometimes called potential transformers. Instrument transformers are used in conjunction with indicating or recording instruments, protective equipment, alarms and control gear. Their usage gives the following advantages.

1. The indicating and protective instrument movements can be standardised e.g. 5 A for the current coils and 110 V for the voltage coils. The arrangement allows a cheap, workable system of instrumentation which is flexible and yet uses meters which are produced on a bulk basis and are comparatively inexpensive.

2. The indicating and protection equipment can be isolated from the high or medium voltage side of a system. The main insulation between the primary and secondary systems can be built into the instrument transformers and the meter system can utilise this to advantage. Thus instruments, relays, alarm units, etc, can be small, with the minimum of insulation and at a potential to earth which is considered to be safe.

3. The connections from the transformer secondaries can be made "light". This allows cheaper and neater wiring and control boards can be located conveniently and at some distance from the heavy power equipment such as the generators, switchboards, transformers and motors.

In relation to instrument transformers some new terms will be introduced such as burden, class, terminal markings, phase-angle and accuracy. These terms will be understood when explained in context but, since the secondaries feed into systems which are isolated from the main system, it is obvious that the total load of these systems must be known and that the transformers must be of a size capable of operating the systems efficiently, accurately and effectively. The ratio of the transformer is the prime requirement but its rating must be suited to the load or burden which it is to carry. It can be mentioned that the term "burden" is special to instrument transformers.

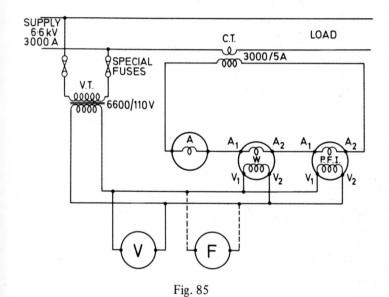

Fig. 85

THE CURRENT TRANSFORMER (C.T.)

The main requirements for a C.T. are ratio accuracy, small angular variation from the 180° phase displacement between primary and secondary currents, sufficient rating to operate the connected burden, high insulation strength, mechanical rigidity and ability to withstand heavy momentary currents.

It will be seen from the diagram of Fig. 85 that the primary forms part of the main series circuit. In this respect, the C.T. is dissimilar to all other transformers where the primary is connected across the system and is subjected to the supply voltage.

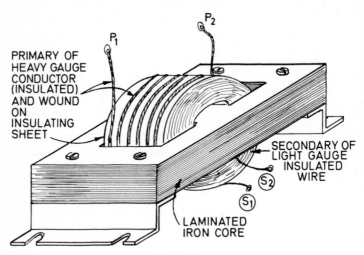

Fig. 86a

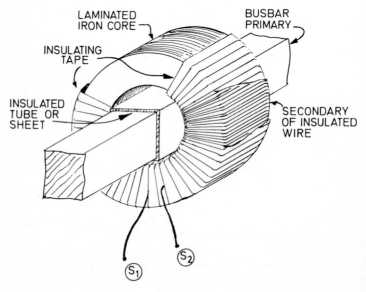

Fig. 86b

Because of the method of connection, the primary must consist of a conductor section which is decided by the magnitude of the main line current and not by the secondary line current. The ampere-turns, necessary for the magnetic circuit are similarly decided by this current, and we have transformer constructions of the types shown in diagrams (a) and (b) of Fig. 86. The former uses a few turns of thick conductor and is known as a "wound primary", whereas the latter uses the main cable or busbar as the primary which, in itself, constitutes a single turn. This arrangement is known as a "bar primary". The transformers are available in variations of these two basic forms of construction and are usually built by instrument makers or specialists rather than by power transformer manufacturers.

THEORY OF OPERATION. From knowledge of fundamental transformer theory it is known that $I_2N_2 = I_1N_1$ and in practice I_2N_2 is approximately equal to I_1N_1. Reference to the phasor diagram (Fig. 87), shows that this approximation is acceptable if I_o. is small. In the design of current transformers every attempt is made to minimise I_o by reducing its components I_m and I_w.

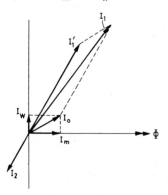

Fig. 87

The iron used for current transformers is chosen for its excellent magnetic characteristics. Thus Mu-metal may be used instead of Stalloy since, at low flux-density values, its permeability value may be seven times that of Stalloy. Modern technology has produced grain-oriented silicon irons such as Unidi and Stantranis which have even more superior characteristics, and if thin laminations are used with good insulation between the laminations, the iron losses can be kept to a minimum since the flux density is always low for associated reasons. These requirements will thus make for a small I_w value.

For any magnetic circuit   Magnetomotive Force = Flux × Reluctance

or $I_m N$ (the energising ampere-turns) $= \dfrac{\Phi l}{\mu A} = \dfrac{BAl}{\mu A}$ or

$$I_m N = \dfrac{Bl}{\mu} \quad \text{and} \quad I_m = \dfrac{Bl}{\mu N}$$

Thus I_m is decided by the factors shown and for a C.T. the working flux density is kept at 1 to $1 \cdot 5 \times 10^{-4}$ T as compared to the normal figure of 14×10^{-4} T for a power transformer. The length of the iron circuit is made as short as possible and, as mentioned above, iron with a high permeability is used. N should be as large as possible, but this is not always possible since frequently a single-turn or "bar-primary" must be used.

Other features of construction are the result of the following requirements. Since the iron circuit must have low reluctance, for the better class C.Ts., the laminations have the minimum of joints or can be complete rings. The windings are put on by hand and leakage reactance is reduced by superimposing or sandwiching the windings.

PRECAUTIONS. Since the secondary low voltage system is isolated from the "mains", it is possible to work on this circuit provided appropriate precautions are taken. One rule must, however, **always** be observed. **Never open the secondary circuit of a C.T. whilst the primary is energised.** The reason for the rule will become obvious if reference is made to the phasor diagram.

With correct functioning $I_2 N_2 = I_1 N_1$ or the primary ampere-turns are nullified by the secondary ampere-turns, the main flux being produced by $I_m N_1$, whose value, by design, is kept to an acceptable figure for the reasons already mentioned. Thus I_m may be in the region of 500 mA for a C.T. of ratio 500/5. Assume the secondary to develop a terminal voltage of 3 V under correct loading conditions. If the secondary circuit was to be opened and the demagnetising turns removed, the primary magnetising current would

now become 500 A or $\dfrac{500}{500 \times 10^{-3}} = 1000$ times greater than the

normal value. The primary magnetising ampere-turns would rise in proportion and, as the magnetic material of the core is operated at a very low value of flux density, it is capable of being raised appreciably. The material is chosen for its magnetic characteristic *i.e.* a steep graph and high saturation value. It is thus possible for the flux density of the transformer to rise in proportion with the new ampere-turn value and the secondary voltage would rise to some 3000 V for the example being considered.

It is now evident that a transformer designed to operate with 3 V across its output terminals may well break-down due to insulation failure. Again if one terminal is earthed then other parts of the secondary system, being subjected to an exceedingly high voltage, may be damaged. The danger to an operator working on this circuit would be considerable — hence the rule regarding open-circuiting the secondary of a C.T.

Unlike the normal transformer, no danger is introduced if the secondary is short-circuited since the maximum current which can flow is 5 A or less, the actual value being decided by the relation Secondary At = Primary At. Under emergency conditions it is possible to take an instrument out of circuit by first shorting the terminals to which the meter is connected. Thus by ensuring that secondary current is still flowing, the instrument can be disconnected, leaving the shorting link in position. Similarly when inserting the current-coil of a meter into circuit, connect it across the shorting link and then open the latter.

It should be apparent from the above that working on the secondary side of a C.T. can be hazardous and should not be undertaken unless one is quite confident of the operation. Marine engineers may have seen specialists undertake this operation during "setting up" or testing procedure and may be tempted to make similar checks. **Proceed with caution.**

It is appropriate, at this point, to note that a special current-circuit switch, such as an ammeter switch, is frequently provided on a switchboard. This would be used to enable the current to be noted in the three lines of a three-phase supply. The meter is used with three C.Ts., one in each line, and the switch is special, in that it shorts out a line C.T. and then disconnects the instrument. It next connects the ammeter across the shorting link of the next line and then breaks the latter connection. It thus carries out the operation described above in correct sequence.

BURDEN AND CLASS. The burden of a current transformer is the output VA value at the rated secondary current. Thus for a 15 VA burden at 5 A, the terminal voltage would be 3 V and the impedance of the instrument or relay circuit must not exceed 0·6 Ω, otherwise errors would be introduced into the readings and operation. The appropriate B.S. specification No. 3938 covers current transformers and sets out the various sizes of burden together with additional useful information.

B.S.S. 3938 also sets out the various classes of current-transformers. Those to be used for precision and industrial instruments are listed, together with the permissible errors for each class. A study of this specification is advised if more detailed information about instrument transformers is required.

TERMINAL MARKINGS. For simple current metering conditions the marking of terminals is not important but for the measurement of power, power factor and, especially for three-phase instruments, the correct sequence of connecting up is essential. To ensure that connections are made in accordance with the requirements of the instruments, the primary terminals of the transformer are marked P_1 and P_2 whilst the secondary terminals are marked S_1 and S_2. The connections would then be such that at the instant when current is flowing in the primary from P_1 to P_2, it is also flowing in the secondary from S_1 through the external circuit to S_2. A colour (red spot) is sometimes used to mark P_1 and S_1. For switchboard work the S_2 side of the secondary is often connected to earth.

PHASE ANGLE. The phasor diagram shows the secondary current phasor to be almost anti-phase *i.e.* almost at $180°$ to the primary current phasor. If the angle was $180°$, no phase-angle error would result. Components I_m and I_w cause the angle to be less than $180°$ and methods of minimising these have been considered. The angle between I'_1 or I_2 reversed and I_1 is the phase angle. The angle is +ve if, when reversed, secondary current leads the primary current. On a low power-factor circuit, this angle could be −ve.

For current measurement, I_2 should be a definite known fraction of I_1 and phase-angle error is not important. For power measurements, current ratio must be accurate and secondary current should be $180°$ out of phase with the primary current. As this condition does not exist, the phase-angle error may introduce an appreciable error into power measurements if appropriate precautions are not taken by the instrument and transformer makers.

Example 51. For the bar-primary type of current transformer, shown in Fig. 85, estimate the secondary turns and the burden if the impedances of the ammeter, wattmeter and power-factor indicator current-coils are 0·15, 0·3 and 0·3 ohms respectively.

Since the ratio of the transformer is $\dfrac{3000}{5}$ or $\dfrac{600}{1}$, it follows that if the primary Ats are to be equal the secondary Ats then 1×3000 = $N_2 \times 5$ or $N_2 = 600$ turns (in inverse proportion to the current ratio).

The 600 turns would be of insulated wire of section sufficient to carry 5 A.

With 5 A flowing, the voltage drop across the ammeter would be $5 \times 0.15 = 0.75$ V. The meter burden would be $5 \times 0.75 = 3.75$ VA.

Similarly the wattmeter and P.F.I. burdens would be $5 \times 0.3 \times 5$ = 7.5 VA. The total burden would be $3.75 + 7.5 + 7.5 = 18.75$ VA and a C.T. of rating equal or greater than this would be required. Thus a 30 VA unit would be suitable. Note that this is a case where instrument burdens have been added arithmetically, since they are due mainly to a resistance load, and any error introduced by the assumption is small.

If a suitably rated unit was not available from the standard range, it would be necessary to use two C.Ts. and two independent instrument circuits. Thus the ammeter could be energised from the first C.T. together with another instrument like an overcurrent relay, and the second C.T. could be used for the wattmeter and power-factor indicator. It is usual however, to keep instrument and protective-gear circuits separate rather than mixed.

THE VOLTAGE TRANSFORMER (V.T. or P.T.)

The theory of this transformer follows that of the power transformer, the main difference being that the secondary current, when referred to the primary, could be of the same order of magnitude as the no-load current. The phasor diagram of Fig. 88 shows that if V_2 is required to be a definite fraction of V_1 then the impedance voltage-drop triangles must be as small as possible. From theory V_2' is to V_1' as the turns ratio, and the desired approximation of V_2 to V_1 can be achieved if the voltage drops, as mentioned, are reduced.

The following precautions are usually taken in the building and use of potential transformers:

1. The copper on the windings is made of much larger section than would be used for a commercially built power transformer. Thus the turns are the same but the area of the wire is much greater. This means a much reduced ohmic value for R_1 and R_2.

2. By careful distribution of the windings, leakage reactances X_1 and X_2 would be much reduced. This precaution results in a further reduction of the voltage drops.

3. An ample, good quality iron core is used with careful construction and suitable insulation between the laminations. The no-load current

components I_m and I_w are thus reduced and I_o would be much smaller than that for a power transformer of equivalent rating.

4. Sensitive instruments should be used since these require a smaller operating current and, if I_2 is small then, the internal voltage drops are reduced.

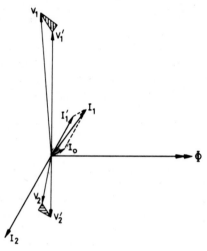

Fig. 88

The point to note regarding the use of voltage transformers is that, unlike the C.T., the primaries are connected across the "mains" and a short-circuit on the secondary would be a short referred to the primary. Short-circuit protection is therefore necessary and high-rupturing-capacity fuses are fitted in the primary circuit. Since this is a high-voltage supply, the fuses must be suitably insulated and a special tool is usually provided to allow their renewal without requiring the "mains" to be switched off. The mid-point of the secondary circuit can be earthed so that the maximum voltage to earth on this system is 55 V. Since the instruments are all in parallel and not in series, as for the C.T., the disconnection or reconnection of a suspect meter presents no difficulty. There is no danger of a high induced voltage, as for the C.T., and short-circuiting of terminals must not be undertaken.

BURDEN AND CLASS. B.S.S. No. 3941 specifies a secondary voltage of 110 V when the rated primary voltage is applied. A range of primary voltages from 110 V to 220 kV and more (line to line) is specified together with the classes and permissible errors. The standard rated burdens, from 15 to some 200 VA, are also listed together with the type of duty for each class. These are usually similar to those mentioned for the current transformers.

TERMINAL MARKINGS. Single-phase. Primary terminals are marked A and B or N. Secondary terminals are marked a and b or n and the polarity is such that, when A is positive with respect to B then a is +ve with respect to b .

Three Phase. Standard P.Ts. are star-connected on the primary and secondary sides. The neutrals may be brought out, especially for the secondary, and other arrangements may be used for special require-ments. The primary terminals are marked A, B and C while the sec-ondaries are a, b, and c. The primary neutral (if provided) is N and the secondary neutral is marked as n. The transformer polarity is such that when A is +ve with respect to N, a is +ve with respect to n.

PHASE ANGLE. Like the current transformer, a voltage transformer can introduce an error into the metering. Such an error can be both of magnitude and phase in terms of the voltage. Ratio and phase-angle errors depend on the relative sizes of the impedance voltage drops, and mention has been made as to how these can be kept to a minimum. Note that phase-angle error is only important when the measuring of power is required.

Example 52. If the voltage transformer shown in Fig. 85 is rated at 15 VA, estimate whether a frequency meter can be added to the installation. The manufacturer's catalogue lists the burden of this meter at 110 V, to be 2 VA and that of the voltmeter is 4 VA. The voltage windings of the wattmeter and power factor indicator are rated at 3·5 VA and 4·5 VA respectively at 110 V.

Total burden of existing instrument voltage windings

$$= \ 4 \ + \ 3{\cdot}5 \ + \ 4{\cdot}5 \ = \ 12 \, \text{VA}$$

The transformer is suitable for a burden of 15 VA, thus an additional $15 - 12 = 3$ VA can be added. It will be possible to add the fre-quency meter in the manner shown.

THE AUTO-TRANSFORMER

FIXED RATIO TYPE

The auto-transformer is a unit employing only one winding which is tapped to provide the appropriate voltage. From the theory of operation set out below, it will be seen that there is no electrical difference between the action of a normal double-wound transformer and that of the auto-transformer, and the related ratio of V and I to the number of turns on the winding still holds good. Because there is only one winding, it is cheaper than a double-wound unit of the same kVA rating. Also, as it contains less copper, the copper loss is lower and the efficiency is higher.

The auto-transformer is used to advantage for certain applications, the chief of these being where the output voltage is not greatly dissimilar from the input voltage. In this form it is operated as a voltage booster and is connected into a transmission line at a point where the load terminal voltage would be below an acceptable value. In this position, its function would be to raise the voltage to the required value. For marine work, its main application is in relation to motor starters. Some forms of starters for induction motors operate on the principle of lowering the voltage to a motor, in order to limit the starting-current surge taken from the ship's mains. As the motor runs up to speed, the voltage is increased by tappings on the transformer, until the full supply is attained when it is switched out of circuit. Since the transformer is used for only a brief period, it is apparent that, the unit should be as cheap as possible and the auto construction satisfies this requirement.

The auto-transformer has an obvious disadvantage in that, the secondary is not isolated from the primary and the benefits of a two voltage system are not realised. This would apply to a marine installation where the main power system was of the order of 440 V and the small-power system at 110 V. An earth fault on one line of either system, would immediately be felt on the other. Of greater importance is the fact that if a break occurred on the common section of the winding, the high voltage would become evident on the low voltage system. This will be seen, if the diagram of Fig. 89b is studied and an open circuit imagined at the point X.

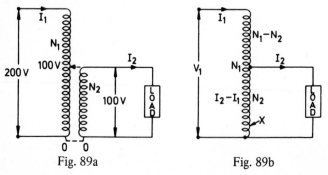

Fig. 89a   Fig. 89b

THEORY OF OPERATION.   Consider the diagrams (a) and (b) of Fig. 89. It will be seen that for a normal transformer, if one primary and secondary terminal is joined − using a pair which are always of the same polarity as each other, then it is possible to find a point on the primary winding which has the same potential as the remaining secondary terminal. These points could be connected and the secondary winding dispensed with. Advantage can now be taken of the fact that I_2 and I_1 are practically in phase opposition. If the secondary is formed at two tappings on the primary, then current in this section is $I_2 - I_1$ and the area of the winding conductor can be reduced, thereby effecting a further saving in cost. The extent of the saving in copper can be investigated thus:

SAVING OF COPPER .   For both the double-wound and auto-transformer it is known that

$$\frac{V_1}{V_2} = \frac{N_1}{N_2} = \frac{I_2}{I_1} = k$$

It can therefore be deduced that

For the double-wound transformer

$$\text{Weight of copper in Primary} \propto I_1 N_1$$
$$\text{Weight of copper in Secondary} \propto I_2 N_2$$
$$\text{Total weight of copper} \propto (I_1 N_1 + I_2 N_2)$$

For the auto-transformer

The top section of $(N_1 - N_2)$ turns carries I_1 so weight of copper
$$\propto I_1 (N_1 - N_2)$$

Bottom section of N_2 turns carries $(I_2 - I_1)$ so weight of copper
$$\propto N_2 (I_2 - I_1)$$

or total weight of copper $\propto I_1 N_1 - I_1 N_2 + I_2 N_2 - I_1 N_2$
$$\propto I_1 (N_1 - 2N_2) + I_2 N_2$$

Thus $\dfrac{\text{Wt. of Cu. in auto-transformer}}{\text{Wt. of Cu. in double-wound transfr.}} = \dfrac{I_1 (N_1 - 2N_2) + I_2 N_2}{I_1 N_1 + I_2 N_2}$

Dividing top and bottom by $I_1 N_2$

$$= \frac{\dfrac{(N_1 - 2N_2)}{N_2} + \dfrac{I_2}{I_1}}{\dfrac{N_1}{N_2} + \dfrac{I_2}{I_1}} = \frac{\dfrac{N_1}{N_2} - 2 + \dfrac{I_2}{I_1}}{\dfrac{N_1}{N_2} + \dfrac{I_2}{I_1}}$$

$$= \frac{k - 2 + k}{k + k} = \frac{2k - 2}{2k} = \frac{k - 1}{k} = 1 - \frac{1}{k}$$

(a) If the primary to secondary voltage is 10:9 then $k = \dfrac{10}{9}$

and the ratio is $\quad 1 - \dfrac{1}{\dfrac{10}{9}} = 1 - \dfrac{9}{10} = \dfrac{1}{10}$

Therefore Wt. of Cu. in auto: Wt. of Cu. in double-wound unit = 1:10

or Wt of Cu. in auto is $\dfrac{1}{10}$ of that required for a normal transformer.

(b) If the primary to secondary voltage is 10: 1 then $k = \dfrac{10}{1}$

and the ratio is $\qquad 1 - \dfrac{1}{10} = \dfrac{9}{10}$

Therefore Wt. of Cu. in auto: Wt. of Cu. in double-wound unit = 9:10

or Wt. of Cu. in auto is $\dfrac{9}{10}$ of that required for a normal transformer.

Examples (a) and (b) show that saving is only effective when k is not far removed from unity.

Example 53. A single-phase, 400/440 V auto-transformer is used to step up the voltage and supply a load of 8·8 kVA, operating at unity power factor. Neglecting losses and the magnetising current, find the output current, the input current and the current in the common section of the winding of the transformer.

$$\text{Load or output current} \ = \ \frac{8800}{440} \ = \ 20 \text{ A}$$

$$\text{Input current} \ = \ \frac{8800}{400} \ = \ 22 \text{ A}$$

$$\text{Current in common section} \ = \ 22 \ - \ 20 \ = \ 2 \text{ A}.$$

Example 54. Determine the core area, the number of turns and the position of the tapping point for a 500 kVA, 50 Hz, single-phase 6·6/5·0 kV auto-transformer, assuming the following values: e.m.f. per turn = 8 V, maximum flux density = 1·3 T.

Using the transformer e.m.f. equation $E \ = \ 4·44 \ \Phi_m fN$ volts

$$\text{Then} \qquad \frac{E}{N} \ = \ 8 \ = \ 4·44 \ \Phi_m \ 50$$

$$\text{or} \ \Phi_m \ = \ \frac{8}{222} \text{ webers} \quad \text{Since} \ \Phi_m \ = \ B_m A$$

$$\therefore \text{Area} \ = \ \frac{8}{222 \times 1·3} \text{ square metres} \ = \ \frac{8 \times 10^6}{222 \times 1·3} \text{ square milli-}$$
$$\text{metres}$$

$$= \ \frac{8\ 000\ 000}{288·6} \ = \ 27720 \text{ mm}^2 \ = \ 0·028 \text{ m}^2$$

$$\text{Primary turns} \ = \ \frac{6600}{8} \ = \ 825 \quad \text{Secondary turns} \ = \ \frac{5000}{8} \ = \ 625$$

Thus the winding must be tapped 625 turns from the end, common to primary and secondary.

VARIABLE RATIO TYPE

This unit is usually called a Variable Transformer or a Variac, the original trade name, and consists of one winding wound on an iron core. The core is in the form of a cylinder, built up from annular lamination stampings and insulated from each other in accordance with good transformer practice. The winding consists of enamelled wire put on in toroidal form, as illustrated by the diagram of Fig. 90. A movable carbon brush is arranged on a radial arm, to make contact with a track formed on the winding by cleaning off the enamel insulation.

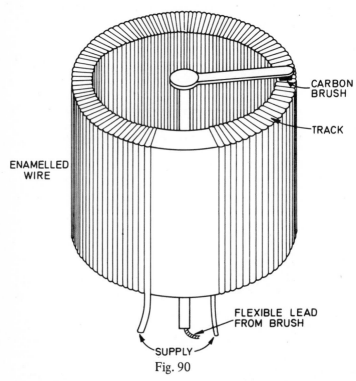

Fig. 90

An evident weakness of construction is that, full voltage is applied along the length of the track where the insulation between adjacent turns is reduced to a minimum, Furthermore, the collection of grime and carbon dust at this very point can lead to insulation break-down between turns. This results in a shorting of the turns with eventual burn out or an insulation failure along the whole length of the track. Modern techniques of construction include the use of durable synthetic enamel and varnish impregnation or "potting", together with a carbon brush of suitable type, grade and shaped tip. These refinements have removed the original causes of failure, and have resulted in this auto-transformer being accepted as a reliable item of equipment, much used in laboratories and the test-rooms of electrical factories. The unit is built in sizes up to 2·5 kVA for 230 V working, and can be arranged in "ganged" form for 440 V working. In the smaller sizes, it is easily operated by a small sensitive motor and in this form can be built into voltage regulators. The author knows of no specific marine applications where Variacs are used for power purposes but the probability of their use in small automatic control and supervisory systems must be expected.

Example 55. A 250/200 V auto-transformer is rated at 2 kVA. What is the economy in copper compared with the equivalent double-wound transformer. Estimate the copper section of the windings.

From the deduction made already

$$\frac{\text{Wt. of Cu. in auto-transformer}}{\text{Wt. of Cu. in double wound transformer}} = 1 - \frac{1}{k} \text{ where } k = \frac{V_1}{V_2}$$

Here $k = \dfrac{250}{200} = \dfrac{5}{4}$ So ratio $= 1 - \dfrac{1}{\dfrac{5}{4}} = 1 - \dfrac{4}{5} = \dfrac{1}{5}$

Therefore Wt. of Cu. in auto is $\dfrac{1}{5}$ of that in the double-wound unit,

making for a saving of $\dfrac{4}{5}$ or 80 per cent in the copper required.

Input current $= \dfrac{2000}{250} = 8$ A   Output current $= \dfrac{2000}{200} = 10$ A

Current in common section $= 10 - 8 = 2$ A

Working on a current density of 1·55 A/mm² the common section should be of conductor at least 1·29 mm² say 1·3 mm². Top section should be 4 times this *i.e.* 5·2 mm².

THREE-PHASE TRANSFORMATION

The production of a three-phase output supply from a three-phase input supply of the same frequency but different voltage value, can be effected in one of two ways: (a) by the use of one composite three-phase unit, (b) by interconnecting three similar but separate single-phase units.

(a) THE THREE-PHASE TRANSFORMER. The magnetic circuit for this unit is built up so as to provide three cores on which are wound the primary and secondary of each phase. The diagram of Fig. 91 shows the arrangement. It should be noted that the three windings are never wound on one single limb, since the resultant flux caused by the three primaries would be zero because, being 120° displaced from one another, the fluxes would cancel. The construction technique is similar to that for the single-phase unit except that the windings for each phase are confined to one core. The most usual arrangement favours the connection of either the primary or secondary in delta to enable the passage of harmonic currents.

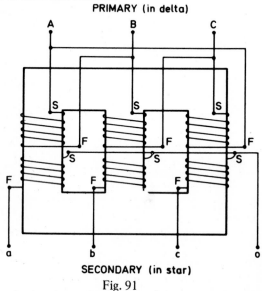

PRIMARY (in delta)

SECONDARY (in star)

Fig. 91

Although, up to now, no mention has been made of the generation of harmonics in transformers, the condition can be summarised briefly thus.

Assume a sinusoidal applied voltage, then since $\Phi \propto V$. The flux waveform will also be sinusoidal but not so the no-load current. This will be distorted for two reasons. 1. Since the B/H curve for the core material is not a straight line then a "peaky" current wave results, especially if the iron is worked near the saturation value. 2. Since due to hysteresis, the B/H curve is dissimilar on the ascending and descending values then the current waveform will be distorted still further, being unsymmetrical about the maximum value. The result is illustrated by the diagram of Fig. 92.

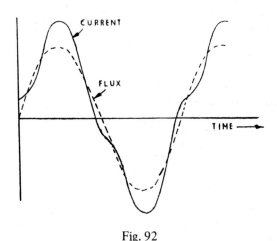

Fig. 92

Fourier's theorem shows that, on analysis, the no-load current waveform can be considered to be made up of a fundamental waveform together with those of 3rd, 5th, 7th and 9th harmonics. These are present in decreasing magnitude values and their effects can be minimised by taking appropriate precautions. A delta connection for a three-phase winding enables the 3rd harmonic e.m.f.s — the most important, as produced in each phase, to circulate 3rd harmonic currents. These currents add to the fundamental to produce the necessary sinusoidal flux wave with the resultant sinusoidal induced e.m.f.s.

(b) THREE SEPARATE SINGLE–PHASE UNITS. Here three similar transformers are interconnected into the three-phase primary and secondary lines. This arrangement is favoured for marine practice, since only one spare unit need be carried. By a system of selector switches, the spare unit can be connected into the circuit in the event of one of the other units failing. The rating of each of the single-phase units will only need to be one-third of that of the total output. See Fig. 93 for the arrangement.

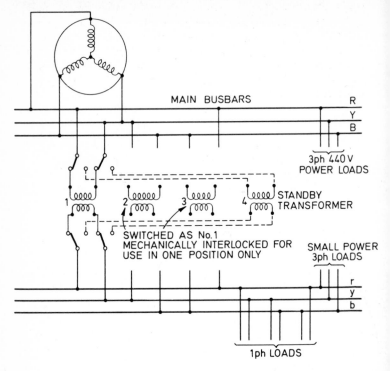

Fig. 93

METHODS OF CONNECTION

1. Star/Star. Diagram (a), of Fig. 94, shows the arrangement and

connection diagram. The phase windings have to withstand only $\dfrac{V}{\sqrt{3}}$

which means that less insulation material is required, a saving in space
and easier design. The windings have to carry full line current and if
one phase winding should fail, the whole three-phase arrangement
would be out of action. For this connection, third harmonic currents
will not flow in the line and distortion of secondary e.m.f. can be
expected. If a neutral line is run from the star point, third harmonic
current will flow along this wire — an undesirable feature.

2. Delta/Delta. Diagram (b) of Fig. 94 shows that with this connection, each phase winding must be insulated to full line voltage value

but the windings need only be designed to carry $\dfrac{1}{\sqrt{3}}$ times line current. In the event of one winding becoming faulty the unit need not be shut down since, by disconnecting the faulty winding, the supply can be maintained but with a reduced output. This method of working is known as "open-delta" or "vee", and is an arrangement favoured by Continental and American marine practice. As no neutral point is available, a three-wire supply only is used. Any third harmonic currents, as set up by third harmonic e.m.f.s in the phases, will circulate round the closed mesh and will not flow out into the lines. As seen earlier, the connection avoids distortion of the secondary e.m.f. waveform by internally generated harmonics.

3. Delta/Star or Star/Delta. Diagram (c), of Fig. 94, will stress the important point that, here the transformation ratio is not merely the transformer turns ratio but involves the type of connection.

Thus for a delta/star arrangement, transformation ratio $= \dfrac{V_1}{V_2}$

$$= \frac{V_{\text{ph}_1}}{\sqrt{3}\, V_{\text{ph}_2}} = \frac{N_1}{\sqrt{3}\, N_2} \quad \text{where} \quad \frac{N_1}{N_2} \text{ is the phase-turns ratio.}$$

Irrespective of whether the delta/star or star/delta connection is used, no trouble is experienced with third harmonic currents since they circulate round the mesh. A star-point is available if a four-wire arrangement is required for earthing purposes. For marine works, star/delta working allows an acceptable arrangement since a star-point on the alternator is usual and a neutral can be introduced at a later date, should this be considered advantageous.

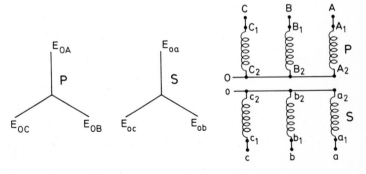

Fig. 94a

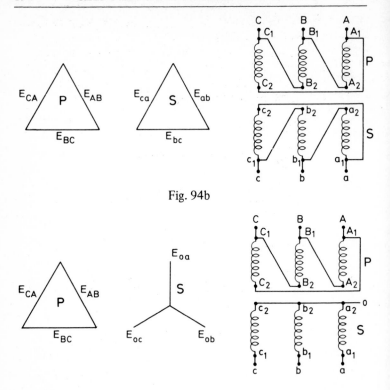

Fig. 94b

Fig. 94c

In relation to the polyphase transformer connections, those considered above are the most common but, it should be noted that, three- to two-phase and three- to six-phase arrangements are practicable although these are only used for special applications. Even for the examples already set out above, variations are possible where the windings per phase are split into two sections and combinations such as star/inter-star, star/double-star, delta/double-delta are used for special purposes. The only marine applications for such special connections are in conjunction with static rectifiers. A mention of one such arrangement was made in Chapter 5 in relation to three-phase, full-wave rectification.

Example 56. The bow docking-propeller of a large liner is driven by a 600 kW, 3·3 kV, three-phase, 60 Hz induction motor which is delta-connected. The motor is supplied from the ship's 440 V auxiliary power mains through a step-up transformer unit, connected in delta, on the L.T. side. Assuming a full-load efficiency and power factor of 85 per cent and 0·8 (lagging) respectively for the motor and an efficiency of 98 per cent for the transformer, find the motor phase and line currents and also the phase and line currents taken by the primary winding of the transformer. If the secondary is in star, find the turns ratio and the transformation ratio.

$$\text{Output power of motor} \; = \; 600 \text{ kW}$$

$$\text{Input power to motor} \; = \; \frac{600}{85} \times 100 \; = \; 705 \cdot 8$$

$$\text{Apparent power input to motor} \; = \; \frac{705 \cdot 8}{0 \cdot 8} \; = \; 882 \text{ kVA}$$

$$\text{Motor current} \; = \; \frac{882\,000}{\sqrt{3} \times 3300} \; = \; \frac{8820}{57 \cdot 2} \; = \; 154 \text{ A}$$

$$\text{Power output from transformer} \; = \; 705 \cdot 8 \text{ kW}$$

$$\text{Power input} \; = \; \frac{705 \cdot 8}{98} \times 100 \; = \; 720 \text{ kW}$$

$$\text{Apparent power input to transformer} \; = \; \frac{720}{0 \cdot 8} \; = \; 900 \text{ kVA}$$

$$\text{Mains current} \; \frac{900\,000}{\sqrt{3} \times 440} \; = \; \frac{90\,000}{76 \cdot 2} \; = \; 1181 \text{ A}$$

Motor line current = 154 A	Phase current = 88·9 A
Primary line current = 1181 A	Phase current = 682 A

The current ratio is given by $\dfrac{682}{154}$. Note these are the transformer

phase currents.

The turns ratio is in inverse proportion to the currents *i.e.*

$$\frac{154}{682} \;=\; \frac{1}{4 \cdot 43} \quad i.e. \;\; 1:4 \cdot 43$$

The transformation ratio is $\dfrac{440}{3300} \;=\; \dfrac{4}{30} \;=\; \dfrac{1}{7 \cdot 5} \quad i.e. \;\; 1:7 \cdot 5$

THE SATURABLE REACTOR

If the B/H curve of the iron material, as used for a reactor core, is examined (Fig. 95a), it will be seen that there is a comparatively straight portion for which $B \propto H$ and the permeability $\mu = B/H$ is large and substantially constant. This part of the curve is used by the reactor and transformer designer. At higher values of H as the "knee" of the curve is reached, the characteristic ceases to be linear and the B/H or μ value varies from point to point as the graph flattens out. The curve shows that the iron has passed into a state of saturation and in its final state becomes almost horizontal. The μ value after decreasing rapidly has now reached a minimum value.

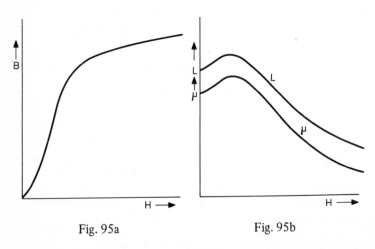

Fig. 95a   Fig. 95b

The variation of the μ value is often referred to as its *incremental permeability* and, if plotted to a base of H, gives a graph of the form shown in Fig. 95b. From the work done in Chapter 6, it was seen that

$$L = \frac{\mu A N^2}{l}$$ or for a given reactor, $L \propto \mu$. Thus for an iron-cored

reactor, inductance varies with μ and it follows that if inductance is plotted against H, the curve would be similar to that shown for μ.

It could well happen that the flux density of a reactor core is carried into the region of saturation by the energising coil being subjected to over-voltage. This could also occur for a badly designed transformer when an increase of the no-load current would result, being accompanied by distortion of the output voltage waveform. For both conditions, the reactance would be reduced if the inductance is reduced by saturation and, in the case of the reactor, control of the choking effect could be used to advantage if the incremental permeability or inductance could be controlled.

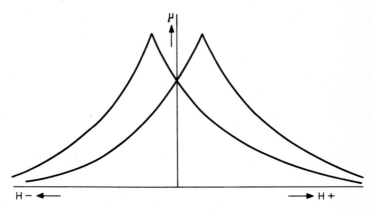

Fig. 96

The usual method of achieving saturation is by providing the iron with an initial biasing flux-density value. This can be done by fitting the reactor with an additional winding fed from a d.c. source. The additional ampere-turns or initial H value would result in biasing the value of B and can also be seen as a movement of the H axis upwards. Irrespective of the polarity or direction of the direct current, variation of this biasing flux density would allow the inductance of the reactor to be varied from a value approaching zero to its maximum value. Thus, when the current of the biasing or control coil is at its maximum value, the reactor inductance would be a minimum and when the former is zero, the inductance would have its normal value. This control can be illustrated by the "butterfly" curve shown by Fig. 96. Note that the two curves are due to the hysteresis effect of the B/H. curve, whilst symmetry about the μ axis is due to the reversal of the direction of current in the d.c. control coil.

If a saturable reactor is connected as shown in the diagram (Fig. 97), the variation of inductance by the control winding affects the reactance and hence the total impedance (Z) offered to the a.c. supply source. The control winding thus produces, by a changing impedance, a change of load current and thus a variation of power as supplied to the load. The arrangement was first used for fading of theatre lights but an obvious disadvantage was that an alternating voltage was induced into the control circuit by transformer action. Refinements have resulted from the development work done to make the unit more effective, and have produced the transductor, as it is now more commonly called.

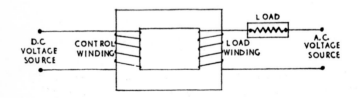

Fig. 97

THE TRANSDUCTOR

This device is similar in construction to a power transformer. It may have one or two magnetic cores fitted with load and control windings connected in the manner to be described. To continue investigations on the theory of operation, we must again consider the B/H curve of the material from which the core is made and assume that a permanent additional B value is provided by the control winding.

The graph of Fig. 98 shows that if the d.c. bias control-current value is zero and no biasing flux exists then, on application of a sinusoidal supply voltage, the waveform of the alternating current can be obtained graphically as shown.

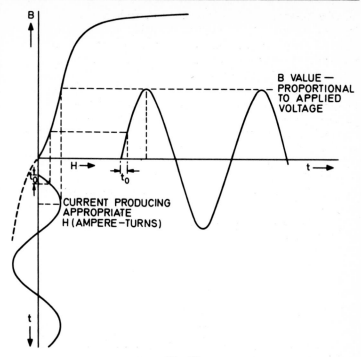

Fig. 98

Next consider the core premagnetised by the bias winding, the amount being shown by the ordinate OA. The corresponding flux density would be OB and the axis of the applied voltage wave would now be along BX (Fig. 99). If the same value of a.c. voltage is applied, it will be seen that the current waveform can be obtained as before, but, as the core is driven into saturation, the right hand half-cycle, shown by the broken lines, has an amplitude limited only by the cir- cuit resistance. There can be no increase of flux and no induced counter e.m.f. once the core has saturated. The current value during this half-cycle is therefore large and independent of the d.c. control current.

By using a high incremental permeability material for the core which also has a "square-loop" type of hysteresis characteristic (Fig. 100), the transductor can be designed to be driven into saturation by a very small change of magnetising current. If the a.c. winding is capable of supporting the supply voltage when unsaturated and the control winding achieves saturation readily, then the device can be regarded as a switch.

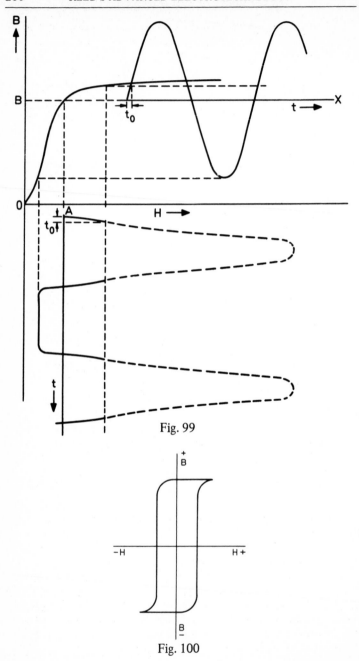

Fig. 99

Fig. 100

To continue, remember that the load current is only dependent on the resistance of the circuit once the device has saturated. A switch if timed to close at an appropriate instant after the start of a cycle and open at the end of a half-cycle, would result in voltage being applied to the load as shown (Fig. 101). The shaded area shows when the voltage is effective, and the mean value of the current passed by the load can thus be controlled by varying the instant of closing the switch.

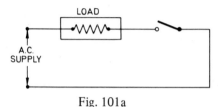

Fig. 101a

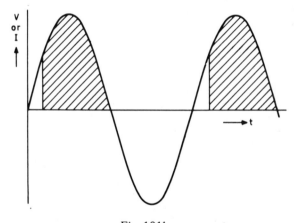

Fig. 101b

The similarity with transductor operation is illustrated by the diagrams of Figs. 98 and 99. With no control current the transductor supports the full a.c. supply voltage and the only current evident at the load is the small no-load magnetising current. With the application of control current, the transductor is driven into saturation on the +ve half-cycle by the sum of the load current magnetising m.m.f. and the control current m.m.f. The instant, at which this occurs, can be varied by adjusting the level of the control current *i.e.* the length of the ordinate OA of Fig. 99. As stated, upon saturation, the device cannot support the a.c. supply voltage and the load current is limited only by the circuit resistance. On the −ve half-cycles the control m.m.f. opposes saturation and the transductor supports the supply voltage.

The disadvantages of the saturable reactor, as already mentioned, are still present, namely: 1. Due to transformer action a voltage is induced into the d.c. control ciruict and because of this an uneven, pulsating direct current is the result. 2. Control is not as effective as desired since it occurs for half-cycles only. These defects are overcome by using two transductor elements connected as shown in the diagram of Fig. 102. The connection allows the induced e.m.f.'s in the control windings to cancel out, and the arrangement can be further developed to give a characteristic nearer that required of an amplifier.

THE MAGNETIC AMPLIFIER

For the arrangement of Fig. 102, the load windings are shown connected in parallel even though the control windings are in series. The load windings could also be arranged in series but the parallel arrangement is easier to understand, and follows the usual practical method.

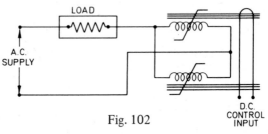

Fig. 102

The load windings are connected in opposition so that on alternate half-cycles of the supply voltage, the control ampere-turns drive an opposite core into saturation. Each load winding is designed to support the full supply voltage when the control current is zero.

Consider core 1. With any given control current condition, for a +ve half-cycle of supply voltage, the flux increases until a point is reached when the core goes into saturation and the winding cannot then support the voltage. Core 2 is, at the same time, driven away from saturation and would support the supply voltage but for the fact that the winding of core 1 is across it. Since core 1 has gone into saturation, its winding effectively short-circuits the load winding of core 2 with the result that the supply voltage is not carried by the transductor and is impressed across the load, to result in a partial half-cycle of current.

For the −ve half-cycle, of the supply voltage the operation is repeated. Core 1 comes out of saturation but its winding is short-circuited by that of core 2 which has gone into saturation. Thus control is obtained over a full cycle by the d.c. control winding with its current maintained continuously in one direction. Reversal of this control current would merely change the order in which the saturation of the cores occurs, and the controlling effect would be as it was originally.

The diagram of Fig. 103 shows the waveforms associated with the magnetic amplifier in its simple form. The output is alternating, though of a pulsing nature, and can be converted for d.c. operation by the use of a suitably rated rectifier.

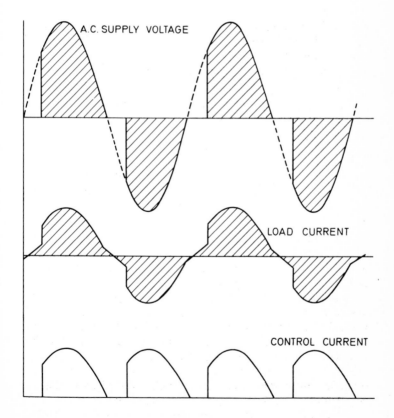

Fig. 103

CURRENT GAIN. This is a measure of the amplifying property of the arrangement and can be simply deduced. Under the controlled condition *i.e.* when one core is saturated and the effect of the other is offset by the short-circuiting of its winding, there can be no change of flux. Any increase of load ampere-turns must be balanced by an equal and opposite increase of control At, so that the resultant m.m.f. for the core remains the same. Since high quality iron is used and worked at high permeability, the m.m.f. necessary to cause saturation is relatively small and can be neglected. Thus, under saturated conditions, we can write

$$I_c N_c \ = \ IN \quad \text{or} \quad \frac{I}{I_c} \ = \ \frac{N_c}{N} \quad \text{where} \quad N_c \ = \ \text{the number of turns of}$$

the control winding and I_c the control current. $N =$ the number of turns of the main winding and I the load current. The relationship shows the current gain to be proportional to the turns ratio of load and control windings.

Also since $\quad I \ = \ \dfrac{I_c N_c}{N}$, load current is dependent on the turns ratio

and the control current. It is independent of the supply voltages or load resistance and the amplifier can be regarded as a constant current source.

Before leaving the current waveforms shown on the diagram of Fig. 103, it is pointed out that, since the control current is fed from a d.c. source its value might be assumed constant being decided by the control circuit resistance. The above deduction however, shows that if flux is constant, $I_c N_c \ = \ IN$ and if I varies, I_c will vary. Hence a reflection of the load-current peaks is evident in the control current.

When the cores are unsaturated, the impedance of the load winding is large and only a small magnetising current will flow, which lags the supply voltage by nearly 90°. When control is ineffective, I is deter-

mined by $\dfrac{\text{Supply voltage}}{\text{Load resistance}}$, and since we have the relation $I \ = \ k I_c$,

the magnetising current being neglected, the characteristic shown by Fig. 104 can be deduced. This indicates that the magnetic amplifier can be used to advantage where a large load current is to be varied in a linear manner by a small control current.

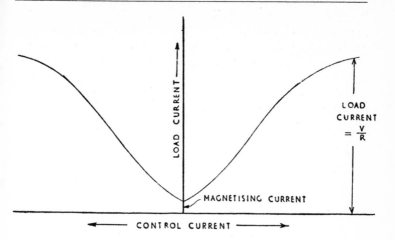

Fig. 104

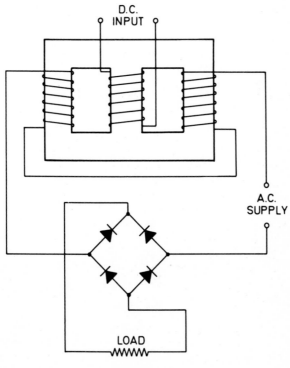

Fig. 105

FEEDBACK.   Another way of showing the construction and diagram of a magnetic amplifier is by Fig. 105. From this it is seen that, the inductance of the a.c. windings is controlled by a d.c. winding which can consist of two sections in series opposition to eliminate the induced alternating voltage. As seen above, by suitably proportioning the ratio of d.c. to a.c. turns, the amplifier is effective with an output practically independent of the load circuit resistance. When d.c. amplification is required, a rectifier is placed in series with the a.c. windings. The amplifier will not however, discriminate between the polarity of d.c. input signals.

To increase the gain of the device and to make it polarity sensitive, positive feedback is employed. For the type of construction just mentioned, a second winding can be placed on the centre limb and connected to the d.c. terminals of a further bridge rectifier in series with the load rectifier (Fig. 106).

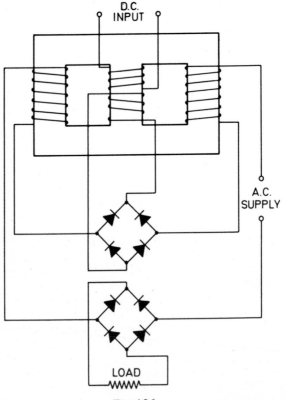

Fig. 106

With the input signal in one direction the current through the feedback winding will be aiding the control winding and increasing the gain. With a reversed signal, the current through the feedback winding will remain in the same direction and will now oppose the control current to reduce the gain. The new characteristic is shown by Fig. 107. Current gains of up to 300 and power gains of 10^4 are possible with this arrangement, which can be developed further for specialised work. Thus a bias winding can be added to move the no-signal operating point and if two amplifiers are used in push-pull, a reversible output with no standing current is possible. The amplifiers can also be used in cascade for increased gain.

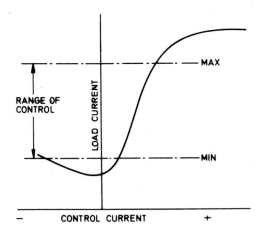

Fig. 107

The magnetic amplifier has been used in various control schemes for many marine applications. It formed the basis of the B.T.H. "Magnastat" and the G.E.C. "Accurex" automatic voltage regulating systems and can be found in association with the monitoring of large d.c. currents and the protection of a.c. systems. It must be stressed however that recent developments in semiconductor technology indicate that usage of the "Mag-amp" will decrease. Devices, such as the Thyristor or S.C.R. (silicon controlled rectifier), can be developed to have characteristics similar to a magnetic amplifier,and, since the disadvantages of the latter include its size and weight then, the electronic alternative is to be preferred once the cost of production is reduced and availability improves.

MISCELLANEOUS EXAMPLES

Example 57. A coil of unknown inductance and resistance is connected in series with a 25 Ω, non-inductive resistor across 250 V, 50 Hz mains. The p.d. across the resistor is found to be 150 V and across the coil 180 V. Calculate the value of coil resistance and inductance. Find also the power factor of the coil.

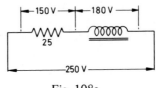

Fig. 108a

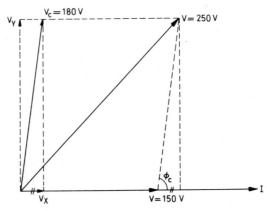

Fig. 108b

From the diagram of Fig. 108 the following mathematical solution is possible.

Let x = the resistive voltage drop of the coil

y = the reactive voltage drop of the coil

Then $\qquad x^2 + y^2 = 180^2$ ... ... (a)

and $(x + 150)^2 + y^2 = 250^2$ ... ... (b)

or, on further simplification and rearranging of (b) and (a), we have

$$x^2 + 300x + 150^2 + y^2 = 250^2$$
$$\text{and} \quad x^2 + y^2 = 180^2$$

Then by subtraction $300x + 150^2 = 250^2 - 180^2$

or $300x = (250 + 180)(250 - 180) - 150^2$

giving $300x = 430 \times 70 - 150^2$

So $300x = 30\,100 - 22\,500 = 7600$ or $x = \dfrac{76}{3} = 25 \cdot 33$ V

Then $y^2 = 180^2 - 25 \cdot 33^2 = 32\,400 - 641 \cdot 6 = 31\,758 \cdot 4$

or $y = 10^2 \sqrt{3 \cdot 176} = 178 \cdot 2$ V

The circuit current $= \dfrac{150}{25} = 6$ A

Resistance of coil $= \dfrac{25 \cdot 33}{6} = 4 \cdot 221 \ \Omega$

Reactance of coil $= \dfrac{178 \cdot 2}{6} = 29 \cdot 7 \ \Omega$

Inductance of coil $= \dfrac{29 \cdot 7}{2 \times 3 \cdot 14 \times 50} = 0 \cdot 0945$ H

Power factor of coil $\cos \phi_L = \dfrac{25 \cdot 33}{180} = 0 \cdot 141$ (lagging).

Example 58. A replacement relay coil for an alarm circuit is obtainable but is rated to operate correctly from a 120 V, 50 Hz supply. It is also stamped 1050 Ω, 1·5 H. The coil is required to replace a burnt-out unit from a 220 V, 60 Hz circuit and, in order to put the coil into operation, it is decided to use a capacitor as a voltage-dropping device. A range of capacitors is available as ship's "radio spares". Estimate the size of capacitors which should prove suitable.

Reactance of new coil on 50 Hz $= 3 \cdot 14 \times 100 \times 1 \cdot 5$
$$= 3 \cdot 14 \times 150 = 471 \ \Omega$$

$$\text{Impedance of new coil on 50 Hz } = \sqrt{1050^2 + 471^2}$$
$$= 100\sqrt{10\cdot5^2 + 4\cdot71^2}$$
$$= 100\sqrt{111 + 22\cdot2}$$
$$= 100\sqrt{133\cdot2} = 1155 \ \Omega$$

$$\text{Current taken by coil } = \frac{120}{1155\cdot5} = 0\cdot104 \text{ A}$$

$$\text{Required impedance on 220 V circuit } = \frac{220}{0\cdot104} = 2120 \ \Omega$$

The reactance of the 220 V circuit would be

$$= \sqrt{2120^2 - 1050^2} = 100\sqrt{21\cdot2^2 - 10\cdot5^2} = 100\sqrt{450 - 111}$$
$$= 100\sqrt{339} \qquad\qquad = 1840 \ \Omega$$

Reactance of new coil on 50 Hz is 471 Ω

$$\text{On 60 Hz it is } 471 \times \frac{6}{5} = 565\cdot2 \ \Omega$$

Now the reactance of the required capacitor arrangement must cancel this inductive reactance and provide the additional reactance for the 220 V circuit, *i.e.* it must be $1840 + 565\cdot2 = 2405\cdot2 \ \Omega$. The catch in the problem is involved with this point. Note it.

The required value of capacitance is given by $X_c = \dfrac{10^6}{2\pi f C}$

$$\text{or } C = \frac{10^6}{2\pi f X_c} = \frac{10^6}{2 \times 3\cdot14 \times 60 \times 2405} = \frac{10^2}{6\cdot28 \times 6 \times 2\cdot405}$$

$$= \frac{100}{90\cdot62} = 1\cdot1 \ \mu\text{F}$$

This could be made up from capacitors of $1\mu\text{F}$ and $0\cdot1\mu\text{F}$ in parallel.

Example 59. A three-phase, 440 V, 50 Hz, 90 kW "circulating water pump" motor has an efficiency of 82 per cent and operates at a power factor of 0·8 (lagging). Calculate (a) the *kVA* input to the motor. (b) the load current. If the motor is connected to the main switchboard by a three-core, 95 mm² cable, 100 m long, calculate the cable line-voltage drop. Take the resistivity of copper as 17 $\mu\Omega$ mm.

$$\text{Motor power input} = 90 \times \frac{100}{82} = 109 \cdot 76 \text{ kW}$$

$$\text{Apparent power input} = \frac{kW}{\cos \phi} = \frac{109 \cdot 76}{0 \cdot 8} = 137 \cdot 2 \text{ kVA} \quad \text{(a)}$$

$$\text{Load current} = \frac{kVA}{\sqrt{3}\,kV} = \frac{137 \cdot 2}{\sqrt{3} \times 0 \cdot 44} = 180 \text{ A} \qquad \text{(b)}$$

$$\text{Resistance of cable core} = \frac{\rho\,l}{A} = \frac{17 \times 10^{-6} \times 100 \times 10^{3}}{95}$$

$$= \frac{17 \times 10^{-1}}{95} = 0 \cdot 018 \ \Omega$$

Then the voltage drop in one cable core $= 180 \times 0 \cdot 018 = 3 \cdot 24$ V

So cable voltage drop in the three-phase line $= \sqrt{3} \times 3 \cdot 24 = 5 \cdot 6$ V

Note the introduction of the $\sqrt{3}$ when considering the three-phase conditions.

Example 60. Whilst in port a tank-ship obtained its "shore-main" supply from a three-phase, 3300/440 V, delta-star transformer. For lighting purposes on board ship the voltage is stepped down by three, 440/110 V, single-phase transformers connected in delta/delta. If the total lighting load, comprised of tungsten filament and fluorescent lamps, is balanced to 15 kW at a power factor of 0·85 (lagging), cal-culate the currents in the respective connecting cables and the phase currents of the transformer windings. What would be the kVA supplied from the high-voltage supply? It is assumed that the trans-former losses are negligible and only the lighting is being supplied.

Lighting transformer.

$$\text{Secondary line current} = \frac{15\,000}{\sqrt{3} \times 110 \times 0 \cdot 85} = 93 \text{ A}$$

$$\text{Secondary phase current} \quad = \quad \frac{93}{\sqrt{3}} \quad = \quad 54 \text{ A}$$

$$\text{Primary phase current} \quad = \quad 54 \times \frac{110}{440} \quad = \quad 13{\cdot}5 \text{ A}$$

$$\text{Primary line current} \quad = \quad 13{\cdot}5 \times \sqrt{3} \quad = \quad 23{\cdot}3 \text{ A}$$

Supply transformer.

Secondary line current = 23·3 A

Secondary phase current = 23·3 A (since the windings are connected in star)

$$\text{Primary phase current} \quad = \quad 23{\cdot}3 \times \frac{440/\sqrt{3}}{3300} \quad = \quad 1{\cdot}76 \text{ A}$$

$$\text{Primary line current} \quad = \quad 1{\cdot}8 \times \sqrt{3} \quad = \quad 3{\cdot}1 \text{ A}$$

Assuming no losses

kVA input from supply = kVA output from lighting transformer

$$= \quad \frac{kW}{\cos \phi} \quad = \quad \frac{15}{0{\cdot}85} \quad = \quad 17{\cdot}65 \text{ kVA.}$$

CHAPTER 6

PRACTICE EXAMPLES

1. A direct-reading wattmeter having a 5 A current-coil and a 110 V, voltage-coil is to be used to measure the power in a single-phase, 6·6 kV circuit, carrying a maximum current of 100 A. State the appropriate ratios for the instrument transformers and calculate the constant by which the wattmeter reading must be multiplied to measure the power consumed.

2. A 75 kW, 415 V, three-phase induction motor has a full-load efficiency of 80 per cent. The input line current is to be measured by a C.T. operated ammeter. Suggest an appropriate ratio for the C.T. if the full-scale deflection of the ammeter is 5 A. State the expected ammeter reading. The power factor of the motor on full load is 0·87 (lagging).

3. A single-phase wattmeter with 5 A and 250 V ranges is used in conjunction with a 25/5 C.T. to measure the power of one phase of a balanced, three-phase, star-connected load. If the load absorbs 12 kW from a three-phase, 415 V, 50 Hz, supply and the power-factor is 0·8 (lagging), calculate (a) the wattmeter reading (b) the impedance per phase of the load (c) if a phase impedance consists of a resistor and reactance in series, calculate their phase values.

4. The output power of a 415 V, three-phase alternator supplying a balanced load is measured by one wattmeter. If the current transformer has a 25/5 ratio and the wattmeter reading is 7 kW, what is the total power?

5. For recording the input to a three-phase, 7·5 kW induction motor, which is rated to take a line current of 14 A at 415 V, the following instruments are to be used.
    Ammeter 0 to 5 A, of resistance 0·08 Ω.
    Voltmeter 0 to 120 V, of resistance 3636 Ω.
    Three-phase Wattmeter (two element type). Current coils 0 to 5 A, of resistance 0·1 Ω. Voltage coils 0 to 120 V, of resistance 4000 Ω.
    Estimate the ratio and burden rating of suitable current and voltage transformers.

6.	The primary and secondary voltages of an auto-transformer are 500 V and 400 V respectively. Find the current distribution in the windings when the secondary current is 100 A and calculate the economy of copper for this arrangement.

7.	The possibility, is to be considered, of using a double-wound, 6 kVA, 250/150 V, single-phase transformer, as an auto-transformer on 400 V mains, to supply a load of 12 kVA at 250 V. By checking the current loading of the windings and a practical connection method, determine the suitability of the arrangement.

8.	Find the values of the currents flowing in the various branches of a three-phase, star-connected auto-transformer loaded with 400 kW at a power factor of 0·8 (lagging) and having a ratio of 440/550 V. Neglect voltage drops, magnetising current and all losses in the transformer.

9.	A 440 V, three-phase induction motor is to be started by the use of a delta-connected auto-transformer provided with a 70 per cent tapping on each phase winding. Find the voltage applied across the motor terminals at the stage of starting when the transformer tappings are in circuit.

10.	A three-phase, 440 V, 40 kW induction motor has an efficiency of 82 per cent and operates at a power factor of 0·85 (lagging). When direct-on started, the motor takes a current of 6 × full-load current and produces a torque of 1·5 × full-load torque. Calculate the current taken from the supply and the ratio of starting to full-load torque if the motor is started through an auto-transformer having a 75 per cent tapping. This transformer is star-connected.

THE A.C. GENERATOR (i)

In Chapter 7 of Vol. 6 the reader was introduced to the basic laws of electromagnetic induction, and the principles of e.m.f. generation by dynamic induction were also considered. The generation of an alternating voltage was seen to be accomplished by a comparatively simple arrangement but, if the a.c. generator was to be functional as a machine, distinct from the d.c. generator then, it was evident that much development work was necessary before it became a commercial proposition. The advantages of a.c. over d.c., from the generation and utilisation point of view, are many and will be outlined later but if the generators are compared, it is apparent that the a.c. generator scores because it has no commutator. Furthermore, since it is merely a matter of moving conductors with respect to the magnetic field then it is immaterial as to whether the field moves and the conductors are stationary or vice versa. Thus we have two methods of operation open to the machine designer but the arrangement of a moving field and fixed armature conductors is usually favoured because of reasons to be mentioned. The basics of machine construction are therefore considered before the main substance of a.c. generator theory is detailed.

THE ALTERNATING-CURRENT GENERATOR

In the interests of standardisation the title of "synchronous a.c. generator" or more simply the "a.c. generator" is now suggested to compare with that of "d.c. generator" but the term alternator will continue to be used by engineers whilst the new terminology is being accepted and established. The main types of a.c. generators in general use are now considered.

ROTATING-ARMATURE TYPE

In the elementary generator, a.c. is always produced. The provision of a reversing switch – the commutator, results in d.c. but, if a.c. is required then slip-rings are necessary. A machine of this type is constructed like a d.c. generator with a fixed-field system and a moving armature. It is known as a "rotating-armature" alternator and is an acceptable arrangement for small machines up to about 40 kVA at a voltage of 450 V. It would be a cheaper machine than an a.c. generator constructed on the rotating-field principle, because a d.c. generator or motor design of suitable size would be used with slip-rings replacing the commutator. A d.c. armature winding would be retained and appropriate tappings brought out to two or three slip-rings, as required for a single-phase or three-phase supply. The number of poles would be chosen to suit the speed of the prime-mover and the required frequency but the construction of the yoke and field system would follow d.c. practice although separate excitation is necessary. The waveform of the generator voltage would not be perfectly sinusoidal but would be acceptable for applications for which this type of alternator is used. A sine-wave machine in the sizes for which this alternator is offered would be "a special" and would be very costly, necessitating a special design. Such "sine-wave sets" are usually only found in the test-rooms and laboratories of large electrical firms.

ROTATING-FIELD TYPE

For alternators of this type, the stationary portion is usually called the stator and carries the armature conductors whereas the field system moves and is usually called the rotor. The magnetic field is produced by energising the pole windings with direct current through slip-rings from a d.c. generator called "the exciter". The exciter is usually driven off the main shaft and the d.c. voltage is usually below 110 V, although 220 V has been used and even 440 V. The arrangement of using the stator to carry the armature conductors has advantages. It can be uniformly slotted and wound for a high working voltage. Since no centrifugal forces are involved, a great thickness of insulation can be used and there is more room for the windings and end-connections than if these were on the rotor. Bracing of the end-connections against the electro-magnetic forces, most evident under short-circuit conditions, is made easier and more effective. The rotor can be made compact, with a construction which allows a low-voltage, high-current winding capable of withstanding the large centrifugal forces which would be encountered.

The stator core is built up from laminations insulated from each other, as for the armature of the d.c. machine, but for small a.c. generators each lamination may be a complete ring. For large machines the laminations are sections of such a ring and in general the slots can be "open" or "semi-enclosed" to suit the type of winding and design requirements. The latter are favoured to reduce "tooth ripple". The rotating-field machine is built in two main forms — chosen to suit the type of the prime-mover. The basic forms may be varied to suit particular requirements but fall under the general category of either the Salient-pole Rotor or the Cylindrical Rotor.

SALIENT-POLE ROTOR. The salient-pole type of field construction is used for slow and medium speeds and is suitable for being driven by steam-engines, water-wheels or turbines and by slow-speed diesel engines. The construction is distinguished by the machine having a large diameter compared to its axial length. The basics of the arrangement are shown by the diagrams of Fig. 109.

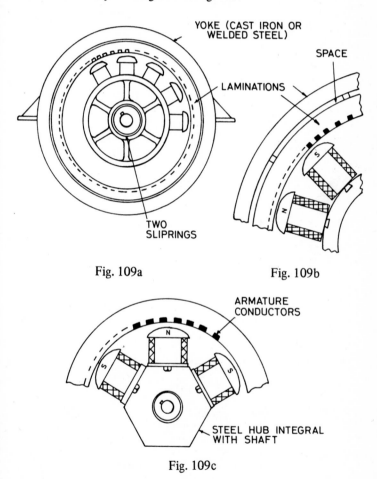

Fig. 109a   Fig. 109b

Fig. 109c

The pole-pieces are similar to those used for the d.c. machines and are bolted or keyed to a flanged magnet wheel of solid or spoked construction. Smaller machines may have their poles fitted to a solid hub.

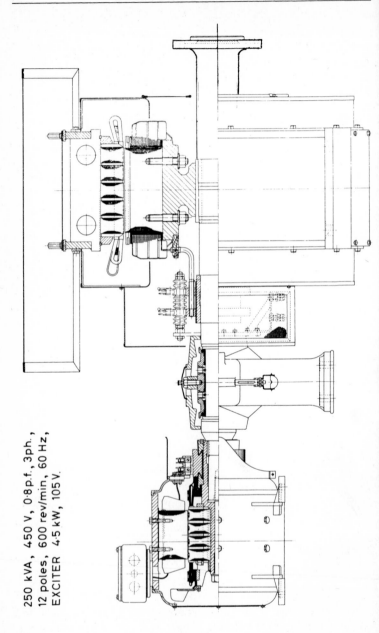

250 kVA , 450 V , 0·8 p.f. , 3ph. ,
12 poles, 600 rev/min , 60 Hz,
EXCITER 4·5 kW, 105 V.

Fig. 110

The number of poles may vary from four to 40 and are usually made up from steel laminations or may be of cast-iron with laminated pole-shoes. The width of a pole-shoe is about $\frac{2}{3}$ of a pole-pitch and its contour is decided by the need for providing a sinusoidal distribution of flux density across the air-gap. The field windings follow the practice used for d.c. machines, except that effective bracing is necessary, and to achieve this, copper strip wound on edge and suitably insulated is frequently used. Fig. 110 is included to illustrate the construction of a typical marine a.c. generator and has been kindly provided by Messrs. G.E.C./A.E.I.

If the rotor construction for a three-phase a.c. generator is examined, it will be seen that the slots are uniformly distributed around the inside periphery of the stator and are so spaced to allow three or multiples of three to be accommodated in a pole-pitch. Assuming a 16-pole machine and three slots per pole per phase, it will be seen that nine slots are required for a pole-pitch or 144 slots in all. The number of slots/pole/phase is decided by the design requirements of voltage and waveform, as is the shape of the stator slots which can be of open-rectangular or semi-enclosed form.

CYLINDRICAL ROTOR. This consists of a steel forging, out of which the slots for the rotor winding are milled. The shaft may be integral with the field system or, for a large rotor, may be made up from two identical parts which are bolted on to the ends of the magnet unit. The number and spacing of the slots is fixed by the number of poles to be provided. The diagrams of Fig. 111 show two- and four-pole

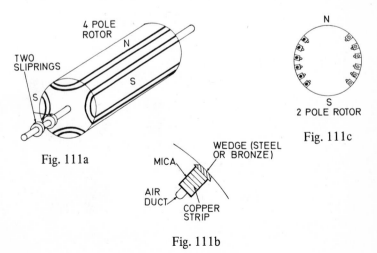

Fig. 111a

Fig. 111b

Fig. 111c

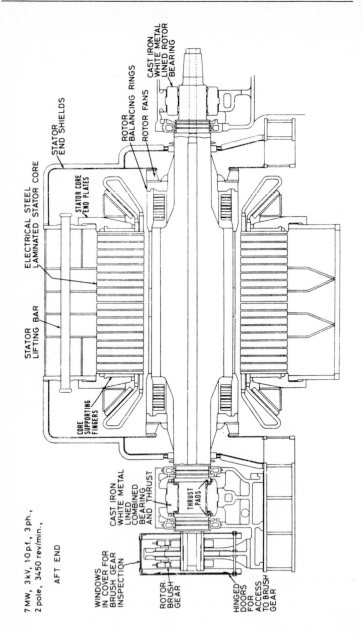

Fig. 112

Fig. 113

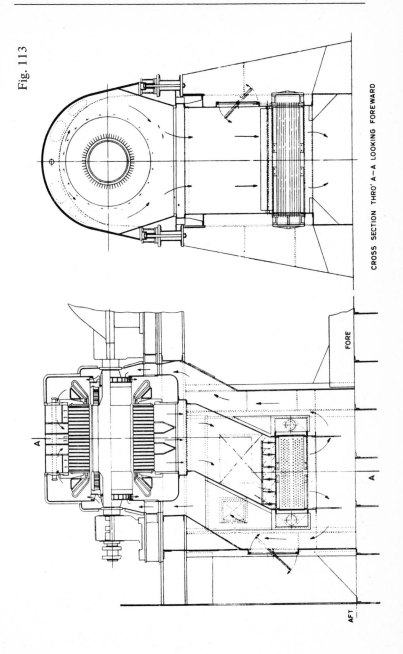

CROSS SECTION THRO' A–A LOOKING FOREWARD

rotor systems, and it will be seen that about $\frac{2}{3}$ of a pole pitch is slotted to leave $\frac{1}{3}$ for the pole centre. The field windings are also shown, and these are made up from coils of copper strip, wound on edge, insulated with mica or micanite and layered into the slots. Wedges of steel or manganese-bronze are driven into the slots to hold the winding against the centrifugal forces and special methods are used to secure the end-connections, since the mechanical stress to which the copper is subjected can be very high. Non-magnetic steel retaining rings or end-caps, arranged to screw on to the rotor, are usually provided to cover the end-connections, once they have been clamped and braced between bakelised-paper and hard-wood blocks.

Because of the smooth surface of the rotor when completed, cooling is achieved by a forced ventilation system, since there is relatively no fanning effect as exists for the salient-pole type of machine. Here the rotor is provided with suitable air-ducts, and the diagrams show how the lower part of the slots is used for this purpose after the upper part has been filled with a winding.

The rotor construction enables it to be driven at high speeds and the generator, in this form, is always associated with a steam or gas turbine. Because of speed considerations, the diameter is limited to 3 or 4 ft. and, to achieve a high electrical output, the axial length must be large, This machine is characterised by its shape since the length is four to five times the diameter.

The stator construction follows that of the salient-pole machine except that because of its length, cooling considerations are more important and influence the design. The laminations are punched and assembled into a fabricated steel frame in such a manner that axial air-vents and radial ducts are formed to allow the air streams, provided by rotor or external fans to circulate. The alternator is enclosed in a steel casing, and the air is usually fed into the air-gap and cooling passage-ways at the ends of the stator, and discharges through the radial ducts into the space between the stator core and the casing. The air is filtered and a closed-circuit system is usual with the air being passed through a cooler. For marine work, the alternator and sea-water cooler unit are usually arranged as a composite unit. Illustrations of this type of a.c. generator and the cooling arrangements are shown by the diagrams of Fig. 112 and 113 which have been kindly provided by Messrs. C.A. Parsons & Co. Ltd.

Total enclosure of the machine reduces noise and for large power-station units, a further development is the use of hydrogen instead of air as the cooling medium. Hydrogen cooling allows a reduction of windage losses and an increase of heat removal which permits of a 1 per cent increase in efficiency and a 20 per cent increase in rating.

Hydrogen cooling also results in a reduction of windage noise, less oxidation of insulation and lower fire risk. Disadvantages are, the need for an explosion-proof construction, the complications of the cooling system and the need for gas-tight shaft-seals.

EXCITATION ARRANGEMENTS

Modern methods of energising the field of an a.c. generator can be classified under the two general headings of (i) Rotary excitation systems and (ii) Static excitation systems. Originally the field current was obtained from a separate d.c. generator but with the development of the semiconductor rectifier, alternative methods of excitation have been devised which can also be made to incorporate desirable characteristics. One such characteristic for a marine alternator, is a fast speed of response to the automatic voltage regulator (A.V.R.). It is now usual practice to "direct-on" start large induction motors and such motors may have starting currents of the order of the full-load current of the supply alternator. Thus transient voltage performance is important and as the voltage dip is a function of the alternator reactance, then this latter aspect must be taken into account, even if at the expense of machine size and cost. In relation to the excitation, due to the high inductance of the field systems of the exciter and main alternator, large time constants can be involved. Thus many designs of excitation use field-forcing techniques to which appropriate reference will be made when systems are described. The two main methods of excitation will now be considered.

(i) ROTARY-EXCITATION SYSTEMS. The d.c. generator or exciter is usually mounted on an extension of the main alternator shaft and is thus driven by the common prime-mover. For the large a.c. synchronous generators as used in power stations, two or more exciters can be used in cascade. Here the first exciter energises the field of the next, the output of which supplies a further larger exciter field or the field of the alternator. Marine machines being smaller do not need the cascade exciter arrangement but a pilot exciter is frequently used. This comprises a permanent-magnet field system rotating within a stator containing the armature coils. The alternating voltage so generated is fed into a rectifier system, the d.c. output of which feeds the field of the main exciter. Reliability of excitation is thus ensured since a d.c. generator by itself may fail to excite because of loss of residual magnetism. With the advent of silicon semi-conductor diodes capable of handling large power values, the d.c. generator type of exciter is rapidly giving way to a machine arrangement which dispenses with commutator, slip-rings and the associated brushgear. Before consider-

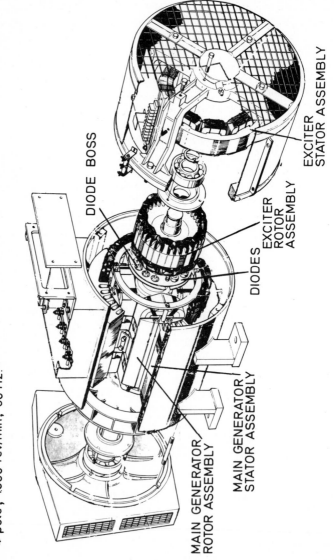

UP TO 250 kVA, 0·8 p.f., 3 ph.,
4 pole, 1800 rev/min, 60 Hz.

DIODE BOSS

EXCITER STATOR ASSEMBLY

EXCITER ROTOR ASSEMBLY

DIODES

MAIN GENERATOR ROTOR ASSEMBLY

MAIN GENERATOR STATOR ASSEMBLY

Fig. 114a

ing such an arrangement, attention is drawn in passing, to the cross-field generator which was in favour for a period. Since this machine had been developed to provide a high degree of amplification and fast control action, it could thus force a rapid change of alternator field in response to a small change of input controlling current. However with the development of the transductor (Chapter 6) and the thyristor or silicon-controlled rectifier (Chapter 12), static excitation systems have come to the fore, if only, because of the fact that machines with their associated maintenance are not involved.

THE BRUSHLESS A.C. GENERATOR

In this machine the exciter consists of an a.c. armature mounted on the main alternator shaft and rotating within a normal d.c. type of field system. The a.c. output of the armature is rectified by semiconductor diodes mounted on and revolving with the shaft. The arrangement is shown in the diagrams of Figs 114a and b. The d.c. output of the diode assembly is fed to the rotating field of the main alternator by leads taken through or secured to the driving shaft. The whole assembly can be made so compact that the length of the original main alternator is not much increased. Both exciter and alternator can be mounted in a common frame and air space to constitute what

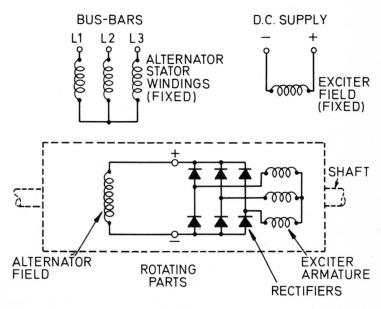

Fig. 114b

can now be described as a brushless generator. A two-bearing arrangement can be used and maintenance requirements are reduced considerably since there is no commutator, slip-rings and associated brushgear.
The illustration of a brushless a.c. generator (Fig. 114a) was kindly
provided by the Brush Electrical Company of Loughborough, England.

The machine can be made self-exciting by ensuring that sufficient
residual magnetism is present. As no commutator is required, the normal interpoles or compoles can be replaced by permanent magnets
fixed within the interpolar spaces. Alternatively the exciter main
poles can be built up with permanent-magnet laminations interleaved
with the normal laminations to follow accepted practice of field-pole
assembly. An adequate residual voltage at the main alternator terminals is thus ensured to overcome the impedance of any components
of the exciter field circuits.

Unlike the conventional alternator-exciter arrangement, since
power supply for the A.V.R. cannot be obtained from the exciter
armature then it must be obtained from a pilot exciter or a reliable
stationary d.c. supply or from the output of the a.c. generator itself
through rectifiers. The disadvantage of this last method is that if a
three-phase "short-circuit" occurs then the main machine voltage
collapses. The alternator current will have an initial high value but
will decay so rapidly that the machine circuit-breaker will not be
"tripped". Under-voltage or instantaneous-current protection must
consequently be fitted but if these are not desirable then auxiliary
short-circuit transformers must be provided. These transformers ensure an additional power supply to the exciter field or A.V.R. and
enable the main alternator to maintain sufficient short-circuit current
to trip its associated circuit-breakers. The need for "field suppression"
also introduces a further complication for brushless machines. The
problem can be solved by using thyristors on the main alternator
rotor but usually suppression of the exciter field is used and although
this results in a longer period to achieve complete and correct machine "shutdown" it has generally proved satisfactory.

The opportunity is taken here to make some mention of the
requirements for the rectifier diodes, the rating of which depends
not only on the full-load excitation current of the alternator but also
on the cooling arrangements i.e. amount of air and heat-sink areas.
Since transient voltages are induced in the rotor during a stator three-
phase short circuit, the peak inverse voltage (P.I.V.) rating must be
sufficient to withstand these prolonged induced voltages and currents.
The diodes are normally arranged in a bridge connection with a combination of units in series or parallel in each arm. This assembly avoids
the need for a "shut down" of the machine since usually a failure of

a diode by internal short-circuit breakdown clears to an open circuit because of the over current. The diodes should be checked periodically but an insulation test on the rotor should be made only with these diodes either disconnected and isolated or with a temporary short circuit being applied across each such unit.

(ii) STATIC EXCITATION SYSTEMS. This heading covers both self-regulating and thyristor methods of excitation. It should be noted that for these methods, the a.c. generator requires slip-rings and brush-gear but no rotating exciter. There are many variations of the systems as developed by different manufacturers to provide the desired voltage regulation performance for "direct-on starting" large motors.

The self-regulating systems can be divided into those which rely on excitation power being taken from the main alternator output and conveyed to the field by a combination of transformers, chokes and rectifiers and those which also use some form of A.V.R. Since close voltage regulation cannot be obtained by transformers and chokes alone a magnetic amplifier or transductor may be employed to control the field current, the control current of the transductor being supplied by an A.V.R. which provides the closer regulation required — in the region of ± 1 per cent. Without the A.V.R., a regulation of say ± 5 per cent could be achieved. The diagram of Fig. 115 shows the type of arrangement under discussion.

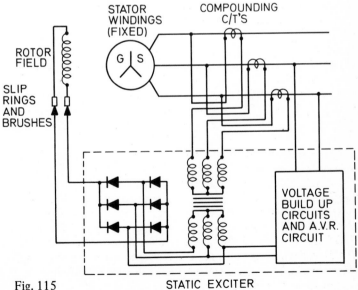

Fig. 115   STATIC EXCITER

Because of the inclusion of compounding transformers or chokes, a field-forcing effect is achieved whenever large alternator load currents are passed. If a transductor is used in the A.V.R. then, if this is in series with the alternator field, the voltage dropped across it can be lowered by the control current and thus an increased voltage is applied across the field to result in an additional field-forcing effect. As for the brushless machine, the need for ensuring sufficient generated terminal voltage under short-circuit conditions is also evident. The use of compounding transformers satisfies this need and currents up to five times normal can be provided to operate the machine's circuit-breaker. Initial voltage build-up when the set is run up to speed can also be provided by the techniques discussed earlier e.g. a pilot exciter, permanent-magnet exciter field inserts or a special starting circuit incorporating a relay which shorts out exciter-circuit impedance until voltage builds up.

Thyristor excitation systems have been developed to exploit the advantages of electronic control. The use of high efficiency silicon rectifiers and thyristors results in a unit of small size and weight with the elimination of large iron cores and rotating exciters. Although the thyristor has yet to be studied, in the context of field excitation it can be considered to be a rectifier for which conduction can be controlled. It has, in addition to the normal anode and cathode of the diode, a third electrode called "the gate". When the thyristor is appropriately forward biased by a half-cycle of applied a.c. voltage, it will conduct only when the gate is pulsed with the correct polarity but will cease to conduct at the end of the half-cycle. If the following half-cycle is to be conducted then a second thyristor is required with its gate pulsed accordingly. The thyristors are arranged in the manner already discussed for diodes, as used in rectification (Chapter 5), although for a bridge assembly only two thyristors need to be used with diodes in series in the opposite arms. The instant at which the thyristor is made to conduct or "fire" can be controlled during the half-cycle and thus where switch-on is delayed, the power level averaged over a number of cycles is less than when no delay is included. Variation of the length of delay varies the average power level. The system described is known as "phase shift" control. An alternative system employs "burst triggering" or "burst fire" control but either system achieves control of the conducted power and when thyristors are used for excitation, irrespective of the control-method employed, the alternator field current is varied by the static exciter.

The diagram of Fig. 116 shows a typical thyristor-controlled excitation arrangement. Automatic voltage regulation is built into the system but as the electronics involved require specialist knowledge, at this stage of study the equipment and its operation is treated in block form. A three-phase voltage sensitive measuring unit is shown together with a transistorised error amplifier, triggering circuit and half-wave thyristor output stage. As for the other excitation systems already described, provision is made for some of the field power to be obtained from excitation current transformers (not shown) in the a.c.

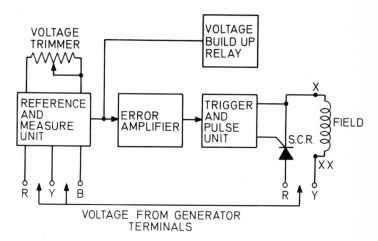

Fig. 116

generator output. This latter arrangement ensures that, in the event of a short-circuit, excitation will be maintained.

Operation of the system can be followed through with the assistance of the diagram. Three-phase a.c. generator terminal voltage is applied to the measuring unit where it is rectified and applied across a resistor/zener-diode reference bridge. The bridge signal output, a measure of the error in the generator voltage, is passed into the amplifier and then fed into a trigger circuit. The electronic circuitry ensures that, the greater the magnitude of the amplified signal then the greater the delay time before the trigger fires the thyristor to allow half-wave rectified current-flow through the field winding. Thus if the alternator terminal voltage tends to rise due to load reduction then the thyristor delay time of firing is extended. The greater the delay time, the smaller the average field current with consequent lowering of machine-generated voltage. Conversely, a reduction of terminal voltage reduces the delay time, firing commences earlier during the rectified half-wave and the average current value rises with consequent increase of generated voltage. A voltage build-up relay unit is also shown on the diagram which operates in a manner already described for the other static-excitation systems. Its inclusion assists voltage build-up by residual magnetism during the initial starting up period.

The above system controls voltage regulation to an accuracy of ± 1 per cent from no-load to full-load. The need for close regulation has already been covered but when an A.V.R. is fitted then B.S. Specification 2949 lays down minimum performance requirements for ships' a.c. generators. The reader is advised to make reference to this specification.

SPEED-FREQUENCY EQUATION

Before considering the conditions for the generation of e.m.f. by an alternator, it is necessary to develop the basic relation which exists between the frequency of voltage alternation, the speed at which the machine is driven and the number of poles. Although the term *synchronous speed* is more usually used in connection with motors, it can be introduced here since it is the speed which is directly related to the frequency, and thus the equation to be developed relates to both alternators and motors alike.

From the basic theory we know that a voltage of alternating waveform is generated as a coil passes through two pole-pitches, and that the frequency of the waveform is the number of cycles generated per second. If f = number of cycles per second or number of *hertz*

then time for one cycle $= \dfrac{1}{f}$ second.

The coil moving through two pole-pitches, has generated in it one cycle of e.m.f. and if driven at a speed of N rev/min then the time for

one revolution $= \dfrac{60}{N}$ second.

The time taken to pass through one pole-pitch $= \dfrac{60}{PN}$ second, if P is

the number of poles. Consequently the time taken to pass through

two pole-pitches $= \dfrac{2 \times 60}{PN}$ second, and thus:

Time taken to move through two pole-pitches = time for one cycle

giving $\dfrac{120}{PN} = \dfrac{1}{f}$ or $f = \dfrac{PN}{120}$ hertz.

Here N would be the synchronous speed for a frequency of f hertz.

Example 61. Find the speed at which an eight-pole, salient-pole a.c. generator must be driven in order to generate voltage at a frequency of 60 hertz.

Since $N = \dfrac{120f}{P}$ then $N = \dfrac{120 \times 60}{8} = 900$ rev/min.

E.M.F. EQUATION

This is deduced on lines similar to that used for the d.c. generator. Consider the diagram of Fig. 117, where the pole system and stator conductors have been laid out in a developed form.

In each revolution one conductor is cut by $P\Phi$ webers. P is the number of poles and Φ the flux per pole (in webers).

The time for one revolution $= \dfrac{60}{N}$ second

So the average rate of cutting lines of force

$$= \frac{\text{Total lines cut}}{\text{Time taken to cut the lines}} = \frac{P\Phi}{\dfrac{60}{N}} = \frac{P\Phi N}{60}$$

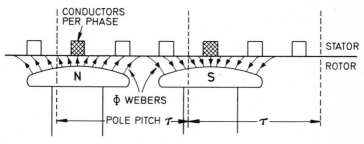

Fig. 117

Now from Faraday's law, the average e.m.f. generated in volts is given by the average rate of cutting lines of force and so

The average e.m.f. generated in a conductor $= \dfrac{P\Phi N}{60}$ volts $= E_{av}$

Now for a.c. work, r.m.s. values are required and if sine-wave working is assumed, then the relation

Form factor $= \dfrac{\text{r.m.s. value}}{\text{average value}}$ can be used.

$$\text{Form Factor} = \frac{^1\!/\!\sqrt{2} \text{ Maximum Value}}{^2\!/\!\pi \text{ Maximum Value}} = \frac{\pi}{2\sqrt{2}} = 1 \cdot 11$$

So r.m.s. value of e.m.f./conductor $= 1 \cdot 11 \times \dfrac{P\Phi N}{60}$ volts

and since $N = \dfrac{120f}{P}$

$\therefore$ r.m.s. value of e.m.f./conductor $= 1 \cdot 11 \times \dfrac{P\Phi}{60} \times \dfrac{120f}{P}$

$$= 2 \cdot 22 \ \Phi f \text{ volts.}$$

It should be noted that, if the generated waveform is not sinusoidal then $1 \cdot 11$ cannot be used for the form-factor and if K is used for a general case then the e.m.f./conductor $= 2K \Phi f$ volts.

CONCENTRATED WINDING. The diagram of Fig. 117 has been drawn to show all the conductors of one phase Z_{ph}, concentrated in one slot per pole. This would be a purely theoretical arrangement and is not used in practice. However, if it is considered, it will be seen that since the conductors are spaced one pole-pitch apart, then for every one, the flux rises and falls simultaneously and the e.m.f.'s rise and fall together i.e. they are in phase. If all the conductors under a pole-pair are connected in series, the total e.m.f. is the phasor sum — here the arithmetical sum of the e.m.f. per conductor. Thus for a concentrated winding: e.m.f. of winding per phase $= 2 \cdot 22 \ Z_{ph} \Phi f$ volts.

DISTRIBUTED WINDING. A winding of one slot per pole per phase, as described above, would need a broad deep slot which would alter the air-gap flux distribution and the generated waveform. The winding would also be difficult to accommodate and the more practical arrangement is the distributed winding which uses more than one slot per pole per phase — usually two or three. Note that too many slots would mean that the teeth would be mechanically weak and, along with other possible faults, would vibrate and contribute to mechanical noise.

The diagram (Fig. 118) shows a working condition, and it will be noted immediately that, the conductors in one slot will have e.m.f.'s generated in them which are not in phase with those generated in the conductors of an adjacent slot. If all conductors per phase are connected in series then the resultant e.m.f. will be the phasor sum, and

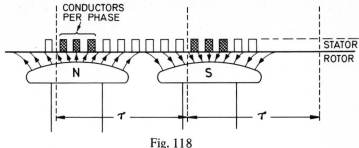

Fig. 118

such a resultant will be slightly smaller in magnitude than if a concentrated winding had been used. This introduces the need for a further constant of multiplication into the e.m.f. equation. This constant is called the *Distribution Factor* or *Breadth Factor*. This term is written as K_D and is defined as the ratio of the e.m.f. in a distributed winding to the e.m.f. in a concentrated winding.

Thus $K_D = \dfrac{\text{e.m.f. in distributed winding}}{\text{e.m.f. in concentrated winding}} = \dfrac{\text{OA}}{\text{OB}}$

OA and OB are shown in the diagram (Fig. 119) and it can be inferred that K_D is always less than 1. If the same number of conductors for each type of winding is considered then, for a Distributed Winding, e.m.f. per phase $= K_D \times$ e.m.f. for a concentrated winding,

or $\qquad E_{ph} = 2 \cdot 22 \, K_D Z_{ph} \Phi f$ volts.

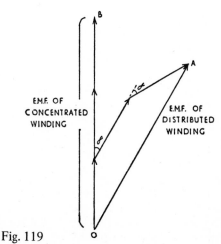

Fig. 119

Summarising we can say that the distribution of the phase winding in several slots for each pole-pitch improves the waveform but reduces slightly the resultant e.m.f. value.

Example 62. Find the no-load terminal voltage of a three-phase, four-pole a.c. generator from the following data. Flux per pole (assumed sinusoidally distributed) = 0·14 webers. Slots per pole per phase = 2. Conductors per slot = 2. Assume a Distribution Factor of 0·966. The machine is driven at 1500 rev/min and is star connected.

Here $f = \dfrac{PN}{120} = \dfrac{4 \times 1500}{120} = 50$ Hz

$$\therefore E_{ph} = 2{\cdot}22\,K_D Z_{ph}\Phi f \text{ volts}$$
$$= 2{\cdot}22 \times 0{\cdot}966 \times (4 \times 2 \times 2) \times 14 \times 10^{-2} \times 50$$
$$= 2{\cdot}22 \times 0{\cdot}966 \times 16 \times 14 \times 10^{-2} \times 50$$
$$= 2{\cdot}22 \times 0{\cdot}966 \times 8 \times 14$$
$$= 240{\cdot}19 \text{ volts/phase}$$

Terminal voltage

$$= \sqrt{3} \times 240{\cdot}19 = 1{\cdot}732 \times 240{\cdot}19 = 416 \text{ V.}$$

DETERMINATION OF BREADTH FACTOR. In order to determine the value of K_D for a particular stator-slot arrangement, the general expression developed here can be used. Reference to Fig. 120 shows

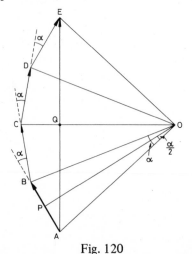

Fig. 120

that the voltages of the conductors in successive slots AB, BC, CD, *etc.* are all equal but with a phase displacement of α (electrical degrees). Points A, B, C, D, *etc.* all lie in a circle of centre O. Draw OP perpendicular to AB. Then angle AOP = angle POB = $\alpha/2$ since angle AOB = α.

Now AP = AO sin $\alpha/2$ and AB = 2AO sin $\alpha/2$

The resultant e.m.f. = AE and angle AOE = $n\alpha$, if there are n slots per pole per phase.

Now $\quad$ AQ = AO $\sin \dfrac{n\alpha}{2}$

$\therefore \quad$ AE = 2AO $\sin \dfrac{n\alpha}{2}$

and $\quad K_D = \dfrac{2AO \sin \dfrac{n\alpha}{2}}{n(\text{voltage/slot})} = \dfrac{2AO \sin \dfrac{n\alpha}{2}}{n \text{ AB}} = \dfrac{2AO \sin \dfrac{n\alpha}{2}}{n \text{ } 2AO \sin \dfrac{\alpha}{2}}$

or $\quad K_D = \dfrac{\sin \dfrac{n\alpha}{2}}{n \sin \dfrac{\alpha}{2}}.$

Example 63. For Example 62 the information was given regarding the Distribution Factor. This could have been determined from the given data thus: Find the no-load terminal voltage of a three-phase, four-pole alternator having two slots per pole per phase with two conductors per slot. The flux per pole is 0·14 Wb, the machine speed 1500 rev/min and the connection of the phases is in star.

As before $\quad f = \dfrac{4 \times 1500}{120} = 50 \text{ Hz}$

Also $\quad K_D = \dfrac{\sin \dfrac{n\alpha}{2}}{n \sin \dfrac{\alpha}{2}} \quad$ Here $\alpha = \dfrac{180}{3 \times 2} = 30°$

and $K_D = \dfrac{\sin \dfrac{2 \times 30}{2}}{2 \sin \dfrac{30}{2}}$

or $K_D = \dfrac{\sin 30}{2 \sin 15} = \dfrac{0 \cdot 5}{2 \times 0 \cdot 2588} = \dfrac{0 \cdot 5}{0 \cdot 5176} = 0 \cdot 966$ (as given)

So $E_{ph} = 2 \cdot 22 \times 0 \cdot 966 \times 16 \times 14 \times 10^{-2} \times 50 = 240 \cdot 19$ volts/phase

Terminal Voltage $= \sqrt{3} \times 240 \cdot 19 = 416$ V.

COIL SPAN, PITCH OR CHORDING FACTOR. Sometimes to improve the waveform of an alternator, the span of a coil is made less or greater than a pole-pitch. Reference to the diagrams of Fig. 121 will show that there is a phase difference between the e.m.f.'s in the coil sides and the resultant e.m.f. is further reduced.

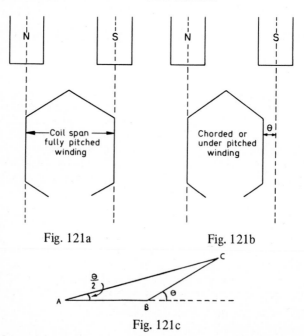

Fig. 121a   Fig. 121b

Fig. 121c

In the example of the diagram, the two sides of the coil are not 180 electrical degrees apart, the span being less than a pole-pitch. From the phasor diagram (c), we see that the resultant e.m.f. is AC or $= 2\,\mathrm{AB}\cos\dfrac{\theta}{2}$ and the Pitch Factor, K_S can be defined as the ratio of the actual e.m.f. to that obtained from a fully pitched coil

or $\quad K_S \;=\; \dfrac{\mathrm{AC}}{2\mathrm{AB}}$

Thus $\quad K_S \;=\; \dfrac{2\mathrm{AB}\cos\dfrac{\theta}{2}}{2\mathrm{AB}} \;=\; \cos\dfrac{\theta}{2}$

The effect of the Pitch Factor is noticeable if the span is $140°$ or less *i.e.* $\theta \;=\; 40°$ or more. For $\theta \;=\; 40°$ $K_S = 0\cdot94$. The term "chorded" is also used in connection with coils which are pitched less or greater than the pole-pitch. Occasionally the term Winding Factor may be encountered. This is used to include the Distribution and Coil-span Factors.

Thus Winding Factor $=$ Distribution Factor $\times$ Coil-span Factor.

Example 64. The e.m.f. generated in a coil on an alternator is 20 V. Calculate the e.m.f. between the ends of two such coils, connected in series, if they are separated on the alternator core by 30 electrical degrees.

The solution is made by reference to the diagrams of Fig. 121.

Since the voltages of two coils are out of phase by $30°$ then the resultant is given by $2\,V\cos\theta/2$. Here θ is the external angle of an isosceles triangle $=$ twice the internal angle.

$\therefore$ Resultant voltage $\;=\; 2\times20\times\cos15° \;=\; 2\times20\times0\cdot9659$
$\qquad\qquad\qquad\qquad\; =\; 4\times9\cdot659 \;=\; 38\cdot64\ \mathrm{V}.$

Example 65. Find the no-load terminal voltage of a four-pole, star-connected, marine, turbo-alternator from the following data: Flux per pole (assumed sinusoidally distributed) $0\cdot12$ Wb; slots per pole per phase 4, conductors per slot 4, coil-span $150°$, connection — star, speed 1500 rev/min.

$$\text{Distribution Factor } K_D = \frac{\sin \dfrac{n\alpha}{2}}{n \sin \dfrac{\alpha}{2}}$$

$$= \frac{\sin \dfrac{4 \times 15}{2}}{4 \sin \dfrac{15}{2}}$$

Note $\alpha = \dfrac{180}{3 \times n} = \dfrac{180}{3 \times 4} = 15°$

So $K_D = \dfrac{\sin 30}{4 \sin 7 \cdot 5} = \dfrac{0 \cdot 5}{4 \times 0 \cdot 1305} = \dfrac{0 \cdot 5}{0 \cdot 522} = 0 \cdot 958$

Pitch Factor $K_S = \cos \dfrac{\theta}{2}$ where $\theta = 180 - 150 = 30°$

or $K_S = \cos \dfrac{30}{2} = \cos 15 = 0 \cdot 966$

Generated Voltage per phase or $E_{ph} = 2 \cdot 22 \, K_D \, K_S \, Z_{ph} \Phi f$

Here $f = \dfrac{4 \times 1500}{120} = 50 \text{ Hz}$

$\therefore E_{ph} = 2 \cdot 22 \times 0 \cdot 958 \times 0 \cdot 966 \times (4 \times 4 \times 4) \times 0 \cdot 12 \times 50$

$= 2 \cdot 22 \times 0 \cdot 958 \times 0 \cdot 966 \times 64 \times 6 = 789 \text{ V}$

Terminal Voltage $= \sqrt{3} \times 789 = 1367 \text{ V}$ or $1 \cdot 367 \text{ kV}$.

WAVEFORM OF GENERATED E.M.F.

For alternating current theory and practice, the ideal conditions of sinusoidal working are assumed. Since the alternator is the source of e.m.f. then, in order to achieve this ideal condition, it is essential that the generated voltage waveform approximates as closely as possible to that of a sine wave. For modern power-stations, synchronous a.c. generators of the order of 300 to 500 MW are in use and, if attention is given to this requirement then, it can be assumed that the "mains" voltage is sinusoidal. For marine work also, although the alternator sizes are very much smaller, every effort is made to achieve sine-wave voltage conditions since the generating capacity is both appreciable and independent. Non-sinusoidal conditions not only give rise to errors in the metering, protection of circuits and the rating of plant but can also introduce adverse effects such as excess voltages due to resonance caused by unwanted harmonics. The aspect of theory now considered, is therefore related to the waveform of alternator voltage, as it is generated.

In Chapter 6, a first mention was made of harmonics in connection with the no-load current of a transformer. Fourier proved that an alternating waveform, of whatever shape, can be analysed into a fundamental sine wave and a number of subsidiary sine waves of a frequency and amplitude different to that of the fundamental. The frequency of a second harmonic is twice that of the fundamental, that of a third harmonic three times that of the fundamental and so on. The subsidiary sine-wave harmonics need not have their zero values at the same instant as the fundamental. The diagrams of Fig. 122 have been drawn to show the resultant of a sine wave combined separately with second and third harmonic sine waves. It will be seen that with an even harmonic, the second half-cycle of the wave is not an exact inverted reflection of the first, whereas with an odd harmonic this condition is fulfilled.

Because, for an alternator, a North magnetic pole followed by a South pole is made to sweep past a conductor and because the e.m.f. generated from instant to instant varies with the flux density, it follows that since the flux-density distribution in the air gap is similar for both poles, even though the South-pole flux is in opposite sense to the North-pole flux, the generated waveform will have a −ve half-cycle which will be exactly the same as the +ve half cycle. The deduction can therefore be made that only odd harmonics can be generated.

Earlier study has shown that for a circuit containing inductance and capacitance, there is some frequency for which the circuit is resonant. For a series circuit this gives rise to a high current and voltages across the inductive and capacitive parts of the circuit which may be very much greater than the supply voltage with unpleasant results. For a parallel circuit the "mains" current is reduced to a minimum but large currents result in the inductive and capacitive branches, again an undesirable effect. Now it may be that a circuit was not resonant at 50 Hz but was resonant at 150 Hz. If a supply voltage having a third harmonic, of amplitude less than that of the fundamental but nevertheless quite appreciable, is applied to the circuit then, due to the strong third harmonic e.m.f., disagreeable effects would result which would not be evident if only a fundamental voltage sine wave of 50 Hz is used. Other effects of generated harmonics are concerned with their disturbing and interfering with associated communication and control circuits through the medium of induction. The methods by which a designer achieves the generation of a sine wave with the minimum of distortion by harmonics, will be considered

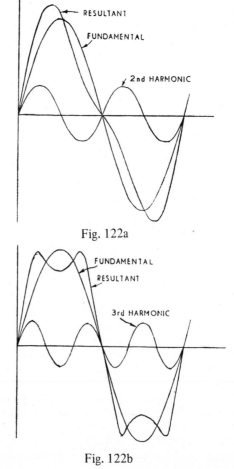

Fig. 122a

Fig. 122b

shortly and will help to explain some of the constructional features of the a.c. generator as described earlier.

THE AMPERE-TURN OR M.M.F. WAVE.   In order to understand a method of treatment which will be used on several occasions, it is of advantage to consider the procedure for drawing out the m.m.f. or ampere-turn wave. Consider the simple three-turn, flat coil with conductors arranged as in the diagram of Fig. 123. If conductors **a** and **d** are considered to form one turn, then the m.m.f. available with a current of one ampere to set up flux is one ampere-turn and is constant in space between the current carrying conductors. Considering the diagram it will be seen that each time a current-carrying conductor is passed over, the m.m.f. reverses and if the resultant m.m.f. wave for the set of three turns **ad, be** and **cf** is drawn from the series of rectangular waveforms, then a stepped m.m.f. distribution wave is obtained. In actual fact the m.m.f. increases gradually across each slot so the ampere-turn wave is more like that shown dotted *i.e.* a trapezium.

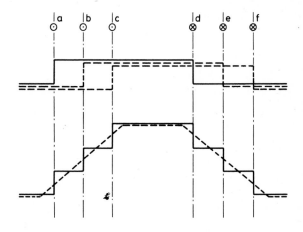

Fig. 123

ALTERNATOR WAVEFORM IMPROVEMENT.   Reference has already been made to the ideal method of obtaining a sine wave of induced e.m.f.; namely, the armature conductors are required to cut or to be cut by a flux whose density is sinusoidally distributed over a pole-pitch. In practice this is not possible but all working methods give close approximations. These methods differ for the cylindrical and salient-pole types of alternators and are now considered separately.

Cylindrical Rotor. Since the rotor of a turbo-alternator is smooth, the machine air-gap is of necessity uniform. The field winding is distributed over the surface of the cylinder and results in a flux wave in space which is more nearly sinusoidal than if the field winding had been concentrated in one slot for each coil side.

Consider the rotor of Fig. 124a to be spread out as shown in the diagrams of Fig. 124b. A two-pole field system has been used for the example. The graphs, drawn to show the magnitude, location and direction of the effective resultant magnetomotive force resulting from the rotor current in the coil sides, allow the following explanation. If a smooth curve is drawn by ignoring the steps, the trapezium is obtained which represents the m.m.f. acting at any point along the surface of the air-gap. The trapezium can be considered to be made up of a number of sine waves — three are shown, the fundamental, the 3rd and the 5th harmonics. Since the flux is directly proportional to the m.m.f., it follows that the flux-density distribution over a pole-pitch, is sinusoidal and that a voltage of sine waveform will be induced if the presence of any harmonic e.m.f.'s is eliminated. This can be achieved by the use of additional corrective measures such as chording of the stator windings, etc.

Salient-Pole Rotor. For this arrangement, since independent poles exist — each with its own field winding, a sinusoidal flux-density waveform in the air-gap across the pole face cannot be achieved by the "distribution of the field winding" method. However, as the poles

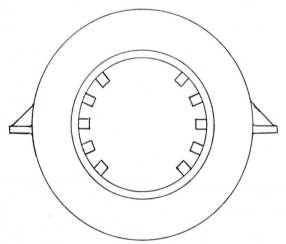

Fig. 124a

are comparatively small and independent they are capable of being shaped and modified to allow one or more of the following methods to be effective.

(a)   Shaping of Pole-shoes. The reluctance of the magnetic circuit is varied by shaping the profile of the pole so that the air-gap is graded, being least at the centre of the pole-shoe and getting larger gradually towards the edges. In practice the profile is determined on

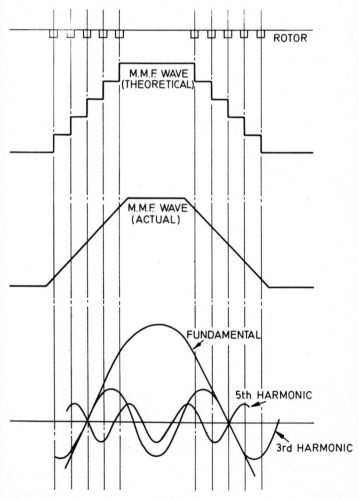

Fig. 124b

a theoretical basis and the final shape is settled emperically, to some extent, by trial and error. A simple approximation is obtained by making the radius of curvature of the pole-shoe equal to 0·7 times the inside radius of the stator core. Fig. 125 illustrates simply the features of the method and here again additional corrective methods are usual *i.e.* chording of the stator winding, etc.

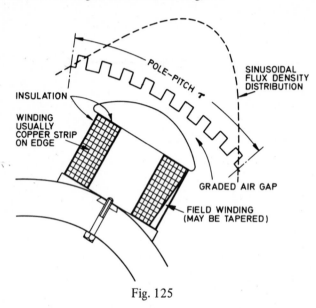

Fig. 125

(b) Skewing of Poles or Armature Slots. If a pole face could be sinusoidally shaped as in Fig. 126, the active lengths of the armature conductors would vary as they are cut by the field, the air-gap being uniform. The arrangement is not practical nor is that shown by the

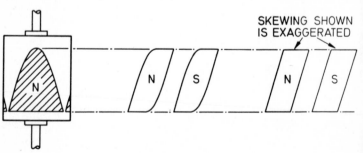

Fig. 126

next diagram. Here the pole edges are sine shaped. A commercial approximation is the third arrangement and an e.m.f. wave of trapezium form (Fig. 127) would result. This, as has been seen, can be corrected to give the required fundamental sine wave. The same result could be obtained by fixing the poles axially and skewing the armature conductors. Method (b) is less favoured than that described under (a) because it does not permit easy modification once a prototype machine is built.

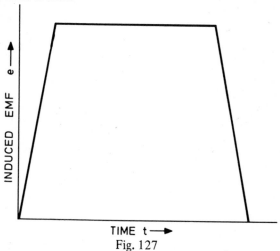

Fig. 127

Example 66. A four-pole alternator, on open circuit, generates 200 V at 50 Hz, when its field current is 4 A. Determine the generated e.m.f. at a speed of 1200 rev/min and a field current of 3 A, neglecting saturation of the iron parts.

This problem has been introduced to provide revision of basics.

The original speed of the alternator is given by $f = \dfrac{PN}{120}$

or $N_1 = \dfrac{f\,120}{P} = \dfrac{5 \times 120}{4} = 1500$ rev/min.

Now generated e.m.f. $E \propto \Phi N$ and assuming Φ is proportional to the exciting current I_f then

$E \propto I_f N$ or we can write $\dfrac{E_2}{E_1} = \dfrac{kI_{f_2}N_2}{kI_{f_1}N_1}$

and $E_2 = \dfrac{E_1 \, I_{f_2} \, N_2}{I_{f_1} \, N_1} = \dfrac{200 \times 3 \times 1200}{4 \times 1500} = \dfrac{200 \times 3 \times 4}{4 \times 5}$

Thus generated e.m.f. $= 40 \times 3 = 120$ V.

STATOR WINDINGS

Mention has been made already of the types of slots into which the stator or armature conductors can be placed. From the constructional point of view, the "open" slot is to be preferred (Fig. 128), since it is possible to preform and insulate the coil side − consisting of one or more conductors, which can then be inserted into the slots. The main disadvantage of the "open" slot is that it results in teeth which encourage the flux to "tuft" as it crosses the machine air-gap. The local magnetic leakage path of this type of slot is high and the *leakage reactance* is low. This is the first occasion in which the term leakage reactance has been used in relation to alternator theory but it will be used in context later on and will be explained fully. Referring back to the "tufting" effect it is seen, from the appropriate diagram, that the flux is caused to jump from tooth to tooth and thus a "ripple" frequency tends to be generated. This is illustrated by the diagram in (Fig. 128) but is exaggerated. The "closed" type of slot overcomes

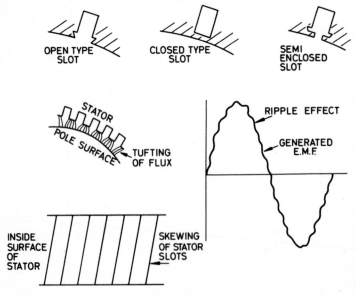

Fig. 128

this effect but makes winding difficult and endows the machine with a high leakage reactance. The semi-enclosed" slot is a compromise. Attention is drawn to the fact that skewing of the conductor slots also helps to reduce the ripple in the e.m.f. wave. The "semi-enclosed" slot increases leakage reactance but reduces the air-gap reluctance and for this reason is preferred for induction-motor construction.

The "semi-enclosed" slot is obviously more suited to the multi-turn coil where, as for a d.c. winding, the conductors of hand-wound or preformed coils can be inserted singly into the space available after the slot insulation has been positioned. The ' open" slot arrangement is particularly suited to bar-type conductors or coils which are insulated with baked-on or heat-moulded insulation, a method much favoured for high-voltage machines.

TYPES OF WINDING. It is not proposed here to consider winding techniques in great detail, since this constitutes a large part of machine design work in itself and requires specialist knowledge. Only basic examples will be described but in general we can say that for a.c. machines "open-type" windings are used in contrast with those for d.c. machines where "closed" windings are necessary. In line with this distinction, it is usual for the smaller and infrequently used rotating-armature a.c. generator to follow d.c. practice and use a closed winding provided with tappings, brought out to the appropriate number of slip-rings. For the more common rotating-field alternator, the armature or stator winding is of the "open" type *i.e.* there are two "free ends" per phase.

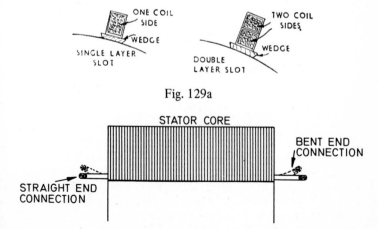

Fig. 129a

Fig. 126b

A.c. windings may be of (a) the "single-layer" type or (b) the "double-layer" type, and reference should be made to the first two small diagrams of Fig. 129.

(a) Single-layer Windings. (Fig. 130.) For a single-layer winding each slot is fully occupied by a conductor or coil-side, an arrangement, as already mentioned, suited to high-voltage machines or coils with a large number of conductors per slot.

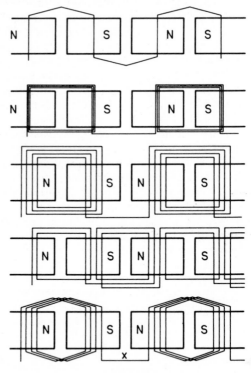

Fig. 130

The third diagram of Fig. 130 shows one phase of a three-phase, single-layer winding made up from concentric, half-coiled or hemitropic coils. The mean coil span equals a pole-pitch and the number of coil groups always equals the number of pole-pairs. The diagrams (Fig. 130) are set out in a basic form to show how the winding explanation can be followed from the simple skeleton bar winding. In each case one phase only is shown and the diagrams are set out logically to follow on. Thus the bar winding is extended next into the second diagram which shows a full-pitched concentrated winding. This in turn

is developed to show a distributed arrangement. In connection with
the distributed winding it should be noted that, since the coils are in
series, the order in which the conductors are connected is not of importance and thus alternative arrangements are possible. Such an
alternative is the use of concentric whole coils as shown by the fourth
diagram of Fig. 130. Here the numbers of coil groups equals the number of poles.

Both the half-coiled and whole-coiled concentric windings use coils
of different spans and it will also be seen that some difficulty is
presented in making the "overhang" or end connections, since these
have to pass each other. The accepted method is to bend the appropriate coils through 90° or 45° where necessary. This has been shown
by the final small diagram of Fig. 129. Another method used is to
"crank" the overhand of the end-connections but a logical alternative
winding sequence is shown by the fifth diagram of Fig. 130. This
makes for a regular arrangement with coils of the same pitch and
shape, the coil-span being equal to a pole-pitch although "chording"
is possible. The necessary "distributed" effect is also achieved and a
lap and wave effect is obtained as the winding progresses round the
stator. The need for the necessary end-connection X between the coil
groups should be noted. The lap-wave arrangement is often called a
"lattice" winding, because of the symmetrical woven appearance, but
as it is better suited to a double-layer assembly, it is most usually used
in this form which is now considered.

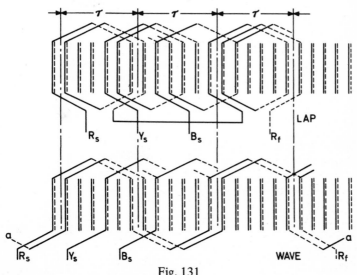

Fig. 131

(b) Double-layer Windings (Fig. 131). For d.c. machines, the two types of winding, lap and wave, are the result of the manner in which the coils per pole-pair are connected. To assist the lay of the coils at the ends and to allow a symmetrical assembly, with all coils being pre-formed, the winding is arranged in two layers with one coil-side in the upper half of a slot and the other coil-side in the lower half of a slot distant one pole-pitch. For an a.c. winding, coils similar to those used and constructed for d.c. armatures, are arranged in either a lap or wave winding as shown by the diagrams of Fig. 131. The coil span is usually equal to a pole-pitch but can be made slightly different if "chording" is desired. Such windings also come under the heading of "lattice" windings, because of their symmetry, and are the most common for medium-sized, salient-pole machines as used for marine work. For each example, only one phase of a three-phase winding has been shown, since the other phase windings are repetitions of the first.

Example 67. An 11 kV alternator provides a three-phase, 50 Hz supply when running at 3000 rev/min. It has 48 slots with twenty conductors per slot. Each coil spans 18 teeth. Find the flux per pole assuming it to be sinusoidally distributed.

$$\text{From } f = \frac{PN}{120} \quad P = \frac{50 \times 120}{3000} = \frac{6000}{3000} = 2$$

Also since a fully-pitched coil would span $\dfrac{48}{2}$ *i.e.* 24 teeth, then 24 teeth would constitute $180°$ (electrical)

$$\text{and 18 teeth} = 180 \times \frac{18}{24} = \frac{540}{4} = 135°$$

$$\text{Here } K_D = \frac{\sin \dfrac{n\alpha}{2}}{n \sin \dfrac{\alpha}{2}} \qquad \text{Also } \alpha = \frac{180}{3n}$$

$$\text{where } n = \text{slots/pole/phase} = \frac{24}{3} = 8$$

$$\text{So } \alpha = \frac{180}{3 \times 8} = 7{\cdot}5° \quad \text{and } K_D = \frac{\sin \dfrac{8 \times 7{\cdot}5}{2}}{8 \sin \dfrac{7{\cdot}5}{2}}$$

Thus $K_D = \dfrac{\sin 30}{8 \sin 3 \cdot 75} = \dfrac{0 \cdot 5}{8 \times 0 \cdot 0654} = \dfrac{0 \cdot 5}{0 \cdot 5232} = 0 \cdot 956$

Also $K_S = \dfrac{\cos(180 - 135)}{2} = \cos \dfrac{45}{2} = \cos 22 \cdot 5 = 0 \cdot 924$

Now $E_{ph} = \dfrac{11\,000}{\sqrt{3}} = 6360$ V (assuming star connection)

and since $E_{ph} = 2 \cdot 22 K_D K_S Z_{ph} \Phi f$

then $6360 = 2 \cdot 22 \times 0 \cdot 956 \times 0 \cdot 924 \times \dfrac{960}{3} \times \Phi \times 50$

Here $Z_{ph} = \dfrac{48 \times 20}{3} = \dfrac{960}{3} = 320$

and $\Phi = \dfrac{6360}{111 \times 0 \cdot 956 \times 0 \cdot 924 \times 320}$

or $\Phi = \dfrac{636}{3552 \times 0 \cdot 883} = 0 \cdot 203$ Wb.

THE ALTERNATOR ON LOAD

As for any source of e.m.f., the terminal voltage of an a.c. generator can be expected to fall when current is supplied. The difference of the terminal voltage between open-circuit and load conditions is dependent on the value of current supplied and on the causes of voltage drop inside the machine. These causes are analysed to be due to:

(a) the impedance of the machine, and (b) the armature reaction effect. It should be noted that the magnitude of the terminal voltage variation is also affected by the power factor at which the machine is operated, and this is illustrated by the curves of Fig. 132 which show the "on-load" characteristics.

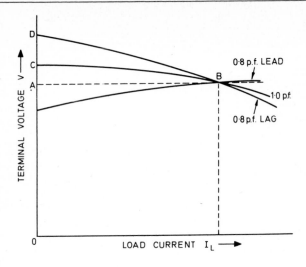

Fig. 132

Before the load condition is covered in more detail, the term *Regulation* is considered. This defines one aspect of the performance of the machine and can be estimated without direct loading, by developing the phasor diagram of the machine and making use of figures obtained from assessing tests.

VOLTAGE REGULATION. The variation of terminal voltage between full load and no load, expressed as a percentage or per unit value of full load voltage, is termed the regulation. Thus:

Voltage regulation =

$$\frac{\text{Rise in terminal voltage when full load is removed}}{\text{Full-load terminal voltage}} \times 100$$

or referring to Figure 132

$$\text{Voltage regulation} = \frac{\text{AC}}{\text{OA}} \times 100 \text{ per cent at unity power factor}$$

$$= \frac{\text{AD}}{\text{OA}} \times 100 \text{ per cent at a power factor of } 0.8 \text{ (lagging)}.$$

Since regulation is influenced by power factor and load current, these must be specified when stating the percentage regulation value. Also, due to the hysteresis effect in the iron of the magnetic circuit, the rise of potential difference when load is thrown off, is less than the fall in p.d. when the load is applied. One method has therefore been standardised and the regulation is defined in terms of a rise in p.d. when load is thrown off. It is sometimes qualified as the regulation "up" or "down".

Example 68. The terminal voltage of a three-phase alternator is set at 440 V, by adjusting the field excitation when the speed is correct and when the full-load current is supplied at a power factor of 0·8 (lagging). When the machine circuit-breaker is opened and the load thrown-off, the terminal voltage is seen to rise to 506 V. Estimate the voltage regulation.

$$\text{Voltage regulation} \quad = \quad \frac{506 - 440}{440} \times 100$$

$$= \quad \frac{66}{440} \times 100 \quad = \quad \frac{66}{4\cdot4} = \frac{6}{0\cdot4}$$

$$= \quad 15 \text{ per cent (up).}$$

PHASOR DIAGRAM. The causes of internal voltage drop have already been mentioned and when considered in relation to the generated e.m.f. and the terminal voltage, the obvious approach must be through a phasor diagram since phase relationships must be examined and appropriate phasor procedures followed. For polyphase machines, a simplification is immediately introduced if, the conditions for one phase only are considered, the appropriate adjustment to line values being made once all the necessary procedures and calculations have been made in phase values. Unless otherwise stated in this book, phase conditions can always be assumed.

When an alternator supplies current, internal voltage drops occur due to the following reasons:

(a) Armature Resistance. The voltage drop *IR,* caused by this factor, is always small and at times can be neglected. It is proportional to the current causing it and is in phase with the current. Note that, for the basic phasor diagram, current has been taken as the reference since the build up of the diagram follows the procedure for a simple series circuit.

(b) Armature Reactance. The voltage drop, due to this factor, is caused by a number of individual inductive effects associated with the stator winding. It is known that, any a.c. winding has self-inductance, when carrying current, but the consequent reactance is frequently attributed to the joint reactance of individual parts and thus mention is made of slot-reactance, end-connection reactance, etc. The overall reactance is proportional to current and is in quadrature with it. This reactance is one constituent part of the X_S shown on the phasor diagram of Fig. 133.

(c) Armature Reaction. The magnetic field due to the armature current modifies the main field and alters the generated e.m.f. For a lagging power-factor condition, the armature reaction effect weakens the main field resulting in a reduction of generated voltage. For a leading power-factor condition, the main flux is strengthened and the generated voltage is raised. The overall effect of the armature reaction can be represented by a reactance voltage drop which is proportional to the current and in quadrature with it. This reactance is the other constituent part of X_S as shown on the phasor diagram.

The armature reactance and the armature reaction produce similar effects and are combined in what is termed the "*synchronous reactance*" X_S. This is not a true reactance, but is considered as such for practical purposes. It is shown on the phasor diagram and when combined with the armature resistance in the usual way, the "synchronous impedance" Z_S is obtained. The effect of Z_S can be represented by a voltage drop on the phasor diagram which illustrates the on-load voltage conditions. Thus for the diagram, I is the load current, OV the terminal voltage and OE the generated voltage. The armature resistance voltage drop (IR), the synchronous reactance voltage drop (IX_S) and the synchronous impedance voltage drop (IZ_S) are also shown.

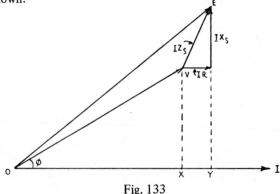

Fig. 133

The phasor diagram can be used to determine the regulation of an a.c. generator if the simple construction is made, as shown dotted for Fig. 133.

In terms of the large triangle, having hypotenuse OE

Horizontal side　$= \mathrm{OX} + IR$　or　$V \cos \phi + IR$

Vertical side　　$= \mathrm{VX} + IX_S$　or　$V \sin \phi + IX_S$

$\therefore$　　　　$\mathrm{OE}^2 = (\mathrm{OX} + IR)^2 + (\mathrm{VX} + IX_S)^2$

or Generated voltage　$E = \sqrt{(V \cos \phi + IR)^2 + (V \sin \phi + IX_S)^2}$

and Voltage regulation　$= \dfrac{E - V}{V} \times 100$ per cent.

The following examples will illustrate the use of the above diagram and expressions, but it should be noted that the regulation value obtained by this method is always higher than the true practical value. The reason for this will be explained in Chapter 9. Here we become familiar with the terms Synchronous Impedance and Synchronous Reactance and with the use of the Voltage Equation as developed from the phasor diagram.

Example 69. An alternator has the following phase values: effective resistance $0 \cdot 2\ \Omega$, synchronous reactance $1 \cdot 1\ \Omega$. Find the generated e.m.f. for a load current of 100 A and a phase terminal voltage of 250 V. The load is operated at a power factor of $0 \cdot 8$ (lagging).

$E_{\mathrm{ph}} = \sqrt{(V \cos \phi + IR)^2 + (V \sin \phi + IX_S)^2}$

$= \sqrt{\{(250 \times 0 \cdot 8) + (100 \times 0 \cdot 2)\}^2 + \{(250 \times 0 \cdot 6) + (100 \times 1 \cdot 1)\}^2}$

$= \sqrt{(200 + 20)^2 + (150 + 110)^2}$

$= \sqrt{(220^2 + 260^2)} = 10^2 \sqrt{2 \cdot 2^2 + 2 \cdot 6^2} = 10^2 \sqrt{4 \cdot 84 + 6 \cdot 76}$

$= 10^2 \sqrt{11 \cdot 60} = 100 \times 3 \cdot 4 = 340$ V.

Example 70. A $53 \cdot 28$ kVA, 440 V, three-phase, star-connected a.c. generator has a stator resistance value of $0 \cdot 2\ \Omega$ per phase and a synchronous reactance of $1 \cdot 6 \ \Omega$ per phase. Estimate the voltage regulation when the full-load current is switched off. The load was being operated at a power factor of $0 \cdot 8$ (lagging).

Load current $I = \dfrac{53\ 280}{\sqrt{3} \times 440} = 70 \text{ A} \quad V_{ph} = \dfrac{440}{\sqrt{3}} = 254 \text{ V}$

From the voltage diagram

$$E_{ph} = \sqrt{(V \cos \phi + IR)^2 + (V \sin \phi + IX_S)^2}$$

$$= \sqrt{\{(254 \times 0 \cdot 8) + 70 \times 0 \cdot 2)\}^2 + \{(254 \times 0 \cdot 6) + (70 \times 1 \cdot 6)\}^2}$$

$$= \sqrt{(203 \cdot 4 + 14)^2 + (152 \cdot 4 + 112)^2}$$

$$= \sqrt{217 \cdot 2^2 + 264 \cdot 4^2} = 10^2 \sqrt{2 \cdot 172^2 + 2 \cdot 644^2}$$

$$= 10^2 \sqrt{4 \cdot 72 + 6 \cdot 98} = 10^2 \sqrt{11 \cdot 7} = 10^2 \times 3 \cdot 42 = 342 \text{ V}$$

$$\therefore \text{ Voltage regulation } = \frac{E - V}{V} \times 100 = \frac{342 - 254}{254} \times 100$$

$$= \frac{8800}{254} = 34 \cdot 6 \text{ per cent.}$$

CHAPTER 7

PRACTICE EXAMPLES

1.   A three-phase, 1000 rev/min a.c. generator supplies a 50 Hz current to a three-phase, 120 rev/min synchronous motor. Calculate the number of poles on both the alternator and motor.

2.   A 440 V, three-phase, eight-pole marine alternator generates a 50 Hz supply when driven at normal speed. If the system is to be converted to 60 Hz working, find the speed at which the prime-mover is to run. Assuming the same excitation value find the new system voltage.

3.   The e.m.f. generated in a coil of an a.c. generator is 20 V. Calculate the e.m.f. between the ends of two such coils connected in series, if they are displaced on the alternator core by 30 electrical degrees.

4.   Calculate the breadth factor for a three-phase, six-pole alternator having 72 slots.

5.   A three-phase, eight-pole alternator has 96 slots. One side of a coil of a two-layer winding lies in slot 1, while the other side lies in slot 9. Calculate the pitch factor for the winding.

6.   A 3·3 kV, 50 Hz, three-phase alternator which is driven at 750 rev/min has a single-layer winding contained in 120 slots. The flux per pole is 0·0448 Wb. If the machine is to have a full-pitched winding and is to be star connected, calculate the number of conductors per slot.

7.   For the machine of Example No. 6, if the stator winding is short chorded to $\frac{3}{4}$ pitch and is to be mesh connected, what would then be the necessary number of conductors per slot to the nearest whole number?

8.   The three-phase, two-pole, star-connected alternator of a turbo-electric vessel supplies the 10 000 kW, mesh-connected, fifty-pole synchronous motor which operates at a power factor of 0·95 (lagging) and an efficiency of 90 per cent at a full-load propeller speed of 110 rev/min. Calculate the phase currents of the alternator and of the motor when the line voltage is 3·3 kV. Find also the *kVA* output of the alternator when

supplying the motor at the above output power. Find the supply frequency of the system on full load.

9.   When a reduced excitation current is applied to a single-phase a.c. generator the O.C. voltage is noted to be 45 V. With the same excitation current and the alternator terminals short-circuited, the current is found to be 30 A. The stator resistance is measured to be 0·4 Ω. Estimate the synchronous reactance of the machine.

10.   A 750 kVA, 3·3 kV, three-phase, star-connected alternator has a resistance of 0·5 Ω per phase and a synchronous reactance of 3 Ω per phase. Find the percentage voltage regulation when full load at a power factor of 0·8 (lagging) is switched off.

CHAPTER 8

ELECTRONICS (i)

An introduction to thermionics was made when the diode valve and its action was described in Chapter 15, Volume 6. This work is summarised here and a continuation made with describing the basic operation and usage of the other multi-electrode valves. Solid-state devices, such as the semiconductor diode, transistor and thyristor (S.C.R.) are fast displacing thermionic valves in all fields of electronic development and evolution but to date, the complete obsolescence of the valve cannot be claimed. It is still appropriate therefore, to continue considering valve types in a general manner, noting their operating features and associated circuitry. Semiconductor devices although functioning in a different way can achieve results similar to those obtained from valves and it is for this reason that the reader is advised to pay attention to the shapes of characteristic curves and the techniques employed in using them. If valves and semiconductor devices are seen to be complementary then study can be advanced on the basis of both common and comparable treatments.

THE VACUUM DIODE (Revision)

The valve cathode when heated, emits electrons which are attracted to the anode provided the latter is at a positive potential with respect to the former. The diagrams (Fig. 134) show the circuit symbols and the position of the valve in a circuit.

The behaviour of the diode can be determined from the test circuit shown (Fig. 135) and the accompanying characteristics illustrate the relation between anode voltage V_a and anode current I_a. The curves for differing heater voltages also stress the effect of cathode temperature on electron emission.

From the characteristics can be deduced the internal resistance of the valve, namely from an appropriate $\dfrac{V_a}{I_a}$ value. This value would vary for each operating point on a curve and since the valve is always used with a load in its anode circuit, a distinction is made between the *static* and *dynamic* characteristic. The test circuit would also show that the diode valve would not conduct when the anode is made negative with respect to the cathode and as a result the device, being unidirectional in its conduction, will act as a rectifier. This is now its main application in valve-type electronic equipment.

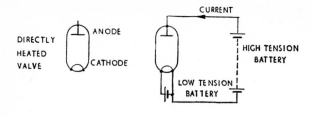

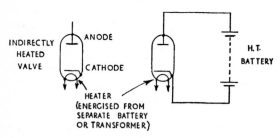

Fig. 134

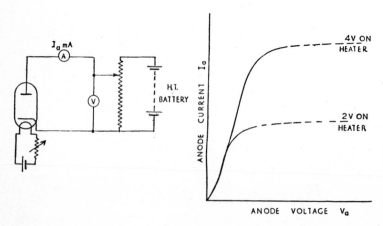

Fig. 135

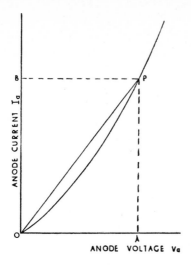

Fig. 136

(a) STATIC CHARACTERISTIC. A curve obtained from the results of a test made in accordance with the circuit of Fig. 135 is termed a static characteristic. If a point P on this curve (Fig. 136) is considered, the *d.c. resistance* is given by $\dfrac{BP}{AP}$ or $\dfrac{OA}{OB}$. This value would also be given by the reciprocal of the slope of the line through P and O. Point P is called a "quiescent" or "operating point" and the reciprocal of $\dfrac{PA}{AO}$ or $\dfrac{OA}{OB}$ gives the valve d.c. resistance value – usually in kilohms, for the quiescent point being considered.

If a regular but fluctuating voltage is applied to the anode, as would occur if a small alternating voltage is superimposed on the h.t. voltage then the circuit current would fluctuate between a maximum and minimum value. The valve thus offers a resistance value different from the d.c. value and the static characteristic can be used to determine the *a.c.* or *slope resistance* (symbol r_a).

Consider the curve of Fig. 137. XZ is drawn as a tangent through the point P, the changes in voltage and current values being obtained by drawing the small triangle XYZ. The slope of the graph *i.e.* the hypotenuse would give a conductance being $\dfrac{\text{current}}{\text{voltage}}$. It is therefore

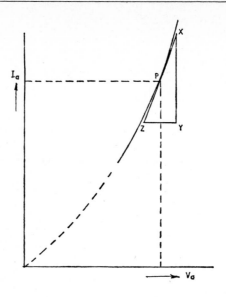

Fig. 137

more practical to use the reciprocal of the slope or $\dfrac{\text{voltage}}{\text{current}}$ i.e. $\dfrac{ZY}{XY}$

Thus $r_a = \dfrac{\text{small change in anode voltage}}{\text{resulting small change in anode current}} = \dfrac{\delta V_a}{\delta I_a}$

The a.c. resistance is less than the d.c. resistance over the working range. The example shows comparative values.

Example 71. The anode current-voltage static characteristic of a diode can be drawn from the following test values. Deduce the values of a.c. and d.c. resistance for anode current values of 5mA and 10mA.

Anode Voltage (V)	0	30·5	46	58	68	77·5
Anode Current (mA)	0	2	4	6	8	10

The V_a/I_a characteristic is plotted and shown by Fig. 138. Consider the 5 mA value. Point A.

D.C. resistance $= \dfrac{52 \cdot 5 \text{ V}}{5 \text{ mA}} = \dfrac{52 \cdot 5}{5 \times 10^{-3}}$ ohms

$$= \dfrac{52 \cdot 5}{5} \times 10^3 = 10 \cdot 5 \text{ k}\Omega$$

A.C. resistance. Assume a small a.c. voltage such as 2·5 V to be superimposed on the d.c. voltage of 52·5 V for a standing current of 5 mA. The anode voltage then varies between 50 V and 55 V and the corresponding currents would be 4·6 and 5·5 mA;

Thus $V_a = \dfrac{55 - 50}{5 \cdot 5 - 4 \cdot 6} = \dfrac{5 \text{ V}}{0 \cdot 9 \text{ mA}} = \dfrac{5}{0 \cdot 9 \times 10^{-3}}$ ohms

$$= \dfrac{50 \times 10^3}{9} = 5 \cdot 55 \text{ k}\Omega$$

Consider the 10 mA value. Point B.

D.C. resistance $= \dfrac{77 \cdot 5 \text{ V}}{10 \text{ mA}} = \dfrac{77 \cdot 5}{10} \times 10^3 = 7 \cdot 75 \text{ k}\Omega$

A.C. resistance. As earlier, assume a superimposed a.c. voltage of 2·5 V. For a standing current of 10 mA, the anode voltage values are $77 \cdot 5 \pm 2 \cdot 5 = 80$ V and 75 V. The corresponding currents would be 10·6 and 9·5 mA.

Thus $V_a = \dfrac{80 - 75}{10 \cdot 6 - 9 \cdot 5} = \dfrac{5 \text{ V}}{1 \cdot 1 \text{ mA}} = \dfrac{5}{1 \cdot 1 \times 10^{-3}}$ ohms

$$= 4 \cdot 55 \text{ k}\Omega \, .$$

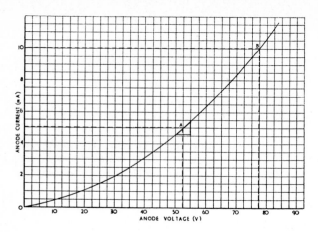

Fig. 138

(b) DYNAMIC CHARACTERISTIC. LOAD LINE. As stated earlier, a valve is required to operate with a load — usually a resistor. Thus even though the supply voltage V is kept constant, when the anode current I_a varies, the voltage drop across the load resistor R will alter and, as a consequence, the h.t. voltage being applied to the anode will vary. This is a dynamic and not a static condition for the valve and must be considered when the operating conditions are being investigated. One method of doing this is by determining the dynamic characteristic by test. The load resistor R is included in the test circuit as shown by the diagram (Fig. 139) and the resulting curve deduced from the results.

An alternative method could also be used to determine the dynamic from the static characteristic. Assume P to be a point on the latter. Then $I_a = $ OB and the valve voltage drop (V_a) is OA. For this current, the load-resistor voltage drop (V_R) is I_aR. The h.t. voltage required for this current would be $V = V_a + V_R$. Thus point P′ is obtained or BP′ = BP + PP′. The procedure is repeated to obtain the complete characteristic which is then used to find the anode current and voltage condition for any supply voltage value. Note that a separate dynamic characteristic must be obtained or deduced for different values of anode load resistor.

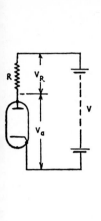

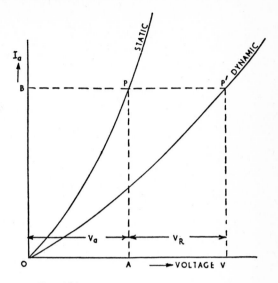

Fig. 139

Load Line. As shown above, the voltage-drop conditions in the anode circuit of a diode with a resistor in circuit can be expressed as $V = V_a + V_R$ or $V_a = V - V_R$. Since the voltage drop across the resistor will be proportional to the anode current, the relationship can be represented by the straight line graph which would result from

the expression $V_a = V - I_a R$ or $I_a = \dfrac{V - V_a}{R} = \dfrac{V}{R} - \dfrac{V_a}{R}$.

If the line is superimposed on the static characteristic (Fig. 140), it is seen to have a "negative slope" *i.e.* it slopes in the opposite direction to the valve characteristic because an increase in I_a means a decrease in anode voltage. Such a line will therefore cut the static characteristic to give a point which shows the respective voltage-drop conditions across the valve and resistor for the indicated current value. The straight-line graph or "load line" will be different for various values of R and can be used to advantage for a specific problem such as that shown below. Thus when the loading condition for one particular value of R at a specified h.t. voltage is required, it is not necessary to construct a dynamic characteristic but merely to draw the load line which is obtained as follows.

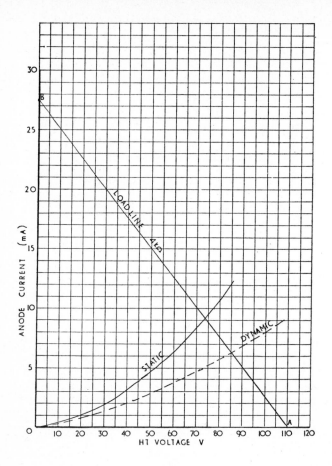

Fig. 140

Referring to Fig. 140, point A is obtained by assuming $I_a = 0$. Under this condition the voltage drop across R would be zero and V_a would be the full h.t. voltage V. Point B is obtained by finding the value of current which would flow if the anode voltage was reduced to zero *i.e.* if the diode developed a short-circuit between anode and cathode. Under this condition anode current would be given by

$$I_a = \frac{V}{R}.$$

The example and Fig. 140 shows the two methods of solution.

Method 1 has involved plotting the static characteristic and the load line to give the required answer.

Method 2 has involved deducing the dynamic characteristic — its values being determined from the table compiled. This method is obviously of more value if the operating conditions for various values of h.t. voltage are to be considered.

Example 72. Using the test values of Example 71, deduce the dynamic characteristic for a load resistor of 4 kΩ. Find the circuit current for an h.t. voltage of 110 V and the voltage drop across the resistor.

Method 1. If the dynamic characteristic had not been required this method would give the required answers. Consider Fig. 140. Plot the static characteristic and obtain the load line thus;

$$\text{Point A} = 110 \text{ V} \quad \text{since } I_a = 0$$

$$\text{Point B} = \frac{110}{4 \times 10^3} = 27 \cdot 5 \times 10^3 = 27 \cdot 5 \text{ mA. Here valve}$$

resistance is assumed to be zero.

The point of intersection gives the answers. Circuit current = 9·1 mA. Voltage drop across resistor = $9 \cdot 1 \times 10^{-3} \times 4 \times 10^3 =$ 36·4 V. Also since $V_R = V - V_a$ then as V_a (read from the graph) = 74 V so $V_R = 110 - 74 = 36$ V.

Method 2. This involves deducing the dynamic characteristic which is shown plotted. One point only is considered. Assume a current of 4 mA. Then $V_a = 46$ V and voltage drop across the resistor = $4 \times 10^{-3} \times 4 \times 10^3 = 16$ V. H.T. voltage = $46 + 16 = 62$ V i.e. a value for the dynamic characteristic. The following table can be deduced.

Anode Current (mA)	0	2	4	6	8	10
Anode Voltage (V)	0	30·5	46	58	68	77·5
Resistor Voltage drop (V)	0	8	16	24	32	40
H.T. Voltage (V)	0	38·5	62	82	100	117·5

For 110 V of h.t., the circuit current is seen to be 9·1 mA. The voltage drop across the resistor = $110 - 74 = 36$ V.

DIODE AS A RECTIFIER . The most common usage of the vacuum diode is as a rectifier valve for a.c. "mains" radio sets. A radio receiving set can be divided into four basic parts: the R.F. (radio-frequency) unit, the Detector unit, the A.F. (audio-frequency) unit and the Power Pack. Up to now the h.t. battery has been considered as the voltage source for the valve anodes but, for "mains" sets, it is possible to obtain the d.c. voltage from a power pack, and for most general purposes a double-diode in one glass envelope is used. The operation of this arrangement is to be considered but, it can be mentioned in passing that, there is a limit for the electrical loading to which the normal vacuum diode rectifier can be operated and that, prior to the more recent developments in the rectifier field, it was possible to extend the working capabilities of the diode valve by the introduction of gas or vapour. The inert gas or mercury vapour, considerably altered the diode characteristic but allowed it to function more efficiently by enabling it to carry a heavier anode loading and reducing the internal voltage drop and thus the valve resistance. Our studies, however, are confined here to the consideration of vacuum valves only, since the use of hot-cathode, gas-filled valves has decreased considerably with the development of semiconductor devices with comparable characteristics.

Forward and Reverse Resistance. These terms are frequently used in conjunction with rectifier theory and are best explained when considered with the graphs of Fig. 141.

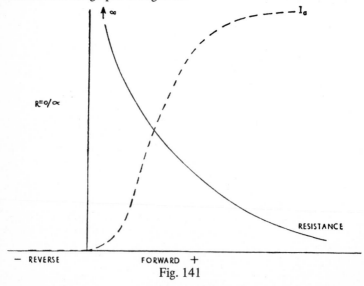

Fig. 141

When the anode is made $+$ ve with respect to the cathode, the diode is biased in the "forward" direction. The resistance value, in this direction, is derived from the static characteristic. It is equal to $\dfrac{V_a}{I_a}$ and is comparatively low when compared to the "reverse resistance" which is (in theory) infinite. When the anode is made $-$ ve with respect to the cathode, the diode is reverse biased and $I_a = 0$. The valve behaves as though it is disconnected and has a high value of Peak Inverse Voltage. The meaning of this term was introduced in Chapter 5.

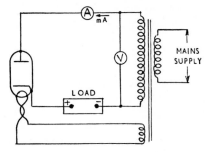

Fig. 142a

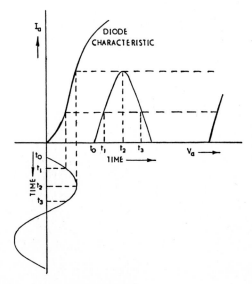

Fig. 142b

Half-wave Rectification. If the diode is connected to the secondary side of a mains transformer, then during the half cycles when the anode is +ve with respect to the cathode, anode current will flow and its magnitude at any instant can be determined from the diode characteristic and the waveform of the a.c. voltage. This is illustrated by the diagrams of Fig. 142. During the half-cycles when the diode is reverse biased, no current flows and, since conduction is in one direction only for alternate half-cycles, then half-wave rectification is obtained. It should be noted that, since a load is included in the circuit then in practice, the dynamic characteristic rather than the static characteristic should be used. Furthermore, if battery-charging is being undertaken then the procedure described at the end of Chapter 5 will be necessary before the actual circuit conditions can be determined.

Full-wave Rectification. This can be obtained by using two diode valves in conjunction with a centre-tapped transformer as illustrated below. The two diodes are usually enclosed in the same glass envelope. When P is + ve to Q, the current flows through diode D_1 to the load. On the next half-cycle R is + ve relative to Q and current flows through D_2 to the load. Thus current flow is through the load for both half-cycles and is in same direction. It continues to be pulsating.

The diagrams of Fig. 143 show the circuit basics and how the rectifier voltage fluctuates from a maximum to a zero value. The effect on the current of these fluctuations can be partially smoothed by connecting capacitor C across the load. Capacitor C is charged to the maximum value of the voltage which appears across the load, and

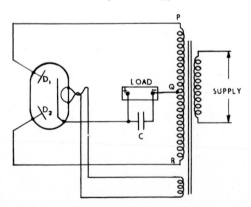

Fig. 143a

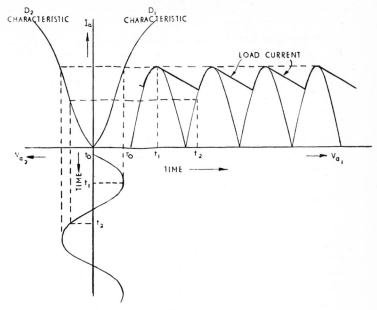

Fig. 143b

when the supply voltage falls below the value of the potential across the plates, the capacitor commences to discharge through the load and so tends to keep the load current and hence the terminal voltage substantially constant. Capacitor C is called a *reservoir* capacitor and its effect can be investigated in greater detail by reference to Fig. 144.

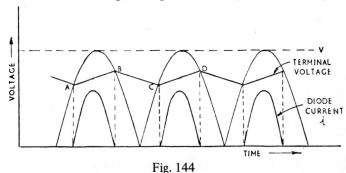

Fig. 144

With no load across the power pack, capacitor C would be charged to the peak value of the supply voltage and its plate potential would remain at this value V. When the circuit is loaded by a resistance R, a current of instantaneous value i amperes would flow from the diode as shown *i.e.* for the short period between points A and B when the capacitor would be charged and the load supplied. When the supply voltage falls below the capacitor p.d. value B, which is determined by the capacitor size and the time of charging, the diode ceases to supply the load and the capacitor starts to discharge through the load, supplying current and allowing the terminal voltage to fall slowly. The discharging continues until time C is reached, when the supply voltage has again risen sufficiently to supply the load and recharge the capacitor. The charging sequence CD is a repeat of AB and it follows that the terminal voltage only fluctuates between the limits represented by the values of A and B. The diode only supplies current for short pulses but the peak current value is now larger than if the capacitor C had not been fitted, because during the charging period both the capacitor and the load have to be supplied.

An approximate value for capacitor C can be found if certain assumptions are made. This is illustrated by the Example which follows. Since the output voltage fluctuates in a regular manner it can be regarded as a steady direct voltage on which is superimposed a small alternating voltage or ripple voltage. The frequency of the ripple would be 100 Hz for a full-wave rectifier and 50 Hz for a half-wave arrangement. The ripple voltage is not sinusoidal but can be assumed to be triangular for practical purposes — this assumes instantaneous charging of the capacitor. This assumption is made for the following example. The diagram of Fig. 145 shows a typical radio receiver power-pack circuit in which is shown the reservoir capacitor C and an additional filter circuit, the purpose of which will be explained after the example.

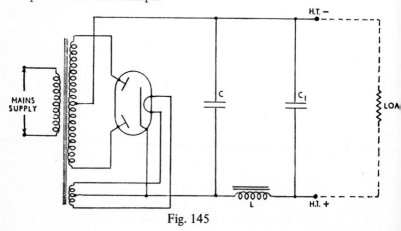

Fig. 145

Example 73. In a 50 Hz mains radio set where the h.t. current is 60 mA, the permissible ripple voltage at the reservoir capacitor is to be 17 V. If full-wave rectification is used, find the value of reservoir capacitor.

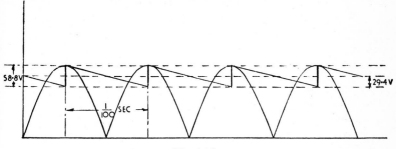

Fig. 146

Note 17 V is assumed to be the r.m.s. value of the ripple voltage which is assumed to be a triangular wave (Fig. 146). The r.m.s. value

of such a wave $= \dfrac{1}{\sqrt{3}}$ Maximum value.

$\therefore$ R.M.S. ripple $= \dfrac{\text{Maximum Value}}{\sqrt{3}}$

or Maximum Value $= \sqrt{3} \times$ r.m.s. value

$$= \sqrt{3} \times 17 = 29\cdot4 \text{ V}$$

The charge Q given out by the capacitor $= I_a t = 60 \times 10^{-3} \times 0\cdot01 = 600 \times 10^{-6}$ coulombs.

Also $Q = CV$ where V is the change in p.d. across the capacitor $= 2 \times 29\cdot4 = 58\cdot8$ V.

Here $C = \dfrac{Q}{V} = \dfrac{600 \times 10^{-6}}{2 \times 29\cdot4} = \dfrac{600}{58\cdot8} \times 10^{-6} = 10\cdot2 \ \mu\text{F}.$

Smoothing. Although Capacitor C does reduce the fluctuation of voltage across the load, additional smoothing is required for a radio set to remove the tendency for "mains hum" from the loudspeaker, which would be detectable when no signal was being received at the aerial. The components L and C_1 constitute a filter arrangement which provides this additional smoothing. The choke/capacitor circuit (L/C_1) is designed so that L has a high impedance and C_1 a low impedance to any 100 Hz a.c. current set up by the ripple voltage. The components L and C_1 are in series across the supply, and allow the small a.c. ripple current to flow and thus to by-pass the load R. Most of the a.c. ripple voltage is dropped across L and very little across C_1. Since R is in parallel with C_1, very little a.c. ripple voltage appears across the load and thus hum due to ripple is eliminated.

Note that for the circuit diagram, a centre-tapped heater winding has been shown. The use of such a centre tap is very common and is favoured because it gives a more balanced arrangement than if the cathode is connected to one side of the heater winding. The centre-tap connection is also found to reduce "mains hum". The load R shown on the diagram is the joint resistance of the various valve h.t. circuits, it being remembered that the other valves in the radio receiver also require a h.t. supply which would have to be provided by a battery, if the power pack was not available.

It should also be noted that the addition of a reservoir capacitor increases the peak inverse voltage as applied to a half-wave rectifier. Without such a capacitor, the maximum value which the diode would have to withstand in reverse would be the peak supply voltage *i.e.* $\sqrt{2}\,V$ where V is the r.m.s. value of the transformer secondary. With a capacitor C charged to the value of $\sqrt{2}\,V$, then at the instant of maximum reverse voltage *i.e.* with no current flow, the voltage which would appear at the inter-electrode space between the anode and cathode would be the sum of the transformer voltage and that to which the capacitor is charged $= \sqrt{2}\,V + \sqrt{2}\,V = 2\sqrt{2}\,V = 2\cdot83\,V$. For a full-wave rectifier, it has already been shown in Chapter 6 that the P.I.V. was of this value *i.e.* $2\cdot83\,V$ and the size of the capacitor C does not alter this value.

Example 74. A full-wave valve rectifier is supplied from a centre-tapped transformer whose secondary is rated at 250—0—250 V. Assuming no smoothing find the average d.c. voltage and the P.I.V. which the valve has to withstand.

The average value of a half-cycle is $\dfrac{2}{\pi}\,V_m$ where V_m is the maximum value of the wave. The r.m.s. value of voltage available from the transformer $= 250$ V.

$\therefore$ The peak supply voltage $= \sqrt{2} \times 250$

$\therefore$ Average value (d.c.) $= \dfrac{2}{\pi} \sqrt{2} \times 250 = \dfrac{500 \times 1\cdot414}{3\cdot14}$

$= 225$ V (d.c.)

The P.I.V. $= 2 \times \sqrt{2} \times 250 = 500 \times 1\cdot414$

$= 707$ V.

Example 75. If the a.c. resistance of each diode section of a thermionic full-wave rectifier is 400 Ω and the circuit provides a d.c. output of 250 V at 50 mA, calculate the turns ratio of the power transformer required. The input a.c. voltage is 240 V at 50 Hz.

The load terminal voltage $= 250$ V $= V_R$ $i.e.$ voltage across load. Now for a diode $V = V_a + V_R$

Also $V_a =$ anode current $\times$ anode resistance $= 50 \times 10^{-3} \times 400$
$= 5 \times 10^{-2} \times 4 \times 10^2 = 20$ V

$\therefore V = 250 + 20 = 270$ V

Also $V = \dfrac{2}{\pi} V_m$ for a half-wave rectifier.

$\therefore V_m = \dfrac{270 \times 3\cdot14}{2} = \dfrac{847\cdot8}{2} = 423\cdot9$ V

and r.m.s. value of $V_m = 0\cdot707 \times 423\cdot9 = 300$ V.

This is the voltage of half the transformer secondary. Full voltage rating of secondary $= 600$ V.

$\therefore$ Turns ratio of primary to secondary $= 240:600 = 1:2\cdot5$.

Example 76. A high-voltage, gas-filled diode has a forward conduction voltage drop of 20 V plus 1 V for every ampere conducted. Plot the diode characteristic and hence estimate the power dissipated in the diode when it is connected in series with a load of resistance 200 Ω to a 5000 V supply of polarity such as to make the diode conduct.

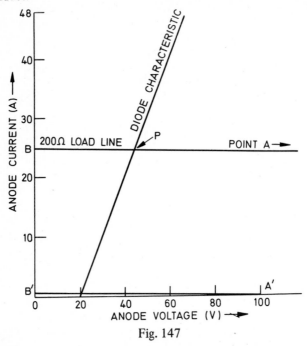

Fig. 147

Although the subject matter covered in this Chapter is concerned with the vacuum diode, this example is included to illustrate the use of a diode characteristic with a load line.

From given data it is evident that the equation $V_a = 20 + (1 \times I_a)$ is satisfied and the following table has been built up for typical anode current values.

Thus if I_a =   5 A V_a = $20 + (1 \times 5)$ = 25 V

similarly if I_a = 10 A V_a = $20 + (1 \times 10)$ = 30 V

"   "   "   = 20 A V_a = $20 + (1 \times 20)$ = 40 V

"   "   "   = 30 A V_a = $20 + (1 \times 30)$ = 50 V

"   "   "   = 40 A V_a = $20 + (1 \times 40)$ = 60 V

This characteristic is plotted as shown in Fig. 147 and the load line for a 200 Ω resistor is drawn thus:

When $I_a = 0$ $V_a =$ the h.t. voltage $= 5000$ V   This gives Point A.

The maximum current condition for the valve is given by

$\dfrac{5000}{200} = 25$ A. This gives Point B.

Note that because of the scales chosen it is difficult to show point A, but the load line can be drawn in, if its slope is determined. Thus by proportionality of similar triangles we have 5000:100 volts as 25:0·5 amperes. Thus points A^1 and B^1 are put in and the line $A^1 B^1$ drawn. Load line AB is drawn through point B by making it parallel to the line $A^1 B^1$. The inter-section point P with the diode characteristic gives a working current of 25 A at an anode voltage of 45 V. Thus the power dissipated at the anode $= 45 \times 25 = 11·25 \times 100 = 1125$ W or 1·125 kW.

THE VACUUM TRIODE

If a third electrode is introduced into the valve, the electron flow can be closely controlled. Such a valve is called a triode and the third electrode is usually in the form of an open spiral of wire wound closely round the cathode. The arrangement is shown in Fig. 148 and this additional electrode is called the *grid*. The circuit symbol for the vacuum triode is also shown. The grid is positioned in the region of the "space charge" and is usually made − ve with respect to the cathode. By altering the grid potential the space charge effect can be modified and the anode current can be controlled.

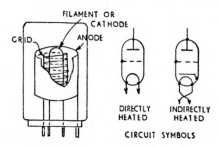

Fig. 148

Consider the diagram of Fig. 149 which shows the potential gradient between the cathode and anode for three different values of grid voltage. For Curve 1, the grid is made only slightly —ve with respect to the cathode. For Curve 2, the —ve grid potential is increased slightly and for Curve 3, the potential has been increased further to make the grid highly —ve with respect to the cathode. By using the term potential gradient, it will be remembered that we can be dealing with a measure of the electric field intensity or the magnitude of the electric force which acts on a charge which is placed in the field between two electrodes.

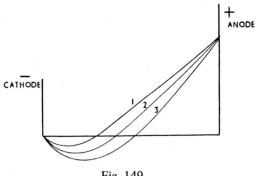

Fig. 149

An electron has a —ve charge and, when emitted from the cathode, it will experience a force which moves it from a point of low potential to one of high potential. The magnitude of the force is directly proportional to the potential gradient and if the latter is +ve, the force is towards the anode and if —ve, the force is towards the cathode. For Curve 1 most of the emitted electrons from the cathode, pass the zero potential gradient value and are then accelerated towards the anode to give a large anode current. For Curve 3 conditions, only a few electrons will have sufficient velocity to pass the point of zero potential, the remainder being urged to return back to the cathode. The few electrons which do pass the point of zero potential gradient will proceed to the anode to give a relatively small anode current. It will now be seen how, by merely altering the potential of the grid, the anode current can be controlled. Furthermore, since the grid is invariably kept at a potential which is —ve to the cathode, electrons do not enter it and therefore no current is drawn from the grid voltage supply source. The power required by the grid to control the anode current is zero and thus, although comparatively large powers can be controlled by this method yet the power required to achieve this control is zero.

STATIC CHARACTERISTICS. The circuit set out in Fig. 150, shows how the triode can be tested and the control of the anode current by, (i) variation of grid voltage and, (ii) variation of anode voltage, can be investigated. The curves when plotted from the test results are divided into two groups of static characteristics known as, (a) the *Mutual Characteristics,* which show the change in anode current I_a for a change in grid voltage V_g, the anode voltage V_a remaining constant and, (b) the *Anode Characteristics* which show the variation of anode current I_a with a change in anode voltage V_a, the grid voltage V_g remaining constant.

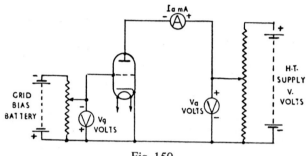

Fig. 150

Mutual or Transfer (I_a/V_g) Characteristics. By keeping the anode voltage constant, the effect on the anode current of a variable grid voltage can be investigated by the test circuit of Fig. 150. If the results of such tests are plotted a family of curves, such as those shown by Fig. 151a are obtained which provide information similar to that available from the anode characteristics although in a more direct form for particular purposes.

Valve manufacturers provide information, on any particular type of valve, in the form of the above curves which also allow three *coefficients* or *parameters* to be determined. These specify the behaviour of a triode when introduced into a circuit and each parameter — sometimes called a "valve constant", represents the relation between the changes in any two of the variables, the third being kept constant.

Anode or Output (I_a/V_a) Characteristics. By keeping the grid at cathode potential *i.e.* $V_g = 0$, a characteristic curve relating anode current to increasing anode voltage can be obtained from the tests. This curve is shown by Fig. 151b. If the grid is next given a small −ve potential and the test is repeated, a curve of similar shape is obtained which is displaced towards the right. Such a curve is marked as

$V_g = -2$ V and the displacement is due to the fact that, as has been explained, the anode must attain a certain minimum +ve potential to overcome the repelling force of the grid on the emitted electrons. The higher the −ve potential on the grid, the more the anode characteristic is moved to the right. A complete family of such curves at various −ve values of grid voltage can be obtained in the manner described.

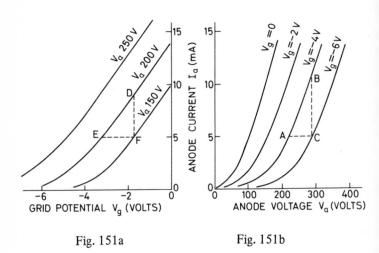

Fig. 151a   Fig. 151b

Valve Parameters. These are derived from the slopes or change conditions of the graph and, as an introduction, are considered as being obtained by constructing small triangles as shown. Since a straight position of a graph is required, the sides of any such triangle are kept as small as possible and are referred to as "a small change of" or the sign δ is used. A mention of this convention has already been made when the diode was being described.

(1) ANODE A.C. RESISTANCE

As for the diode, the slope of an anode characteristic, such as $\dfrac{BC}{AC}$ (Fig. 151b) is seen to give a conductance, being $\dfrac{\text{current}}{\text{voltage}}$. In practice it is more convenient to use the resistance of the valve and the reciprocal of this slope $\dfrac{1}{\dfrac{BC}{AC}} = \dfrac{AC}{BC}$ is known as the *anode slope* or

anode a.c. resistance. Another alternative term is the *internal resistance.* The symbol used is r_a and if a definition is considered we have:

Slope or a.c. resistance (r_a) is the ratio of the small change of anode voltage to the small change of anode current produced, the grid voltage being kept constant.

Thus $r_a = \dfrac{\text{small change of anode voltage}}{\text{resulting small change of anode current}}$

or $r_a = \dfrac{\delta V_a}{\delta I_a}$, V_g being constant.

In the limit, mathematically the above can be written as

$$r_a = \dfrac{dV_a}{dI_a}.$$

The anode a.c. resistance of a triode for any particular operating condition can be found from either the anode or mutual characteristics and is illustrated by the examples which follow.

(2) MUTUAL CONDUCTANCE

The influence of the control-grid voltage on the anode current can be measured most readily by determining the slope of a mutual

characteristic curve. Thus for example, $\dfrac{DF}{EF}$ (Fig. 151a) gives the

second parameter, namely $\dfrac{\delta I_a}{\delta V_g}$. This is termed the *mutual conductance.*

The symbol g_m is used and the definition would be

Mutual conductance is the ratio of the small change of anode current to the small change of grid voltage producing it, the anode voltage being kept constant.

Thus $g_m = \dfrac{\text{small change of anode current}}{\text{small change of grid voltage}}$

or $g_m = \dfrac{\delta I_a}{\delta V_g}$, V_a being constant.

Here again either an anode or mutual characteristic can be used to determine g_m. The examples shows the unit as $\dfrac{\text{Milliamperes}}{\text{Volts}}$ or millisiemens. Since the latter is a little used unit, the usage of milliamperes per volt or mA/V is more general.

(3) AMPLIFICATION FACTOR

This is the third parameter and is used to represent the maximum theoretical *voltage gain* which can be obtained from any particular valve. Note that as it is a ratio of two voltages, a unit is unnecessary. The symbol used is μ and it is defined thus:

Amplification factor is the ratio of the change in anode voltage to the change in grid voltage when both are concerned with the same change of anode current.

$$\text{Thus } \mu = \frac{\text{change in anode voltage}}{\text{change in grid voltage}}$$

$$\text{or } \mu = \frac{\delta V_a}{\delta V_g}, \; I_a \text{ being constant.}$$

As for the two previous parameters, amplification factor can be determined from either the anode or mutual characteristics. A relationship connecting r_a, g_m and μ will be deduced shortly and thus the third constant can always be determined if the other two are known. The examples extend practice in the use of the parameters and characteristics and are instructive.

Parameter Relationships. Mention has been made of the relation between the three valve parameters and this is now considered.

$$\text{Since } \mu = \frac{\delta V_a}{\delta V_g} \text{ equality still exists if written as } \mu = \frac{\delta V_a}{\delta V_g} \times \frac{\delta I_a}{\delta I_a}$$

$$\text{Then } \mu = \frac{\delta V_a}{\delta I_a} \times \frac{\delta I_a}{\delta V_g} \text{ giving } \mu = r_a \times g_m \text{ or}$$

Amplification factor = Anode a.c. resistance × Mutual conductance.

Example 77. In a certain triode, the anode current is 5 mA with an anode potential of 220 V and a grid potential of -3 V. When the anode potential is increased to 260 V, the current rises to 7 mA and a change of grid potential to -4 V restores the current to its original value. Determine the valve constants.

$$\text{Here } r_a \;=\; \frac{\delta V_a}{\delta I_a} \;=\; \frac{(260-220)\,V}{(7-5)\times 10^{-3}\,A} \;=\; \frac{40\times 10^3}{2}$$

$$=\; 20\,000 \text{ ohms or } 20\text{ k}\Omega$$

$$g_m \;=\; \frac{\delta I_a}{\delta V_g} \;=\; \frac{(7-5)\text{ milliamperes}}{-3-(-4)\text{ volts}} \;=\; \frac{2}{1} \;=\; 2 \text{ mA/V}$$

$$\text{or 2 millisiemens}$$

$$\mu \;=\; \frac{\delta V_a}{\delta V_g} \;=\; \frac{(260-220)\text{ volts}}{-3-(-4)\text{ volts}} \;=\; \frac{40}{1} \;=\; 40$$

or using the relationship

$$\mu \;=\; r_a \times g_m$$

$$=\; (20\times 10^3) \times (2\times 10^{-3}) \;=\; 40.$$

Example 78. The static characteristic test values for a triode are as follows.

V_g = 0 V	Then for V_a	50	100	150	200	(volts)
	I_a	3	6·5	10·1	13·7	(milliamperes)
V_g = -3 V	Then for V_a	150	200	250	300	(volts)
	I_a	1·4	5	8·5	12·1	(milliamperes)

Plot the anode characteristic and determine r_a, g_m, and μ.

The small lengths of anode characteristics are shown by Fig. 152. Where data is given in this form, the most convenient method of solution is to erect a triangle which makes contact with an adjacent curve. Solution is then as follows.

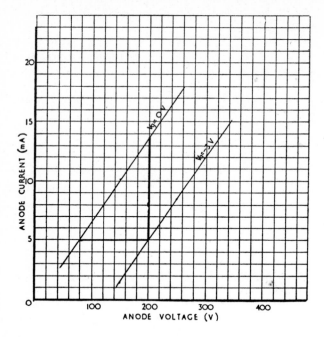

Fig. 152

$$r_a = \frac{\delta V_a}{\delta I_a} = \frac{200 - 80}{(13 \cdot 7 - 5)10^{-3}} = \frac{120 \times 10^3 \text{ volts}}{8 \cdot 7 \text{ amperes}} = 13\ 793\ \Omega$$

$$= 13 \cdot 793\ k\Omega$$

$$g_m = \frac{\delta I_a}{\delta V_g} = \frac{(13 \cdot 7 - 5)10^{-3}}{0 - (-3)} = \frac{8 \cdot 7 \times 10^{-3} \text{ amperes}}{3 \text{ volts}}$$

$$= 2 \cdot 9 \text{ mA/V or } 2 \cdot 9 \text{ millisiemens}$$

$$\mu = \frac{\delta V_a}{\delta V_g} = \frac{200 - 80}{0 - (-3)} = \frac{120}{3} = 40$$

Check $\mu = r_a g_m = 13 \cdot 793 \times 10^3 \times 2 \cdot 9 \times 10^{-3} = 39 \cdot 999 = 40$

Example 79. The following test results were obtained on a triode:

Anode Voltage 100 V

Grid Voltage (V)	−2·0	−1·5	−1·0	−0·5	0	0·5
Anode Current (mA)	1·5	2·7	4·0	5·6	7·4	9·5

Anode Voltage 150 V

Grid Voltage (V)	−2·0	−1·5	−1·0	−0·5	0
Anode Current (mA)	4·0	5·7	7·6	9·6	11·8

Plot the curves and deduce, (a) the anode slope resistance, (b) the amplification factor, (c) the mutual conductance for the zero grid voltage condition.

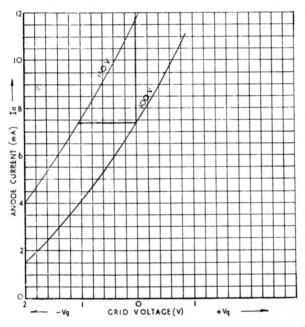

Fig. 153

The data given is shown by Fig. 153 and is seen to be points on the mutual characteristics. A triangle constructed as shown gives the necessary answers

Thus $g_m = \dfrac{\delta I_a}{\delta V_g}$ or slope $= \dfrac{(11·8 - 7·4)10^{-3}}{0 - (-1·05)} =$

$$= \frac{4 \cdot 4 \times 10^{-3} \text{ amperes}}{1 \cdot 05 \text{ volts}} = 4 \cdot 2 \text{ mA/V}$$

$$\mu = \frac{\delta V_a}{\delta V_g} = \frac{(150 - 100)}{0 - (-1 \cdot 05)} = \frac{50 \text{ volts}}{1 \cdot 05 \text{ volts}} = 47 \cdot 6$$

The values used for this ratio are the changes in grid and anode voltages required to maintain a constant anode current of 7·4 mA.

$$r_a = \frac{\delta V_a}{\delta I_a} = \frac{150 - 100}{(11 \cdot 8 - 7 \cdot 4) 10^3} = \frac{50 \times 10^3 \text{volts}}{4 \cdot 4 \text{ amperes}} = 11 \cdot 36 \text{ k}\Omega$$

Check μ = $r_a \times g_m$ = $11 \cdot 36 \times 10^3 \times 4 \cdot 2 \times 10^{-3}$ = 47·7.

The use of valve characteristics is also illustrated by the next example but it should be noted that the constants can be obtained from either the mutual or static characteristics and that all three parameters vary with the operating conditions. For this reason, when quoting those values in published valve data, the measuring conditions are specified.

Example 80. A thermionic triode valve has the characteristics shown. Use these to show (a) the variation of anode current against grid voltage, the anode voltage being constant at 200 V, (b) the variation of anode voltage against grid voltage, the anode current being constant at 2·5 mA. Use the curves to determine, (c) the anode slope resistance when the anode voltage is 200 V and the anode current is 2·5 mA, (d) the amplification factor.

The associated graphs are shown by Fig. 154. The anode characteristics as given have been set out on the right and (a) a mutual characteristic has been derived by the construction shown which is self-explanatory.

For (b) a horizontal line against 2·5 mA has been drawn and the points where it cuts the curves noted thus:

For V_g = 0 − 1   − 2   − 3   − 4   − 5 (V)

then V_a = 40   85   150   220   265   315 (V)

These values are plotted to give the desired graph for 2·5 mA.
(c) The easiest way of determining the anode slope resistance for the conditions specified is to determine point P and since the graphs are parallel at this point, draw triangle ABC with AB as a parallel through P to the appropriate curve.

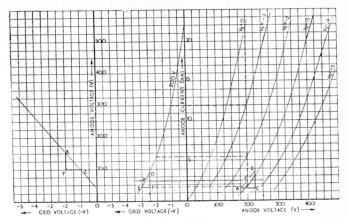

Fig. 154

Then $r_a = \dfrac{\delta V_a}{\delta I_a} = \dfrac{AC}{BC} = \dfrac{(225 - 180) \text{ volts}}{(3 \cdot 7 - 1 \cdot 4) \times 10^{-3} \text{ amperes}}$

$= \dfrac{45 \times 10^3}{2 \cdot 3} = 19\,600 \; \Omega \quad \text{or} \quad 19 \cdot 6 \text{ k}\Omega$

(d) The amplification factor can be obtained from the slope of the special graph deduced *i.e.*

$\mu = \dfrac{\delta V_a}{\delta V_g} = \dfrac{XY}{YZ} = \dfrac{155 - 100}{-1 \cdot 2 - (-2)} = \dfrac{55}{0 \cdot 8} = 69$

Or alternatively find the value of g_m from the deduced mutual characteristic at point P

Thus $g_m = \dfrac{\delta I_a}{\delta V_g} = \dfrac{DE}{EF} = \dfrac{(3 \cdot 4 - 1 \cdot 7) \text{mA}}{-2 \cdot 5 - (-3) \text{V}} = \dfrac{1 \cdot 7}{0 \cdot 5}$

$= 3 \cdot 4 \text{ mA/V}$

So $\mu = r_a \times g_m = 19 \cdot 6 \times 10^3 \times 3 \cdot 4 \times 10^{-3} = 67.$

FUNCTIONS OF THE TRIODE. This valve can be operated in one of two modes (a) as an on-off switch (b) as a voltage controller. Since the vacuum triode is only capable of handling a comparatively small amount of power, its most common application is in the form of an a.c. voltage amplifier which is a variation of mode (b) where it functions as a voltage controller. The other aspects of operation will therefore be considered only briefly and the rest of the chapter will be devoted to study of the voltage amplifier.

(a) On-off Current Control. The signal potential source in the diagram of Fig. 155 is usually a sensing device which would be small and accurate but of small power output. Such a sensor could be a photo-electric cell or thermocouple.

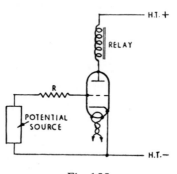

Fig. 155

The usual arrangement is that when the input to the grid is +ve, conduction in the valve occurs and anode current flows to actuate a small-current device such as a relay. When the input to the grid is —ve, the valve is non-conducting and the relay is inoperative or the triode is in the "cut-off" condition. The h.t. voltage and value of R are chosen that although there is grid current when the triode is conducting, ie +ve input signal, nevertheless the voltage drop in R is such that grid voltage is substantially zero. A typical marine application for such an arrangement is one type of "flame-failure" protection equipment, as used in conjunction with an oil-fired boiler. Failure of the flame reduces the output of the photo-electric cell. This in turn reduces the input to a triode valve to the extent of driving it into the cut-off condition. Anode current then ceases and the relay opens to de-energise a spring-loaded, oil-control solenoid valve which stops the flow of fuel to the burners.

(b) Voltage Control — d.c. amplifier. The test circuit of Fig. 150 showed how the anode current of a triode can be controlled by varying the grid voltage. If a load resistor R is connected in series with the valve then the voltage drop across this resistor will be varied. The valve thus operates as a simple controller or even as a voltage amplifier as may be required for computer or cathode-ray oscilloscope work.

Although a d.c. amplifier invariably consists of much more elaborate arrangements to ensure operating stability we can infer, without diverting from basics, that if the valve conditions are so chosen that a linear portion of a mutual characteristic is operated upon, then variation of V_g results in a direct variation of I_a, which in turn gives a proportional variation of voltage drop across the load resistor. If the latter is a 5 kΩ resistor in the 150 V h.t. line and the anode current is varied as for Example 79 through 4·4 mA by a grid voltage variation of 1·05 V, then the voltage drop across the load resistor would be

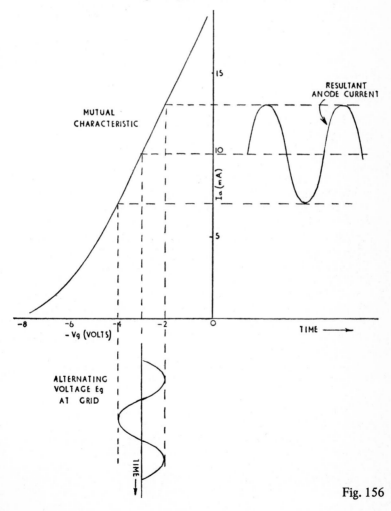

Fig. 156

varied by $5 \times 10^3 \times 4\cdot4 \times 10^{-3} = 22$ V. Thus a 22 V change of output voltage can be effected by a grid voltage change of $1\cdot05$ V and it is not difficult to imagine this input variation being obtained, for example, from a Wheatstone bridge network of a resistance-thermometer arrangement and the valve-load output voltage variation being used to operate a large indicating or recording instrument.

Voltage Control — a.c. amplifier. For this important application the grid is given a —ve direct voltage, *i.e.* it is *biased* negatively and a small a.c. signal voltage is injected into the grid circuit so that the alternating voltage adds to or subtracts from the d.c. bias voltage and operates the valve over a linear part of the characteristic. As explained for the d.c. amplifier, if a load resistor is included in the anode circuit, a voltage drop is developed across it and controlled in a manner which reflects the input signal. Since the output voltage variation is greatly in excess of the input signal voltage, voltage amplification is obtained. This principle is used extensively for amplifiers in electronic equipment such as sound amplifiers and radio sets. It is sufficiently important to warrant more detailed study.

THE TRIODE AS A VOLTAGE AMPLIFIER . The effect on anode current of a grid-voltage variation has been introduced, and if now an alternating voltage is applied to the grid, the effect on I_a is as illustrated by Fig. 156. If the alternating voltage E_g, of peak value 1 V, is applied at the — 3 V point on the V_g axis then the grid is caused to swing between — 2 V and — 4 V. The resulting anode current now consists of an alternating current superimposed on the original direct "standing current" of 10 mA, or an a.c. of 3 mA peak value is being produced.

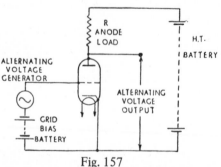

Fig. 157

The valve can now be considered to be a generator of a.c., in which case it needs a load. The simplest load would be a resistor connected in the anode circuit, as is shown by the diagram of Fig. 157. The anode current, flowing through the anode load resistor R, results in a voltage drop across it which varies in magnitude with the current. Provided a suitable value of R is used, the alternating voltage output *i.e.* between anode and cathode will be many times greater than the alternating input voltage, as applied between grid and cathode. The input voltage E_g has thus been amplified and the circuit of Fig. 157 is that of a voltage amplifier. Note that, with respect to the a.c. circuit, the h.t. battery resistance is assumed to be negligible and the existence of the d.c. voltage source can be neglected.

Note also that, the conditions of the circuit of Fig. 150 are not exactly the same as those represented by the diagram of Fig. 157. For the latter arrangement the voltage at the valve anode is not constant because of the varying voltage drop in the load resistor R. The Static Characteristic *i.e.* with fixed anode voltage, if used for circuit investigation would give incorrect results and therefore, as for the diode, dynamic conditions should be considered if the valve is to be used with varying anode voltages. The procedure for deducing the appropriate Dynamic Characteristic will be considered after a practical arrangement for introducing an alternating or varying input signal has been discussed. It should also be noted that up to now the grid bias voltage has been assumed to be provided by a battery but in practice circuits are employed which allow such a battery to be discarded. The method of obtaining this grid bias automatically will be considered in due course.

PRACTICAL ARRANGEMENTS.   Up to now the alternating voltage being applied to the grid of the triode has been assumed to be provided by an alternator but in practice a number of such alternating voltages may be involved. These voltages may be of different amplitudes and frequencies resulting in a complex voltage wave being applied to the grid. As an example the voltage wave may be received from a carbon microphone connected as in Fig. 158a and the voltage at the secondary of the transformer will correspond to the sound waves which reach the microphone. The transformer secondary can replace the original a.c. generator in the grid circuit and Fig. 158b will result in amplified electrical signals appearing at the triode which correspond to the original sound signals. The microphone transformer has two functions: (a) it has a high step-up ratio (1:100) and thus converts and amplifies the minute current variations of the primary circuit into larger voltage variations to be applied to the valve grid circuit, (b) a practical microphone is a low impedance instrument, whereas the valve grid input circuit presents a high impedance. The step-up transformer helps to "match" the two impedances to each other.

Such "matching" is essential for effective operation and, although space does not allow further consideration of this aspect of theory, it is stressed that this is a fundamental requirement for correct circuit conditions. As stated the carbon microphone is a low impedance instrument and a step-up matching transformer is required to feed its output into the high impedance grid/cathode arrangement of a thermionic valve. A moving-coil microphone also needs such a step-up transformer but a crystal microphone gives a relatively high voltage output, is a high impedance instrument and is connected directly between grid and cathode of the valve. Converse conditions apply for a transistor amplifier. A transistor is a current-operated device with a low input impedance. It can be driven directly by a carbon or moving-coil microphone whereas, a crystal microphone will need a step-down transformer before matching is effected.

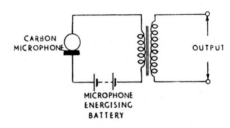

Fig. 158a

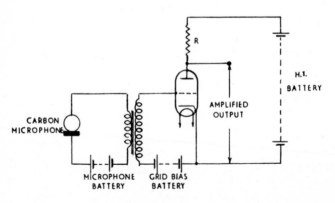

Fig. 158b

CHAPTER 8

PRACTICE EXAMPLES

1.   Tests show that to increase the anode current of a diode valve from 12 mA to 22 mA, the anode voltage must be increased from 75 V to 129 V. Assuming a linear characteristic, find the anode a.c. resistance of the valve.

2.   From the data given for Q1, find the d.c. resistance of the diode for the two specified current conditions.

3.   The I_a/V_a curve of a diode valve is given by

V_a (volts)	5	10	15	20	25	30
I_a (milliamperes)	0·6	2·0	4·2	7·25	10·5	11·8

Plot the curve and find the a.c. resistance over the straight line region.

4.   A diode has the following forward characteristic.

Anode-cathode Voltage (V)	0	10	20	30	40	50
Anode Current (mA)	0	25	60	100	145	200

This diode is placed in series with a 300 Ω resistor load and the combination is fed from a 60 V d.c. supply, connected to make the diode conduct. Determine the current that flows and the power dissipated in the load resistor.

5.   The current in a certain thermionic vacuum diode is related to its anode-cathode voltage by the expression $I_a = V_a^{3/2}$ milliamperes where V_a is in volts. The diode is connected via a 1 kΩ resistor to a d.c. supply. Plot the dynamic characteristic showing the variation of anode current against supply voltage for the range 0–10 V. What is the value of anode current when the p.d. across the circuit is 8 V.

6.   For a triode valve, if the anode voltage is kept constant at 250 V and the grid voltage is varied from −3 V to −1·5 V, the anode current is found to rise from 12 mA to 17·7 mA. Find the mutual conductance of the valve.

7.   A triode valve is supplied from variable anode and grid voltage supplies which together with the anode current are metered to give the following set of readings.

V_g (volts)	V_a (volts)	I_a (milliamperes)
− 1	200	8
− 2	200	3
− 1	140	3

Deduce the three valve constants for this triode, using both the anode and mutual characteristics.

8.   A triode valve has an anode current of 4 mA at a grid voltage of − 2 V and an anode voltage of 100 V. The anode a.c. resistance is measured to be 20 kΩ. Find the amplification factor of the valve and also the value of grid bias which will just cut off the anode current. Assume that the characteristics are linear.

9.   The following test values were obtained for a triode valve.

For V_g = − 0 V
Then when V_a (volts) = 50   100   150   200
I_a (milliamperes) = 3   7   11·5   17

For V_g = − 1 V
Then when V_a (volts) = 50   100   150   200   250   300
I_a (milliamperes) = 1   3   7   11   16   21

For V_g = − 2 V
Then when V_a (volts) = 100   150   200   250   300   350
I_a (milliamperes) = 1   2·5   5·5   9·5   13·5   18

Plot the anode characteristics and determine the valve parameters.

10.   Plot the mutual characteristics of a triode valve from the table of values given below and hence evaluate the valve constants.

V_g (volts)	−7	−6	−5	−4	−3	−2	−1	0
I_a (milliamperes) for V_a = 200 V	0·5	1·2	2·7	5·3	8·5	12	15·5	19
I_a (milliamperes) for V_a = 150 V	−	0·2	0·6	1·7	3·5	6·75	10·0	13·5

CHAPTER 9

THE ALTERNATOR (ii)

Before continuing the study of the alternator it is necessary, at this stage, to consider the phenomenon of the rotating magnetic field which is produced when polyphase currents flow in a related polyphase winding. For the induction motor and synchronous motor, the rotating magnetic field is due to currents which result from the application of externally applied polyphase e.m.fs. For the alternator, this field is due to the currents through the machine itself—such currents are the result of the load circuit being completed and energised by the generated voltage of the alternator. The passage of such currents through the machine windings, can be expected to produce an armature reaction effect similar to that which has been described for the d.c. machines. In this instance, since the polyphase machine is being described, it follows that a fresh examination should be made of the conditions which exist and the overall effect.

THE ROTATING MAGNETIC FIELD

The phenomenon to be described can best be introduced by the simple diagram of Fig. 159. This shows three coils displaced from each other by 120° in space and energised by three-phase currents, which themselves are displaced from each other by 120° in time. If a compass needle is mounted at the centre of the coil system, it will be seen to rotate, and if two connections to the supply are interchanged the direction of rotation of the needle will be reversed.

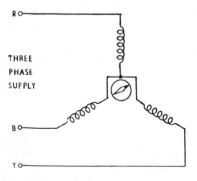

Fig. 159

The rotation of the needle shows that a rotating magnetic field is the result of the combined effect of the three separate single-phase alternating magnetic fields of the coils. In practice the three single-phase windings are laid out along the circumference of the stator and are arranged in a manner as already described for the alternator. By choosing a suitable coil pitch, the machine can be wound for any number of poles, provided the number is even. The speed of rotation of the magnetic field is dependent on the number of poles and the supply frequency. Explanations of the rotating magnetic field effect are now given by different methods of treatment. One of these should suit the reader, but comprehension of all three will allow a better understanding of armature reaction effects when these are introduced.

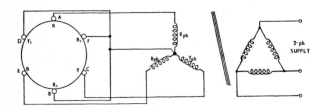

Fig. 160

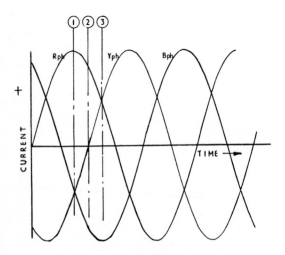

Fig. 161

Method 1. By Consideration of the Resultant Magnetic Field Distribution.

A basic approach to the understanding of the combined magnetic field produced by a three-phase winding, can be made by considering a simple two-pole stator with one conductor per pole per phase. In the diagram of Fig. 160, conductors A and B constitute the "go" and "return" conductors of a single turn and are energised from one phase of a three-phase transformer. Similarly conductors C and D are energised from the yellow phase and EF from the blue phase.

The currents flowing in the stator windings are illustrated by the waveform diagram of Fig. 161, and our investigation can proceed by considering various conditions at different instants of time.

For a suitable time of commencement, consider Instant 1, when the R_{ph} current is a maximum. Then the conditions as shown by the diagram of Fig. 162a exist and the resultant magnetic field will be as

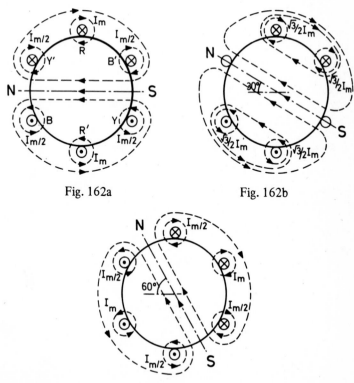

Fig. 162a   Fig. 162b

Fig. 162c

shown. Note that the currents in the Yellow and Blue phases are −ve with respect to the Red phase current and are represented by dots in the respective "go" conductors to illustrate their direction. They will be half their maximum value. The stator will function as a solenoid with its axis along the horizontal and the resultant magnetic field will be as depicted.

For Instant 2, the time is 30° later. The current in the Red phase will still be +ve but will have fallen to $\dfrac{\sqrt{3}}{2}$ of its maximum value.

The current in the Yellow phase will have fallen to a zero value, whilst that in the Blue phase will have risen to $\dfrac{\sqrt{3}}{2}$ maximum value but will still be negative. The resultant magnetic field will be as shown by Fig. 162b and its axis will have rotated through 30°. For Instant 3, the Blue phase current will be at maximum value but still negative. For the Red and Yellow phases the current will be +ve but half of the maximum value. The resultant magnetic field will be as shown (Fig. 162c) and its axis is seen to have rotated through a further 30°.

On examining the results of steps 1, 2 and 3 we can conclude that the three-phase currents flowing through coils, suitably displaced from each other by 120 electrical degrees, will produce a resultant magnetic field which rotates in time and in phase with the currents *i.e.* it rotates at a speed decided by the frequency. This is its synchronous speed. These conclusions are borne out further if alternative methods of treatment, as set out below, are considered.

Method 2. By Mathematical Derivation of the Rotating Field Conditions.

For the mathematically minded student, the following method of considering the phenomenon of the rotating magnetic field will be appreciated.

The diagram of Fig. 163 is similar to that of Fig. 162a, except that three-turn phase windings are shown. For simplicity a two-pole arrangement is treated, the three individual windings producing fluxes Φ_R, Φ_Y and Φ_B each varying in magnitude according to a sine law, but not rotating. Each is a purely alternating flux.

We can then write:
$$\Phi_R = \Phi_m \sin \omega t$$
$$\Phi_Y = \Phi_m \sin (\omega t - 120°)$$
$$\Phi_B = \Phi_m \sin (\omega t - 240°)$$

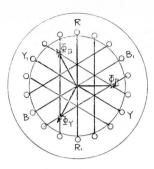

Fig. 163

The resultant can now be found by resolving the appropriate flux vectors into horizontal and vertical components.

$$\text{Thus } \Phi_H = \Phi_R - \Phi_Y \cos 60 - \Phi_B \cos 60$$
$$= \Phi_m \sin \omega t - \Phi_m \sin (\omega t - 120) \cos 60$$
$$- \Phi_m \sin (\omega t - 240) \cos 60$$
$$= \Phi_m [\sin \omega t - \cos 60 \{\sin(\omega t - 120)$$
$$+ \sin(\omega t - 240)\}]$$
$$*= \Phi_m \sin \omega t - \cos 60 \sin(\omega t - 180)$$

Note. The trigonometrical formula as set out below

$$\sin S + \sin T = 2 \sin \frac{S + T}{2} \cos \frac{S - T}{2}$$

continuing $\Phi_H = \Phi_m \{\sin \omega t + \cos 60 \sin \omega t\}$

$$= \Phi_m \left\{\sin \omega t + \frac{\sin \omega t}{2}\right\} = \frac{3}{2} \Phi_m \sin \omega t$$

$$\text{Also } \Phi_V = \Phi_B \cos 30 - \Phi_Y \cos 30$$
$$= \Phi_m \sin(\omega t - 240) \cos 30 - \Phi_m \sin(\omega t - 120) \cos 30$$
$$= \Phi_m \frac{\sqrt{3}}{2} \left\{\sin(\omega t - 240) - \sin(\omega t - 120)\right\}$$
$$*= \frac{\sqrt{3}}{2} \Phi_m \left\{2 \cos(\omega t - 180) - \sin 60\right\}$$

Note. The trigonometrical formula is set out below

$$\sin S - \sin T = 2 \cos \frac{S+T}{2} \sin \frac{S-T}{2}$$

or $\quad \Phi_V = \dfrac{\sqrt{3}}{2}\Phi_m\left\{2\left(-\cos\omega t\right)\left(-\dfrac{\sqrt{3}}{2}\right)\right\} = \dfrac{3}{2}\Phi_m\cos\omega t$

Resultant Flux $\quad \Phi = \sqrt{\Phi_H{}^2 + \Phi_V{}^2} = \dfrac{3}{2}\Phi_m\sqrt{\sin^2\omega t + \cos^2\omega t}$

$$= \frac{3}{2}\Phi_m \ .$$

The resultant flux is seen to have a magnitude one and a half times the maximum value produced by any one phase winding. If θ is the angle the resultant field makes with the horizontal, then

$\tan\theta = \dfrac{\frac{3}{2}I_m\sin\omega t}{\frac{3}{2}I_m\cos\omega t} = \tan\omega t$. Thus $\theta = \omega t$ which means that

the resultant field is rotating at a constant angular velocity $i.e.$ at synchronous speed, since $\theta = \omega t = 2\pi f t$.

Method 3. By Deduction of the Magnetomotive Force or Ampereturn Field-producing Patterns.

Consider the diagram of Fig. 164a to show one pole of a threeturn phase winding. The procedure for drawing out the m.m.f. distribution patterns associated with a winding has already been

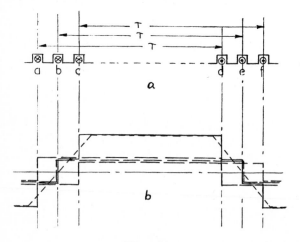

Fig. 164

considered earlier. For brief revision here, consider the "go" and "return" conductors of a single-turn phase winding. The m.m.f. available to set up flux is constant in space between the current-carrying conductors **a** and **d**, **b** and **e**, **c** and **f**. Each time one such conductor is passed over, the m.m.f. reverses and, if the distribution for the set of phase conductors is drawn out as in the second diagram (Fig. 164b) we get a series of rectangular waveforms, which when added give the resultant stepped distribution waveform. In actual fact, the m.m.f. increases gradually across each slot, so the distribution pattern is more like that shown dotted.

The above procedure is now used to find the m.m.f. distribution over one pole pitch due to three-phase currents in their respective windings. Three instants in time are considered and the associated phasor time diagram has been drawn to illustrate the current magnitudes and relationships occurring for each particular consideration. The relevant diagrams are Figs. 165 a, b and c.

(a) Instant when the current in the R_{ph} is a maximum.

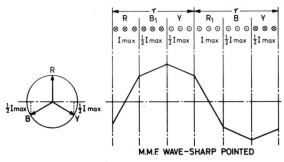

Fig. 165a

(b) Instant when the current in the Y_{ph} is zero.

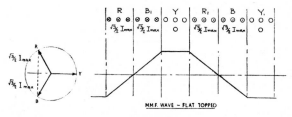

Fig. 165b

(c) Instant when the currents in the R_{ph} and Y_{ph} are +ve and half the maximum value.

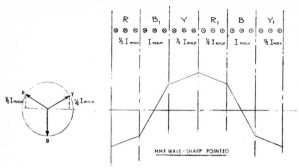

Fig. 165c

Examination of the three m.m.f. distribution waveforms shows the following notable features. 1. The pole pitch is constant. 2. Movement in time − say 30° of the phasors, corresponds to a 30° space movement of the wave. Thus this m.m.f. wave moves at synchronous speed and in the same direction as the phase rotation. 3. The m.m.f. distribution wave is in time phase with the current for any particular phase winding. If the current is of a definite value at any instant in such a winding, the position and form of the m.m.f. wave at that instant can be found.

The two waveforms are extreme shapes (Fig. 166). Both can be approximated to by a fundamental constant height sine wave accompanied by 3rd and 5th harmonics. The latter are not of great importance and, if sufficient care is taken in design to minimise all harmonics, it is sufficient to assume that the m.m.f. wave is sinusoidal, having a maximum value which is a mean between the maximum values of the two limiting cases. It should be noted that as the reluctance of the magnetic path is fairly uniform, especially for the cylindrical rotor alternator and the induction motor, the flux and m.m.f. distribution waveforms are similar.

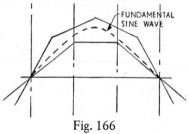

Fig. 166

The following point has already been made above and is stressed. The rotating m.m.f. or flux wave is always in a position where it can combine with maximum linkages. This is illustrated by the diagrams of Fig. 167.

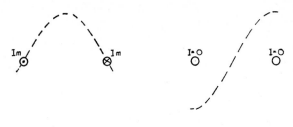

Fig. 167a   Fig. 167b

Thus for (a) if for example, the current in the red phase is a maximum, the m.m.f. and flux wave is symmetrically distributed across that phase winding and maximum flux-linkages occur. Similarly for (b) when the current is zero the m.m.f. or flux distribution is such as to have its zero value at the centre of that winding — zero flux-linkages occur *i.e.* the peak value is displaced by $90°$. Reference will be made to this fact when armature reaction effects are considered.

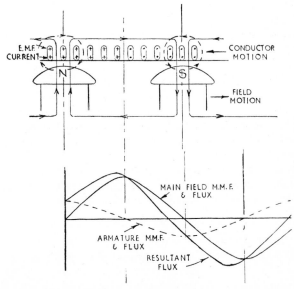

Fig. 168

ARMATURE REACTION

The work done above, in connection with the rotating magnetic field, shows that the armature currents produce a field of constant strength which rotates at synchronous speed. In the diagrams of Figs 168, 169 and 170, the e.m.f.'s are generated in the stator conductors by the movement of the main field poles. The conductors of only one phase winding are shown, those above the N pole constituting a three-turn coil with those above the S pole. The path of the main flux is shown by full lines and that of the armature current produced flux by dotted lines. When such current flows, an additional magnetic effect is produced which modifies the main field, this effect being known as armature reaction. The exact effect of such armature reaction depends upon the phase of the current relative to the e.m.f.'s generated by the main field.

CASE 1. CURRENT IN PHASE WITH GENERATED E.M.F. When the phase winding coil is in the position shown by Fig. 168, the flux linking with it is changing at the greatest rate and the maximum e.m.f. is being generated.

The armature currents are seen to produce flux which passes across the pole shoes without directly effecting the total pole flux. The effect is purely cross-magnetising, the leading pole-tip being weakened and the trailing pole-tip strengthened. The effective pole centre is moved backwards. Since the armature m.m.f. is cross-magnetising, the field flux is reduced only slightly due to saturation effects of the field system. Thus the generated e.m.f. is reduced by only a small amount but the waveform may be distorted. Note that in the diagram, use has been made of both e.m.f. and current conductor "bands". Since E and I are in phase, these bands are in line and similar.

CASE 2. CURRENT LAGGING GENERATED E.M.F. BY 90°−ZERO P.F. CONDITION. (Fig. 169.) Here the current bands lag a quarter cycle behind the e.m.f. bands and produce fluxes which have a direct demagnetising effect on the main poles.

The armature m.m.f. is entirely de-magnetising and the flux produced is large since the coil centre is opposite the poles. The resultant flux in the air-gap is reduced considerably and the generated e.m.f. is accordingly much smaller.

CASE 3. CURRENT LEADS GENERATED E.M.F. BY 90°−ZERO P.F. CONDITION. (Fig. 170.) Here the diagram will be similar to the above except that the current in the armature conductors will be in the opposite direction and the armature flux will be reversed. Thus the current bands lead the e.m.f. by a quarter cycle and produce fluxes

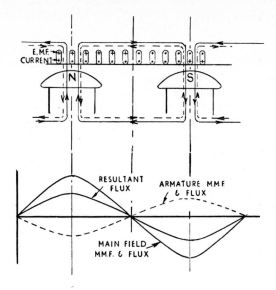

Fig. 169

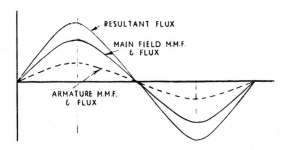

Fig. 170

which have a direct magnetising effect on the main poles. The armature m.m.f. is entirely magnetising but the flux increase is not as great as the decrease in Case 2, due to the saturation effect of the iron circuit. The resulting increase in flux gives a rise in the generated voltage.

THE PHASOR DIAGRAM (Continued).

An introduction has already been made to the simple phasor diagram of the alternator (Fig. 133) and the causes of internal voltage drop have been mentioned. These will now be considered in greater detail and the full diagram evolved.

ARMATURE RESISTANCE (R). The "effective resistance" includes the actual ohmic value and an extra component, resulting from the uneven current distribution in the conductors as caused by the irregular and alternating flux across the slots. This extra resistance is usually taken as ten per cent of the ohmic value and is added to the true resistance to give the effective resistance. It is stressed that even after making this allowance, the resistance voltage drop is small and is frequently neglected in calculations.

ARMATURE REACTANCE (X_S). The reactance, as shown on the simple phasor diagram of Fig. 133 and repeated below in Fig. 172, is termed the synchronous reactance which, from investigation of voltage regulation, is found to be greater than that due to the inherent inductance of the machine. This reactance can be considered to consist of two components (1) true leakage reactance and (2) a reactance effect credited to armature reaction.

(1) Leakage Reactance (X_L). The armature currents give rise to a magnetising field which varies with the frequency and cuts the conductors to generate a small back e.m.f. This is the normal conception of an inductance effect and the machine is known to have this leakage inductance which results in a reactance value. This leakage reactance is considered by designers under two main headings (a) Slot and tooth-tip reactance. (b) End-connection reactance. These are illustrated by the diagrams of Fig. 171 and are approximately of equal value. The former varies with the coil position relative to the poles and a mean value is taken.

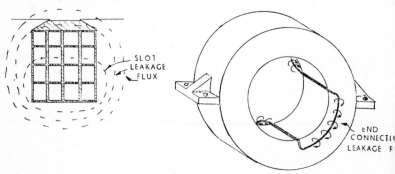

Fig. 171

(2) Armature-Reaction Reactance (X_a). This is not a true reactance but is considered as such for convenience. In this context it must be stressed that armature reaction does not cause reactance, It is responsible, as has been seen earlier, for a change in generated voltage, which change can be likened to a voltage drop. Since this voltage drop is proportional to current, it can be represented by a reactance value multiplied by a current value. Further explanation is made in considering the complete phasor diagram.

THE SIMPLE PHASOR DIAGRAM. This, as already evolved, is redrawn to assist in the solution of the revision example, (Fig. 172).

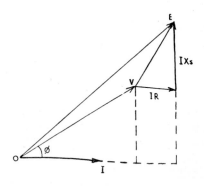

Fig. 172

Example 81. A 50 kVA, 500 V, single-phase alternator has an effective resistance of 0·2 Ω and a synchronous reactance of 1·688 Ω. Find the synchronous impedance and the voltage regulation when a load of 100 A at a power factor of 0·8 (lagging) is removed.

$$Z_S = \sqrt{R^2 + X_S^2} = \sqrt{0\cdot2^2 + 1\cdot688^2} = \sqrt{0\cdot04 + 2\cdot85}$$

$$= \sqrt{2\cdot89} = 1\cdot7 \ \Omega$$

Synchronous impedance $= 1\cdot7 \ \Omega$

From the diagram of Fig. 172. The generated voltage E is represented by the phasor OE.

Thus $E = \sqrt{(OV \cos \phi + IR)^2 + (OV \sin \phi + IX_S)^2}$
$= \sqrt{\{(500 \times 0\cdot8) + (100 \times 0\cdot2)\}^2 + \{(500 \times 0\cdot6)}$
$\overline{\quad + (100 \times 1\cdot688)\}^2}$

$$= \sqrt{(400 + 20)^2 + (300 + 168 \cdot 8)^2} = \sqrt{420^2 + 469^2}$$

$$= 100\sqrt{4 \cdot 2^2 + 4 \cdot 69^2} = 100\sqrt{17 \cdot 64 + 21 \cdot 996}$$

$$= 100\sqrt{39 \cdot 64} = 100 \times 6 \cdot 3 = 630 \text{ V}$$

$$\text{Voltage regulation} = \frac{630 - 500}{500} \times 100 = \frac{130}{5} = 26 \text{ per cent}$$

This can be described as 26 per cent "up", since a voltage rise is involved.

It is of interest to note that the synchronous reactance, resistance and impedance are often given in percentages rather than as definite figures. This practice has already been mentioned in relation to transformers and the following example illustrates the usage. Note that the percentage value refers to a voltage drop and not to an ohmic figure.

Example 82. A 600 kVA, 3·3 kV, 50 Hz, three-phase alternator is star connected and operated at a full-load power factor of 0·8 (lagging). If the resistance and synchronous reactance values are 1 per cent and 6 per cent respectively, find the voltage regulation value for the full-load condition described.

$$\text{Load terminal voltage per phase} = \frac{3 \cdot 3}{\sqrt{3}} = 1 \cdot 91 \text{ kV}$$

$$\text{Resistive voltage drop per phase} = 1 \cdot 91 \times \frac{1}{100} = 0 \cdot 019 \text{ kV}$$

If E is the generated voltage per phase and V is phase terminal voltage then

$$E = \sqrt{(V\cos\phi + IR)^2 + (V\sin\phi + IX_S)^2}$$

$$= \sqrt{\{(1 \cdot 91 \times 0 \cdot 8) + 0 \cdot 019\}^2 + \{(1 \cdot 91 \times 0 \cdot 6) + 0 \cdot 1146\}^2}$$

$$= \sqrt{(1 \cdot 528 + 0 \cdot 019)^2 + (1 \cdot 146 + 0 \cdot 1146)^2}$$

$$= \sqrt{1 \cdot 547^2 + 1 \cdot 261^2} = \sqrt{2 \cdot 39 + 1 \cdot 59}$$

$$= \sqrt{3 \cdot 98} = 1 \cdot 995 \text{ kV}$$

$$\text{Voltage regulation} = \frac{1 \cdot 995 - 1 \cdot 91}{1 \cdot 91} \times 100 = \frac{0 \cdot 085 \times 100}{1 \cdot 91}$$

$$= \frac{8 \cdot 5}{1 \cdot 91} = 4 \cdot 45 \text{ per cent.}$$

THE COMPLETE PHASOR DIAGRAM.   In line with the treatment of
a.c. circuits and machines, a phasor diagram can be derived to illustrate the operating conditions for the alternator. An introduction has
already been made to such a diagram, where reference was made to
the causes of internal voltage drops such as those due to resistance R
and the combined leakage reactance and armature reaction effect —
the term "synchronous reactance", X_S being used. Since "on load"
the main field is altered in magnitude and phase by the armature
flux, it is useful to show this effect on the diagram and also illustrate
the related effects on the generated e.m.f. Thus the diagram can be
drawn to show not only voltage and current phasors but also flux, or
more properly, the m.m.f. vectors. In this form the diagram (Fig. 173)
depicts more completely the operation of the machine as described.
Since the m.m.f. vectors are additional to those shown on the diagram
as deduced up to now, these warrant attention first.

Flux in the air-gap.   When "on-load", in the air-gap of the alternator,
there are two sinusoidally distributed fluxes caused by the m.m.f. distribution waves. These fluxes rotate at synchronous speed in the
same direction. They remain in the same position relative to each
other and produce a resultant flux which can be assumed sinusoidal.
The fluxes and their producing m.m.f.'s are shown on the left-hand
side of the diagram. Thus Φ_0 is the no-load flux set up by F_0 the
main field m.m.f. or ampere-turns. Φ_a is the armature-reaction flux
and the corresponding m.m.f. or ampere-turns are F_a. Φ is the resultant flux caused by F, the resultant m.m.f. or corresponding ampere-
turns.

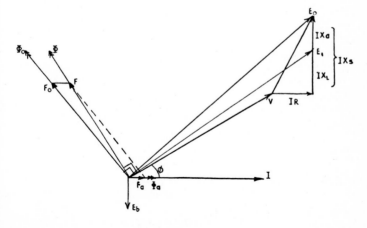

Fig. 173

Since the "on-load" or resultant flux is the original flux altered by the armature-reaction flux, the e.m.f. E_o as originally generated, can be considered to be weakened and retarded by an internally generated e.m.f. E_b as shown. The e.m.f. as generated finally, can now be considered to be E_1 and since E_b is proportional to armature flux which in turn is proportional to current I, the effect is similar to a reactance voltage drop, in quadrature with the current. As seen earlier, the machine can thus be credited with X_S — a larger reactance than the leakage reactance X_L, the additional amount X_a taking care of the armature reaction e.m.f. The term "synchronous reactance", X_S has now been treated in depth and should be appreciated by the reader. It is also shown on the phasor diagram.

Note that the phasor diagram is drawn for one phase only and although the armature-reaction effect is due to the combined action of all the phases in a polyphase machine, its representation by a phasor on a diagram for one phase can be justified. From the diagrams of Fig. 165 it was shown that the armature m.m.f., due to all three phases, produces its maximum value, relative to any one phase, when the current in that phase is a maximum. It can therefore be represented by a phasor in phase with the current in that phase. Note also that the generated e.m.f. is considered as being produced by a change in flux linked with the associated coils, and the e.m.f. phasors are shown lagging the fluxes by $90°$.

Summarising the diagram explanation, we see that it has been drawn for a load current I lagging the terminal voltage V by an angle ϕ. The main field ampere-turns are shown by F_o, and the e.m.f. generated by the associated no-load flux Φ_o is E_o. The armature ampere-turns F_a cause the reaction effect or an associated e.m.f. E_b which is equivalent to a voltage drop IX_a. The on-load flux Φ, caused by F ampere-turns — the resultant of F_o and F_a, is responsible for the on-load generated e.m.f. E_1. The resistance voltage drop is IR in phase with the current and the leakage reactance voltage drop is IX_L in quadrature with the current. IX_a and IX_L are added to give IX_S the synchronous reactance voltage drop from which the synchronous impedance voltage drop IZ_S can be derived. As stated earlier

$$Z_S = \sqrt{R^2 + X_S{}^2}$$

PREDICTION OF VOLTAGE REGULATION

Because of the difficulties of loading and driving large alternators, performance on load is usually determined from assessed tests. It must be appreciated that modern power-station generators have reached a size of 500 MW and marine machines of 300 to 500 kW rating are common. Although the latter are small when compared with the land type units, direct loading is seldom undertaken until the machines are installed on board ship, when performance can be checked against the predicted value.

Two methods of predicting performance will be considered but both require an open-circuit and short-circuit test to be made on the machine.

Open-circuit Test. This is made like that for obtaining the magnetisation or open-circuit characteristic for the d.c. generator. The machine is driven at correct speed and the terminal voltage noted for various values of field current. The test results when plotted would give a curve of the type associated with the O.C.C. of a generator and a check is made to ensure that some 50 per cent overvoltage can be obtained without saturation of the field system. This precaution should cover the requirement for the machine to generate full voltage under over-load and poor power-factor conditions.

Short-circuit Test. This is made with the terminals of the machine short-circuited, a suitable current measuring instrument being included in the circuit. The machine is run at its normal speed and a reduced excitation is applied. Readings of line current are noted for the appropriate field current values, the former being raised to as large a value as possible, whilst keeping in mind the danger of damage to the windings by overheating. The test results when plotted should give a straight line, hence only two or three readings are necessary. The O.C. and S.C. test results are frequently plotted to the same base of field current or field ampere-turns since they are used jointly for making the assessment tests as described below.

THE SYNCHRONOUS-IMPEDANCE METHOD. This is also referred to as the Behn Eschenberg or Pessimistic method — for the reason set out below. The effect of armature reaction is ignored and the alternator is credited with a greater reactance value than it actually has. In relation to the phasor diagram, the effect of F_a is taken into account by the X_a value credited to the windings. Since the terminal-voltage value V is zero then it is assumed that all the voltage generated is utilised in overcoming the voltage drops IR and IX_S when current circulates through the machine. The synchronous impedance can thus be obtained and used in conjunction with the phasor diagram and the regulation formula. The reasoning appears acceptable but for the fact that Z_S is measured from a S.C. test and, if its values, as determined by the procedure being considered, are plotted to a base of excitation current then it is seen to decrease as the excitation is strengthened due to saturation of the field system. For calculations a value of Z_S is chosen as near as possible to the excitation under load conditions.

It should be appreciated that for the S.C. test, field current is low and the field is weak *i.e.* the magnetic circuit is unsaturated. On full-load the field current is high and the iron is nearly saturated. Thus under the latter condition greater opposition is offered to the flux causing armature reaction than under the former or test condition. For the S.C. test, more flux-linkages can be associated with the arma-ture than can occur on load and thus X_S, as obtained from the S.C. test, is larger than its value under normal load conditions. A larger voltage drop than occurs in practice, is the result of the assessment test giving a pessimistic figure for voltage regulation.

The diagram of Fig. 174 shows the shapes of the curves which would be obtained from the O.C. and S.C. tests and illustrates how the results of the tests are used.

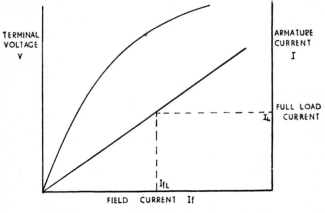

Fig. 174

The voltage generated, when making the S.C. test, is used to drive current through the impedance of the armature. Assume for the full-load current value I_L the field current to be I_{f_L} as shown. Then from the O.C.C., for I_{f_L} a voltage of V_L value is seen to be generated. Thus a current of I_{f_L} in the field gives a voltage of V_L generated and an armature current of I_L results.

Therefore $Z_S = \dfrac{V_L}{I_L}$ for that field current value. R is measured

for a hot armature condition and X_S is then found from $X_S = \sqrt{Z_S^2 - R^2}$.

From the basic phasor diagram E is obtained, as seen already, and the voltage regulation is deduced. The procedure is illustrated by the examples.

Example 83. A single-phase alternator required an excitation current of 1·9 A to produce a short-circuit full-load current of 100 A. The same excitation current results in an open-circuit voltage of 430 V. The armature resistance is measured to be 0·77 Ω. Find the full-load voltage regulation at unity power factor when the terminal voltage is 500 V.

$$Z_S = \frac{430}{100} = 4\cdot3 \ \Omega \quad X_S = \sqrt{4\cdot3^2 - 0\cdot77^2} = \sqrt{18\cdot49 - 0\cdot59}$$

$$= \sqrt{17\cdot9} = 4\cdot23 \ \Omega$$

$$E = \sqrt{(V\cos\phi + IR)^2 + (V\sin\phi + IX_S)^2}$$

$$= \sqrt{\{(500 \times 1) + (100 \times 0\cdot77)\}^2 + \{(500 \times 0) + (100 \times 4\cdot23)\}^2}$$

$$= \sqrt{(500 + 77)^2 + (0 + 423)^2} = \sqrt{577^2 + 423^2}$$

$$= 100\sqrt{5\cdot77^2 + 4\cdot23} = 100\sqrt{33\cdot29 + 17\cdot89} = 100\sqrt{51\cdot18}$$

$$= 100 \times 7\cdot15 = 715 \ V$$

$$\text{Voltage regulation} = \frac{715 - 500}{500} \times 100 = \frac{215}{5} = 43 \text{ per cent}$$

Example 84. A three-phase, 5 MVA, 6·6 kV alternator is star connected and has O.C. and S.C. test results as set out in the following table.

Field current (A)	Terminal voltage (V)	Line current (A)	Phase voltage (V)
100	4800	690	2770
150	6500	1020	3750
200	7400	—	4270
250	7900	—	4560

Estimate the voltage regulation at full load, 0·8 (lagging) power factor by the synchronous impedance method. Assume the stator resistance to be negligible.

The graphs of Fig. 175 show the appropriate values plotted to a common base of field current. Note that the figures appearing in the last column are not test results but are deduced from column 2.

Thus $\dfrac{4800}{\sqrt{3}} = 2770$ V. This is a star-connected machine and the problem has been worked in phase values. The largest measured short-circuit current has been used to obtain Z_S, since this gives the maximum field strength leading up towards saturation and gives a reduced error in the value of Z_S.

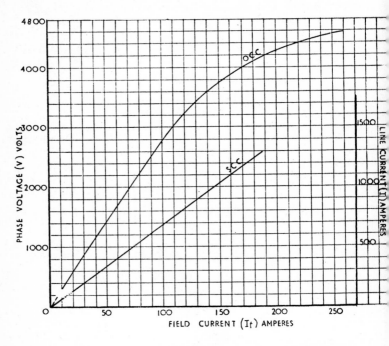

Fig. 175

$$\text{Full-load current} = \frac{5000 \times 1000}{6 \cdot 6 \times 1000 \times \sqrt{3}} = \frac{5000}{6 \cdot 6 \times 1 \cdot 732}$$

$$= 437 \cdot 4 \text{ A}$$

$$\text{Normal phase voltage } V = \frac{6600}{\sqrt{3}} = 3820 \text{ V}$$

The generated voltage per phase $= E \qquad R = 0\Omega$

$$\text{Also } X_S = \frac{6500}{\sqrt{3} \times 1020} = 3 \cdot 68 \ \Omega \quad \text{and} \quad Z_S = X_S$$

Then $E = \sqrt{(V \cos \phi)^2 + (V \sin \phi + IX_S)^2}$

$= \sqrt{(3820 \times 0 \cdot 8)^2 + \{(3820 \times 0 \cdot 6 + 437 \times 3 \cdot 68)\}^2}$

$= \sqrt{3056^2 + (2292 + 1610)^2}$

$= \sqrt{3056^2 + 3902^2} = 10^3\sqrt{3 \cdot 056^2 + 3 \cdot 902^2}$

$= 10^3\sqrt{9 \cdot 3 + 15 \cdot 2} = 10^3\sqrt{24 \cdot 5} = 10^3 \times 4 \cdot 95$

$= 4950$ V

Voltage regulation $= \dfrac{4950 - 3820}{3820} \times 100 = 29 \cdot 6$ per cent

Example 85. Up to now, no example concerning a leading power-factor condition has been considered and here is shown how the terminal-voltage equation is modified to suit. The simple phasor diagram has been redrawn (Fig. 176) and it will be seen that here the reactance voltage drop is treated as a negative value.

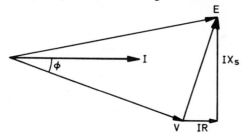

Fig. 176

A 95 kVA, 440 V, three-phase, star-connected alternator is fully loaded and its excitation current is adjusted to give full voltage when the power factor is 0·8 (lagging). What will be the value of the terminal voltage when the power factor changes to 0·8 (leading), if the load current and excitation current remains unchanged.

During tests on the machine it was found that on short circuit, a field current of 40 A was required to circulate full-load current through the stator and that on open circuit, this field current produced a generated line terminal voltage of 120 V. The resistance of the armature measured between two line terminals was 0·2 Ω

Full-load current $= \dfrac{95 \times 1000}{440 \times \sqrt{3}} = 125$ A

Synchronous Impedance per phase $= \dfrac{120}{\sqrt{3} \times 125} = 0.55\ \Omega$

Resistance per phase $= \dfrac{0.2}{2} = 0.1\ \Omega$

Synchronous Reactance per phase $= \sqrt{0.55^2 - 0.1^2}$
$$= \sqrt{0.3025 - 0.01} = 0.54\ \Omega$$

On full load at a power factor of 0·8 (lagging), phase terminal voltage $= \dfrac{440}{\sqrt{3}} = 254\ V$

Therefore E per phase

$= \sqrt{(V \cos \phi + IR)^2 + (V \sin \phi + IX_S)^2}$

$= \sqrt{\{(254 \times 0.8) + (125 \times 0.1)\}^2 + \{(254 \times 0.6) + (125 \times 0.54)\}^2}$

$= \sqrt{(203 + 12.5)^2 + (152.5 + 67.5)^2} = \sqrt{215.5^2 + 220^2}$

$= \sqrt{46\,440 + 48\,400} = \sqrt{94\,840} = 308\ V$

If the excitation current remains unaltered then the generated e.m.f. of the machine at a power factor of 0·8 (leading) will still be 308 V and under the new full-load condition, the terminal voltage will alter to V_1 where

E per phase $= \sqrt{(V_1 \cos \phi + IR)^2 + (V_1 \sin \phi - IX_S)^2}$

Note the $-$ve sign.

$308 = \sqrt{\{(V_1 \times 0.8) + (125 \times 0.1)\}^2 + \{(V_1 \times 0.6) - (125 \times 0.54)\}^2}$

$308 = \sqrt{(0.8\,V_1 + 12.5)^2 + (0.6\,V_1 - 67.5)^2}$

$308 = \sqrt{0.64\,V_1^2 + 20\,V_1 + 156.25 + 0.36\,V_1^2 - 81\,V_1 + 4556.25}$

Thus:

$308^2 = V_1^2 - 61\,V_1 + 4712$ or $V_1^2 - 61\,V_1 - 90\,152 = 0$

Giving $V_1 = \dfrac{61 \pm \sqrt{61^2 + (4 \times 1 \times 90\ 152}}{2}$

$= \dfrac{61 \pm \sqrt{3721 + 360\ 608}}{2} = \dfrac{61 \pm \sqrt{364\ 329}}{2}$

$= \dfrac{61 \pm 603 \cdot 6}{2} = 332 \cdot 3\ V \quad \text{or} \quad -271 \cdot 3\ V.$

The second answer is obviously unreal so the alternator terminal voltage on full load at a power factor of $0 \cdot 8$ (leading) will be $\sqrt{3} \times 334 \cdot 6 = 575\ V.$

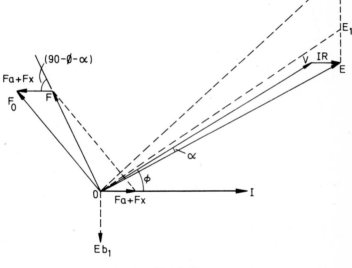

Fig. 177a

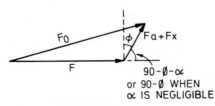

Fig. 177b

THE AMPERE-TURN METHOD.   This is also known as the M.M.F., Rothert or Optimistic Method. Here armature reactance is ignored and its effect is replaced by crediting the armature with an additional fictitious reaction. The leakage-reactance voltage is replaced by the m.m.f. value which would be required to produce an equal voltage and the m.m.f. part of the diagram (Fig. 173) is therefore worked upon. Since the replacing m.m.f. will be in quadrature with the reactance voltage drop IX_S or the equivalent generated voltage E_b, it is in phase with the existing armature-reaction m.m.f. F_a. It can therefore be combined with F_a, as is shown on the modified diagram (Fig. 177), giving the reaction vector, labelled $F_a + F_x$. Since $F_a + F_x$ is equated to a reactance voltage drop due to full load current, it is determined from the short-circuit test. Use of this value can be justified for prediction purposes because it is obtained from a zero power-factor condition; the impedance of the short-circuit being due wholly to reactance, the resistance being negligible.

For the modified diagram (Fig. 177a) OV is the terminal voltage, with the current lagging by angle ϕ. IR is the resistance voltage drop, which is often negligible, and OE shows the required e.m.f. to be generated. To produce E, an m.m.f. F is required which is the original no-load m.m.f. F_o weakened by the armature reaction demagnetising effect as is now being credited to $F_a + F_x$. Since the m.m.f. effects are proportional to the energising currents, the relative magnitudes can be obtained from the O.C. and S.C. characteristics, any vector resolving being undertaken in terms of such currents. Thus the demagnetising value $(F_a + F_x)$ is obtained from the S.C. graph by taking the load current, for which the regulation, is required and noting the corresponding field current. From the O.C. curve the field current (F) for the terminal voltage E is obtained and the resultant m.m.f. is derived from the triangle and vector relationships (Fig. 177b) given by

$$F_o^2 = F^2 + (F_a + F_x)^2 + 2 F(F_a + F_x) \cos (90 - \phi).$$

Note that as the same field system is concerned, then there is no need to use the actual ampere-turn values and that the calculation can be made in terms of field current. Under the conditions used for the calculation, the values are obtained at no saturation. To overcome the same impedance voltage drop, a smaller generated voltage and hence a smaller value of excitation is required than would be used under normal working conditions. This method of prediction will therefore give an optimistic result.

As an alternative to using the mathematical expression set out above and evolved from the Cosine Rule, the value of F_o can be determined by a graphical construction. This is illustrated by Example 86 and since basic procedures are used, the solution should be readily followed.

Once the value of F_o has been found, the O.C.C. is used to determine the open-circuit voltage E_o and hence the voltage regulation is deduced.

The above methods of predicting voltage regulation are not accurate in themselves, but the average of the two results is sufficiently accurate for most practical purposes.

Example 86. The test figures and graphs of Example 84 are repeated here and the voltage regulation is determined by the ampere-turn method.

Field current	Terminal voltage	Line current	Phase voltage
(A)	(V)	(A)	(V)
100	4800	690	2770
150	6500	1020	3750
200	7400	—	4270
250	7900	—	4560

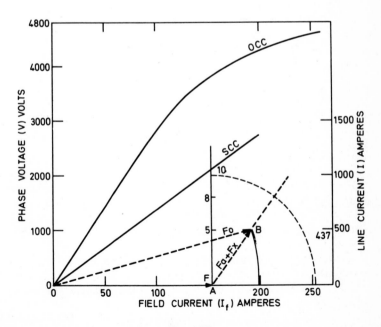

Fig. 178

For a full-load current of 437·4 A *i.e.* 5 MVA, the field current on short circuit is noted to be 65 A (Fig. 178). Therefore the demagnetising effects of the armature-reaction m.m.f. ampere-turns are proportional to this current or $(F_a + F_x)$ can be represented by 65 A.

Again since the amount of voltage appearing across the terminals of the machine is $\dfrac{6600}{\sqrt{3}} = 3820$ V per phase, then the field current, deduced from the O.C.C. is 155 A. This value is represented by F

Thus
$$F_o^2 = 155^2 + 65^2 + 2 \times 155 \times 65 \times \cos(90 - \phi)$$
$$= 155^2 + 65^2 + (2 \times 155 \times 65 \times 0·6).$$

The voltage regulation is to be estimated at a power factor of 0·8 (lagging).

Thus $\cos\phi = 0·8$ and $\cos(90 - \phi)$ must equal 0·6

$$\therefore \ F_o^2 = 24\,025 + 4225 + (310 \times 65 \times 0·6)$$
$$= 24\,025 + 4225 + 12\,090 = 40\,340 = 4·034 \times 10^4$$

or $F_o = 2·008 \times 10^2 = 200·8$ say 201 A

From the O.C.C. the generated voltage E_o would be 4270 V

Voltage regulation $= \dfrac{4270 - 3820}{3820} \times 100 = 11·8$ per cent

If resistance is to be taken into account a further approximation can be made by developing the phasor diagram. A further reference to this evaluation is illustrated by Example 87 (Fig. 180b).

The Graphical method of obtaining F_o is shown dotted on the graphs of Fig. 178. It is seen that $F_a + F_x = 65$ A. A power-factor quadrant can be made as shown *i.e.* read off ten units vertically above A and draw the quadrant. Number the units 1, 2, 3 etc. Then the horizontal through 8, to cut the quadrant, will give a point which determines the slope of AB. Note $\cos\phi = \dfrac{8}{10} = 0·8$.

It has been deduced that $F = 155$ A. Thus OA is made this length. Draw a vertical above A, construct the quadrant and measure off $F_o = 200$ A. Reference to the O.C.C. shows the generated voltage would be 4270 V.

Voltage regulation $= \dfrac{4270 - 3820}{3820} \times 100 = 11·8$ per cent

Average of the Synchronous Impedance and Ampere-turn Methods

would be $\dfrac{29{\cdot}6 + 11{\cdot}8}{2} = \dfrac{14{\cdot}14}{2} = 20{\cdot}7$ per cent

Example 87. This example takes into account the resistance value of the alternator since it is given.

A three-phase, 100 kVA, 400 V, star-connected alternator gave the following O.C. and S.C. test results.

Field current (A)	2	4	6	8	10
Terminal voltage (V)	204	370	452	496	520
Line current (A)	101	198	300		
Deduced phase voltage (V)	118	213	261	286	300

The ohmic resistance of the machine as measured between line terminals is $0{\cdot}16\ \Omega$. Estimate the percentage voltage regulation of the machine for a full load, $0{\cdot}8$ (lagging) power factor condition.

Synchronous-Reactance Method. Terminal voltage per phase

$= \dfrac{400}{\sqrt{3}} = 230{\cdot}6$ V

Full-load current $= \dfrac{100\ 000}{\sqrt{3}\ 400} = \dfrac{250}{\sqrt{3}} = 144$ A

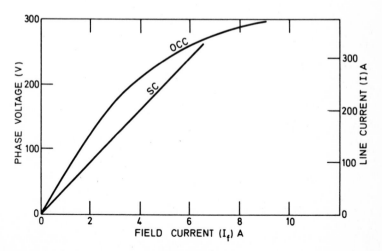

Fig. 179

From the graphs of Fig. 179 for a current of 300 A (largest test current) $I_f = 6$ A and the generated voltage per phase $= 261$ V

$$\therefore \quad Z_{ph} = \frac{261}{300} = 0.87 \ \Omega \quad \text{Now} \ R_{ph} = \frac{0.16}{2} = 0.08 \ \Omega$$

Thus $X_{ph} = \sqrt{0.87^2 - 0.08^2} = \sqrt{0.757 - 0.0064} = \sqrt{0.7506}$

$$= 0.866 \ \Omega$$

Using diagram (a) of Fig. 180 we have

$$E = \sqrt{(V \cos \phi + IR)^2 + (V \sin \phi + IX_S)^2}$$

$$= \sqrt{\{(230.6 \times 0.8) + (144 \times 0.08)\}^2 + \{(230.6 \times 0.6) + (144 \times 0.866)\}^2}$$

$$= \sqrt{(184.48 + 11.52)^2 + (138.36 + 124.3)^2}$$

$$= \sqrt{196^2 + 262.66^2} = 10^3 \sqrt{0.196^2 + 0.263^2}$$

$$= 10^3 \sqrt{0.0385 + 0.069} = 10^3 \sqrt{0.1075} = 10^3 \times 0.327$$

$$= 327 \text{ V}$$

Voltage regulation $= \dfrac{327 - 230.6}{230.6} \times 100 = \dfrac{96.4}{2.306} = 41.8$ per cent

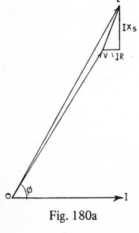

Fig. 180a

Fig. 180b

Ampere-turn Method. IR voltage drop/ph $= 144 \times 0\cdot08 = 11\cdot52$ V

Here $X = 11\cdot52 \cos \phi = 11\cdot52 \times 0\cdot8 = 9\cdot216$ V

Here $Y = 11\cdot52 \sin \phi = 11\cdot52 \times 0\cdot6 = 6\cdot912$ V

As an approximation OP $=$ OV $+ X = 230\cdot6 + 9\cdot216 = 239\cdot82$ V

$\therefore$ OE $= \sqrt{239\cdot8^2 + 6\cdot9^2} = 10\sqrt{575 + 0\cdot476} = 10\sqrt{575\cdot5}$

$\qquad = 100\sqrt{5\cdot755} = 100 \times 2\cdot399 = 239\cdot9 = 240$ V

From the graphs of Fig. 179

O.C. field current for 240 V $= 5$ A
S.C.   "   "   "   144 A $= 2\cdot9$ A

$\therefore F_o^2 = 5^2 + 2\cdot9^2 + \{2 \times 5 \times 2\cdot9 \cos (90 - \phi)\}$

$\qquad = 25 + 8\cdot4 + (10 \times 2\cdot9 \sin \phi)$

$\qquad = 33\cdot4 + 29 \times 0\cdot6 = 33\cdot4 + 17\cdot4 = 50\cdot8$

Thus $F_o = \sqrt{50\cdot8} = 7\cdot127$ A

Using the graph for a field current of $7\cdot127$ A, the voltage would be 278 V.

Voltage regulation $= \dfrac{278 - 230\cdot6}{230\cdot6} \times 100 = \dfrac{47\cdot4}{2\cdot306} = 20\cdot6$ per cent.

Average value $= \dfrac{41\cdot8 + 20\cdot6}{2} = \dfrac{62\cdot4}{2} = 31\cdot2$ per cent

SYNCHRONISING TORQUE

Although the space available in this chapter does not allow treatment of synchronising and load-sharing procedures for the alternator, it would be helpful to consider the conditions which arise when such a machine is connected in parallel with "live" busbars. The problem first requires examination with respect to the system to which the machine is to be connected. Thus the "incoming" alternator may be paralleled onto *infinite busbars i.e.* a system so large — such as the "grid" system, whose stability is unaffected by the speed and size of the incoming machine. Alternatively,this alternator may be connected to a machine of equivalent rating,which is already on load. Since the procedure of synchronising involves bringing the machine up to the correct voltage value and frequency, it is assumed that this can be done accurately by noting the voltmeter, tachometer and frequency-meter readings. However, a further condition to be satisfied for correct synchronising is that the phase sequence and phase angle of the incoming alternator should be correct at the instant of closing the machine circuit-breaker. The syncroscope or synchronising lamps will indicate correctly the phase-sequence requirement, but for the phase-angle requirement some error can exist due to sluggishness of the instrument or slow action by the operator.

In the event of the machine being synchronised when partially out of phase with the busbar voltage, the incoming alternator will experience a "pull-in" torque which will either instantaneously speed it up or slow it down in relation to the system. If paralleled onto an infinite busbar system, the incoming machine alone will be affected but if it is connected to a small independent system, such as would occur on board ship, then the frequency and stability of the system may be altered. If the incoming machine is lagging, the system will sustain a shock, in that the machine will be pulled up into step and the loaded machine momentarily slowed down. Once synchronised, the incoming alternator will run light or "float" on the busbars. Similarly if the incoming machine is leading it will sustain a momentary shock whilst being pulled back and the system frequency may tend to rise until the arrangement attains steady conditions with the machine floating on the bars.

During the transient conditions described above, a circulating current will flow between the incoming machine and the busbars. The magnitude of this current is decided by the degree of "out of phase" when the paralleling circuit-breaker is closed. This current is additional to that being supplied by the system and may well "trip out" circuit-breakers which are closed, by causing their over-current or reverse-power protective devices to operate. Such a current can be considered to be a danger, since the synchronising torque which it produces can result in damage to the incoming machine or to the system. The factors affecting this current and the synchronising torque are now considered and illustrated by an example.

SYNCHRONISING POWER AND TORQUE. The position of the incoming and "running" machines can be regarded from the common circuit point of view *i.e.* the busbars and the phasor diagram, shown by Fig. 181, can be built-up.

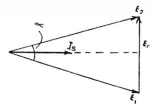

Fig. 181

Let E_1 be the generated voltage of the machine (No. 1) or machines already on the "bars" and E_2 the e.m.f. of the incoming machine (No. 2). E_2 is in advance by an angle of α or its equivalent mechanical displacement. Due to this displacement, a resultant e.m.f. E_r is created which causes the synchronising current I_S to flow. Since I_S is limited mainly by the reactance of the closed circuit, the resistance of the machines being negligible. Then I_S lags E_r by $90°$.

$$\text{Now } \frac{\frac{E_r}{2}}{E_2} = \sin\frac{\alpha}{2} \quad \therefore \quad E_r = 2E_2 \sin\frac{\alpha}{2}$$

$$\text{and } I_S = \frac{E_r}{2X_{S/ph}} = \frac{2E_2 \sin\frac{\alpha}{2}}{2X_{S/ph}} \text{ where } X_{S/ph} \text{ is the synchronous}$$

reactance per phase of each machine. These can be assumed to be identical.

$$\text{Power flow from machine No. 2} = E_2 I_S \cos\frac{\alpha}{2}$$

$$\text{Synchronising Power/ph} = \frac{E_2 E_2 \sin\frac{\alpha}{2} \cos\frac{\alpha}{2}}{X_{S/ph}} = \frac{E_2{}^2 \sin\frac{\alpha}{2} \cos\frac{\alpha}{2}}{X_{S/ph}}$$

Example 88. A single-phase, 10 kV, 50 Hz, alternator runs at 1500 rev/min and has a synchronous reactance of 4 Ω. Assuming the machine is paralleled on to an infinite busbar arrangement, calculate the synchronising power for 1 mechanical degree of phase displacement.

$$\text{Here } P = \frac{120f}{N} = \frac{120 \times 50}{1500} = 4 \text{ poles, Total number of}$$

electrical degrees contained on rotor circumference $= \dfrac{4}{2} \times 360 =$

$720°$. Number of equivalent mechanical degrees $= 360°$

For this machine 1 mechanical degree $= \dfrac{720}{360} = 2$ electrical degrees.

Synchronising Current $I_S = \dfrac{E_r}{2X_{S/ph}}$ for two small machines but

for infinite busbars $I_S = \dfrac{E_r}{X_{S/ph}}$. The $X_{S/ph}$ of the busbar system

being negligible.

Here $E_r = 2E_2 \sin \dfrac{\alpha}{2} = 2 \times 10\ 000 \times \sin \dfrac{2}{2} = 350$ V

$I_S = \dfrac{350}{4} = 87 \cdot 5$ A

Synchronising Power $= E_2 I_S \cos \dfrac{\alpha}{2} = \dfrac{10\ 000 \times 87 \cdot 5 \times 0 \cdot 9998}{1000}$

$= 874$ kW

Note. If this had been a three-phase machine, the total power would have been three times this single-phase power.

Synchronising Torque. This is derived from the relation

Synchronising Power $= \dfrac{2\pi NT}{60}$

or $T = \dfrac{\text{Synchronising Power} \times 60}{2\,\pi\,N}$ newton metres.

Hunting or Phase-swinging. For small phase differences, the synchronising current I_S and hence the synchronising torque are proportional to the phase-difference angle. The arrangement of parallel working can thus have a free period of oscillation about the maintained speed and such oscillation is known as "hunting" or phase swinging. After synchronising, on light load or when load is thrown off, the oscillation may increase in amplitude until the machine falls out of step when, due to the large circulating current, it will be "tripped off" the busbars. All alternators designed for parallel operation are provided with a *damping winding* set into the pole faces. When hunting occurs, relative motion takes place between the damping-winding bars and the field flux. This results in heavy currents in the bars producing a damping or stabilising effect. The damping windings are also referred to as *damping grids* and since hunting occurs, more particularly with salient-pole machines which are engine-driven, where cyclic oscillations are prone to be present, an arrangement such as that shown by Fig. 182 is usual. For extreme cases, special precautions are necessary and the short-circuiting damping rings may be extended round the periphery of the rotor.

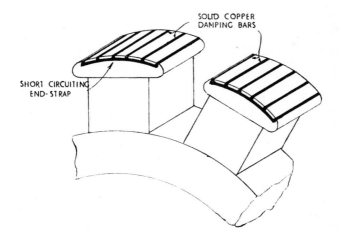

Fig. 182

For the cylindrical rotor, the retaining wedges may be made of brass and bonded at the ends to serve as damping grids. Considerable damping already exists however, due to the fact that the rotor is solid and this is augmented by the effect of the metal wedges in the rotor slots. For special cases, sufficient damping can be provided by an additional bar in the slots, connected to end rings to constitute a damping winding.

The following two examples are included for interest and revision purposes. They complete the number allocated to each chapter.

Example 89. In a 1 MVA, 11 kV, three-phase, star-connected alternator the resistance of the winding per phase is 2 Ω. The iron core loss is 15 kW, the friction and windage loss is 10 kW and the input required to the field excitation circuit amounts to 5 kW. Estimate the efficiency of the machine when operating at full load and a power factor of 0·8 (lagging).

Output power at 0·8 power factor $= 1000 \times 0·8 = 800$ kW

$$\text{Full-load current} = \frac{1000 \times 1000}{\sqrt{3} \times 11\,000} = \frac{1000}{1·732 \times 11} = \frac{1000}{19·05}$$

$$= 52·48 \text{ A}$$

Copper loss per phase $= 52·48^2 \times 2 = \dfrac{52·48 \times 104·96}{1000} = 5·51$ kW

Three-phase copper loss $= 16·53$ kW

Total loss $= 16·53 + 15 + 10 + 5 = 16·53 + 30 = 46·53$ kW

Input power $= 800 + 46·53 = 846·53$ kW

Efficiency $= \dfrac{800}{846·53} \times 100 = \dfrac{800}{8·465} = 94·5$ per cent.

Example 90. An industrial load of 4 MW has a supply voltage of 11 kV and a power factor of 0·8 (lagging). A synchronous motor is required to meet an additional 1130 kW and at the same time raise the power factor to 0·95 (lagging). Find the *kVA* rating of the motor, the power factor at which it must be operated and the line current of the motor if it is star-connected. Assume that the efficiency of the motor is 93 per cent.

Original condition.

Active power of basic load $= 4000$ kW

Apparent power of basic load $= \dfrac{4000}{0·8} = 5000$ kVA

Also $\cos \phi = 0·8$ so $\sin \phi = 0·6$

Reactive power of basic load $= 5000 \times 0·6 = 3000$ kVAr

Additional load.

Motor active power output $= 1130$ kW

$$\text{Motor input power} = \frac{1130}{0.93} = 1215$$

Final condition.

Total active power of load $= 4000 + 1215 = 5215$ kW

$$\text{Apparent power of combined load} = \frac{5215}{0.95} = 5500 \text{ kVA}$$

Also $\cos \phi = 0.95 \quad \therefore \quad \phi = 18.11 \quad$ so $\quad \sin \phi = 0.312$

Reactive power of combined load $= 5500 \times 0.312 = 1715$ kVAr

By deduction:

Reactive power rating of motor $= 3000 - 1715 = 1285$ kVAr

Note that the motor is to operate at a leading power factor in order to raise the power factor of the combined load. The original load was working under a lagging power-factor condition.

$$\begin{aligned}
\text{Apparent power rating of motor} &= \sqrt{1285^2 + 1215^2} \\
&= 10^3\sqrt{1.285^2 + 1.215^2} \\
&= 10^3\sqrt{1.65 + 1.48} \\
&= 10^3\sqrt{3.13} = 1770 \text{ kVA}
\end{aligned}$$

$$\text{Power factor of motor} \cos \phi = \frac{1215}{1770} = 0.687 \text{ (leading)}$$

$$\text{Line current of motor} = \frac{1770 \times 1000}{\sqrt{3} \times 11\,000} = \frac{1770}{1.732 \times 11} = 92.5 \text{ A.}$$

CHAPTER 9

PRACTICE EXAMPLES

1.   A four-pole alternator, on open circuit, generates 200 V at 50 Hz when its field current is 4 A. Determine the generated e.m.f. at a speed of 1200 rev/min and a field current of 3 A, neglecting saturation of the iron parts.

2.   A single-phase alternator has an effective resistance of 0·2 Ω and a synchronous reactance of 2·2 Ω. Find the generated e.m.f. when the load current is 50 A at a terminal voltage of 500 V. The load operates at a power factor of 0·8 (lagging).

3.   Find the synchronous impedance and reactance of an alternator in which a given field current produces an armature current of 250 A on short-circuit and a generated e.m.f. of 1500 V open-circuit. The armature resistance is 2 Ω. Calculate the terminal voltage when a load of 250 A at 6·6 kV and power factor 0·8 (lagging) is switched off.

4.   A 250 kVA, 3000 V, three-phase, star-connected alternator has a synchronous reactance of 20 Ω per phase. The armature resistance measured between two terminals is 2·8 Ω. Determine the value to which the terminal voltage will rise when full load at (a) 0·8 (lagging) power factor, (b) 0·9 (leading) power factor is switched off.

5.   A marine type 9 kVA, 230 V, single-phase alternator has an armature resistance of 0·25 Ω. The results of the O.C. and S.C. tests are as follows:

Field current (A)	1	2	3	3·5	4	5	6	7
Terminal voltage (V)	78	144	198	220	237	265	284	296
Short-circuit current (A)	11	22	34	40	46	57	69	80

Estimated, by means of the synchronous-impedance method, the voltage regulation of the machine on a full-load, 0·8 (lagging) power-factor condition.

6.   Using the data of Example 5, estimate the voltage regulation of the machine by means of the ampere-turn method.

7.   A 500 kVA, 440 V, three-phase, star-connected alternator has a resistance of 0·01 ohms.per phase. The results of open-circuit and short-circuit tests are:

Field current (A)	0	20	40	60	80	100	120	160	200
Terminal voltage (V)	0	130	253	336	393	440	478	524	545
Line current (A)		260	500	−	1000				

Calculate by means of both the ampere-turn and the synchronous-impedance methods the value to which the terminal voltage will rise when full load at a power factor of 0·8 (lagging) is switched off. What would be a more likely value of the voltage for this machine?

8.   A marine-type, 30 kVA, 440 V, three-phase, star-connected alternator has a resistance of 0·15 ohms per phase. Results of open-circuit and short-circuit tests are as follows:

Field current (A)	2	4	6	7	8	10	12	14
Terminal voltage (V)	156	287	396	440	474	530	568	592
Short-circuit current (A)	11	22	34	40	46	57	69	80

Find the percentage voltage regulation of the alternator on a full-load and power-factor condition of 0·8 (lagging), by the synchronous-impedance and ampere-turn methods.

9.   A three-phase, 3·3 kV, 50 Hz alternator having an equivalent armature reactance of 5 ohm per phase and negligible resistance, is connected to busbars to which an identical alternator is already connected. The circuit-breaker was closed at an instant when the rotors of the two machines were running at 1500 rev/min but out of phase by 1 mechanical degree. Find the value of (1) the synchronising current and (2) the synchronising torque at this instant.

10.   A 2 MVA, three-phase, eight-pole alternator runs in parallel with other machines on 6 kV busbars. If full load at a power factor of 0·8 (lagging) is being supplied, calculate the synchronising power per mechanical degree of displacement. The synchronous reactance of the machine is 6 ohm per phase.

CHAPTER 10

ELECTRONICS (ii)

An introduction has already been made in Chapter 8, to the use of the triode valve as an amplifier of small varying signal voltages and, by way of example, a practical method of using such a valve in a sound reproduction system was illustrated. It is of value, at this stage, to differentiate between the d.c. and a.c. voltages which are jointly present in many sections of the circuits. The d.c. voltage sources and consequent currents are necessary to enable the valve to function effectively as an electronic device but, it should be noted that, although a steady current flows in the anode circuit when no signal is applied at the grid, which itself is being correctly biased, nevertheless this current fluctuates when a varying signal is applied at the grid. Since the anode current does not reduce to zero but varies between limits, it is considered to consist of a steady d.c. current on which is superimposed a varying current. In describing arrangements used to exploit the amplifying property of the triode, electrical components will be introduced which block the passage of d.c. currents but allow alternating currents, of appropriate values, to pass with the minimum of opposition at the frequencies involved. The operation of any one valve-stage of an amplifier is concerned with the passage of the alternating currents but the technique is to convert such currents into voltage drops by means of resistors, inductors or transformers. It is for this reason that the method, of treating the valve as an alternating-voltage generator, can be used for the development of equivalent circuits.

ALTERNATING-VOLTAGE AMPLIFICATION

Such amplification is achieved simply by introducing a load resistor into the anode circuit. Dynamic operating conditions now exist and will be considered.

EQUIVALENT CIRCUIT.   This ignores the d.c. voltages and currents and considers the valve as an a.c. generator. The e.m.f of the generator is μE_g and its internal resistance is r_a. The arrangement is shown by Fig. 183 and the generator circuit is seen to be completed through R and the h.t. source, — which is considered to have negligible impedance.

The alternating current I, which would flow is given by $\dfrac{\mu E_g}{r_a + R}$ and

the voltage drop E_a across R would be IR or $\dfrac{\mu E_g}{r_a + R} \times R$

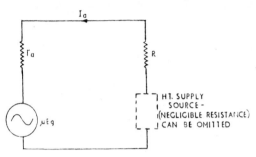

Fig. 183

The voltage amplification of the valve stage or the *stage gain A* is the ratio of the signal output voltage to that applied at the grid/cathode or input voltage. Thus:

$$A = \frac{E_a}{E_g} = \frac{\mu E_g R}{r_a + R}\bigg/ E_g \quad \text{or} \quad A = \frac{\mu R}{r_a + R}$$

Note that E_g, E_a and I are r.m.s. values.

Example 91. A triode valve has an anode resistance of 16 kΩ, an amplification factor of 33 and is employed as a voltage amplifier with a resistive anode load. If the stage gain is to be 25, determine the value of the load resistor required.

Since $A = \dfrac{\mu R}{r_a + R}$ and $A = 25$

$\therefore \quad 25 = \dfrac{33R}{16 + R}$ or $25(16 + R) = 33R$

Whence $33R - 25R = 16 \times 25$ or $8R = 16 \times 25$
Giving $R = 50\,\text{k}\Omega$.

DYNAMIC OPERATION. The behaviour of a triode valve under load conditions can be checked by a test circuit similar to that used for determining the static characteristics (Fig. 150). The new arrangement is shown by the diagram of Fig. 184, from which it is seen that a load resistor R has been included in the anode circuit and an extra voltmeter, to read the supply voltage V, has been added.

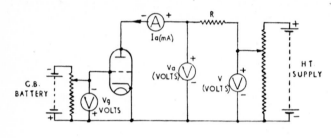

Fig. 184

The results can be plotted to give anode and mutual characteristics as before but, since these will be different for differing values of load resistor R, it is more convenient to obtain the static characteristics from tests as described and then to deduce the dynamic characteristic for any given value of R. Since the dynamic mutual or transfer characteristic is the most useful, its derivation only will be described.

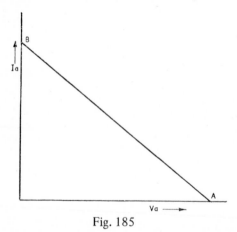

Fig. 185

THE DYNAMIC CHARACTERISTIC—LOAD LINE. The mutual or transfer dynamic characteristic can be deduced by drawing on the family of static output I_g/V_a curves, a "load line" representing the voltage drop across load resistor R. Procedure would be as follows and reference is made to Fig. 185.

Point A is obtained by considering the condition when I_a is 0 *i.e.* the valve is at "cut-off". There is no p.d. across R and hence $V_a =$ the full h.t. voltage V, which is known. OA is therefore marked off to equal the h.t. supply voltage.

Point B represents the maximum value of current which would flow if the valve resistance was ineffective *i.e.* at this point $V_a = 0$ and the full h.t. voltage V is dropped across the load resistor R

$\therefore I_a = \dfrac{V}{R}$. Thus OB is marked off, having a value given by

$$\frac{\text{H.T. supply voltage}}{\text{Load resistance}}.$$

Dynamic Mutual Characteristic — by deduction. A load line for the given load resistance value is drawn on the static anode characteristics as illustrated by Fig. 186. Assume that a grid-bias voltage of -3 V is applied, then the working point is at the intersection of the curve of $V_g = -3$ V and the load line *i.e.* at Q. Q is called the "quiescent point" and OY is the "standing current".

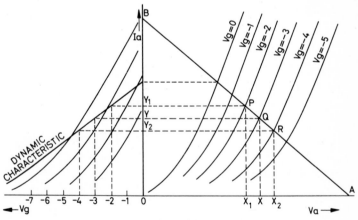

Fig. 186

Suppose now an a.c. voltage of amplitude 1 V is fed to the grid-cathode in series with the grid-bias voltage by means of an input signal device. The grid potential would vary between -2 V and -4 V as shown, and when the potential is -2 V, the working point will move to P and when -4 V to R. The anode voltage would vary between OX_1 and OX_2, having a mean value of OX, while the anode current varies between OY_1 and OY_2 having a mean value of OY.

If the values of anode current, obtained from the intersection of the load line and the different anode characteristics, are plotted against the corresponding values of V_g, the dynamic I_a/V_g characteristic is obtained. If this is superimposed on the static mutual characteristics it is seen that its slope is less, showing that the value for mutual conductance, under practical conditions when the valve is loaded, is less than the value obtained from the static conditions. In considering the triode as an amplifier, it is evident that the dynamic mutual characteristic should be used to obtain the working value of g'_m — the dynamic mutual conductance. The example shows how the value of g'_m can be determined graphically or alternatively how the stage gain can be found. Note that the dynamic amplification factor μ' is also less than the static value and is in fact given by

$$\frac{\mu R}{r_a + R} = A$$

Example 92. The following are the test values obtained on a triode valve which is to be used as a resistance-loaded amplifier.

V_a (volts)		0	50	100	150	200	250	300	350	400	450
I_a (mA) when V_g in volts =	0	0	3	7·8	14	22					
	-1			3	7	12	19				
	-2				2·7	5·5	9·5	16	22·5		
	-3					1·8	4·5	8·5	13·5	20	
	-4						2	4·5	8	13	18·5
	-5							2	4·5	8	17
	-6								2	4·5	7·7

If the load resistor is 20 kΩ and the supply voltage is 400 V, plot the variation of anode current with grid voltage for this amplifier.

If the amplifier is biased at − 2·5 V and a signal of 1 V peak value is applied to the grid/cathode, find the stage gain of the arrangement.

The curves as plotted and deduced are shown on Fig. 187. Points on the dynamic mutual characteristic are obtained in the manner described above and are shown by the dotted construction lines. A static mutual characteristic for 400 V h.t. has been plotted for comparison purposes.

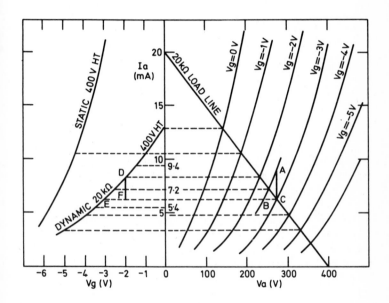

Fig. 187

From the dynamic characteristic we see that for a grid voltage of − 2·5 V the standing current is 7·2 mA.

If the grid voltage, with an input signal of 1 V, swings between − 3·5 V and − 1·5 V, the anode current changes between 5·4 mA and 9·4 mA.

The anode current thus swings through 9·4 − 5·4 = 4 mA and this is equivalent to an alternating current of 2 mA (peak value). The peak value of voltage developed across the load = $2 \times 10^{-3} \times 2 \times 10^{4}$ = 40 V. The peak value of input voltage = 1 V

Thus the stage gain = $\dfrac{40}{1}$ or $A = 40$.

As an alternative solution and check we could proceed as follows. Draw in a portion of the static anode characteristic for a 2·5 V grid-bias value. Such a characteristic would be midway between the -2 V and -3 V curves. On this portion of the characteristic, we can construct the triangle ABC and obtain

$$r_a \; = \; \frac{BC}{AC} \; = \; \frac{275 - 245}{(9 - 6 \cdot 1)10^{-3}} \; = \; \frac{30}{2 \cdot 9 \; \times \; 10^{-3}} \; = \; 10 \cdot 34 \; k\Omega$$

$$\text{also} \; \mu \; = \; \frac{275 - 245}{-2 \cdot 5 \; -(- 3)} \; = \; \frac{30}{0 \cdot 5} \; = \; 60$$

$$\therefore \; A \; = \; \frac{\mu R}{r_a + R} \; = \; \frac{60 \times 20}{10 \cdot 34 + 20} \; = \; \frac{1200}{30 \cdot 34} \; = \; 39 \cdot 5$$

Thus stage gain, as before $= 40$.

The fact that the dynamic mutual characteristic is less steep than the static mutual characteristic is explained by the inclusion of the anode-load resistance which is now part of the circuit.

Thus for the static mutual characteristic $g_m \; = \; \dfrac{\mu}{r_a}$ but for the

dynamic mutual characteristic $g'_m \; = \; \dfrac{\mu}{r_a + R}$

Since $A \; = \; \dfrac{\mu R}{r_a + R}$ then $A \; = \; g'_m R$. This can be confirmed

from the graph of the example or a further alternative solution is possible if the slope of the dynamic mutual characteristic is obtained as $\dfrac{DF}{EF}$.

$$\text{Thus} \; g'_m \; = \; \frac{(8 \cdot 1 - 6 \cdot 1)10^{-3}}{-2 \; -(-3)} \; = \; \frac{2 \; \times \; 10^{-3}}{1} \; = \; 2 \; mA/V$$

Then $A \; = \; 2 \times 10^{-3} \times 2 \times 10^4 \; = \; 40$ (as shown previously).

Distortion. Deduction of the dynamic mutual characteristic for any particular value of load resistor R enables us to understand more fully the process of amplification. Since this characteristic is not straight then, it is important that, the grid-bias value should be so chosen that working should be over a linear section of the curve, otherwise distortion would result. The diagram of Fig. 188 has been drawn to exaggerate the distorting effect obtained by a poor choice of grid-bias value. A bias of -4 V is used and a signal of 2 V peak is assumed. It is seen that as the grid swings less $-$ve, the anode current rises and as the grid becomes more $-$ve, the anode current falls. If the waveform of the incoming signal and that of the anode current is plotted to a time base, the resulting voltage output, due to anode current passing through the load resistor R, is seen to be appreciably distorted. To ensure that the output voltage is a true enlarged replica of the input signal, such distortion should be avoided by, as stated earlier, a correct selection of grid-bias voltage to give operation over a linear portion of the characteristic.

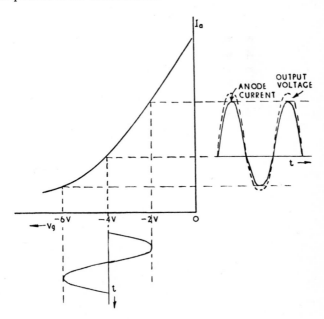

Fig. 188

Linearity of the dynamic mutual characteristic can be encouraged by the use of a load line, so located, to give a uniform ratio of grid-voltage change to anode-current change. This is ensured by drawing a load line so that the intercepts made on it by the various static curves are equal. The point, being made, is illustrated by the following example.

Example 93. The output characteristics of a thermionic triode have been set out in Fig. 189. The triode is to be used in a resistance-loaded amplifier with (a) a load of 16·6 kΩ and a h.t. supply of 200 V and (b) a load of 5·42 kΩ and a h.t. supply of 130 V. Construct the

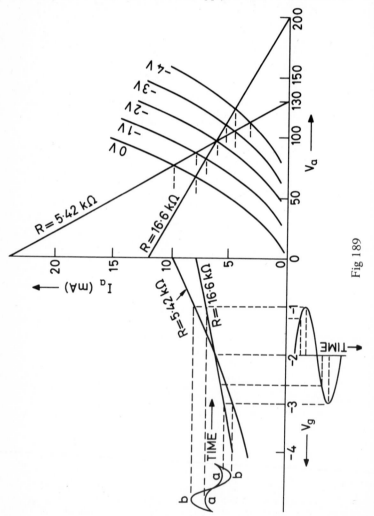

Fig 189

dynamic characteristic of the amplifier for each condition and suggest a suitable bias voltage when the input voltage to the grid is a sine wave of peak value 1 V. With the suggested bias, sketch the resultant anode-current waveform.

The object of the example is to illustrate distortion by poor choice of load resistance and h.t. voltage. Procedure for determining the dynamic mutual characteristics is known and is therefore not described. The results show both characteristics are the result of loading which suggests a standing current of 6mA and a bias voltage − 2 V. When the sinusoidal signal voltage of 1 V is applied to the characteristics, it is noted that, the resulting anode current waveforms appear sinusoidal. On closer examination waveform (a) for the 16·6 kΩ load is seen to be symmetrical, peaking to 0·8 mA but for the 5·42 kΩ load the following is evident. The waveform of (b) peaks to 2 mA for the +ve half-cycle and to only 1·5 mA for the −ve half-cycle. Thus distortion is occurring even though amplification is being achieved. The current waveform is drawn to a small scale and amplification is not apparent but for the 16·6 kΩ load it is

$$0·8 \times 10^{-3} \times 16·6 \times 10^3 \ = \ \frac{13·28 \text{ V}}{1 \text{ V}} \text{ or } 13·3 \text{ times.}$$

For the 5·42 kΩ load it is $\ 2 \times 10^{-3} \times 5·42 \times 10^3 \ = \ \dfrac{10·84 \text{ V}}{1 \text{ V}} \text{ or}$

11 times for the + ve half-cycle and $\ 1·5 \times 10^{-3} \times 5·42 \times 10^3 \ = \ \dfrac{8·13 \text{ V}}{1}$

or 8 times for the −ve cycle. The distortion is thus evident.

PHASE REVERSAL. An effect which may have been noted by the reader when considering the amplifying action of the triode valve, but to which attention is now drawn, is the reversal of phase between the output load voltage and the incoming signal. Because a grid voltage reduction results in an increase of anode current and a rise of load-resistor voltage drop, it follows that the output-voltage wave is in phase opposition to the input-signal waveform. Such a reversal of phase is not in itself a disadvantage but in the design of amplifiers, employing feed-back circuits, due attention must be paid to such phasing. The diagrams of Fig. 188 and Fig. 189 show the reversal of phase to which attention is being drawn.

AUTOMATIC GRID BIAS

For indirectly heated valves, in which the cathode is insulated from the heater, an alternative method of providing bias for the grid is possible. Consider the diagram of Fig. 190a. It will be seen that the cathode is connected to the h.t. negative line through a resistor R_c.

When standing anode current I_a is flowing, it passes from the +ve of the h.t. supply, through the valve, and returns via R_c. There is thus a voltage drop across R_c equal to $I_a \times R_c$. The cathode will thus be at a +ve potential with respect to the −ve h.t. line and since the grid is connected to this −ve line, it will be biased negatively with respect to the cathode by a value equal to $I_a \times R_c$. When I_a varies in response to an incoming signal, in accordance with normal operating conditions, it will vary the voltage drop across R_c and thus the bias voltage. Since this would be a disadvantage, a steady bias being needed, a capacitor C_c is connected in parallel with R_c, This capacitor, as shown in the diagram of Fig. 190b, smooths out the current variations in R_c because, as the voltage across R_c falls, C_c discharges through R_c and when the voltage rises, C_c charges up again. The voltage drop across R_c is thus kept almost constant.

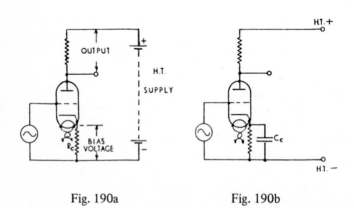

Fig. 190a   Fig. 190b

The method is simple and self-adjusting. If I_a falls due to reduced anode voltage, the voltage drop across R_c falls thus reducing the bias voltage. The method of biasing, as described, is known as "cathode bias" and is usual for all types of indirectly-heated valves. The bias capacitor is of a large value, some 20 to 30 μF and is usually of the electrolytic type since the working voltage is low. Attention must be given to the correct connection of this component as regards its polarity.

Example 94. For a standing current of 30 mA a triode vlave, supplied with the correct h.t. voltage through a load resistor, is found to require a bias voltage of −6 V. Find the value of the grid-bias resistor.

A simple solution is possible since

$$\text{bias voltage} \quad = \quad \text{valve current} \times \text{grid resistor}$$

$$\text{or } R_c \quad = \quad \frac{6}{30 \times 10^{-3}} \quad = \quad \frac{10^3}{5} \quad = \quad \frac{1000}{5} \quad = \quad 200 \text{ ohms}$$

As stated the smoothing capacitor value will be some 20 to 30 μF — decided by practical tests.

MULTI-STAGE A.C. AMPLIFIERS

When more alternating voltage amplification is required than a single triode can provide, then a second triode can be added to amplify the alternating voltage appearing at the anode load of the first. The anode load of the first valve can be a resistor but alternative methods of loading are possible and are also discussed.

Valve Loads. The introduction of a load resistor into the anode circuit of a triode results in a simple effective method of producing an amplified signal voltage. Such a resistor must have an appreciable ohmic value in order to achieve the desired voltage drop and this in turn means that the voltage, produced by the h.t. "power pack", must be high. Since the current through the load is fluctuating, it is apparent that a.c. circuit techniques can be employed and a reactor substituted for the load resistor. Such a reactor or choke would mean a lower ohmic resistance value in the circuit and thus a reduction of h.t. voltage — a beneficial result. Furthermore, since its reactance value is dependent on frequency, then the a.c. voltage drop across it would be large, and the amplification would be effective for the frequency range normally encountered. In the same way a transformer with a step-up turns ratio and a low resistance primary winding could be considered as a substitute for the load resistor. Such choke and transformer methods of achieving voltage amplification will be mentioned as the multi-stage amplifier is being described.

VOLTAGE AMPLIFIERS IN CASCADE. The diagram of Fig. 191 shows two triode valve coupled to give increased amplification. The output of V_1 is required to provide the input of V_2 through a capacitor-resistor arrangement. Here resistance-capacitance (R-C) coupling is employed and since, only the alternating voltage at the anode of

valve V_1 is to be fed to the grid of V_2, operation is effective because
the coupling capacitor C is of a value which offers a low impedance
at the frequencies being encountered. It should be noted that direct
coupling of V_1 anode to V_2 grid would not be possible because of
the steady voltage at the anode of V_1 which would destroy the biasing conditions for the grid of V_2. Capacitor C isolates the h.t. of V_1
from the grid circuit of V_2 which, in the diagram of Fig. 191, is shown
with its own grid-bias battery. This battery is connected to the grid
of V_2 through R_g, a high resistance, of value about 1 MΩ, which is
often referred to as a "grid-leak" resistor. The function of the grid
resistor R_g is to provide conditions similar to those for the grid of V_1.
The grid of the second triode must be tied down to the −ve line and
correctly biased. If it was connected directly to the bias battery, the
impedance of this grid circuit would be negligible in relation to that
of the a.c. voltage source supplying it, and correct amplification
would be impossible. By inserting R_g, conditions are provided similar
to those arising from an alternating-voltage source being inserted
between the −ve terminal of the bias battery and the grid of V_2. Grid
resistor R_g, although in parallel, being of a very large value would not
alter the requirement of a high-impedance voltage source − the voltage
drop across R_g being matched into the high impedance input of V_2.
Note that any permanent charge on the grid of V_2, stimulated by the
presence of C under standing conditions, is dissipated by the electrons
leaking away through R_g − hence the term, grid-leak resistor.

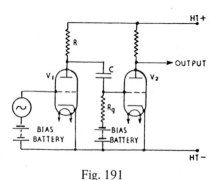

Fig. 191

The R-C Coupled Amplifier. Although this has already been introduced, a more detailed explanation of its working is probably desirable.
The circuit has also been rearranged for Fig. 192 to show the use of a
common grid-bias source. Referring to this diagram, we see that, since
R_g is connected to C then, as the anode voltage fluctuations of V_1

are applied to C, this capacitor can charge or discharge through R_g as required. Suppose the anode of V_1 swings +ve, the potential across C will rise and the charge on C will increase, resulting in an electron flow through R_g in the direction YX. Since electron flow is from —ve to +ve potential, this results in point X and thus the grid of V_2 being driven +ve relative to point Y. Thus the grid of V_2 swings with the anode of V_1. Similarly, if anode V_1 swings —ve, C discharges producing electron flow from X to Y. The grid thus becomes more —ve relative to the —ve terminal of the battery.

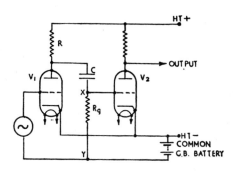

Fig. 192

Equivalent Circuit. This method of treatment has already been introduced and achieves a clearer understanding of the circuit operating conditions. The R-C circuit is redrawn for simplification in Fig. 193 and the Equivalent Circuit diagram is illustrated by Fig. 194. We can summarise the circuit representation as follows, whilst remembering that it applies to the a.c. working conditions only. The valve is considered to be an a.c. generator of voltage output μE_g and internal resistance r_a. The generator circuit is completed through R and the h.t. source, which is considered to have negligible impedance. The voltage developed across R by the input signal is transferred as fully as possible to the grid of the second valve. The voltage drop across C must be as low as possible, so its reactance should be kept small.

The a.c. current flowing would be given by $I = \dfrac{\mu E_g}{r_a + R}$ and the

voltage drop across R would be $\dfrac{\mu E_g R}{r_a + R}$. The stage gain or voltage

amplification, as already shown, is the ratio of the output voltage to

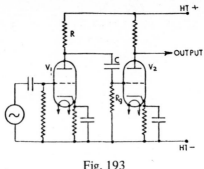

Fig. 193

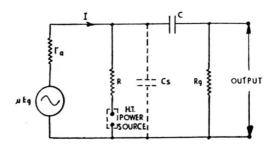

Fig. 194

that applied at the grid-cathode input. Thus $A = \dfrac{\mu E_g R}{r_a + R} \bigg/ E_g$ or

$A = \dfrac{\mu R}{r_a + R}$. Note that this is the same expression as that used for

the dynamic amplification factor $i.e.$ $A = g'_m R$ where g'_m is obtained from the dynamic mutual characteristic.

$R\text{-}C$ coupling gives uniform amplification over a wide range of frequencies, but this tends to fall off at the upper and lower ends of this range. Stray capacitance, due to valve inter-electrode capacitance, capacitance to earth and that of the wiring are shown by C_S (dotted) on Fig. 194. At high frequencies, the reactance of C_S may be low enough to shunt R and reduce amplification, while at low frequencies, the reactance of C may be large enough to decrease the output voltage and thus lower the stage gain.

Example 95. The measured voltage gain of a triode amplifier with a 10 kΩ resistive load is 20. With a 15 kΩ resistive load the gain increases to 25. Calculate the values of the three triode parameters.

From the expression $A = \dfrac{\mu R}{r_a + R}$ we have by substitution

$20 = \dfrac{10\mu}{r_a + 10}$ and $25 = \dfrac{15\mu}{r_a + 15}$. Solving these two

equations then $\dfrac{25}{20} = \dfrac{15\mu}{r_a + 15} \bigg/ \dfrac{10\mu}{r_a + 10} = \dfrac{15}{10} \times \dfrac{r_a + 10}{r_a + 15}$

or $\dfrac{5}{4} = \dfrac{3}{2} \times \dfrac{r_a + 10}{r_a + 15}$ giving $\dfrac{10}{12} = \dfrac{r_a + 10}{r_a + 15}$

$\therefore 10r_a + 150 = 12r_a + 120$ or $2r_a = 30$

Whence $r_a = 15 \text{ k}\Omega$

Substituting in $20 = \dfrac{10\mu}{r_a + 10}$ we have $20 = \dfrac{10\mu}{15 + 10}$

or $20(15 + 10) = 10\mu$ whence $\mu = 2 \times 25 = 50$

Again since $\mu = r_a \times g_m$ $\therefore g_m = \dfrac{\mu}{r_a} = \dfrac{50}{15}$

or $g_m = 3{\cdot}3 \text{ mA/V}$.

Example 96. What percentage of a 50 Hz signal voltage, as developed across the load resistor of the first stage of a two-valve amplifier, is passed to the following stage if the coupling capacitor is 1838 pF and the grid resistor is 1 MΩ.

The reactance of the capacitor, $X_c = \dfrac{1}{2\pi f C}$

$= \dfrac{10^{12}}{2 \times 3{\cdot}14 \times 50 \times 1838} = \dfrac{10^7}{3{\cdot}14 \times 1{\cdot}838} = \dfrac{10^7}{5{\cdot}77}$

$$= \frac{10\ 000}{5 \cdot 77} \times 10^3 = 1740 \text{ k}\Omega$$

The impedance of the resistor-capacitor series circuit is given by $Z = \sqrt{R^2 + X^2}$

$$\text{or } Z = 10^3 \sqrt{1000^2 + 1740^2} = 10^6 \sqrt{1^2 + 1 \cdot 74^2} = 10^6 \sqrt{1 + 3}$$

$$= 10^6 \sqrt{4} = 10^6 \times 2 \quad \text{or} \quad Z = 2 \text{ M}\Omega$$

Let I = the series circuit a.c. current

$\therefore$ voltage drop E_g across grid resistor $= I \times R_g$ volts

But $I = \dfrac{E_a}{Z}$ where E_a = the signal e.m.f. So $I = \dfrac{E_a}{2 \times 10^6}$

$$\therefore E_g = \frac{E_a \times 1 \times 10^6}{2 \times 10^6} = \frac{E_a}{2} \quad \therefore E_g = 0 \cdot 5 E_a$$

Thus 50 per cent of the signal voltage is developed across the grid resistor *i.e.* 50 per cent of signal voltage is passed on or more briefly,

$$\text{Percentage of input signal} = \frac{1 \text{ M}\Omega}{2 \text{ M}\Omega} \times 100 = 50 \text{ per cent}$$

The Choke-Capacitance Coupled Amplifier. In order to avoid the large voltage drop in the load resistor of the *R-C* amplifier, an inductor L can be used with a similar coupling capacitor C and grid-leak resistor R_g. The arrangement is shown by Fig. 195. The equivalent circuit can also be deduced and is illustrated by the diagram of Fig. 196.

Inductor L will have a comparatively low resistance and hence a low d.c. voltage drop. It will however offer a high impedance to the a.c. component of I_a. Amplification is thus better than for the *R-C*

amplifier. Stage gain $A = \dfrac{\mu Z}{r_a + Z}$. It should be noted that in the

expression for the denominator $r_a + Z$, a phasor additon is required, Z being the load impedance.

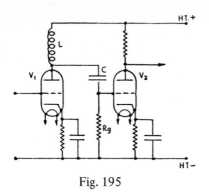

Fig. 195

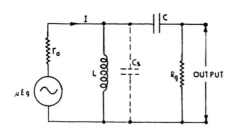

Fig. 196

Example 97. The anode-load impedance of an amplifier is a choke of inductance 10 mH and resistance 10 kΩ. Find the stage gain at 300 kHz, if the a.c. resistance of the valve is 20 kΩ and the amplification factor is 10.

Here $X = 2\pi f L = 2 \times 3\cdot14 \times 300 \times 10^3 \times 10 \times 10^{-3}$

$= 6\cdot28 \times 3 \times 10^3 = 18\cdot84 \times 10^3 = 18\,840\,\Omega \text{ or } 18\cdot84\,k\Omega$

$\therefore Z \text{ (in kilohms)} = \sqrt{10^2 + 18\cdot84^2} = 10\sqrt{1^2 + 1\cdot884^2}$

$= 10\sqrt{1 + 3\cdot53} = 10\sqrt{4\cdot53} = 10 \times 2\cdot13 = 21\cdot3\,k\Omega$

Similarly $r_a + Z = \sqrt{(10 + 20)^2 + 18\cdot84^2} = \sqrt{30^2 + 18\cdot84^2}$

$= 10\sqrt{3^2 + 1\cdot884^2} = 10\sqrt{9 + 3\cdot53}$

$= 10\sqrt{12\cdot53} = 10 \times 3\cdot54 = 35\cdot4\,k\Omega$

Then $A = \dfrac{10 \times 21\,300}{35\,400} = \dfrac{2130}{354} = 6.$

The Transformer Coupled Amplifier. The arrangement is shown by
the diagram of Fig. 197 and the equivalent circuit is illustrated by
the diagram of Fig. 198. It will be seen that the primary of the transformer is inserted as an inductive load in the anode circuit of V_1 and
as no current flows in the secondary, the primary can be regarded as
acting like an inductor of high value.

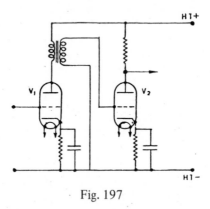

Fig. 197

Since the resistance of the primary is small enough to be neglected
then the method is similar to that used for the choke-capacitance
amplifier, except that because of the transformer action, the ultimate
stage gain will be further increased. The voltage V_p is developed
across the primary by the flow of alternating current, produced by
the generating property of the valve V_1. The secondary voltage V_s,
stepped up by the transformation ratio, is applied across the grid and
cathode of the valve V_2.

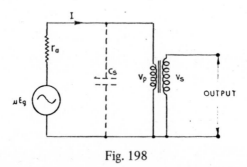

Fig. 198

Transformer coupling was formerly frequently used for sound
reproducing equipment but because of the d.c. magnetisation of the

core caused by the standing current, the transformers besides being large and heavy, required special constructional techniques and were expensive. Modern techniques of valve miniaturisation offer high degrees of amplification and, for the earlier stages in a valve amplifier, there is little advantage in using an inter-valve coupling transformer, which may introduce distortion, instead of the simple R-C coupling.

Example 98. A transformer-coupled amplifier uses a valve with an a.c. resistance of 10 kΩ and an amplification factor of 14. If the transformer has a ratio of 1:3 and a primary inductance of 7 H find the stage gain at a frequency of 500 Hz.

From first principles and using the equivalent circuit we have

$$V_p = IX \text{ and } I = \frac{\mu E_g}{Z_c}. \text{ Here } Z_c \text{ is the anode circuit impedance}$$

i.e. the phasor sum of r_a and X.

$$\text{Thus } I = \frac{14 E_g}{\sqrt{10^2 + X^2}}. \text{ Here } X = \text{ the transformer primary}$$

reactance in kilohms — the resistance is negligible. Here the value of E_g would be in kilovolts but this is unimportant, since it cancels out in the amplification expression.

$$\text{So } X = 2 \times 3 \cdot 14 \times 500 \times 7 = 3 \cdot 14 \times 7 \times 10^3$$
$$= 21 \cdot 98 \times 10^3 = 21 \cdot 98 \text{ k}\Omega$$

$$\text{and } I = \frac{14 E_g}{\sqrt{10^2 + 21 \cdot 98^2}} = \frac{14 E_g}{10 \sqrt{1^2 + 2 \cdot 198^2}} = \frac{14 E_g}{10 \sqrt{1 + 4 \cdot 84}}$$

$$= \frac{1 \cdot 4 E_g}{\sqrt{5 \cdot 84}} = \frac{1 \cdot 4 E_g}{2 \cdot 42} .$$

$$\therefore V_p = \frac{1 \cdot 4 E_g \times 21 \cdot 98}{2 \cdot 42}$$

Since the input is E_g then $A = \frac{1 \cdot 4 \times 21 \cdot 98 E_g}{2 \cdot 42} \bigg/ E_g$ or $A = \frac{1 \cdot 4 \times 21 \cdot 98}{2 \cdot 42}$

This is the stage gain due to the coupling only.

The overall stage gain with transformer action $= 1 \cdot 4 \times 9 \cdot 1 \times 3$

$$= 38 \cdot 22$$

From the above deduction, we see that the stage gain of a transformer coupled amplifier can be written as $A = \dfrac{\mu Z N}{r_a + Z}$.

Here $r_a + Z$ is a phasor summation and Z is the load impedance *i.e.* that of the transformer primary, and N is the secondary to primary turns ratio.

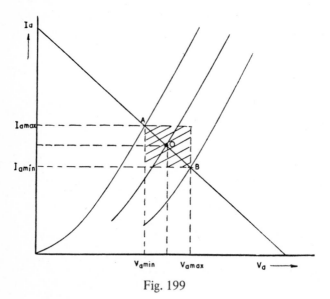

Fig. 199

POWER AMPLIFICATION

Up to now the valve has been considered as an amplifier of alternating voltage *i.e.* as a device used for obtaining an output voltage which is greater than the input voltage but which is otherwise similar in waveform and frequency. The input signal is usually in the form of a complex or sinusoidal alternating voltage and controls a small amount of d.c. power taken from the h.t. supply to produce the necessary output voltage. This output voltage is fed into a high impedance load and thus the output current and power is small. When the output at the final stage of an amplifier is required to drive a load such as a loudspeaker, mechanical indicator or control device then a certain definite power will be required, and the output valve must be capable of dealing with this power. The valve can thus be classed as either a voltage or power amplifier, dependent on the function to be fulfilled, and since the former has been considered in detail, attention is now given to the operating requirements of the latter.

For power amplification both large current and voltage changes are required and it should be noted that power gain is equal to the product of the voltage amplification and the current amplification. Thus the output alternating power developed depends as much on the change of I_a as on the change of V_a. The product gives the area of the rectangle shown by the diagram of Fig. 199, which illustrates the dynamic operation of a valve loaded with resistance. O is the quiescent point and the valve operates along the load line AB.

It should be noted that each side of the rectangle represents the swings of voltage and current and thus each side is twice the peak value which, assuming sinusoidal working, is $\sqrt{2}$ times the r.m.s. value.

Thus area of power rectangle $= 2\sqrt{2} \times 2\sqrt{2} \times P = 8P$ where P is the a.c. power.

$$\text{or } P = \frac{\text{Area of rectangle}}{8}$$

P is in watts and is thus equal to $\dfrac{(V_{a_{max}} - V_{a_{min}})(I_{a_{max}} - I_{a_{min}})}{8}$

The alternating power can also be found without the use of characteristics, provided the appropriate valve constants are known. The method is illustrated by the following example.

Example 99. A power triode, having an a.c. resistance of 2 kΩ and an amplification factor of 6 is loaded with a 4 kΩ resistor. Determine the alternating power output with a signal of 30 V peak.

$$E_s = 0.707 \times 30 = 21.21 \text{ V (r.m.s.)} \quad \text{Also } E_o = \mu\, 21.21$$
$$= 6 \times 21.21 = 127.26 \text{ V}$$

$$I_a = \frac{127.26}{(2+4)10^3} = 21.21 \times 10^{-3} \text{A} \quad \text{or} \quad 21.21 \text{ mA}$$

$$P = I_a^2 R = 21 \cdot 21^2 \times 10^{-6} \times 4000$$
$$= 21 \cdot 21^2 \times 4 \times 10^{-3}$$
$$= 1 \cdot 8 \text{ W.}$$

LOAD COUPLING. To allow an output valve to drive a loudspeaker it is necessary to ensure that the impedance presented to the anode is such that a satisfactory amount of power is developed. The diagrams of Fig. 200 show methods of operating a loudspeaker but for reasons, not detailed here, a power amplifier valve offers best performance in terms of efficiency and low distortion when a particular load impedance is included in the anode circuit. The optimum load impedance is listed in the manufacturer's technical information.

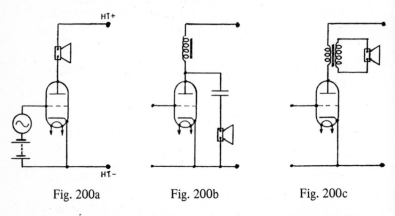

Fig. 200a   Fig. 200b   Fig. 200c

Assume a triode which requires an anode-load impedance of 5 kΩ is being used and a moving-coil loudspeaker, having an impedance of 3 Ω is to be matched. No difficulty is involved in coupling the units provided a transformer of suitable ratio is used. The transformer is coupled as in Fig. 200c, and it can be deduced from the theory of mutual inductance and coupled circuits that, the impedance presented by the primary is equal to the secondary load impedance multiplied by the square of the primary to secondary turns ratio. If this ratio is N

then $Z_1 = N^2 Z_2$ or $N = \sqrt{\dfrac{Z_1}{Z_2}} = \sqrt{\dfrac{5000}{3}} = 41.$

The "speaker" transformer should therefore have a primary/secondary turns ratio of 41:1 but a standard 40:1 ratio would be suitable. Mention has been made earlier of the need for a specially manufactured transformer, when such a component is included in a valve anode circuit. A "speaker" transformer is such a component since the d.c. anode current flows through it together with the a.c. component.

Thus the primary must be wound with wire sufficiently thick to pass the total anode current and since the direct current will produce some core magnetisation, this is minimised by butt-jointing the laminations and introducing a small air-gap. Although a pentode valve is frequently used instead of a triode for the output power stage, such a "speaker" or output transformer would still be necessary.

CLASSES OF AMPLIFICATION. Although it is not intended to give this aspect of theory any detailed attention, it is necessary to mention that three types of valve amplification are defined by the region of the mutual characteristic over which the valve is required to operate. These are illustrated by the diagrams of Figs. 201, 202 and 203. These types are:

Class A—when the valve is operated on the straight line portion of the I_a/V_g curve.

Class B—when the valve is operated on the curved portion of the I_a/V_g curve.

Class C —when the valve is biased beyond the cut-off value.

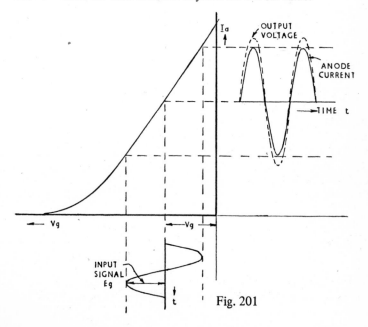

Fig. 201

Class A Amplification. All the work done up to now has assumed this type of operation. The alternating voltage applied to the grid causes an alternating current to appear in the anode circuit and this variation of I_a in turn, causes an alternating output voltage. Thus the peak a.c. output voltage may be 20 V whilst the change in grid voltage may be 2 V. The amplification or stage gain is thus $\dfrac{20}{2} = 10$. If the straight line portion of the I_a/V_g curve is not used, distortion of the signal will occur. The operation is illustrated by Fig. 201.

Class B Amplification. Here the valve is biased near its cut-off value and typical operating conditions are shown by Fig. 202. The output voltage is virtually only a half cycle. This output is therefore distorted as no anode current flows for almost half a cycle and although the Class B amplifier is much more efficient than the Class A amplifier, because of the very small value of standing current, the distortion of output signal is a serious problem.

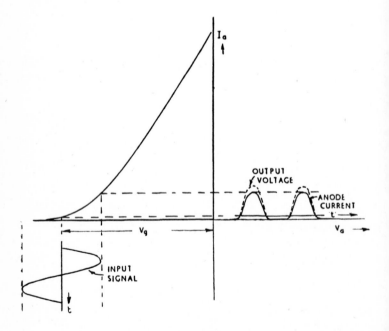

Fig. 202

Class C Amplification. For this the valve is biased beyond its cut-off value. No anode current will flow for the greater part of the cycle as shown in the diagram of Fig. 203. The use of Class C amplification is very limited, such usage being generally for radio frequencies when the turned circuits, which are used, are sufficiently selective to filter out the harmonics which occur as a result of distortion.

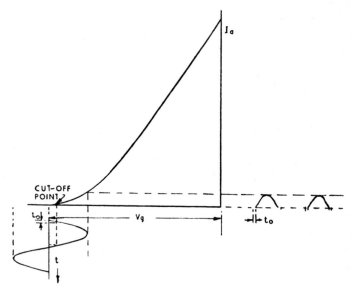

Fig. 203

Because the valve is biased to beyond the cut-off point, anode current flows for a small part of each cycle, forming a succession of short pulses. This current can be resolved into a steady component and other a.c. components, one of these latter is at the input frequency and the others at harmonic frequencies. By the use of tuned circuits, a considerable amount of power can be developed in the load at the input frequency and negligible power at other frequencies. Thus Class C amplification is suitable and efficient for a single frequency or narrow band of frequencies.

PUSH-PULL OPERATION

The power available from the output stage of an amplifier can be increased by using two valves, (a) in series, (b) in parallel and, (c) in push-pull. Series connection introduces difficulties of connection and operation and is therefore not used. Parallel operation can be arranged but is seldom used because, if increased power is required then, the push-pull arrangement with its inherent advantages is a better alternative.

Push-pull operation uses two valves to give a power output greater than twice that of a single valve since distortion is minimised. The diagram of Fig. 204 shows the arrangement which operates as described below.

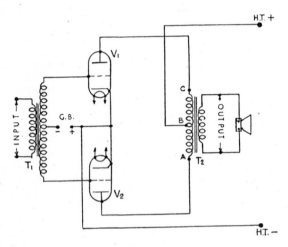

Fig. 204

The input signal is converted into two voltages which are fed to the grids of valves V_1 and V_2. These voltages are equal and in antiphase *i.e.* 180° out of phase, since the grids are connected to the opposite ends of the secondary of the "driver" transformer T1. Because of the phase opposition of the grid voltages, the anode currents of the valves are also in phase opposition. On the application of an input signal, the anode current of V_1, consists, as can be expected, of a steady d.c. value with an a.c. component superimposed on it. This current flows through the primary winding of the "output" transformer T2 in the direction B to C. Similarly for valve V_2, the anode current consists of a steady d.c. value and a superimposed a.c. component which flows in the direction B to A. Since the phase of the grid signal voltage on V_2 is antiphase to that of V_1 the anode current fluctuations are anti-phase or the alternating current, flowing through the transformer primary, is the difference between the two opposite phase a.c. components with the result that their combined effect is twice that of one of them alone. This effect can also be viewed alternatively for the primary of the output transformer where the fall of anode current of V_1 assists the rise of anode current in V_2 to produce an a.c. flux in the core. Thus although the d.c. currents in the two halves of the windings are in opposite directions and cancel out their magnetic effects, the combined effect of the two alternating components, in each half winding, is such as to be double that of one.

Either Class A or Class B operation can be used for the push-pull arrangement but the latter is favoured because the quiescent current is almost zero and the current drain on the h.t. supply is less than for Class A operation. Also, since second and other even harmonics can be shown to be eliminated, harmonic distortion is reduced by push-pull working and the valves can be worked over a larger range of their characteristics thus allowing greater power output.

MULTIELECTRODE VALVES

Since the electrodes of the triode valve constitute the plates of a capacitor then interelectrode capacitance effects can be expected. Although the capacitance values are only of the order of a few pico-farads, at radio frequencies the reactance is very low and a by-pass effect is introduced between the grid and anode electrodes. Inter-electrode capacitance can be reduced by introducing additional electrodes, which are kept at fixed potentials.

THE TETRODE OR SCREEN-GRID VALVE

An electrode to be considered, additional to those used in the triode, is called the "screen grid" and is positioned between the control grid and the anode. It acts as an electrostatic shield to prevent coupling between the anode and control grid. The circuit symbol and arrangement are illustrated by Fig. 205.

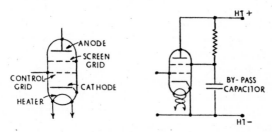

Fig. 205

Since some form of connection to the cathode is required to make the shielding effective, this usually takes the form of a voltage-dropping resistor, connected to the h.t. +ve line and a capacitor, called a by-pass capacitor, which is connected to h.t. —ve and has a very low reactive value at the frequencies involved. Thus although the screen grid is at a +ve d.c. potential, it is also at earth potential for the alternating current circuit and no alternating voltage can be fed back into the input circuit across the anode-grid capacitance.

The introduction of a screen-grid electrode has the following effect on valve action. When a +ve potential — approaching that of the anode, is applied to the screen, electrons from the cathode are attracted. Most of these electrons are travelling so fast that they pass through the mesh of the screen and reach the anode. Some electrons do however strike the screen and thus there is a screen current.

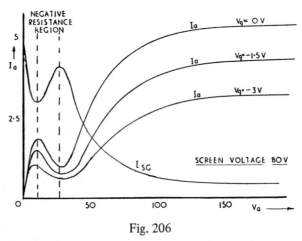

Fig. 206

CHARACTERISTICS. These can be determined by using a test circuit similar to that used for the triode, except that an additional voltage potentiometer resistor across the h.t. supply, is needed to supply the screen grid. The diagram of Fig. 206 shows the typical curves obtained by plotting the variation of I_a with V_a for various V_g values, the voltage on the screen-grid V_{SG} being kept constant. The screen current for one test condition is also drawn, and it will be noted that the situation can be reached where this is greater than the anode current. The irregularity of the I_a/V_a curve is caused by "secondary emission" and is a region of negative resistance $i.e.$ as voltage V_a is increased, I_a falls. A brief explanation for the shape of the characteristic at the section being discussed is as follows. As V_a is increased during tests, some electrons strike the anode with sufficient velocity to liberate surface electrons. These liberated electrons move into the space between the anode and screen grid and, because the latter is at a higher potential than the anode, the electrons move to the screen to cause a screen current. The emission of secondary electrons from the anode also results in repulsion of electrons arriving at the anode from the cathode. Thus there is an increase of screen current and a decrease of anode current until the anode potential is raised to a value which is greater than that of the screen. The electron flow then returns to the anode and I_a rises while screen current falls.

Valve Constants. The important part of the characteristic is the almost horizontal portion. The anode is collecting most of the electrons and screen current is low. The control grid, as for the triode, exercises control over the space current and the anode characteristics are widely spaced and are similarly shaped. The tetrode arrangement is thus seen to provide extra amplification and a greater mutual conductance value. The large +ve potential provided by the anode and screen-grid electrodes ensures that electrons are drawn off as soon as they are emitted and anode current is comparatively independent of anode voltage over the horizontal section of the characteristic. This indicates a high anode resistance value. The vertical spacing of the characteristics, being similar to that for a triode, indicates a similar mutual conductance value. Appropriate values for r_a and g_m are 300 kΩ and 2 mA/V respectively, giving a μ value of $300 \times 10^3 \times 2 \times 10^{-3} = 600$.

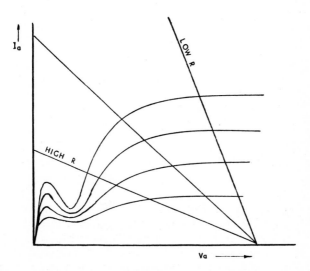

Fig. 207

The diagram of Fig. 207 has been drawn to show that the advantage of a high μ value cannot be used to a great extent since load resistance R should be high compared to r_a. The two load lines show that for a lower R value, equal intersections and satisfactory working is possible but for high R values, the negative resistance region being avoided, the load line intersects the rising part of the characteristic to give uneven intercepts and distortion results. R must thus be a fraction of r_a and the overall amplification, though higher than for a triode, is lower than could be expected.

THE PENTODE VALVE

The irregularity of the tetrode characteristic causes distortion which can be reduced, if the effects of secondary emission are overcome. The simplest way of achieving this is by the use of a "suppressor grid" introduced between anode and screen grid. The diagrams of Fig. 208 show the circuit symbol and the anode characteristic, deduced in a manner similar to that used for the tetrode.

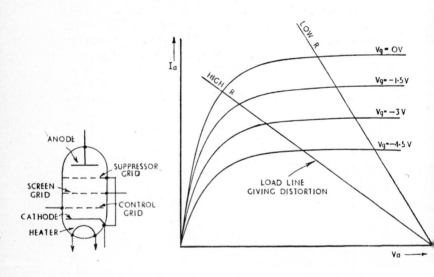

Fig. 208a   Fig. 208b

The suppressor grid is connected to the cathode and is thus − ve with respect to the anode. Any electrons liberated from the anode by secondary emission, pass into the space between anode and suppressor grid, but since the latter is at a −ve potential, the electrons are repelled and return to the anode. In this valve, the additional electrode overcomes secondary emission effects and increases the effective working range of the characteristic since the negative resistance region is eliminated.

Valve Constants. These are of the same order of magnitude as for the tetrode. The vertical spacing of the curves is similar and the g_m value is little changed and reasonably constant. The valve screening is assisted by the suppressor grid and V_a is higher than for the tetrode. Thus the amplification factor tends to be greater than for the tetrode. It must be apparent however, that it is not possible to obtain, from a pentode valve, an overall amplification approaching the μ value because a load resistor of several megohms would be required and the steady anode voltage would be reduced to a low value. For practical purposes an R value of 250 kΩ is usual which allows a gain of 100-200 in the audio-frequency range. I_a is almost constant over the working range and μ depends mainly on the extent of the variation in anode current which can be brought about by control-grid variations. Thus the mutual conductance g_m really determines the performance of this valve as a voltage amplifier. Fig. 208 shows that a load-resistance value which gives a load line passing through the "knee" of the $V_g = 0$ characteristic is a good guide. R values greater than this would give distortion and lower values would reduce the amplification possible. The R value necessary for this working condition would be a fraction rather than a multiple of r_a but would be some two or three times the value used for an equivalent triode valve and an increased

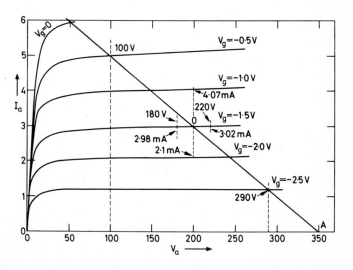

Fig. 209

amplification is thus the overall result of using a pentode. The following example illustrates the use of techniques described when the triode was being considered.

Example 100. The anode characteristics of a typical amplifying pentode are illustrated by the graphs of Fig. 209. Estimate the valve constants at the point $V_a = 200$ V, $V_g = -1.5$ V, $I_a = 3$ mA. If the valve is required to operate at the quiescent point defined by these values with a load resistor of 50 kΩ, determine, (a) the h.t. voltage required, (b) the rise and fall of anode potential for a signal voltage of peak value 1 V, (c) using the valve constants, calculate the amplification and compare the results with those obtained from the load-line construction.

For Point O

With $V_g = -1.5$ V $\quad r_a = \dfrac{\delta V_a}{\delta I_a} = \dfrac{220 - 180}{(3.02 - 2.98)10^{-3}}$

$$= \frac{40}{0.04 \times 10^{-3}} = \frac{40}{4 \times 10^{-5}}$$

$$= 10^6 \, \Omega \quad \text{or} \quad 1 \text{ M}\Omega$$

With $V_a = 200$ V $\quad g_m = \dfrac{\delta I_a}{\delta V_a} = \dfrac{(4.07 - 2.10)10^{-3}}{-1 - (-2)}$

$$= \frac{1.97 \times 10^{-3}}{1}$$

$$= 1.97 \times 10^{-3} = 1.97 \text{ mA/V}$$

$$\mu = r_a \times g_m = 1 \times 10^6 \times 1.97 \times 10^{-3}$$

$$= 1970.$$

Note: μ is difficult to obtain directly from the anode characteristics because a straight section must be used and since this is almost horizontal, the V_g for the corresponding current change is difficult to assess.

For the other answers the load line for 50 kΩ is drawn in thus:

With 3 mA flowing the voltage drop in $R = 3 \times 10^{-3} \times 50 \times 10^3$ = 150 V

(a) $\therefore$ The h.t. voltage $= 200 + 150 = 350$ V. The load line is drawn through points A and O $\therefore$ Answer (a) $= 350$ V

(b) A signal voltage of 1 volt peak would vary the grid voltage from -0.5 V to -2.5 V. Using the load-line intersection, the anode voltage is seen to vary between 290 V and 100 V

(c) Since $A = \dfrac{\mu R}{r_a + R} = \dfrac{1970 \times 50 \times 10^3}{(1000 + 50)10^3} = \dfrac{98 \cdot 5 \times 10^3}{1050}$

$= \dfrac{98 \cdot 5}{1 \cdot 05} = 94$

Using the load line $A = \dfrac{290 - 100}{2} = \dfrac{190}{2} = 95$

CHAPTER 10

PRACTICE EXAMPLES

1.   During a test on a vacuum triode valve, the following constants were obtained $r_a = 16$ kΩ, $g_m = 2.5$ mA/V. If this valve was then used with a resistance load of 80 kΩ, calculate the stage gain.

2.   A triode has the following constants, $r_a = 10$ kΩ, $g_m = 2.5$ mA/V, $\mu = 25$. If it is loaded with a resistor of 40 kΩ and coupled to the following stage through a capacitor, of negligible reactance at the signal frequency, and a grid resistor of 100 kΩ, determine the stage gain.

3.   The anode a.c. resistance of a triode valve is 10 kΩ and its amplification factor is 12. Calculate the input voltage that would give 3 V (r.m.s.) output across an anode-load resistance of 5 kΩ.

4.   A triode resistance-loaded amplifier has a gain of 10 when the anode-load resistor is 10 kΩ and a gain of $\dfrac{40}{3}$ when the anode-load resistor is changed to 20 kΩ. Deduce the value of the valve amplification factor (μ) and the anode-slope resistance (r_a), assuming these to be constant.

5.   An audio-frequency output stage uses a triode valve with an anode-slope resistance of 2 kΩ, an amplification factor of 7 and an anode-load resistor of 3 kΩ. What is the value of the mutual conductance of the valve used? What value of audio-frequency voltages must be applied to the grid of the valve if the power output is to be one watt.

6.   The results of tests on a vacuum triode are as follows

Anode voltage V_a(volts)	50	75	100	125	150	175	200	225
Anode Current I_a (mA) for $V_g = -2$ V	0·2	0·7	1·5	2·7	4·3	6·2		
Anode Current I_a (mA) for $V_g = -3$ V		0	0·3	0·9	2	3·4	5	6·8

Plot the anode characteristics and from them determine the value of the three parameters of the valve when operated at the point $V_a = 175$ V, $V_g = -3$ V and $I_a = 3.4$ mA. If the valve is used as an amplifier with an anode-load resistor of 75 kΩ, determine the stage gain of the amplifier.

7.   The following values of anode current were obtained with a triode

Anode voltage V_a (volts)				25	50	75	100
Anode current I_a(mA) with	$V_g = 0$			0.4	2.8	6	9.5
,,	,,	,,	$V_g = -1.5$ V	0	0.6	3	5.7
,,	,,	,,	$V_g = -3.1$ V	0	0	1.0	2.9

Plot the anode-current/anode-voltage characteristic. If a 150 V battery and anode-load resistor of 20 kΩ are connected across the valve, determine the anode current at the above values of grid voltages. Select a suitable operating point and calculate the stage gain when operating at this point.

8.   A triode valve, having an amplification factor of 60 and a mutual conductance of 2 mA/V, is employed as a voltage amplifier with an anode-load resistor of 47 kΩ. If the triode grid-bias voltage is -3 V and the valve operates under Class A conditions, calculate the standing or no-signal anode current.

9.   A triode has the following characteristics

Anode voltage V_a (volts)

	0	25	50	75	100	125	150	175	200	225	250	275	300

Anode current I_a (mA) with
$V_g =$

V_g	0	25	50	75	100	125	150	175	200	225	250	275	300
-2 V		0.2	0.7	1.5	2.7	4.3	6.3						
-3 V		0	0.3	0.9	2.0	3.4	5	6.8					
-4 V				0.1	0.5	1.3	2.5	4.0	5.6				
-5 V					0.3	0.8	1.7	3.0	4.5	6.2			

Plot the I_a/V_a characteristics and determine the values of mutual conductance and amplification factor when $V_a = 200$ V and $V_g = -4$ V. For this same quiescent point determine the value of the h.t. supply voltage for an anode-load resistor of 60 kΩ. Determine, using this load resistor, (a) the change in anode voltage and anode current when the voltage changes from -2 V to -5 V, (b) what is the voltage gain, (c) plot the anode dissipation against anode volts for each increment of grid voltage, (d) determine the a.c. power output for a signal of 1 V peak. Assume the same quiescent point as stated earlier.

10.   A pentode amplifier stage has an h.t. supply of 250 V. The anode-load resistor is 22 kΩ and the anode voltage is 140 V. The screen-grid decoupling-resistor is 150 kΩ and the cathode-bias resistor is 500 Ω. Draw a practical circuit diagram, inserting the appropriate capacitors and given the cathode voltage as 3 V and a mutual conductance for the valve as 6 mA/V. Calculate

(a) the cathode current, (b) the anode current, (c) the screen-grid current, (d) the screen-grid voltage, (e) the control-grid voltage, (f) estimate the stage gain.

THE INDUCTION MOTOR (i)

This motor is frequently referred to as an asynchronous machine because it operates at a speed below that of the rotating magnetic field set up within it. It is this asynchronous feature which determines the action of the motor which is both simple and robust in construction. It is the most common of the a.c. motors and since alternating current working is almost universal for shore work and is fast becoming accepted for marine practice, the importance of this machine as a motive unit should be appreciated. The prime disadvantage in its usage lies in the fact that its speed cannot be varied when used in its basic or *cage* form and that, if variable speed is a requirement of the driven machinery then, a special type of a.c. motor or an induction motor with built-in speed control characteristics must be used. The latter machines are more expensive and an alternative is to use the basic single-speed cage motor and provide the power output adjustment by mechanical means. Thus for pumps and fans such adjustment can be achieved by valves and dampers. Other practical methods use gears and/or adjustable pulleys with belt drives. Such methods are well-known to the marine engineer and attention will therefore only be given to the forms of induction motor which allow limited speed control.

PRINCIPLE OF OPERATION

In its basic form the machine is known as a "squirrel-cage" or more simply a "cage" motor. The theory of the rotating magnetic field has already been introduced in Chapter 9 and it can now be appreciated how, if the stator of a polyphase motor is suitably wound with the appropriate number of conductors and is energised from a suitable polyphase supply then, a rotating magnetic field is produced. It is the effect of this field on the rotor of the motor which is next to be considered in order to understand the theory of action.

Although it can be shown that the rotating magnetic field will drag round a rotatable solid iron cylinder when this is placed concentrically in the field; in the interests of efficiency, copper conductors are embedded in the face of such a cylinder and thus we have two forms of rotor, known as, (a) the cage type and, (b) the wound-rotor type. The cage type is more simply constructed, consisting of copper bars inserted, without insulation, into slots spaced evenly round the circumference of a rotor built up from iron laminations.

The bars are connected to copper rings at each end of the rotor and the winding, when considered alone, looks like a cage – hence the name. The winding thus, in effect, consists of single turns all connected in parallel. The wound-rotor construction employs insulated conductors put on, in coil form, in a manner similar to that used for the stator. There are thus three insulated phase windings, correctly positioned and spaced relative to each other. The phase windings are usually star-connected and the three free ends are brought out to slip-rings which are insulated from the shaft. Contact is made on to these rings by means of brushes mounted on a brush-arm and the external circuit is completed through three star-connected variable resistors. Control gear, built into the starter, allow these resistors to be cut out and the slip-rings to be "shorted". Thus the winding, when the motor is running, is short-circuited and in effect resembles the cage winding except that there are more turns per phase coil. More details of machine construction will be given in Chapter 13 but for simplicity, the theory set out below will assume conditions for a wound-rotor machine.

THEORY OF ACTION.   The diagram (Fig. 210), drawn for an instant in time, shows conductors on a rotor situated in a rotating magnetic field which is produced by a suitably wound stator fed with poly-phase currents. Some of the stator conductors are shown but current direction is not given since the rotating flux is the result of the effect of currents of different magnitudes in three phases. The rotor conductors can be considered to constitute a winding which is short-circuited.

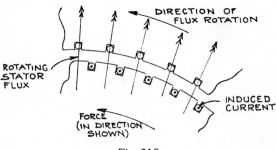

Fig. 210

If the field is considered to move, the relative cutting between flux and rotor conductors induces e.m.f.s in the latter producing currents in the direction shown. This can be checked if the "right-hand generator rule" is applied, with the conductors assumed to move in the direction opposite to that as shown for the flux and the latter is

considered to be stationary. Currents circulate in paths provided by
the end-rings and the conductors under the opposite stator poles.
Applying first principles for the action between the flux and the
rotor currents, we see that a force results which is in the direction
shown *i.e.* the rotor conductors are made to follow the rotating field
because of the torque produced. The rotor therefore commences to
revolve and accelerates up to a speed N_2 approaching the synchronous speed — the speed of the rotating field N_1. There must always be
a difference in speed between N_1 and N_2 otherwise there would be
no relative cutting, no rotor e.m.f.s induced, no rotor currents and
no torque.

Further consideration of the action will show that transformer
action is also involved. Consider again the "standstill" condition *i.e.*
rotor stationary. It has been shown that the rotating magnetic field is
basically of sinusoidal form and thus, as it rotates, the resulting
induced e.m.f. in the rotor and stator conductors are sinusoidal. The
rotor currents are also of sine-wave form and, if transformer working
is kept in mind, the rotor acts as a short-circuited secondary to the
stator which constitutes the primary. Keeping the transformer simile
we see that if the secondary *i.e.* rotor is open-circuited, then the
rotating flux cuts the stator conductors to produce a back e.m.f.
almost equal in value to the supply voltage and thus negligible current
is taken — as for the transformer on no-load. If the rotor winding is
completed to form a short-circuit, rotor currents flow and transformer
on-load conditions are similated. A secondary (rotor) flux is produced
which interacts with the stator flux to produce a resultant which is
smaller than the original. The stator induced back e.m.f. is reduced
and the supply voltage forces a larger current through the stator *i.e.*
stator current is dependent on rotor current.

Up to now standstill conditions have been considered but if the
rotor is considered to rotate, then the speed with which the field cuts
the rotor conductors is reduced, and hence the rotor e.m.f. is reduced
in magnitude. Rotor current is reduced in consequence and the stator
or supply current is reduced. Thus we have the condition of a large
starting current which falls as the machine runs up to speed. Further
details of the comparison with transformer working will be considered
as the theory is evolved but since reference has been made earlier to
the necessity for a difference in speed between rotor and the magnetic
field, then the first term associated with the induction motor, namely
slip or *slip speed* can now be introduced.

Slip speed. This is the speed at which the rotating field cuts the
rotor conductors and it can be expressed as a fraction or percentage
of synchronous speed. Thus if N_S is the slip speed then $N_S = N_1 - N_2$.

The ratio $\dfrac{N_S}{N_1}$ or $\dfrac{N_1 - N_2}{N_1}$ is called the slip s

As a fraction we have

Fractional slip $\quad s \;=\; \dfrac{N_S}{N_1} \;=\; \dfrac{N_1 - N_2}{N_1}$

or Percentage slip $s \;=\; \dfrac{N_1 - N_2}{N_1} \times 100$

Example 101. A four-pole, 400 V, 50 Hz induction motor operates with 4 per cent slip. Find the motor speed.

Synchronous speed N_1 is given from the relation $f \;=\; \dfrac{PN_1}{120}$

$\therefore \; N_1 \;=\; \dfrac{50 \times 120}{4} \;=\; 1500 \text{ rev/min}$

$\therefore \; 4 \;=\; \dfrac{1500 - N_2}{1500} \times 100 \text{ giving } 4 \times 15 \;=\; 1500 - N_2$

or $N_2 \;=\; 1500 - 60 \;=\; 1440 \text{ rev/min}.$

Alternatively, for 100 rev/min the slip is 4 revolutions
   or rotor speed $=$ 96 rev/min

Therefore for 1500 rev/min the slip is 4×15 revolutions
   or rotor speed $=$ 96×15 $=$ 1440 rev/min.

When the motor is running light *i.e.* unloaded, the torque is only that required to overcome friction and windage and the speed is nearly synchronous. The stator draws little current from the supply. As the load torque is increased *i.e.* motor is loaded, the speed falls slightly. This causes an increase in the relative cutting of the rotor conductors by the field and hence the rotor e.m.f. and current is increased. This effect gives the increased torque required and the speed is determined by this condition when power developed equals the power required.

The fall in speed from no-load to full-load is about 4 to 5 per cent for small motors and 1·5 to 2 per cent for large motors. The speed-load characteristic is thus similar to that of the d.c. shunt motor.

ROTOR TO STATOR RELATIONSHIPS

These can be readily deduced and summarise the basic facts of induction motor theory. The final result for each condition should be memorised.

1. FREQUENCY. For both the standstill and running conditions, the stator frequency is the supply frequency or $f_1 = f$. At standstill the rotating field also cuts the rotor at synchronous speed, so $f_2 = f$.

When running at speed N_2, the speed at which the rotor conductors are cut by the field is $N_1 - N_2$. Therefore $*f_2' = \dfrac{(N_1 - N_2)P}{120}$

But fractional slip $s = \dfrac{N_1 - N_2}{N_1}$ Therefore $sN_1 = N_1 - N_2$

and $f_2' = \dfrac{sN_1 P}{120} = sf$ or $f_2' = fs$

Thus Rotor Frequency = Supply Frequency × Fractional Slip.

*The introduction of the dash, as for f_2' should be noted. Such a dash will be used to distinguish "running" conditions as against those for stand-still when the dash is omitted.

Example 102. If the supply voltage to an eight-pole induction motor is of 50 hertz frequency and the induced e.m.f. in the rotor is of 1·5 Hz, find the speed at which the motor is running and the slip.

Since $f_2' = sf$ then $s = \dfrac{1·5}{50}$ or $\dfrac{3}{100}$

i.e. Fractional slip is 0·03 or Percentage slip is 3 per cent

$$\text{Synchronous speed } N_1 = \frac{120f_1}{P} = \frac{120 \times 50}{8} = 30 \times 25$$

$$= 750 \text{ rev/min}$$

$$\text{and } N_2 = 750 - \frac{750}{100} \times 3 = 750 - 22\cdot5 = 727\cdot5 \text{ rev/min}.$$

2. E.M.F. The rotating magnetic field cuts the stator and generates and e.m.f. given by the formula $E_1 = 2\cdot22 \; K_{D_1} \, K_{S_1} \, Z_{ph_1} \; \Phi f_1$ volts. Neglecting the voltage drops due to resistance and reactance then E_1 can be considered as a back e.m.f. being approximately equal to the supply voltage V.

For a stationary rotor the rotating field cuts the rotor at the same speed as for the stator and an e.m.f. E_2 is induced where $E_2 = 2\cdot22 \, K_{D_2} K_{S_2} Z_{ph_2} \Phi f_2$ volts. Here $f_2 = f$ the supply frequency.

With the rotor revolving the induced e.m.f. is variable depending on the rotor frequency since $f_2' = sf.$

$$\therefore \; E_2' = 2\cdot22 \, K_{D_2} K_{S_2} Z_{ph_2} \Phi sf = E_2 s$$

Rotor induced e.m.f. at any speed = Standstill e.m.f. × slip.

3. REACTANCE. As can be expected, the rotor has reactance as well as resistance. This reactance is due to the inductance of the circuits formed by the cage conductors being embedded in iron. The characteristics of the machine can be altered by the degree to which the conductors are embedded in the iron i.e. the depth of slot, but irrespective of this, we know that reactance is directly proportional to frequency and thus the rotor reactance varies as the slip changes.

Accordingly we can write $X_2' = 2\pi f_2' L_2 = s2\pi f_2 L_2$
$$= X_2 s$$

Rotor reactance at any speed = Standstill reactance × slip.

From the above relationships the following can be deduced:

For any speed. Rotor Impedance $= Z_2' = \sqrt{R_2^2 + (sX_2)^2}$

$$\text{Rotor Current } I_2' = \frac{E_2'}{Z_2'} = \frac{sE_2}{\sqrt{R_2^2 + (sX_2)^2}}$$

and Rotor Phase Angle $\cos \phi_2' = \dfrac{R_2}{Z_2'} = \dfrac{R_2}{\sqrt{R_2{}^2 + (sX_2)^2}}$

Example 103. A six-pole, three-phase, 50 Hz induction motor is running at full load with a slip of 4 per cent. The rotor is star-connected and its resistance and standstill reactance are 0·25 Ω and 1·5 Ω per phase respectively. The e.m.f. between slip-rings at standstill is 100 V. Find the full-load conditions, (a) the e.m.f. induced in each rotor phase, (b) the rotor impedance per phase, (c) the rotor current and power factor assuming the slip-rings are short-circuited.

The rotor phase e.m.f. at standstill $E_{ph_2} = \dfrac{100}{\sqrt{3}} = 57\cdot7$ V

The rotor phase e.m.f. at full-load $= E_{ph_2}' = sE_{ph_2}$

$$= 0\cdot04 \times 57\cdot7 = 2\cdot31 \text{ V}$$

Rotor reactance per phase on full load $X_{ph_2}' = sX_{ph_2}$

$$= 0\cdot04 \times 1\cdot5$$

$$= 0\cdot06 \ \Omega$$

∴ Rotor impedance per phase on full load $Z_{ph_2}' = \sqrt{0\cdot25^2 + 0\cdot06^2}$

$$= 0\cdot257 \ \Omega$$

$$\text{Rotor current} = \dfrac{2\cdot31}{0\cdot257} = 9 \text{ A}$$

$$\text{Rotor power factor } \cos \phi_2' = \dfrac{0\cdot25}{0\cdot257} = 0\cdot97 \text{ (lagging).}$$

RELATION BETWEEN ROTOR LOSS, ROTOR INPUT POWER AND ROTOR OUTPUT

This deduction is important and summarises much of motor theory. The power taken from the supply is utilised in creating a rotating magnetic field or the flux (for the transformer simile). Copper losses occur in the stator winding, as do iron losses. Thus the power put into the air-gap flux or the stator output = stator input − stator losses. Since no losses occur in the air-gap we can assume that the stator output power is the rotor input power whence

$$\boxed{\text{Stator output} = \text{Rotor input}}$$

The fact that the power output of a machine is given by the

expression $\dfrac{2\pi NT}{60}$ watts, is well known to the marine engineer. It

must be remembered however that the torque T is in newton metres.

Reverting to the induction motor relationships, we know that losses occur in the rotor, consisting of copper and iron losses. On load, because of the very low frequency, the latter are usually negligible and we can write

Rotor input   =   rotor output + rotor copper loss
or rotor copper loss   =   rotor input − output
                    =   stator output − rotor output

$$= \frac{2\pi N_1 T}{60} - \frac{2\pi N_2 T}{60} = \frac{2\pi T (N_1 - N_2)}{60}$$

T is in newton metres and N is in revolutions per minute.

Thus $\dfrac{\text{Rotor copper loss}}{\text{Rotor input}} = \dfrac{2\pi T(N_1 - N_2)}{2\pi N_1 T} = \dfrac{N_1 - N_2}{N_1} = s$ ...... (a)

$$\therefore \text{ Rotor Copper Loss} = s \times \text{Rotor Input}$$

Since the above relation can also be written as

$$\text{Slip} = \frac{\text{Rotor Cu Loss}}{\text{Rotor Input}}$$

$$\therefore \ s = \frac{3\,I_2{}^2 R_2}{\text{Rotor Input}} \text{ from which it follows that}$$

if the rotor resistance is large, the slip will be large and vice versa.

A further deduction is:

$$\frac{\text{Rotor copper loss}}{\text{Rotor output}} = \frac{2\pi T(N_1 - N_2)}{2\pi N_2 T} = \frac{N_1 - N_2}{N_2}$$

Since $\dfrac{N_1 - N_2}{N_1} = s$ $\ \therefore sN_1 = N_1 - N_2$ and $N_2 = N_1(1-s)$

Thus $\dfrac{\text{Rotor copper loss}}{\text{Rotor output}} = \dfrac{N_1 - N_2}{N_2} = \dfrac{sN_1}{N_1(1-s)} = \dfrac{s}{1-s}$ ...... (b)

or Rotor Copper Loss $= \dfrac{s}{1-s} \times$ Rotor Output

From (a) and (b) we have

Rotor input : Rotor copper loss $= 1 : s$
and Rotor copper loss: Rotor output $= s : 1 - s$

Rotor Input : Rotor Copper Loss : Rotor Output $= 1 : s : 1 - s.$

Example 104. The input power to a three-phase induction motor is measured to be 50 kW. The stator losses amount to 800 W. Find the rotor copper loss per phase and the mechanical power developed, if the slip is three per cent.

Stator input 50 kW Stator output $= 50 - 0.8 = 49.2$ kW
Rotor input $= 49\ 200$ W and Rotor Cu Loss $= s \times$ Rotor input
$$\text{or Rotor Cu Loss} = 0.03 \times 49\ 200$$
$$= 492 \times 3 \text{ watts}$$
Rotor Cu Loss per phase $= \dfrac{492 \times 3}{3} = 492$ W

Mechanical power developed $=$ Rotor input $-$ Rotor Cu loss
$$= 49\ 200 - (492 \times 3) = 49\ 200 - 1476$$
$$= 47\ 724 = 47.72 \text{ kW.}$$

TORQUE CONDITIONS

Earlier theory has considered how torque is produced by the rotor but it is evident that further examination of the conditions resulting in such torque is necessary. A "brake test" on the machine will produce a mechanical speed/torque characteristic of the type shown by the diagrams of Fig. 211, where the normal operating region is shown, for each type of motor, as a shaded area. The comparison with that of a d.c. shunt motor is given and from these it is seen that the torque of an induction motor is limited to a maximum value. Since the torque and slip are seen to be directly related — the deductions which follow show this; it is usual to give the characteristic of the machine as a torque/slip or torque/speed curve as shown for Fig. 212. The actual value of the maximum torque is decided by the physical dimensions of the motor and this cannot be exceeded, even by overloading the machine but it can be adjusted, to some degree, to occur at a desired value of slip. Such adjustment can be achieved by altering the resistance of the rotor-cage winding and is exploited fully in the wound-rotor machine.

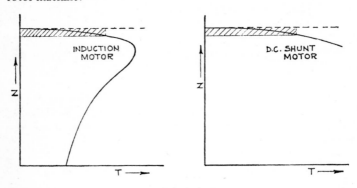

Fig. 211

STANDSTILL TORQUE. For a stationary rotor condition, let T_S be the standstill torque in newton metres. This torque is produced on the rotor by the magnetic field, which is rotating at a speed of N_1 rev/min. Since the motor is not running, there is no mechanical output and it follows that the rotor input power must be converted into rotor electrical power *i.e.* in this instance — a copper loss. Note that the rotor iron loss, being small is neglected for this assumption.

Thus   Rotor Input = Rotor Copper Loss

or $\dfrac{2\pi N_1 T_S}{60}$ $=$ $3I_2^2 R_2$ or $= 3E_2 I_2 \cos \phi_2$

$$= \frac{3E_2^2 R_2}{Z_2^2} = 3E_2^2 \ \frac{R_2}{R_2^2 + X_2^2}$$

Thus $T_S = \dfrac{3E_2^2 \times 60}{2\pi N_1} \times \dfrac{R_2}{R_2^2 + X_2^2}.$ Here E_2 and N_1 are constants whose magnitudes are dependent on the motor design.

We thus have $T_S = K \dfrac{R_2}{R_2{}^2 + X_2{}^2}$ or $T_S \propto \dfrac{R_2}{R_2{}^2 + X_2{}^2}$

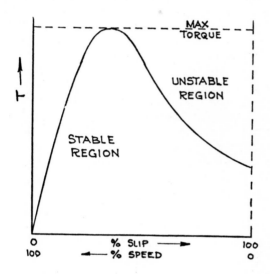

Fig. 212

Starting. Since the rotor is usually coupled to machinery, then maximum starting torque is a necessary feature, and the requirements for obtaining this need to be investigated. The expression deduced above shows that torque is affected by the rotor resistance, and the following mathematical deduction indicates the direction in which control can be made. The result and not the method of this deduction is of importance to the engineer but it is set out here for the interest of the mathematically-minded student.

The torque expression has a maximum value for a value of R_2. This can be found by differentiating the expression with respect to R_2 and equating to zero. Since $T_S = \dfrac{KR_2}{R_2{}^2 + X_2{}^2}$

then $\dfrac{dT_S}{dR_2} = K \dfrac{(R_2{}^2 + X_2{}^2) - R_2(2R_2)}{(R_2{}^2 + X_2{}^2)^2} = 0$

or $K \dfrac{X_2{}^2 - R_2{}^2}{(R_2{}^2 + X_2{}^2)^2} = 0$ from which $X_2{}^2 = R_2{}^2$ or $X_2 = R_2$.

This is the condition for obtaining maximum starting torque and explains why a wound-rotor may be used for ensuring a good starting torque. At starting, rotor frequency is f, so X_2 is large compared with R_2. Normally the rotor-circuit power factor is also low and T_S would be small. Improvement can however be effected, if $\cos \phi_2$ is raised by (1) decreasing X_2 —this is not readily feasible, (2) by increasing resistance — this is achieved by putting in external resistance, as for the wound-rotor machine, or by using a high resistivity material for the rotor cage. The latter arrangement does not allow for good running conditions and high efficiency, and as a result special cage forms have been evolved to give an improved overall performance. As already mentioned for the wound-rotor motor, as the rotor speeds up, the external resistance is cut out.

Before considering running conditions, it is important to point out the effect of supply voltage on starting torque.

The expression $T_S \propto \dfrac{R_2}{R_2{}^2 + X_2{}^2}$ was deduced on the assumption

that E_2 and N_1 were constant. It should be noted that E_2 is dependent on the machine flux Φ and this flux is in turn proportional to the supply voltage V. The torque expression could well be written as

$$T_S = \frac{CV^2 R_2}{R_2{}^2 + X_2{}^2}. \quad \text{Here } C \text{ is a constant.}$$

We now see that a change of supply voltage will considerably alter the starting torque. This fact must be kept in mind when considering the various methods used for motor starting.

Example 105. An induction motor with star-connected stator takes 45 A from a 110 V supply with the rotor stationary, the power factor being 0·3 (lagging). If the ratio of stator to rotor turns per phase is 2, determine the resistance and reactance per phase of the rotor winding. What additional resistance per phase would be required to make the starting torque a maximum.

Stator voltage per phase $= \dfrac{100}{\sqrt{3}} = 63 \cdot 5 \text{ V}$

Rotor voltage per phase $= 63 \cdot 5 \times \dfrac{1}{2} = 31 \cdot 75 \text{ V}$

Stator current per phase $= 45 \text{ A}$

Rotor current per phase $= 45 \times 2 = 90 \text{ A}$

Impedance Z_2 of rotor per phase $= \dfrac{31 \cdot 75}{90} = 0 \cdot 353 \ \Omega$

$$\cos \phi_2 = 0 \cdot 3$$

$\therefore R_2 = Z_2 \cos \phi_2 = 0 \cdot 353 \times 0 \cdot 3 = 0 \cdot 106 \ \Omega$

$X_2 = \sqrt{Z_2{}^2 - R_2{}^2} = \sqrt{0 \cdot 353^2 - 0 \cdot 106^2}$

$\qquad = \sqrt{0 \cdot 1246 - 0 \cdot 01124} = \sqrt{0 \cdot 11336} = 0 \cdot 337 \ \Omega$

For maximum starting torque R_2 must equal X_2

$\therefore R_2$ must be $0 \cdot 337 \ \Omega$ per phase or $0 \cdot 337 - 0 \cdot 106 = 0 \cdot 231 \ \Omega$ must be added in each phase.

RUNNING TORQUE. With the rotor up to speed, as seen earlier, the rotor e.m.f. is variable depending on the rotor frequency. Thus $E_2{}' = sE_2$.

Also now that mechanical power is produced at the shaft, it follows that, rotor input = rotor output + rotor copper loss.

$$\therefore \frac{2\pi N_1 T}{60} = \frac{2\pi N_2 T}{60} + 3 E_2{}' I_2{}' \cos \phi_2$$

$$\text{or } \frac{2\pi N_1 T - 2\pi N_2 T}{60} = 3 E_2{}' I_2{}' \cos' \phi_2 \ \text{ or } = 3 \frac{E_2'^2}{Z_2{}'} \times \frac{R_2}{Z_2{}'}$$

$$\frac{2\pi T(N_1 - N_2)}{60} = 3\,s^2 E_2{}^2 \times \frac{R_2}{R_2{}^2 + (s\,X_2)^2}$$

$$\frac{2\pi T(s\,N_1)}{60} = 3\,s^2 E_2{}^2 \; \frac{R_2}{R_2{}^2 + s^2 X_2{}^2}$$

$$\text{or}\; T = \frac{3\,s\,E_2{}^2 \times 60}{2\pi N_1} \times \frac{R_2}{R_2{}^2 + s^2 X_2{}^2} = K \frac{s\,R_2}{R_2{}^2 + s^2 X_2{}^2}$$

Whence $\;T \alpha \dfrac{s\,R_2}{R_2{}^2 + (s\,X_2)^2}$

A fuller understanding as to the shape of the torque/slip characteristic is now possible. From the work done above, it is seen that, on examining the expression $\;T \alpha \dfrac{s\,R_2}{R_2{}^2 + (s\,X_2)^2}\;$, when the rotor is just starting to move, the slip is large and the reactance X_2' is large. In comparison the resistance R_2 is small and the expression can be modified to $\;T \alpha \dfrac{s}{s^2\,X_2{}^2} = \dfrac{1}{s\,X_2{}^2}\;$. Thus at starting, the torque varies inversely with the slip at low speeds. This means that when slip is high, torque is small and the relation explains the right-hand side of the torque/speed curve as shown on Fig. 210. This is the unstable region.

At high speeds *i.e.* small slip, the rotor frequency is low and the reactance is small compared with the resistance. Again with reference to the expression $\;T \alpha \dfrac{s\,R_2}{R_2{}^2 + (s\,X_2)^2}\;$, for approximation under the condition of small slip $\;T \alpha \dfrac{s\,R_2}{R_2{}^2}\;$ or $\;T \alpha \dfrac{s}{R_2}\;$.

Thus $\;T \alpha s\;$ or torque is directly proportional to slip and this explains the left-hand side or stable region of the characteristic (Fig. 212).

Summarising, the torque/slip curve first rises steeply showing an increase of load will produce an increased torque. This linearity continues until the maximum torque is reached when the curve flattens and turns back. Beyond this the curve falls to follow an inverse law. Here an increase in load results in a decrease in torque and the slip will increase to result ultimately in a "stalled" condition.

Maximum Running Torque. The technique for determining this condition follows that used for the standstill condition. Thus by differentiating with respect to slip and equating to zero, we have

$$\text{Since } T = \frac{KsR_2}{R_2{}^2 + (sX_2)^2}$$

$$\text{then } \frac{dT}{ds} = \frac{\{R_2{}^2 + (sX_2)^2\} KR_2 - \{KsR_2 \times 2sX_2{}^2\}}{\{R_2{}^2 + (sX_2)^2\}^2} = 0$$

For the maximum condition

$$\therefore KR_2{}^3 + Ks^2X_2{}^2R_2 = 2KR_2s^2X_2{}^2$$
$$\text{or } R_2{}^2 + s^2X_2{}^2 = 2s^2X_2{}^2$$
$$\text{and } R_2{}^2 = s^2X_2{}^2$$
$$\text{or } R_2 = sX_2.$$

The above shows that maximum torque occurs when the slip

$s = \dfrac{R_2}{X_2}$. R_2 and X_2 are normally constants but by increasing R_2

as in the wound-rotor machine, the maximum torque value can be made to occur at any slip value. Thus maximum starting torque can be arranged for starting against heavy loads and this value can be maintained through the accelerating period by cutting out the external series resistance. The value of maximum torque is not affected by the resistance, but it is nevertheless cut out when the motor gets up to speed to reduce losses and ensure a small slip. The diagram of Fig. 213, shows the effect of rotor resistance on the torque/speed or torque/slip curve. Note that here the curves are drawn with speed as the base and they appear reversed with respect to those shown on Fig. 212. Either method of representation is acceptable although slip as a base is the more usual.

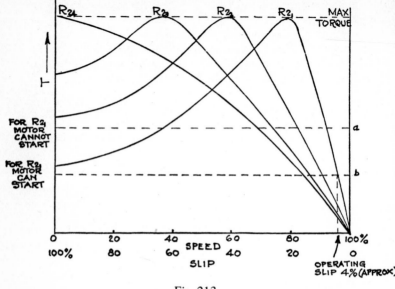

Fig. 213

For Fig. 213, the following condtions are considered for load torque. Thus for line (b) and curve R_{2_1}, the starting torque produced by the motor is greater than the load torque, hence the motor can start and accelerate the load. The machine will thus run up to speed and operate at about 4 per cent slip. For line (a) and curve R_{2_1}, the load requires a greater starting torque than that produced by the motor. The motor cannot therefore start and accelerate unless the load torque is reduced. However, if resistance is introduced into the rotor circuit to operate on curve R_{2_2}, then starting is possible and the motor will run up to speed and operate at about 6 per cent slip.

Example 106. The rotor current per phase of an induction motor is 50 A, with the rotor stationary; the power per phase being 5 kW. The angle of lag in the rotor is 75°. Find the resistance of a rotor phase, then calculate the current per phase and the angle of lag when the slip is 10 per cent. At what slip and speed will the torque be a maximum. Assume a four-pole machine and a 50 Hz supply.

Since $P = I_2^2 R_2$ $\therefore R_2 = \dfrac{5000}{50^2} = 2\ \Omega$

Also $P = E_2 I_2 \cos \phi_2$ $\therefore 5000 = E_2\,50 \cos 75°$

So $E_2 = \dfrac{5000}{50 \times 0.259} = 386$ V/ph (at standstill)

Then $Z_2 = \dfrac{386}{50} = 7.72\ \Omega$

$X_2 = \sqrt{7.72^2 - 2^2} = \sqrt{59.7 - 4} = \sqrt{55.7} = 7.47\ \Omega$

At 10 per cent slip, e.m.f. $E_2' = 386 \times \dfrac{10}{100} = 38.6$ V

Impedance $Z_2' = \sqrt{2^2 + 0.747^2} = \sqrt{4 + 0.557} = \sqrt{4.557}$
$$= 2.13\ \Omega$$

Note. $X_2' = s\,X_2 = 7.47 \times \dfrac{10}{100} = 0.747\ \Omega$

Current $I_2' = \dfrac{38.6}{2.13} = 18.1$ A

$\cos \phi_2' = \dfrac{2}{2.13} = 0.939$ (lagging) $\therefore\ \phi_2' = 20°$

Maximum Torque occurs when $sX_2 = R_2$ $\therefore\ s = \dfrac{2}{7.47}$

$$= 0.268 \text{ or } 26.8 \text{ per cent.}$$

Synchronous speed $N_1 = \dfrac{50 \times 120}{4} = 1500$ rev/min.

$\therefore$ Motor speed $= 1500 - 0.268 \times 1500 = 1500 - 402$
$$= 1098 \text{ rev/min.}$$

THE PHASOR DIAGRAM

Since the motor functions like a transformer, with the stator as the primary and the rotor as the secondary, then the phasor diagram will be very similar to that of the transformer and the associated reasoning can be applied, provided the following points are noted.

(1).   Transformer action is set up by a rotating magnetic field.

(2).   Because of the air-gap between the stator (primary) and the rotor (secondary), the magnetic circuit has a high reluctance and the magnetising component of current I_m is larger than for a transformer. Due to the large I_m, the no-load current I_o is at a low power factor. The motor should not therefore be run light, but should be switched off.

(3).   Since the rotor circuit is closed there is no secondary terminal voltage V_2. The rotor e.m.f. E_2' is responsible for the rotor current I_2' and overcoming the rotor resistance and reactance voltage drops. The secondary impedance voltage triangle will therefore be larger than for the transformer and will be constituted by all the secondary voltage drops.

(4).   Standstill conditions relate directly with the transformer but for running conditions slip must be considered, since it influences the induced secondary e.m.f. and secondary reactance.

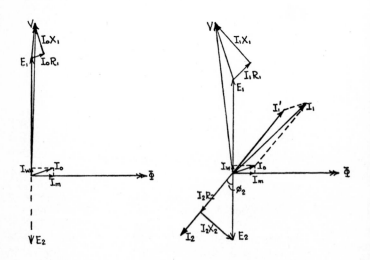

Fig. 214a   Fig. 214b

The diagram of Fig. 214a, shows the motor circuit with the rotor open-circuited. The condition resembles the transformer on "no-load". Fig. 212b, is drawn for rotor standstill conditions *i.e.* rotor locked and circuit complete. This condition is similar to the transformer "on load" and phasor relations, as evolved earlier for the transformer, apply here. Thus for the "locked rotor" condition, the main flux generates stator and rotor e.m.f.'s of E_1 and E_2 respectively. I_o is the no-load current consisting of components I_m and I_w. I_m is the current component which produces the ampere-turns resulting in the main flux Φ, and I_w is the energy component supplying the iron loss. The stator resistance voltage drop is $I_1 R_1$ and the reactance voltage drop is $I_1 X_1$ and these when added by phasors with the back e.m.f. E_1 equal the supply voltage V. The rotor e.m.f. E_2 produces a rotor current I_2 which is limited by the resistance R_2 and reactance X_2 of this circuit. The rotor secondary resistance and reactance voltage drops, $I_2 R_2$ and $I_2 X_2$ respectively, by phasor addition equal the induced e.m.f. E_2 or $E_2 = I_2 Z_2$.

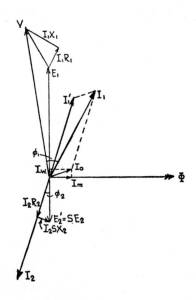

Fig. 214c

When the rotor is allowed to rotate, we get the conditions shown by the phasor diagram of Fig. 214c. Here the secondary e.m.f. is now reduced to E_2' or sE_2 and the secondary reactance falls to X_2' or sX_2. Since the rotor resistance is unaffected by slip, the value of R_2

remains unchanged and the power factor of the circuit rises as the value of sX_2 falls. Note that, as mentioned above under point (2), the power factor is low when the motor is running light since component I_o is large in relation to I_1'. When the motor is loaded I_2 and hence I_1' increases, I_o being constant, the result is that not only does I_1 rise, but the power-factor condition improves.

Some uncertainty may be felt by the student at this stage at the prospect of relating rotor or secondary conditions at slip frequency to those of the stator or primary which are at the "mains" frequency. The practice is seen to be in order however, if it is noted that, when running, although the rotor frequency is low, the field produced by the rotor current is rotating at a speed relative to the rotor itself and is in the same direction as the stator field. Thus the rotor field relative to the stator is at synchronous speed, and the rotor quantities can be combined and related to the stator quantities, provided appropriate adjustment is made for the turns ratio and slip.

Example 107. Full-load torque is obtained from a 3 kV, twenty-four pole, 50 Hz star-connected induction motor at a speed of 240 rev/min. The slip-ring motor has a resistance of 0·02 Ω per phase and a stand-still reactance of 0·27 Ω per phase. Calculate (a) the ratio of maximum to full-load torque (b) the speed at maximum torque.

$$\text{Since } f = \frac{PN_1}{120} \quad \therefore N_1 = \frac{120 \times 50}{24} = 250 \text{ rev/min}$$

$$\therefore \text{Slip } s = \frac{250 - 240}{250} = \frac{10}{250} = 0\cdot04 \quad \text{or} \quad 4 \text{ per cent}$$

Let T = the full load torque

$$\text{Then } T = \frac{KsR_2}{R_2{}^2 + (sX_2)^2} = \frac{K \times 0\cdot04 \times 0\cdot02}{0\cdot02^2 + (0\cdot04 \times 0\cdot27)^2}$$

$$= \frac{K \times 8 \times 10^{-4}}{0\cdot0004 + (0\cdot0108)^2} = \frac{8K\,10^{-4}}{10^{-4}(4 + 1\cdot1664)} = \frac{8K}{5\cdot1664}$$

Also let T_m = maximum torque.

Now T_m occurs when $s_m X_2 = R_2$ or when $s_m \, 0 \cdot 27 = 0 \cdot 02$

$\therefore$ For maximum torque, $s_m = \dfrac{0 \cdot 02}{0 \cdot 27} = 0 \cdot 074$

Then at this slip s_m, $\quad T_m = \dfrac{K \times 0 \cdot 074 \times 0 \cdot 02}{(0 \cdot 02)^2 + (0 \cdot 074 \times 0 \cdot 27)^2}$

$\therefore T_m = \dfrac{K \times 14 \cdot 8 \times 10^{-4}}{0 \cdot 0004 + (0 \cdot 01998)^2} = \dfrac{14 \cdot 8 \, K \, 10^{-4}}{10^{-4}(4 + 3 \cdot 992)} = \dfrac{14 \cdot 8 \, K}{7 \cdot 992}$

$\therefore \dfrac{T_m}{T} = \dfrac{\dfrac{14 \cdot 8 \, K}{7 \cdot 992}}{\dfrac{8 \, K}{5 \cdot 1664}} = \dfrac{14 \cdot 8 \times 5 \cdot 1664}{8 \times 7 \cdot 992} = \dfrac{76 \cdot 45}{63 \cdot 94}$

or $\dfrac{T_m}{T} = \dfrac{1 \cdot 2}{1}$ (approx)

For maximum torque, slip $= 0 \cdot 074$ or $7 \cdot 4$ per cent.

$\therefore 0 \cdot 074 = \dfrac{250 - N_2}{250} \quad \therefore N_2 = 250 - 250 \times 0 \cdot 074$
$$= 250(1 - 0 \cdot 074) = 250 \times 0 \cdot 926$$
$$= 250 \times 0 \cdot 2315 \times 4 = 231 \cdot 5 \ \text{rev/min}.$$

Note that the information relating to motor-voltage rating and connection, (3 kV, star-connected), is superfluous and is not necessary for the solution.

THE CIRCLE DIAGRAM

Locus diagrams are frequently used in electrical engineering for determining the working conditions of a circuit when its variables are altered in a known manner. Although space here does not allow the examination of other examples of such locus diagrams nevertheless, the basics upon which the induction motor circle diagram are based should receive the appropriate consideration.

This circle diagram is used to predict motor performance under different operating conditions and can be developed to show current, efficiency, power factor, slip, speed and torque. The diagram can be constructed from the results of simple tests which do not involve loading the machine and this method of assessing performance is similar to that used for the transformer and alternator. It is used in industry as the basis for making acceptance tests, as agreed between motor manufacturer and the customer, and enables a large machine to be installed on site once it is deemed to be satisfactory, as indicated by the circle-diagram test results. These results and the behaviour of the motor are then checked by investigating its performance under the actual working conditions.

Since the circle diagram is developed from the phasor diagram, operations appropriate to the latter can be applied to the former, and the results of any measurements made are dependent entirely on the accuracy of the draughting and graphical constructions. To this end the largest convenient scale should be used.

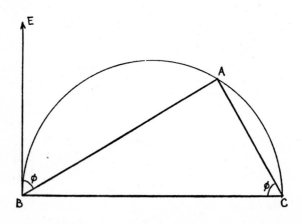

Fig. 215

Consider the diagram of Fig. 215 which shows any circle of diameter BC. Since A is a point on the circumference, the angle BAC is a right angle and AB = BC sin ϕ or AB = K sin ϕ where K is the diameter of the circle.

Next consider the rotor expression where the rotor current is given by $I_2 = \dfrac{sE_2}{\sqrt{R_2{}^2 + (sX_2)^2}}$. If we multiply top and bottom by X_2

and rearrange, we have

$$I_2 = \frac{sE_2}{\sqrt{R_2{}^2 + (sX_2)^2}} \times \frac{X_2}{X_2} = \frac{sX_2}{\sqrt{R_2{}^2 + (sX_2)}} \times \frac{E_2}{X_2}$$

$$= \frac{E_2}{X_2} \sin \phi_2 \text{ (from the rotor-circuit relationships)}$$

$$= K \sin \phi_2, \text{ since } E_2 \text{ and } X_2 \text{ alter in proportion.}$$

Comparing this with the expression $AB = K \sin \phi$, it is seen that the locus of I_2 is on a circle with a diameter of constant value $\frac{E_2}{X_2}$.

This diameter is at right angles to E_2 as shown by the diagrams of Figs. 216a and b. Note that I_1' is the reflection of I_2 in the primary or stator circuit. It follows that I_1' lies on a circle of appropriate diameter. Again, since the primary current is the phasor sum of I_o and I_1', then the relevant parts of the primary diagram can be adapted to give the Circle Diagram. It is not necessary to know the actual value of the diameter, since the circle can be built up from current values obtained from simple motor tests. The circle diagram, once deduced, can be used for assessment of performance as will be illustrated by examples.

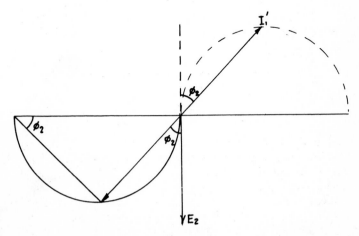

Fig. 216a

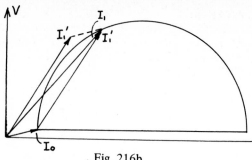

Fig. 216b

EXPLANATION OF THE CIRCLE DIAGRAM

Reference here is made to the diagram of Fig. 217, which is the circle diagram as extracted from the phasor diagram. It should be noted immediately that both stator and rotor conditions are represented, the latter being in terms of appropriate stator values. Thus OB is the no-load current, OA is a stator-load current value, made up of the no-load current OB and the rotor-load current BA is reflected into the stator circuit – this would be I_1'. Thus current phasors, originating from B and lying on the semi-circle, are rotor values referred to the stator, whilst phasors originating from point O are stator values. Note also that all vertical components of these current phasors are in phase with the voltage, are energy components and are a measure of input power, *i.e.* output power plus losses. These will be explained as the diagram is detailed.

The semi-circle can be constructed if two phasors originating from O are known in length and also in their angular relation to an axis. Such current phasors are obtained from simple tests on the machine, performed under conditions requiring the minimum of loading arrangements.

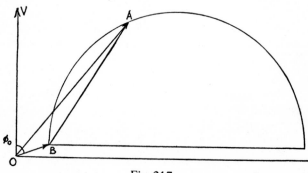

Fig. 217

TESTING PROCEDURE

Since the machine is a three-phase motor, three-phase measurements are made. The data can be used in this form but sometimes it is more convenient to work in phase values. This will become evident from the problem being worked.

NO-LOAD TEST. The machine is run light at normal voltage and frequency. Measurements are made of line voltage, line current and total power. Referring to the diagram of Fig. 218, the length OB is obtained, also the phase angle ϕ_o.

LOCKED-ROTOR OR STAND-STILL TEST. A reduced voltage is applied to the stator and this is raised until full-load current is passing. The rotor is clamped meanwhile so as to prevent rotation. The line voltage, line current and total power are measured. Referring to Fig. 218, the length OA_S is found, also the phase angle ϕ_S. It should be noted that the data from this test must be adjusted to full voltage values. Thus if V_S is the reduced voltage and V the normal voltage then,

Standstill current (at normal voltage) $=$ measured current $\times \dfrac{V}{V_S}$

Power (at normal voltage) $=$ measured value $\times \left(\dfrac{V}{V_S}\right)^2$

Students are warned not to forget this operation.

CONSTRUCTION OF CIRCLE DIAGRAM

A power-factor quadrant is useful for laying off and measuring power factors. The procedure for doing this has been described earlier but is repeated here. Along the vertical axis (Fig. 218), starting from point O, a scale of ten units is laid off. This is made of convenient length to give accuracy and a quadrant is drawn. Scale the vertical 0 to 1. As an example, if a line is drawn horizontally through the 0·2 mark to cut the quadrant and this point of intersection is joined to the origin then the line so obtained must subtend with OV, an angle whose cosine is 0·2.

No-load Test. This test gives I_o and ϕ_o. Thus no-load power factor

$$\cos \phi_o = \frac{P_o}{\sqrt{3}\, VI_o} \quad \text{where } P_o = \text{ the no-load three-phase power. To}$$

start the circle diagram, using a suitable scale, lay off OB equal to I_o at an angle ϕ_o.

Load Test. The power factor for this test is given by $\dfrac{P_S}{\sqrt{3}\ V_S I_S}$

$= \cos \phi_S$. Next adjust the locked-rotor or short-circuit test current to the correct current value at full voltage. This deduced value of current gives OA_S, which is drawn to the chosen current scale at an angle ϕ_S.

The construction can now proceed by joining BA_S and bisecting it perpendicularly. The bisector is drawn to cut the horizontal through B at point M, which now becomes the centre of the circle.

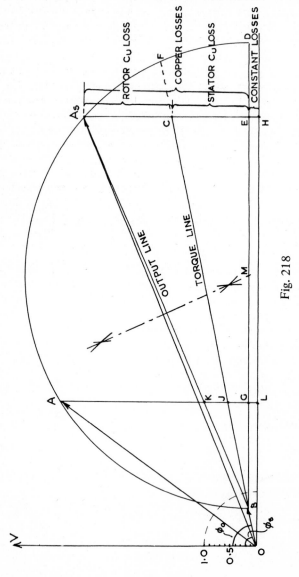

Fig. 218

The diagram can now be used to estimate the power factor for any load current and the optimum power-factor condition. It can also be extended to estimate power output, losses, slip, etc. Thus consider any load-current value OA. Then AL will be the power component of OA and the input power to the motor will be $\sqrt{3} \; V \times$ AL. This input power supplies the output power, as developed at the motor shaft, and also the losses in the machine. These losses comprise the Iron (hysteresis and eddy current) losses, the Rotational (friction and windage) losses and the Copper (stator and rotor) losses.

Consider the length GL. This is seen to be the power component of OB and is therefore the no-load power input which goes to make up the rotational loss, the stator iron loss and a very small stator copper loss. The last can be neglected but $GL = EH$, can be taken as a measure of the constant losses of the machine.

For the locked-rotor test all input power is converted into losses *i.e.* copper and iron losses. Here there is no rotational loss but since the rotor frequency is now at full "mains" value, it can be assumed that the constant loss component EH, under this condition, is made up of the iron losses, *i.e.* rotational loss is nil but the rotor iron loss increases to replace it. Component $A_S E$ is a measure of the rotor and stator copper losses under the locked-rotor condition.

Next consider any value of stator current such as OA. Now since the stator current is proportional to the rotor current then

Total copper loss α (rotor current)^2 *i.e.* $\alpha \, (AB)^2$

Now $\cos ABD = \dfrac{AB}{BD}$ also $\cos ABG = \dfrac{BG}{AB}$

$\therefore \; \dfrac{AB}{BD} = \dfrac{BG}{AB}$ or $AB^2 = BD \times BG$

But BD is a constant $\therefore \; AB^2 = K \times BG$ or $AB^2 \, \alpha \, BG$

or the copper loss $\alpha \, BG$

We can thus deduce $\dfrac{\text{Copper loss for any current}}{\text{Copper loss at standstill}} = \dfrac{BG}{BE} = \dfrac{KG}{A_S E}$

(Since similar triangles are involved).

Thus a line joining B to A_S cuts off the copper losses (to scale) for the current considered *i.e.* KG, to scale, is a measure of the total copper losses.

Output Line. Consider the power component AL of any current such as OA.

The length GL then represents the constant losses.
The length KG represents the copper losses,

and AK must then represent the Power Output from the motor. The line BA_S is called the *Output Line* and any length of a power component above this line is a measure of the output (in watts). Note that the current must be multiplied by the voltage and $\sqrt{3}$.

Maximum output is obtained by drawing a tangent to the circle parallel to BA_S or by producing bisector of BA_S to cut the circle to give this tangent point. The length of the vertical between this tangent point and the output line gives the maximum output possible.

Example 108. The following test data refers to a 500 V, three-phase, 50 Hz induction motor. No-Load Test: With 500 V applied, the current was 18 A and the power 1200 W. Locked-Rotor Test: With 250 V applied, the current was 100 A and the power 11 kW.

Estimate the maximum power output, the current and power factor under this condition.

The problem is worked for a three-phase condition and thus the motor connection is immaterial.

Thus on no load $\sqrt{3} \; VI \cos \phi_o \; = \; 1200$

or $\cos \phi_o \; = \; \dfrac{1200}{\sqrt{3} \; \times \; 500 \; \times \; 18} \; = \; 0{\cdot}077$

Stand-still test $\cos \phi_S \; = \; \dfrac{11\,000}{\sqrt{3} \; \times \; 250 \; \times \; 100} \; = \; 0{\cdot}254$

Note. The solution shown, has been drawn using a scale of 10 mm = 10 A. This is not apparent in the printed form due to photographic reduction. The student should therefore draw out this solution to benefit from the problem.

Using the origin O, draw the power-factor quadrant, (5 mm has been used for 1 division). Draw OB = 18 A *i.e.* 18 mm at 0·077 power factor. The horizontal line cutting the quadrant is shown.

Draw OA_S = 200 A *i.e.* 200 mm at 0·254 power factor. The horizontal line cutting the quadrant is shown. Note that the locked-rotor current value has been deduced for full-voltage value *i.e.*

$$OA_S = 100 \times \frac{500}{250} = 200 \text{ A}$$

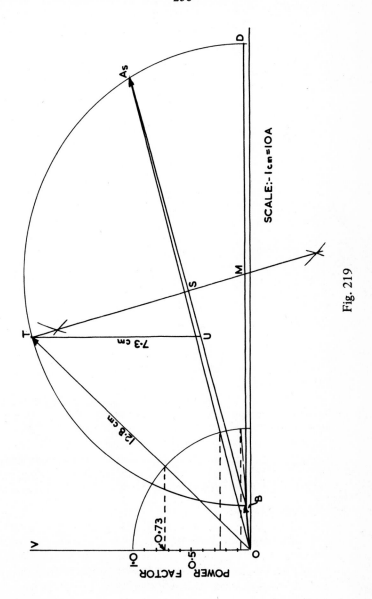

Fig. 219

Join BA_S. Bisect this line and produce bisector to give point M on BD. Draw the circle. Produce MS to cut the circle at T. Drop TU —this is the output component. Thus $TU = 73$ mm $= 73$ A

and Output Power $= \dfrac{\sqrt{3} \times 73 \times 500}{1000} = 63 \cdot 2$ kW

Joint OT. Then the estimated line current given by OT is 128 mm *i.e.* 128 A at a power factor of 0·73 (lagging).

Torque Line. Further work on the circle diagram (Fig. 218) allows the determination of other values for any machine. To draw the *Torque line,* stator and rotor copper losses are separated. For the locked-rotor test, since the voltage is reduced, the iron losses are negligible and the test is assumed to give the copper losses — deduced for the full voltage condition. Furthermore, if R_S is the stator resistance per phase then the stator copper loss during the test is $3 I_S{}^2 R_S$ and, Rotor copper loss $= P_S - 3 I_S{}^2 R_S$.

Thus the length $A_S E$ can be divided in the ratio

$\dfrac{A_S C}{CE} = \dfrac{\text{Rotor copper loss}}{\text{Stator copper loss}}$ and if BC is joined, then we have the

Torque Line, which is explained as follows

AL is a measure of the power input to the machine.
GL is a measure of the constant loss.
GJ is a measure of the stator copper loss for the line current OA being considered. It is deduced from the locked-rotor condition, using the proportion given by similar triangles.

AJ is a measure of the power passed across the air-gap from stator to rotor *i.e.* is the rotor input power and the torque can be deduced thus; any power intercept up to the torque line such as AJ, when multiplied by the appropriate factors, will give the torque in "synchronous watts". Since we are concerned with the power put into the rotor *i.e.* produced by the rotating magnetic field then synchronous speed N_1, must be substituted for N when using the expression $\dfrac{2\pi N T}{60}$.

Other data can be obtained from the circle diagram such as estimated slip, overall efficiency, optimum power factor etc. The following examples show how these can be deduced and here again the student should undertake the practical solution himself since the scales used are not apparent.

Example 109. A four-pole, 56 kW, 440 V, 50 Hz, three-phase induction motor gave the following figures on test.

No-load Test.
Applied voltage 440 V. Line current 18·25 A. Input power 3·19 kW.

Locked-rotor Test.
Applied voltage 110 V. Line current 55·2 A. Input power 3·17 kW.

Construct the circle diagram and from it find (i) the line current, (ii) the power factor, (iii) the full-load percentage slip and (iv) the efficiency. Assume at standstill the rotor loss and stator loss to be equal.

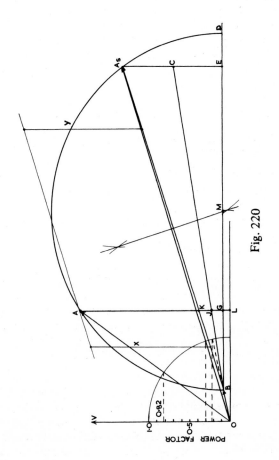

Fig. 220

For this solution (Fig. 220) a scale of 10 mm = 10 A is used. The construction is commenced by drawing the power-factor quadrant, using 5 mm for 1 division.

From the No-load Test.

$$\text{Power factor } \cos \phi_o = \frac{3190}{\sqrt{3} \times 440 \times 18 \cdot 25} = 0 \cdot 23$$

Then using the power-factor quadrant, draw in the line for an angle whose cosine is 0·23. Next on this line measure off OB = 18·25 mm.

From the Locked-rotor Test.

$$\text{Power factor } \cos \phi_S = \frac{3170}{\sqrt{3} \times 110 \times 55 \cdot 2} = 0 \cdot 3$$

Using the power-factor quadrant draw in the OA_S line by adjusting the locked-rotor current to full voltage value *i.e.*

$$OA_S = 55 \cdot 2 \times \frac{440}{110} = 220 \cdot 8 \text{ A} = 221 \text{ mm}.$$

Join A_S to B giving the output line, bisect this so that the bisector cuts BD at M. M is the centre of the circle, draw this in.
For an output of 56 kW

$$\text{the power component} = \frac{56\,000}{\sqrt{3} \times 440} = 73 \cdot 5 \text{ A} = 73 \cdot 5 \text{ mm}$$

From the output line erect two verticals X and Y = 73·5 mm. and draw the parallel to cut the circle at point A.
Then OA = 114 mm or line current = 114 A
The operating power factor would be 0·82 (lagging)
The next construction involves drawing in the Torque Line
Since Rotor Loss = Stator Loss. Bisect $A_S E = 62$ mm
∴ $A_S C = 31$ mm = CE. Join BC,
Draw in the vertical from point A, marking in the points K, J, G and L. Then

Rotor copper loss = $s \times$ Rotor input

or in terms of the proportions of the power component of load current

$$\text{Slip} \quad = \quad \frac{\text{length KJ}}{\text{length AJ}} = \frac{0.8}{8.1} = 0.099 = 9.9 \text{ per cent}$$

$$\text{Efficiency} = \frac{\text{length AK}}{\text{length AL}} = \frac{7.3}{9.3} = 0.785 \text{ or } 78.5 \text{ per cent}$$

The answers indicate that the machine is overated *i.e.* the slip is excessive and the efficiency low. The line current would also be excessive and the power factor should be higher. A rating of 37 kW would be more appropriate.

Example 110. A 45 kW, 440 V, 60 Hz, six-pole, three-phase induction motor was tested and the following results were obtained.

No-load Test.
Applied voltage 440 V. Line current 18 A. Power input 2·06 kW.

Locked-rotor Test
Applied voltage 110 V. Line current 55·5 A. Power input 2·9 kW.

At standstill the rotor copper losses were assessed to be two-thirds of the stator copper losses.
Draw the circle diagram and estimate for full-load conditions, the line current, power factor and efficiency.
Determine also the speed of the motor when the torque is maximum. Estimate also the value of this torque.

No-Load Test.
Line current = 18 A represented by 18 mm. Scale used 10 mm = 10 A.

Thus OB (Fig. 221) = 18 mm

and power factor $(\cos \phi_o) = \dfrac{2060}{\sqrt{3} \times 440 \times 18} = 0.15 \text{ (lagging)}$

Locked-rotor Test.

Line current at full voltage $= 55.5 \times \dfrac{440}{110} = 222 \text{ A}$

To scale $OA_S = 222 \text{ mm}$

and power factor, $\cos\phi_S = \dfrac{2900}{\sqrt{3} \times 110 \times 55\cdot5}$

or $\cos\phi_S = 0\cdot275$ (lagging).

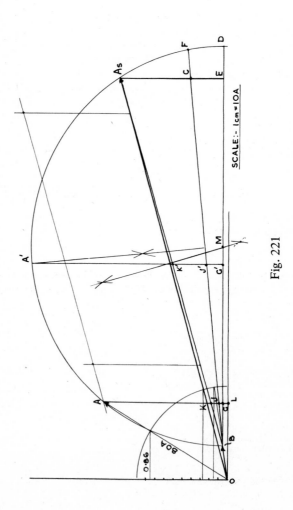

Fig. 221

Basic features of Circle Diagram construction.

OB and OA_S are drawn to scale and correct phase angles by using a power-factor quadrant. (Scale 5 mm = 1 unit). B and A_S are joined to give the Output Line. BA_S is bisected to give the centre of the circle at M.

Vertical $A_S E$ is divided in the ratio 2:3 to give point C. BC being the Torque Line.

$$\text{Full-load power component of current} = \frac{45\,000}{\sqrt{3} \times 440} = 59 \text{ A}$$

or AK = 59 mm.

Note. Point A is obtained by drawing a parallel to the output line, spaced 59 mm and cutting the circle at A.

Line current OA (80 mm) = 80 A

Power factor, 0·86 (lagging).

$$\text{Efficiency} = \frac{\text{Output}}{\text{Input}} = \frac{\text{length AK}}{\text{length AL}} = \frac{59 \text{ mm}}{68 \text{ mm}} = \frac{59}{68} = 0.867$$

Thus $\eta = 86.7$ per cent.

To obtain the maximum torque condition, draw the tangent to the circle which is parallel to the torque line. This is done most readily by bisecting BF and extending this to cut the circle. Dropping the vertical to give A', K' and J' we can deduce

$$\text{Slip} = \frac{\text{Rotor Cu loss}}{\text{Rotor input}}$$

Proportional lengths $\dfrac{K'J'}{A'J'} = \dfrac{19}{96.5} = 0.196$ or 19·6 per cent.

$$\text{Synchronous speed } N_1 = \frac{120f}{P} = \frac{120 \times 60}{6} = 1200 \text{ rev/min.}$$

$$\therefore \text{ Motor speed } N_2 = N_1 - sN_1$$
$$= 1200 - 0.196 \times 1200 = 1200 - 235.2$$
$$= 964.8 \text{ say } 965 \text{ rev/min.}$$

Power input to motor

$$= A'J' \times 440 \times \sqrt{3} \text{ watts } = \frac{96 \cdot 5 \times 440 \times \sqrt{3}}{1000} \text{ kilowatts}$$

Thus $\dfrac{2\pi NT}{60} = \dfrac{96 \cdot 5 \times 762 \cdot 1}{1000}$

or $T = \dfrac{96 \cdot 5 \times 762 \cdot 1 \times 60}{6 \cdot 28 \times 1200}$

$\qquad = \dfrac{96 \cdot 5 \times 762 \cdot 1}{6 \cdot 28 \times 20}, = 585 \cdot 5 \text{ newton metres}$

CHAPTER 11

PRACTICE EXAMPLES

1.   The input to a three-phase induction motor is 50 kW. The stator losses amount to 800 W. Calculate the rotor copper loss per phase and the total mechanical power developed, if the slip is 3 per cent.

2.   A 500 V, six-pole, 50 Hz, three-phase induction motor develops 20 kW inclusive of mechanical losses, when running at 990 rev/min, the power factor being 0·85 (lagging). Calculate, (a) the percentage slip, (b) the frequency of the rotor e.m.f., (c) the rotor copper loss, (d) the input to the motor if the stator loss is 2·5 kW, (e) the line currents.

3.   A four-pole, three-phase, 60 Hz induction motor is rated at 15 kW. Its full-load slip is 4 per cent. The stator loss is measured as 950 W and the mechanical losses of the rotor as 830 W. Find the rotor copper loss and the efficiency of the machine.

4.   A six-pole, 50 Hz, three-phase, wound-rotor induction motor running on full load develops a useful torque of 162·7 Nm. A moving-coil ammeter in the rotor circuit fluctuates at 90 complete "beats" per minute. The mechanical torque lost in friction is 13·56 Nm and the stator losses total 750 W. Calculate, (a) speed of motor, (b) brake power input, (c) rotor copper loss (watts), (d) motor input watts, (e) motor efficiency.

5.   A 75 kW, 440 V, three-phase, wound-rotor induction motor has a rotor resistance and standstill reactance of 0·02 and 0·27 Ω/ph respectively. The stator to rotor phase turns ratio is 3:1 and the stator windings are connected in delta.
    If the motor is started by means of a resistance starter having a resistance of 0·25 Ω/ph, calculate the current taken by the motor from the supply, (i) at starting, (ii) under full-load running conditions if the full-load slip is 4 per cent. What would be the current taken from the supply if the motor was accidently started with the starting resistance in the "run" position? Neglect the no-load current and the resistance and reactance of the stator windings. Assume the rotor to be star-connected.

6.   Full-load torque is obtained with a 440 V, twelve-pole,
60 Hz, three-phase, delta-connected induction motor, when
driving a main circulating water pump at a speed of 576 rev/min.
The slip-ring rotor has a resistance of 0·02 Ω/ph and a standstill
reactance of 0·27 Ω/ph. Calculate, (a) the ratio of maximum
to full-load torque, (b) the speed at maximum torque.

7.   A 440 V, three-phase, ten-pole, 50 Hz, boiler induced-draught
fan motor has a delta-connected stator winding and a wound
rotor. The resistance per phase of the rotor winding is 0·018 Ω
and the ratio of the stator conductors per phase to rotor conductor per phase is 4 to 1. If the motor develops maximum
torque when running at a speed of 540 rev/min, calculate the
value of the rotor current and power factor for this machine
when running at full load, with a slip of 4 per cent and the slip-
rings short-circuited. Also calculate the developed full-load output power, assuming the mechanical losses to be negligible.
What value of starter resistance per phase must be connected
into the rotor circuit to give maximum torque at start?
        Neglect the resistance and reactance of the stator winding.

8.   A 37 kW, twelve-pole, three-phase, 50 Hz squirrel-cage
induction motor as used aboard a tankship for driving a main
circulating water pump, gave the following test results

No-load Test.
Applied voltage 440 V. Line current 19 A. Input power 2·17 kW

Locked-rotor Test.
Applied voltage 100 V. Line current 70 A. Input power 3·88 kW.

        The ratio stator/rotor copper loss is 4:5. Construct a circle
diagram and find for full-load conditions
(a) Input line-current and power factor, (b) percentage slip,
(c) percentage efficiency.

9.   During tests on a three-phase, 440 V, 60 Hz, eight-pole,
22 kW induction motor, the following figures were obtained

No-load Test.
Applied voltage 440 V.at 60 Hz. Input line current 10·5 A
Input power 1·82 kW.

Locked-rotor Test.
Applied voltage 113 V at 60 Hz. Input line current 50 A
Input power 3·92 kW.

Estimate for full-load conditions, (a) Input line current and power factor, (b) percentage slip, (c) percentage efficiency and, (d) developed torque of the motor.

For this motor the rotor copper loss under standstill conditions was found to be the same as the stator copper loss.

10.   Draw the circle diagram for a three-phase, 415 V, 37 kW induction motor from the following test results

No-load Test.
Applied voltage 415V. Input current 40 A. Input power 3·6 kW.

Locked-rotor Test.
Applied voltage 83 V. Input current 80 A. Input power 3·45 kW.

Under standstill conditions the rotor copper loss and the stator copper loss are equal. From the circle diagram deduce (a) The full-load current and power factor, (b) the full-load efficiency, (c) the maximum output, (d) the ratio of full-load torque to maximum torque.

CHAPTER 12

ELECTRONICS (iii)

Semiconductor theory was introduced in Chapter 15 of Volume 6 when a start was made by considering the basics of atomic physics together with the concept of covalent bonding, doping and the formation of n-type and p-type materials. The junction diode was also described and, before further solid-state devices are considered, a brief revision is made of its rectifying action and operating characteristics.

THE SEMICONDUCTOR DIODE

The junction diode consists of a section of p-type material and a section of n-type material grown together to form a continuous crystal. At the junction electrons from the n-type region diffuse across into the p-type region to combine with the holes. Similarly some holes from the p-type region diffuse across the junction into the n-type region to combine with electrons. Thus $I = I_F - I_R = 0$. This action results in a narrow *depletion layer* on either side of the junction and due to both sides losing some of their majority carriers, the p-side will exhibit a −ve charge and the n-side a +ve charge. Note that the +ve charge of the n-side relative to the p-side tends to attract electrons so that migration finally ceases. Similarly the −ve charge of the p-side relative to the n-side attracts holes back to limit migration. The final charge distribution results in a potential barrier being set up across the junction which can be likened to a battery connected in the sense shown by the diagram (Fig. 222). The potential barrier is normally referred to as the *junction barrier* and the diode behaviour is now investigated for the condition when an external electromotive force or electric potential is applied.

Forward Bias. (Fig. 223.) This is applied by connecting the p-type region to the +ve pole of a d.c. supply and the n-type region to the −ve pole of the supply. As the applied potential is increased, the majority carriers of holes and free electrons are repelled towards the junction. The first effect is to neutralise the charged section or junction barrier. The voltage at which this occurs is 200 mV for germanium and 600 mV for silicon. Further increase of voltage drives the holes and electrons across the junction where they combine. The

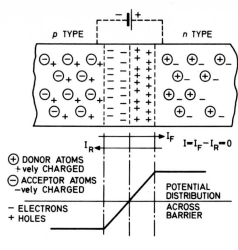

Fig. 222

barrier potential has thus been reduced, the effective resistance decreases and a current of several milliamperes can flow for a voltage of a little above 200 mV for germanium and 600 mV for silicon. Note that for each combination of holes and electrons at the junction, a co-valent bond breaks down at the +ve connection and holes are liberated. Simultaneously breakdown at the −ve connection occurs and electrons enter the crystal lattice. Thus the majority carriers in the p-type material are reinforced by the holes and in the n-type material, the electrons are reinforced. This action constitutes a forward current (I_F) which is considerably larger than the reverse current (I_R) due to the minority carriers or $I = I_F - I_R$. For working conditions I_R can be disregarded and the current I can be seen as being due to the applied voltage V. Figure 225 shows the test circuit and typical semiconductor diode characteristics. The forward part of the curve shows a sharp rise of current for increasing voltage but this only occurs for voltages above 0·6 V for silicon and 0·2 V for germanium.

Whilst still considering the semiconductor diode when operating in the forward direction, it will be seen that the slope of the forward characteristic is $\dfrac{\delta I}{\delta V}$ *i.e.* the a.c. conductance of the diode. The reciprocal of this slope is the a.c. resistance which decreases in value as the current increases. Similarly the d.c. resistance value reduces as the working current increases. Further examination of these values will be introduced when rectifier working is considered.

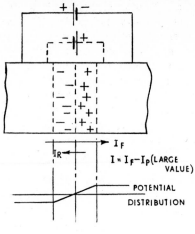

Fig. 223

REVERSE BIAS (Fig. 224). This is applied by connecting the p-type region to the −ve pole of the supply and the n-type region to the +ve pole of the supply. The barrier potential is increased as more

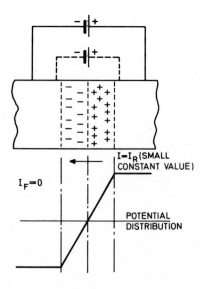

Fig. 224

holes and electrons are assisted in diffusing into the opposite type
regions. The effect is to widen the depletion layer strength of the
junction barrier whose effective resistance is thus increased. Under
this condition no forward current can flow due to no mobile majority
carriers being available to cross the junction $I_F = 0$. In fact a small
reverse current I_R (called a *leakage current*) does flow because of the
minority carriers or electrons released by intrinsic conductivity, i.e.
thermally generated electron-hole pairs in the semiconductor materials.
This current, between 1 μA and 100 μA for a silicon diode and 10 μA
and 1 mA for a germanium diode, is affected by temperature and
remains fairly constant as the voltage is increased. This is known as
the *Reverse Saturation Current.* Here $I = I_R$, $I_F = 0$.

If the reverse voltage continues to be increased, a value will be
reached on the characteristic where the current is seen to rise appreciably. This point of turn on the characteristic is called the *Reverse
Breakdown Voltage.* Note that the ability of a junction diode to withstand a high reverse potential is an important property, especially for
rectifiers. The Peak Inverse Voltage for any commercial diode is
always specified and will have a value slightly less than the Reverse
Breakdown Voltage (*R.B.V.*) to avoid damage if used "on the limit".

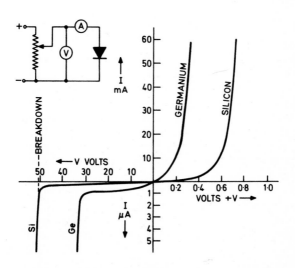

Fig. 225

The sudden increase in current at the R.B.V. point is due to two causes:

1.   The Zener effect. This is due to electrons being torn from their co-valent bonds by the strength of the electric field *i.e.* the applied potential.

2.   The Avalanche effect. Here the minority carriers *i.e.* the electrons, are given sufficient velocity to dislodge other electrons from their bonds.

Provided the current build-up is limited at the Reverse Breakdown Voltage and the diode is not damaged by overheating, the effect is reversible and such Zener diodes have applications for voltage reference and stabilisation, for meter protection and other specialised purposes. The use of the Zener diode is treated in some depth later in this chapter.

RECTIFIER OPERATION

This has already been described in Volume 6, Chapter 15 and the reader is asked to pay particular attention to the Examples which illustrate the use of a characteristic and load line in rectifier operation. For convenience and emphasis this area of work is repeated here and the student is also referred to the treatment on rectification already covered in Chapters 5 and 8.

Since the junction diode requires no heater supply, it can be introduced directly into circuit arrangements to allow either half-wave or full-wave rectification. The characteristic shows the device to be suitable for a rectifier and if an alternating voltage of about 1 volt (peak to peak) is applied to a *p-n* junction, the potential barrier is alternatively strengthened and weakened to allow the rectifying action. The reverse current can be considered as negligible. Silicon is used in preference to germanium for power rectifiers since it can carry large currents and can operate at higher temperatures. Its reverse current is also lower than that for germanium with similar forward current values.

The advantages of semiconductor rectifiers over the earlier type of "metal" rectifiers such as copper-oxide and selenium, lie chiefly in their smaller size, longer life and greatly improved regulation and efficiency. The latter features are the result of the low forward resistance and voltage drop and the fact that fewer elements are required in series to handle a given voltage.

Example 111. The following values refer to a germanium diode.

Forward current (mA)	0	0·1	0·18	0·22	0·4	0·6	0·8	1·1	1·6	2·3	3·45
Forward voltage (mV)	100	140	160	180	200	220	240	260	280	300	320

Reverse current (μA)	−0·2	−0·3	−0·4	−0·4
Reverse voltage (mV)	−200	−400	−600	−800

Plot the anode characteristics for the above diode and determine from it (a) the "forward" d.c. resistance when the current is 3 mA. (b) the forward anode voltage when the d.c. resistance is 200 Ω.

(a) From the graph of Fig. 226 the forward d.c. resistance

$$= \frac{OB}{OA} = \frac{315 \times 10^{-3}}{3 \times 10^{-3}} = 105 \ \Omega$$

(b) Construct the 200 Ω d.c. resistance lines thus. Assume a current of 1 mA. Then the applied voltage would be $1 \times 200 = 200$ mV. Plot this point P. Draw the voltage drop line to cut the curve. The forward anode voltage would be 272 mV.

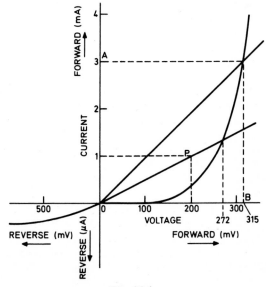

Fig. 226

STATIC AND DYNAMIC OPERATION.   It has been seen from the I_a/V_a curve that, as for the thermionic diode, there is a definite ratio of voltage to current for any particular anode current value. As shown by the preceding example, this is the anode d.c. resistance, determined for any one point on the characteristic by dividing the anode-voltage value by the corresponding anode-current value. Operation of the semiconductor diode for a.c. conditions, as when used as a R.F. demodulator or detector, will not be considered here but, mention is made in passing that, the slope or a.c. resistance value for any working range can be determined as for the vacuum diode, by the ratio

$$\frac{\text{small change in anode voltage}}{\text{resulting small change in anode current}}$$

When used as a rectifier, the diode is loaded or is operating dynamically. It is therefore apparent that a dynamic characteristic should be obtained either by direct testing or by deduction in a manner similar to that used for the diode valve. Since a separate dynamic characteristic is required for each value of R — the load resistance, then use of a "load line" would be often more appropriate for solving some problems. This is illustrated by the following example.

Example 112. The characteristic of a germanium diode is shown by Fig. 227. If the value of load resistance is to be 100 Ω and the average value of the applied voltage is to be 2 V, find the average value of the terminal voltage and the load current.

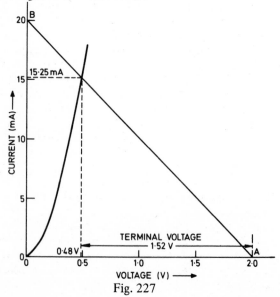

Fig. 227

The characteristic is plotted as shown and the load line drawn for 100 Ω thus:

With no current flow, the full 2 V would be applied across the diode and point A is obtained. With a diode d.c. resistance of zero,

the maximum current which could flow would be $\dfrac{2}{100} = 0.02$ A

or 20 mA. Thus point B is obtained. The point of intersection between the load line and characteristic shows that the terminal voltage would be about $2 - 0.48$ V $= 1.52$ and the load current would be 15.25 mA.

THE ZENER DIODE

As has been described earlier, a normal semiconductor diode when reverse biased passes a small but constant current until the Reverse Breakdown Voltage is attained. The current then rises rapidly to a comparatively large value and can damage the diode if maintained. The Zener diode, or more correctly the voltage reference or voltage regulator diode, due to controlled impurities injected during manufacture, can operate in the reverse biased condition, provided the current after breakdown is limited by a series resistor. The diagrams (Fig. 228a and b) show the circuit symbol and a typical characteristic curve. AB is the region where little current flows and the device has a resistance of some 50 kΩ. Region BC is the breakdown voltage V_2 — usually a very sharp knee. Region CD, above the breakdown voltage, gives large increases in current for small increase in voltage. Thus the voltage across the diode remains nearly constant although the current may vary considerably. Manufacturers data, for zener diodes, gives a definite value for current $I_{Z(min)}$ for the breakdown voltage and lies in the conductory region CD. The device has a finite resistance R_Z —its a.c. resistance after breakdown which can be determined from

$\dfrac{\delta V}{\delta I}$, The maximum current $I_{Z(max)}$ which can safely be carried for

a maximum continuous power dissipation, $P_{C(max)}$ is also quoted. Zener diodes are readily obtained for breakdown voltages between 2 and 100 volts and at different power ratings. It should be noted from the diagram that the symbol is drawn with the n-type material at the pointed end. On the actual component, a painted dot or ring marks the cathode *i.e.* the end at which current leaves the device when used as a rectifier or the end opposing current flow when connected to the +ve supply terminal when used as a zener or voltage reference diode.

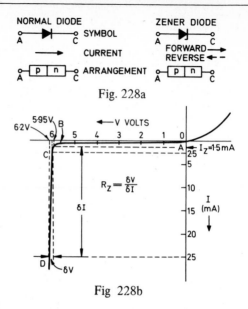

Fig. 228a

Fig 228b

USE AS A VOLTAGE STABILISER.   The circuit (Fig. 229) shows the typical arrangements used to produce a constant output voltage V_O for a varying load current I_O or a varying input/supply voltage V_I. These conditions are considered in turn.

1. If load is varied so that I_O rises then I_I will tend to rise with a consequent larger R_S voltage drop, V_Z will tend to fall but a small decrease of V_Z causes a large decrease in I_Z and this compensates for the increased I_O. I_I tends to remain constant so does the R_S voltage drop and thus the output voltage V_O.

If load current falls, I_I will tend to fall reducing the voltage drop across R_S. A small increase of V_Z will cause a large increase of I_Z which counters the effect of a reduced I_O. The R_S voltage drop tends to keep constant and thus V_O will remain substantially constant. Note that when no load current is taken, maximum load current value plus $I_{Z(min)}$ flows through the zener diode, while on full load the zener conducts only $I_{Z(min)}$.

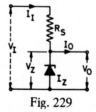

Fig. 229

2. If the supply voltage varies so that V_I rises then V_O will tend to rise. V_Z will thus rise causing a large increase of I_Z. The R_S voltage drop will get larger and this compensates for the rise of supply voltage to keep V_O substantially constant.

If the supply voltage V_I falls, then V_O and V_Z will tend to fall. This causes a large decrease of I_Z resulting in a smaller R_S voltage drop. This voltage drop offsets the fall of V_I and thus V_O tends towards a minimum of change. Note that V_Z will only remain constant provided that $I_{Z(min)}$ is allowed to flow continuously.

Example 113. From the zener diode characteristic shown (Fig. 228b) find the a.c. resistance. If used in a stabilising circuit (Fig. 229) to provide 25 mA at 6·0 V, find the value of the series resistor. The input supply voltage is 12 V. If the input supply voltage changes by 1 V, find the change in the output voltage.

From the characteristic, the a.c. resistance

$$R_Z = \frac{(6 \cdot 2 - 5 \cdot 95) \text{ volts}}{(25 - 1 \cdot 5) \times 10^{-3} \text{amperes}} = \frac{0 \cdot 25 \times 10^3}{23 \cdot 5} = 10 \cdot 64 \ \Omega$$

Also from the characteristic I_Z at 6 V = 2·5 mA. I_O = 25 mA

Thus I_I = 25 + 2·5 = 27·5 mA.

Voltage across R_S = 12 − 6 = 6 V

$$\therefore R_S = \frac{6}{27 \cdot 5 \times 10^{-3}} = \frac{6000}{27 \cdot 5} = 218 \cdot 18 \ \Omega$$

The power rating of this resistor would be

$$(27 \cdot 5 \times 10^{-3})^2 \times 218 \cdot 18 \text{ watts}$$

$$= 0 \cdot 756 \times 10^{-3} \times 218 \cdot 18 = 0 \cdot 165 \text{ W}.$$

The nearest preferred values for the series resistor would be 200 Ω of 0·25 W rating.

$$\text{The load resistance} = \frac{6}{25 \times 10^{-3}} = \frac{6000}{25} = 240 \ \Omega$$

Deducing the equivalent circuit (Fig. 230) we have,

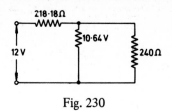

Fig. 230

Here $\dfrac{1}{R} = \dfrac{1}{10\cdot64} + \dfrac{1}{240} = \dfrac{240 + 10\cdot64}{10\cdot64 \times 240}$

or $\quad R = \dfrac{2554}{250\cdot64} = 10\cdot19 \ \Omega$

Thus change in output voltage

$=$ change in supply voltage $\times \ \dfrac{10\cdot19}{10\cdot19 + 218\cdot18}$

$= \ 1 \ \times \ \dfrac{10\cdot19}{228\cdot37} \ = \ 1 \times 0\cdot0446$

The output voltage changes by 0·045 V or 45 mV.

VOLTAGE REGULATION. The rise in terminal voltage when load is removed is termed the voltage regulation. It can be expressed as a "per unit" or as a "percentage". The rise in voltage is caused by the load current I_o passing through the zener path when the load circuit is opened.

$$\text{Per Unit Voltage Regulation} \ = \ I_O \times R_Z$$

and $\quad$ Percentage Voltage Regulation $= \dfrac{I_O R_Z}{V_O} \times 100$

For the example

Per Unit Voltage Regulation $\quad = \ 25 \times 10^{-3} \times 10\cdot64 = 0\cdot266$ V

Percentage Voltage Regulation $= \dfrac{0\cdot266}{6} \times 100 \ = \ 4\cdot43$

STABILITY.   The following assumptions can be made if V_O is kept substantially constant by zener action. If V_I is fairly constant then I_I is constant and is given by $I_I = \dfrac{V_I - V_O}{R_S}$. Also if V_I rises by 1 volt then I_I rises by $\dfrac{1}{R_S}$ amperes. This rise in current value is diverted through the zener diode, consequently the rise of I_Z equals the increase of I_I and the rise of voltage of $V_o = I_Z R_Z = \dfrac{1}{R_S} \times R_Z = \dfrac{R_Z}{R_S}$.

Stability can be defined by the rise of output voltage when the input voltage rises by 1 volt. This stability value can be obtained from the ratio $\dfrac{R_Z}{R_S}$.

For the example it is $\dfrac{10\cdot64}{218\cdot10} = \dfrac{0\cdot048}{1}$ *i.e.* 0·048 volts per volt.

Before continuing our studies of the junction solid-state devices, it is worth summarising the history of investigation, research and development of partial conductors. The various physical properties of certain materials were of interest to the earlier scientists; thus Michael Faraday in 1833 had discovered that silver sulphide had a negative temperature coefficient of resistance. Austrian physicist Braun, about 1874, investigated the resistance properties of contacts between metals and their oxides. He used a wire for a point contact and observed that resistance depended on the polarity of the applied voltage. He also noted the rectification effects of selenium. Point contact rectifiers were used in the early days of wireless at the commencement of the century and up to the 1920's, the crystal detectors were in common use until displaced by the thermionic valve. Metal or plate rectifiers, consisting of their semiconductor layers such as cuprous oxide or selenium between metal electrodes, were accepted by industry between the mid 1920's and the late 1930's even though little was generally known about their functioning theory. However, interest in the properties of semiconductor materials was aroused sufficiently, at about this period, to initiate research in depth. With the advent of the second World War, the silicon point-contact detector was developed for use in radar equipment and other "spin-offs" included the discovery of the properties of germanium and the sensitising of crystals by "doping". By the end of the war research into semiconductors was well advanced, the junction diode had been evolved in 1941 and the first discovery of transistor action was made at the Bell Laboratories, New York on Christmas Eve 1947 by J. Bardeen and W. Brattain, part of a team working under the direction of Dr. W. Shockley. The first public demonstration was in June 1948 and this was on a *point-contact* transistor which substantiated Shockley's theories. This was followed by the *junction* transistor which was devised by him to be practicable and yet performed in accordance with a predicted action.

The point-contact transistor was obsolete by 1956 and the junction-alloy transistor had come into general manufacture and usage. The diagram of Fig. 231a shows how this device is constructed. Physical electronic theory had also progressed to the stage where a range of semiconductor junction possibilities was opened up for exploration. Shockley also proposed the four-layer *p-n-p-n* device and arrangements for a different class of uniplanar devices based on the "field-effect" principle. The present day situation is such that an enormous range of solid-state devices are available for various circuit requirements. This range covers various families of such devices, but the chief still comprises the junction transistor which will be the main subject of our studies. This family can be split into two groups covering *p-n-p* and *n-p-n* types. Another family covers the field-effect devices which again can be divided into two main groups — namely junction field-effect transistors (f.e.t. or j.f.e.t.) and insulated-gate field-effect transistors (MOSFET or MOST — MOS stands for metal-oxide semiconductor and refers to construction). Still further families cover three-junction devices such as the thyristor, diacs, triacs etc whilst the developments in techniques relating to diffusion, metal-oxide processing and photographic etching have created the integrated circuit block with transistors, resistors and capacitors being formed during manufacture. The field of knowledge relating to semiconductor devices is thus being continually extended. Study of these would require to be in depth and we would be taken further along the road of specialisation. We must therefore return, at this stage, to the object of our immediate attention, namely the junction transistor.

THE JUNCTION TRANSISTOR

This device in its basic form consists of a sandwich of three layers, the layers being either of *p*-type material *i.e.* rich in +ve charge carriers (holes) or *n*-type material *i.e.* rich in electrons.

Adulterated germanium was the material first used but it has since been displaced by silicon because the latter results in lower *leakage* currents in the device and it has the ability to withstand higher temperatures (180°C) as against that for germanium (75°C). It can also have higher voltage ratings. Most present day junction transistors are *bipolar* and of the *planar-epitaxial* type. Bipolar means that current is conducted by the two types of carrier; electrons and holes. Planar-epitaxial refers to the construction. Although the junction transistor will be treated in terms of the simple diagram shown (Fig. 233), it is stressed that practical assemblage is more in accordance with the diagrams of Fig. 231. Diagram (Fig. 231a) shows a section of the earlier form of construction to which reference has already been made. The diagram (Fig. 231b) shows a section of the planar-epitaxial transistor where *n*-type silicon disc is oxidised, a window is etched photo graphically, the whole cleaned chemically and an impurity such as boron is fused in at the window to form a region of *p*-type silicon. The window is then closed by oxide and a smaller window etched in. The chemical cleaning process is repeated and phosphorus is fused into the *p*-region so as to convert it into *n*-type. The window is finally closed by oxide and the disc cut into section so as to provide numbers of transistor chips. Connections are bonded in by aluminium fusing into the silicon and each chip is enclosed in a can. Note a *n-p-n* transistor has been described. A *p-n-p* transistor is made in a similar manner except that different doping elements would be used. The impurity fusion is achieved by "vapour diffusion". This can be very accurately controlled and even though mass production methods are employed some 10,000 transistor chips can be produced from the original 50 mm diameter disc.

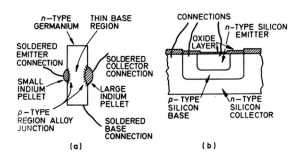

Fig. 231

OPERATION. The diagrams (Fig. 232) shows a basic diode and a basic *p-n-p* transistor arrangement, the latter consisting of a zone of *n*-type material flanked on either side by zones of *p*-type material. The *n*-type material, called the *base*, is rich in electrons and kept very thin (0·005 mm to 0·05 mm). The outer zones, called *emitter* and *collector*, are rich in +ve charge carriers or holes. Note. A *n-p-n* transistor would have the materials reversed. Amplifying action is obtained because current between emitter and base influences current, at a higher power level, between collector and base. Before describing this action in detail, we revise the explanation of junction diode operation by considering that on one side there are only holes (Fig. 232a) and on the other side there are only electrons. The application of a voltage across the junction, so that the *p*-end is +ve and the *n*-end is —ve, results in current flow since holes move from left to right and electrons from right to left. A reversal of applied potential results in no current since there are no holes to move from right to left or electrons to move from left to right. Note that the junction provides an easy path in one direction and is practically an open circuit in the other direction. With a +ve potential to the *p*-side and a —ve potential on the *n*-side, the diode is forward biased and current is readily passed provided the voltage is above the barrier potential; 0·6 V for silicon and 0·2 V for germanium.

Consider next the diagram of Fig. 232b. The arrangement shows two *p-n* junctions close together in a single piece of semiconductor. The current through one junction is found to influence that through the other junction and the arrangement constitutes a transfer-resistor (transistor). The action is best followed by taking the theory in two stages.

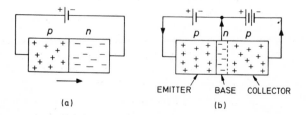

(a) (b)

Fig. 232

Consider the collector-base junction and disregard the other emitter-base diode. If the collector is made −ve to base, the arrangement is that of diode with reverse bias. The only current will be due to minority carriers *i.e.* electrons from the *p*-type collector and holes from the *n*-type base. The numbers of such minority carriers are small and the current would be small. Consider next the emitter-base diode. If the emitter is made +ve to base, the arrangement is that of a diode with forward bias and the current will be large, being due to majority carriers. For the modern bipolar transistor both the *p* and *n* type materials are heavily doped and current is conducted by both types of carrier, electrons and holes. The holes injected into the base from the emitter collect in a high concentration at the junction and because of the thinness of the base are caused to diffuse away towards the collector junction. If now a large −ve potential is applied across the collector-base junction, any holes that reach this barrier are swept across to reach the collector and increase collector current. Note that all the injected holes from the emitter do not reach the collector because some combine with electrons in the base. By keeping the base thin; a few millionths of a metre − called microns, the electrons available for such recombination are reduced and so most of the holes leaving the emitter can reach the collector. For a typical transistor for each 1 mA leaving the emitter, the collector current is about 0·99 mA and amplification occurs because of the different impedances in which these currents flow. The emitter current is in the direction of easy flow, so the impedance may be less than 50 Ω, whilst the collector current is in the direction of difficult flow and the resistance may be of the order of 1 MΩ or more. It should be noted also that to make up for the electrons lost in the base by recombination with holes, the base requires a small electron flow, viz a conventional current flow out of the base is required. The current flow out of the base can also be regarded as a flow of holes. The result of such a flow is a reduction of the base positive charge and the greater such a flow, the more the base-collector junction depletion layer is reduced. This is equivalent to reducing the resistance of this junction and in effect the collector current is increased. Thus we see that an increase of base current results in an increase of collector current. This condition should be kept in mind when the operation of the transistor in the various configurations is being considered. These configurations will be described shortly.

The diagrams of Fig. 233 show the connection arrangement for the *p-n-p* transistor, the circuit symbol and an illustration symbolising the hole flow. The conditions for a *n-p-n* transistor are also shown (Fig. 234).

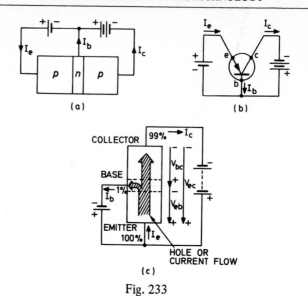

Fig. 233

Fig. 234

CURRENT TRANSFER RATIO. As already stated, all transistors behave like a control unit in that current, in one part of the device, can be made to change the value of current in another part of the circuit. Early research showed that the effectiveness of such control action was dependent on the thinness of the base region. This thinness also ensured that the base current value is only a very small fraction of the total emitter current and that the ratio of these currents is substantially constant. This fact relates to any one transistor sample but the ratio value does vary quite considerably for different transistors of the same type because of the tolerances allowed in mass production manufacture.

Consideration of the diagrams (Figs. 233 and 234) leads to the obvious deduction that $I_e = I_c + I_b$. Here I_e is the emitter current, I_b the base current and I_c the collector current. Since the ratio

$\dfrac{I_c}{I_e}$ can be taken as a measure of the control effectiveness of the

device it is given the modern term *current-transfer ratio* and the symbol h_F. Originally this ratio was loosely but incorrectly called the "current gain" — symbol α. The up-dating of the terms and

symbols is explained thus. If $\alpha = \dfrac{I_c}{I_e}$ then since I_c is less than I_e,

α will be less than unity. Again any manufacturing improvements resulting in a smaller I_b value leads to a reduction in the difference value between I_c and I_e and α approaches unity even closer. It is

evident that a more useful ratio would be $\dfrac{I_c}{I_b}$ — symbol β. A simple

relation between α and β is easily deduced but it should be noted that, as for the valve, characteristic curves can be obtained and the linear parts of these used to give the values of α and β. It is also useful to consider small changes of current instead of actual current values. We can write

$$\alpha = \frac{\delta I_c}{\delta I_e} \quad \text{and} \quad \beta = \frac{\delta I_c}{\delta I_b} \quad \text{Also since } I_e = I_c + I_b \text{ then}$$

$$\delta I_e = \delta I_c + \delta I_b \quad \text{and} \quad \delta I_b = \delta I_e - \delta I_c$$

$$\textbf{Substituting in } \beta = \frac{\delta I_c}{\delta I_b} = \frac{\delta I_c}{\delta I_e - \delta I_c}$$

$$= \frac{\delta I_c}{\delta I_e} \Big/ \frac{\delta I_e}{\delta I_e} - \frac{\delta I_c}{\delta I_e} = \frac{\alpha}{1 - \alpha}$$

Thus $\beta = \dfrac{\alpha}{1 - \alpha}$

Mention was made earlier of the modern and accepted term for α *viz* the "forward current transfer ratio" $h_{FB} = \dfrac{\delta I_c}{\delta I_e}$. Similarly β is also given the term "forward current transfer ratio" or $h_{FE} = \dfrac{\delta I_c}{\delta I_b}$.

These current ratios are defined by the B.S. specification as "the static value of the short-circuit forward current ratio" which specifies the symbol h_F. More will be said about this term when the various transistor modes and characteristics are considered. Another subscript is added to indicate the configuration, hence the above h_{FB} and h_{FE}.

Transistor usage soon showed that any value of α had to be closely specified and this necessitated a symbol which would state within limits, under what conditions the value was measured. Thus to show α was measured in "grounded emitter" mode it was primed α'. To show it was measured at zero frequency (d.c.) it was written α'_o. If measured at large current values and not small values, it would be written as α_o and so on. The use of h with subscripts to define the modes of operation gives clearer information but it should also be noted that other h values are also given with the manufacturer's data; example h_{fe} (small-signal gain *i.e.* change of I_c for unit change in I_b). Such h values (hybrid parameters) are termed small-signal parameters and their usage is related to circuit theory involving specialised treatments, equivalent circuits etc. The symbols α and β will continue to be used here for basic study considerations with reference being made to the h representation at appropriate points.

Example 114. For an increment of base current of 0·1 mA in a certain transistor, the corresponding collector current increment under short-circuit conditions was observed to be 1 mA. What is the common-base short-circuit current gain of the transistor?

Here $\delta I_b = 0{\cdot}1$ mA and $\delta I_c = 1$ mA $\therefore \delta I_c = 1{\cdot}1$ mA

$$\therefore \alpha = \frac{1{\cdot}0}{1{\cdot}1} = 0{\cdot}909.$$

Note that δI_e is deduced from the fact that emitter current equals collector current plus base current.

TRANSISTOR CHARACTERISTICS.

The performance of a transistor can be investigated by making suitable tests and from the results can be drawn the characteristics of each configuration. These are (i) the Input Characteristic, (ii) the Output Characteristic and (iii) the Transfer Characteristic. The last can be deduced from the first two but a test is useful in that an average value for the forward current transfer ratio can be directly obtained.

Before proceeding to consider the tests and characteristics of the different configuration or modes, attention is drawn to the circuit details of Fig. 236. It should be noted that the emitter arrow indicates the direction of conventional current *i.e.* the direction in which +ve holes are injected into the base material. For the *n-p-n* transistors, which operate in a similar fashion except that, the current carriers which constitute the collector current, consist of electrons flowing from the *n*-type emitter into the *p*-type base. Conventional current flow is in the opposite direction, hence the reverse of arrow on the emitter which points away from the base. The polarities to the device are reversed in that the collector is made +ve to base and the emitter is −vely biased. It is of interest to note that usage of the *n-p-n* transistor has increased appreciably of late, as manufacturing techniques have improved with consequent reduction in cost. An advantage, of major consideration, is that we have now "got our circuit polarities right" *i.e.* they are in line with accepted valve practice.

CIRCUIT CONFIGURATION. The electrodes of a transistor are analogues to those of a thermionic triode. The emitter corresponds to the cathode, the base to the grid and the collector to the anode. Like the valve, the device may be used in one of three different circuit configurations but as for the valve, attention will only be given to the most common form of usage.

The diagrams of Fig. 235 illustrate the three different circuit configurations and the corresponding triode valve connections are given for comparison purposes. The first and most used is shown by (a) and is called the "common or grounded emitter" mode. This corresponds to the most usual arrangement in valve practice where the cathode is "earthy", the grid being the input electrode and the anode the output electrode. The second configuration, as shown by (b), is the "common or grounded base" mode which, as for the triode valve, is infrequently used and then only for special applications *e.g.* matching a low-input source to amplifying stages. For mode (c) is shown the "common or grounded collector" usage. This again is mainly adopted for impedance matching or as a buffer stage. A common use is for coupling an amplifier to a low-impedance load, since this mode has a high input impedance and low output impedance.

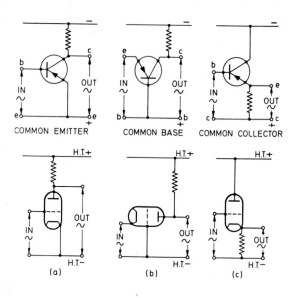

Fig. 235

The features of the various transistor modes are

(a) Common or grounded emitter. Medium impedance for both input and output circuits. The battery +ve line is common and, as the capacitors—see Fig. 245b, connect emitter to +ve line for a.c. signals then, the emitter is also common to both input and output circuits. There is a phase shift of 180° and both the a.c. current gain and voltage gains are high.

(b) Common or grounded base. Low input impedance to emitter, high output impedance from collector. Capacitors—Fig. 239 allow a.c. signals while blocking d.c. The base is thus grounded to a.c. and is common to both input and output signal circuits. The phase shift is zero and the a.c. current gain is less than 1. The a.c. voltage gain is high.

(c) Common or grounded collector. High input impedance and low output impedance. The arrangement is often called an emitter follower. It should be remembered that the battery offers negligible impedance to a.c. signals and thus the input is applied between base and collector and the output is taken from the emitter and collector. The phase shift is zero and the a.c. current is high. The a.c. voltage gain is less than 1.

COMMON-BASE CONFIGURATION. The circuit for making tests on a transistor connected in the common-base mode is shown by the diagram of Fig. 236.

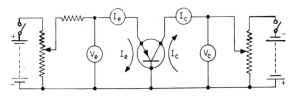

Fig. 236

The "input characteristic" is obtained by maintaining the base-collector voltage V_c constant and noting emitter current I_e for various values of emitter-base voltage V_e. The voltage adjustments are made by means of the circuit potentiometer and graphs, of the form shown by Fig. 235, can be plotted from the test results. It will be seen that the graphs can be plotted as I_e to a V_e base or more commonly as V_e to I_e base. The curves allow the input impedance to be determined for any working condition. Thus the slope of the graph in Fig. 237a

is $\dfrac{\delta I_e}{\delta V_e}$ and the input impedance is the reciprocal of this slope or

$r_{in} = \dfrac{1}{\text{slope}}$. Thus input impedance $= \dfrac{\delta V_e}{\delta I_e}$. The slope of the

graph of Fig. 237b gives the value of the input impedance directly.

Input impedance $r_{in} = \dfrac{\delta V_e}{\delta I_e}$.

The "output characteristic" is obtained by holding the values of I_e constant and, over a range of values, readings of I_c are obtained for various values of base-collector voltage V_c. Control would be by means of the potentiometer resistors provided. A family of curves, as illustrated by Fig. 238b can be obtained which are seen to resemble the I_a/V_a curves of a thermionic pentode. The important difference is that, here each curve is for a separate value of emitter current

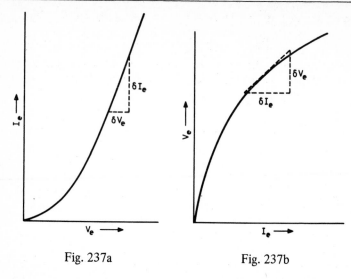

Fig. 237a   Fig. 237b

instead of grid voltage. Thus we confirm that the transistor is a current-operated device and must be thought of in these terms when considering the associated circuits.

The explanation for the shapes of the curves is that the collector will always collect holes as long as it has a −ve bias. The current I_C is almost independent of the voltage applied to it, because it is a reverse current, and is dependent only upon the numbers of minority carrier +ve holes which are made available. It thus depends on the magnitude of I_e.

It should be noted that double subscripts are sometimes used to designate the base-collector voltage as V_{BC}, the emitter-base voltage as V_{EB} etc. Capital letters are used for the d.c. condition when no signal is applied.

The "transfer characteristic" can be found by direct test by holding the base-collector voltage at a fixed value and by varying I_e. The effect on I_C is noted, a graph as shown by Fig. 238b is plotted and its slope is a measure of α or h_{FB} —the forward current transfer ratio (current gain). A value for α could also be deduced directly from the output characteristic. Thus for this mode $\alpha = \dfrac{\delta I_C}{\delta I_e}$, so when, for a fixed value of V_C, the change in value for I_C is noted for a change in I_e, the current gain can be deduced. For example (Fig. 238a) for a constant V_C value of − 4 V, when the emitter current is raised from 2 mA to 4 mA (difference 2 mA), the collector current rises approximately from 1·9 mA to 3·8 mA(1·8 mA differenc). Thus

$$\alpha = \frac{1\cdot8}{2} = 0\cdot9 \text{ (approximately)}.$$ It will be seen that difficulties

arise in achieving the required accuracy for α, when working by this method, and a more reliable result is achieved by plotting the transfer characteristic.

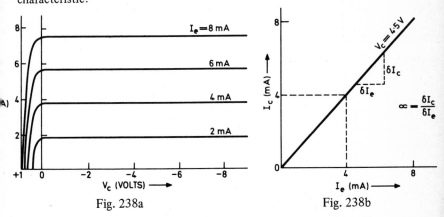

Fig. 238a   Fig. 238b

Amplification. The arrangement for a single-stage amplifier, connected in common-base configuration, is shown by the diagram of Fig. 239. The battery polarities are important. A *p-n-p* transistor is assumed.

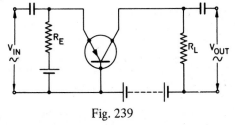

Fig. 239

Assume that a supply voltage of 10 mV, applied to the emitter, produces a current of say 1 mA of which 0·98 mA pass to the collector and 0·02 mA to the base. Transistor supply voltages are usually multiples of 1·5 V and even the smallest voltage unit would give too high a current, if applied direct to the emitter. R_E is therefore included to limit the emitter current to a standing value of a few milliamperes. By inserting a load resistor R_L in the collector circuit, amplifying action is obtained. This is shown below.

With no a.c. input signal steady emitter current will flow, limited to a convenient value (1 mA) by R_E. If the input signal varies by

5 mV to produce a fluctuation of emitter current of 0·5 mA, then collector current will vary by about 0·5 mA (actually some 0·5 × 0·98 = 0·49 mA). If R_L is 1000 Ω, the a.c. fluctuation of the output voltage is 1000 × 0·5 × 10⁻³ = 0·5 V. Thus a voltage amplification

of output voltage to input voltage of about $\dfrac{0·5}{5 \times 10^{-3}}$ = 0·1 × 10⁻³

= 100 is obtained.

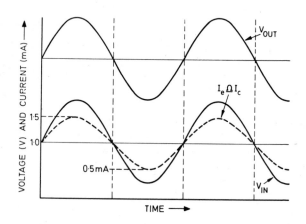

Fig. 240

The diagram of Fig. 240 shows the phase relationship between the input and output voltages. They are in phase and this is the reverse of that for a valve where the input and output voltages are 180° out of phase.

It will be seen that for the common-base mode, the full capabilities of the transistor are not exploited in that, the control of the base current on the collector current is not used. The configuration gives a current gain of less than one and amplification is only achieved by the different impedance levels in emitter and collector circuits. It may be convenient to deal in terms of voltage gain, but it must be remembered that true amplification should be considered in terms of a power gain and this aspect is important for transistor work. Thus for the circuit of Fig. 239, the emitter-base diode of the transistor is biased in the forward direction and is of low impedance—say 15 Ω. If a current change of 0·5 mA takes place, the power change = 0·5² × 10⁻⁶ × 15 = 3·75 μW. In the collector circuit, a corresponding change of 0·49 mA takes place in an impedance of say 1000 Ω. Thus power change is $0·49^2 \times 10^{-6} \times 10^3 = 0·49^2 \times 10^{-3} = 0·24 \times 10^{-3}$ or 240 μW.

There is thus a power gain of $\dfrac{240}{3·75} = 64$ (approximately).

To achieve such a gain, it is necessary to match the signal source to the input impedance and the load to the output impedance of the transistor.

We continue our consideration of the common-base mode by stating that current amplification cannot be obtained since α is less than unity. Voltage amplification is possible as was shown earlier because the emitter-base junction has a low resistance and, if a large load resistance value is used in the collector circuit, a high voltage gain results. The power gain can also be large, as is shown above. The arrangement is now chiefly used for R.F. circuits.

Example 115. For the common-base arrangement shown in Fig. 239 assume a R_L of 50 kΩ and a transistor current gain of 0·95. Find the voltage gain if the input signal causes an emitter current change of 100 μA. The transistor input resistance can be taken as 100 Ω.

Since $\alpha = \dfrac{\delta I_c}{\delta I_e}$ then $0·95 = \dfrac{\delta I_c}{\delta I_e}$ $\therefore \delta I_c = 0·95 \times 100 = 95 \ \mu$A

The input-signal voltage variation $= 100 \times 10^{-6} \times 100 = 10^{-2}$

$$= 0·01 \text{ V}$$

Output-voltage variation $= 95 \times 10^{-6} \times 50 \times 10^3 =$

$$= 475 \times 10^{-2} = 4·75 \text{ V}$$

$\therefore$ Voltage gain $= \dfrac{4·75}{0·01} = 475.$

In the circuit of Fig. 239, an emitter resistance R_E is shown. An explanation for its presence is useful here, in case it is confused with the value of 100 Ω as given in the above example for the input resistance. This input resistance is a transistor value and does not refer to R_E. Tests would show that the value of voltage to be applied between emitter and base to cause normal current to flow is about 100 mV. If a fixed voltage is applied, then as a result of temperature variation, the emitter current would alter since the effective emitter-base resistance decreases with a rise of temperature. To overcome this, a resistor R_E is included in the emitter circuit. Its value is made much greater than the emitter-base resistance and thus the current will not alter to any extent when the temperature changes. This technique comes under the general heading of "stabilisation" and prevents "thermal run away".

COMMON-EMITTER CONFIGURATION. The diagram of Fig. 241 shows the test circuit for obtaining the characteristic curves for a transistor connected in common-emitter configuration. Potential adjustments for the input and output circuits are made by means of the resistance potentiometers.

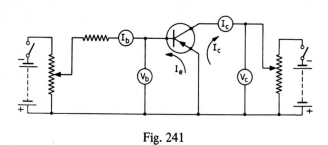

Fig. 241

Fig. 242a

Fig. 242b

One set of characteristics can be obtained for a constant value of collector voltage V_c, by varying the base-emitter potentiometer to give a series of different values of V_b, the current I_b being noted for each value of V_b. A typical curve for this "input characteristic" is shown by Fig. 242a. Note this can be plotted as I_b to a base of V_b or as V_b to a base of I_b. The latter form is the more common — Fig. 242b.

Example 116. The "input characteristic" of Fig. 243 is that of a typical transistor when connected in the grounded-emitter mode. Find the input impedance when $I_b = 40\,\mu A$ and V_c is kept constant at -4.5 V.

Point P is determined on the characteristic and a tangent is drawn in order to determine the slope.

The input impedance or a.c. resistance is given by this slope.

$$\therefore \; r'_{in} \; = \; \frac{\delta V_b}{\delta I_b} \; = \; \frac{(175 - 135)\,10^{-3}}{(60 - 20)\,10^{-6}} \; = \; \frac{40 \times 10^3}{40}$$

$$= \; 1 \; \times \; 10^3 \; \text{ohms or 1 k}\Omega.$$

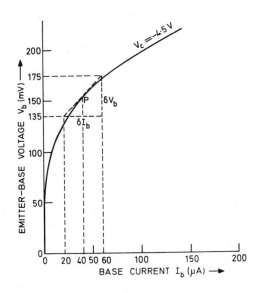

Fig. 243

A further useful set of graphs are the "output characteristics". The collector current/collector voltage curve is plotted by obtaining corresponding readings for I_c and V_c, I_b being held constant for each set of readings. Typical curves, as shown by Fig. 244a, would be obtained.

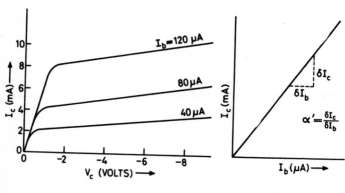

Fig. 244a

Fig. 244b

A "transfer characteristic" can also be obtained by plotting I_c to a base of I_b, V_c being kept constant. This can be deduced from the output characteristic and is shown by Fig. 244b.

The curves obtained from the tests are static characteristics. When a small a.c. signal is applied between base and emitter, the effect is shown on the input characteristic—Fig. 242a. The varying signal causes a small swing of I_b and V_b over a part of the curve, as shown by PQ. The reciprocal of the slope of PQ will give the input a.c. resistance as already described. Assume a working point of 93 mV and the voltage input to vary so as to give points P and Q. If US is a straight line through PQ, the a.c. resistance for the working points can be obtained as the reciprocal of the slope.

$$\text{Thus } \frac{UT}{TS} = \frac{(150 - 50) \times 10^{-3}}{(145 - 35) \times 10^{-6}} = \frac{100}{110} \times 10^3 = 909 \text{ ohms.}$$

Note that the input resistance depends upon the working d.c. bias value because the gradient of the curve changes with the V_b value.

The output resistance can be obtained from Fig. 244a. It is given by the slope of the appropriate curve. The output a.c. resistance value will be high and nearly constant for a given value of I_b as long as the working conditions are beyond the turn-over point of the curve. When V_c is small *i.e.* below the turn-over value, the output resistance is much reduced.

The attention of the reader is drawn to the deduction already made earlier, namely

$$\beta \text{ or } h_{\mathrm{FE}} = \frac{\alpha}{1-\alpha} \text{ or } \frac{h_{\mathrm{FB}}}{1-h_{\mathrm{FB}}}$$

Example 117. If the change of collector current is 2 mA when the emitter current (in a common-base circuit) is changed by 2·2 mA (the collector voltage being constant), calculate the current gain of the transistor when connected in a common-emitter circuit.

For the common-base mode $\alpha = \dfrac{\delta I_{\mathrm{c}}}{\delta I_{\mathrm{e}}}$

where $\delta I_{\mathrm{c}} = 2\,\mathrm{mA}$ and $\delta I_{\mathrm{e}} = 2\cdot2\,\mathrm{mA}$

$$\therefore \alpha = \frac{2}{2\cdot2} = \frac{1}{1\cdot1} = 0\cdot91$$

also $\beta = \dfrac{\alpha}{1-\alpha} = \dfrac{0\cdot91}{1-0\cdot91} = \dfrac{0\cdot91}{0\cdot09}$

Thus $\beta = \dfrac{91}{9} = 10\cdot1.$

Amplification. The arrangement for a single-stage amplifier connected in common-emitter configuration is shown by the diagrams of Fig. 245. The base and collector are made −ve with respect to the emitter. In Fig. 245b the base bias is obtained by the voltage drop along R_{B} since R_{B} forms a bias potentiometer with the base-emitter resistance in series.

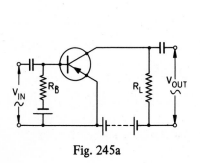

Fig. 245a

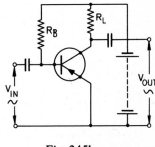

Fig. 245b

The common or ground-emitter arrangement is most commonly used to take advantage of the high current gain that can be achieved with a transistor. Here $\dfrac{I_c}{I_b}$ is the current gain since the base current is now the input current. As I_b is very small, this ratio is large and has a value of about 50. The diagrams of Fig. 245, show how the supply voltage is applied between emitter and collector and current, in this circuit, is controlled by the base current. As was stated earlier the symbol α' or β was used for the common-emitter "current-gain" but h_{FE} is now recommended *i.e.* the "forward current transfer ratio". The value of this parameter can range from 20 to 100. The slope of the transfer characteristic (Fig. 244b) will give this value.

Reverting to fundamentals, it is easy to explain the high gain for this mode. Two separate current paths are seen for Fig. 245. For one, a current flows into the emitter and out through the base. For the other, current flows out from the collector and returns via the supply to the emitter. The collector current is thus less than the emitter current by the very small base current or alternatively, we can say, base current is the difference between emitter and collector currents, namely $I_b = I_e - I_c$. Any change occuring in I_e or I_c leads to a change in I_b equal to the difference. Thus for a transistor with a common-base gain factor α of 0·98, when I_e is increased by 1 mA, I_c changes by 0·98 mA and the resulting change in I_b is 0·02 mA. Here a certain change in base current is associated with one in collector current 49 times as large, and there is a current gain between the circuits concerned.

Example 118. The "output characteristic" of Fig. 246 is that of the typical transistor mentioned in Example 116 and connected in the grounded-emitter mode. Find the output impedance when $V_c = -4\cdot5$ V and $I_b = -40\,\mu\text{A}$.

Also find the current gain for this point.

Point P is determined on the $-40\,\mu\text{A}$ characteristic and the slope is found.

$$\text{Slope} \ = \ \frac{\delta I_c}{\delta V_c}$$

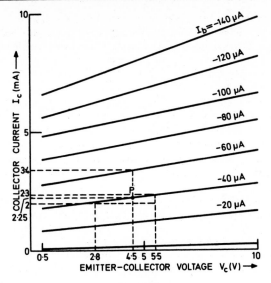

Fig. 246

Since the output impedance is the reciprocal of the slope then output impedance

$$r_o' \;=\; \frac{1}{\text{slope}} \;=\; \frac{\delta V_c}{\delta I_c} \;=\; \frac{5 \cdot 5 - 2 \cdot 8}{(2 \cdot 3 - 2 \cdot 0)\,10^3} \;=\; \frac{2 \cdot 7 \times 10^3}{0 \cdot 3}$$

$$=\; 9 \times 10^3 \text{ ohms or } 9 \text{ k}\Omega$$

Current gain $\alpha' \;=\; \dfrac{\delta I_c}{\delta I_b}$

Then for a voltage value of $-4 \cdot 5$ V we see that a base current change of $(60 - 40) = 20\ \mu$A or 20×10^{-6} A is associated with an I_c current change of $(3 \cdot 4 - 2 \cdot 25) = 1 \cdot 15 \times 10^{-3}$ mA.

$$\therefore \;\alpha' \;=\; \frac{\delta I_c}{\delta I_b} \;=\; \frac{1 \cdot 15 \times 10^{-3}}{20 \times 10^{-6}} \;=\; \frac{0 \cdot 115}{2} \times 10^3$$

$$=\; 0 \cdot 0575 \times 10^3 \;=\; 57 \cdot 5$$

or current gain $= 57 \cdot 5$.

Example 119. If the base current of the transistor shown in the common-emitter amplifier, (Fig. 245b) is to be 50 μA and the collector supply voltage is 10 V, find the value of the single resistor R_B used to provide the bias current. Ignore the base-emitter forward conduction voltage drop.

Since I_b is to be 50 μA and the base-emitter voltage is neglected then the base is 10 V positive with respect to the —ve line *i.e.* the full 10 V is dropped across R_B.

$$\therefore \ R_B \ = \ \frac{10}{50 \times 10^{-6}} \ = \ 2 \times 10^5 \ = \ 200 \, k\Omega \, .$$

The amplification action is similar to that considered for the common-base arrangement. Here the operation is applied to an arrangement as for the circuit of Fig. 241b. A change of emitter-base signal voltage of say 10 mV will produce a change of base current δI_b, say 0·05 mA and because of the high β value for this mode of operation, a relatively large change in the current I_c will occur. Suppose this current change δI_e, to be 2·5 mA, then almost all of this current appears in the collector current I_c and if this passes through a high value of load resistor R_L, say 1 kΩ, then the a.c. output voltage change appearing across this resistor will be

$$\delta I_c \times R_L = 2\cdot 5 \times 10^{-3} \times 10^3 \ = \ 2\cdot 5 \, V$$

A voltage amplification of about $\dfrac{2\cdot 5}{10 \times 10^{-3}} \ = \ 250$ is obtained.

From the waveforms of Fig. 247, an inversion of the output to input voltage is seen. This is similar to the common triode valve operation where input and output voltages are 180° out of phase. The reason for the transistor inversion is explained as follows. Upon the application of a small alternating voltage signal across the base-emitter junction, the base current will fluctuate, such variation resulting in similar variations of the collector current. Thus a rise of input voltage will produce an increase in collector current. The voltage drop across the load resistor R_L will increase and the amplifier output voltage will fall as a result. Similarly a fall in input voltage produces a fall of collector current and a rise of output voltage.

Power gain for the arrangement being considered would be
Power in $= 10$mV $\times 0\cdot 05$ mA $= 0\cdot 5 \, \mu$W

Power out $= \ I_c{}^2 R_L \ = \ 2\cdot 5^2 \times 10^{-6} \times 1000 \ = \ 6\cdot 25 \times 10^{-3}$
$= \ 6250 \, \mu$W.

So power gain $= \dfrac{6250}{0\cdot5} = 12\,500.$

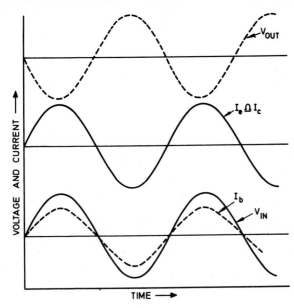

Fig. 247

COMMON-COLLECTOR CONFIGURATION.   As stated earlier this
arrangement is little used and will therefore be given no further
attention in this book.

LOAD LINES

The use of a load line, in conjunction with the transistor characteristics for any particular mode, follows thermionic valve procedure
and the example shows how this can be constructed and used.

Example 120. The diagram (Fig. 248) shows the typical common-
emitter characteristic of a transistor. From the information available
draw the load line for a collector load resistance of 4 kΩ and a supply
voltage of 15 V. Assume that it is possible to swing a sinusoidal signal
over the full range of the load line, state a suitable value of I_c and
calculate (a) the d.c. power dissipated in the load resistor, (b) the a.c.

power available (c) the $\dfrac{I_c}{I_b}$ gain. Estimate the saturating value of V_c.

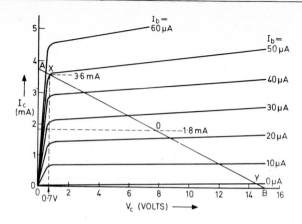

Fig. 248

The load-line is drawn using the expression $V_C = V_S - I_c R_L$. Here V_C is collector-emitter voltage, V_S is the supply voltage and R_L the load resistance.

When $V_C = 0$, then $I_c = \dfrac{V_S}{R_L} = \dfrac{15}{4000} = \dfrac{15}{4} \times 10^{-3}$

$$= 3.75 \text{ mA}.$$

Mark Point A on the diagram for $V_C = 0$ and $I_c = 3.75$ mA. Again when $I_C = 0$ then $V_C = V_S$ or the voltage on the collector is 15 V.

This gives point B on the diagram.

Join A and B to give the load line. The saturating value of V_C for this condition is seen from the diagram to be about 0.7 V. The collector current is seen to be 3.6 mA (approx).

A suitable value for the quiescent point O would be $\dfrac{3.6}{2} = 1.8$ mA.

(a) The d.c. power dissipated would be

$$I_c^2 R_L = (1.8 \times 10^{-3})^2 \times 4000 \text{ watts.}$$
$$= 1.8^2 \times 4 \times 10^3 = 12.96 \text{ mW}$$

(b) The power available

$$= \frac{(V_{c(max)} - V_{c(min)})\,(I_{c(max)} - I_{c(min)})}{8}$$

$$= \frac{(15 - 0 \cdot 7)(3 \cdot 6 - 0) \times 10^{-3}}{8} = \frac{14 \cdot 3 \times 3 \cdot 6 \text{ milliwatts}}{8}$$

$$= \frac{12 \cdot 87}{2} = 6 \cdot 44 \text{ mW}.$$

(c) The $\dfrac{I_c}{I_b}$ gain. This is determined for working about the operating

point O; that is between the 30 μA and 20μA base current characteristics

we have β or h_{FE} $= \dfrac{2 \cdot 3 - 1 \cdot 5}{(30 - 20)10^{-3}} = \dfrac{0 \cdot 8 \times 10^3}{10} = 80.$

Note. An expression for the a.c. power available has already been deduced for the thermionic valve.—Chapter 10, Fig. 197. This will be seen to apply to transistor working.

Referring to the example, it is seen that the saturating value for V_c is required. The load line for the conditions specified shows the limits of operation for the appropriate d.c. potentials and a line passing through point X indicates that an increase of I_b (along OX) produces no further increase of I_c and the transistor is said to be "bottomed" or "saturated". At the other extreme at point Y, the collector current is almost zero and the condition is one of "cut-off".

LEAKAGE CURRENT.

It will be noted from the diagram (Fig. 248) Example 120 that collector current values are shown for the $I_b = 0$ characteristic *i.e.* a current of increasing magnitude flows for rising values of V_c. This is the normal reverse current of a diode junction and for most purposes its effect can be neglected, especially for silicon transistors. Manufacturers data gives this figure as I_{ceo} *i.e.* collector-emitter leakage current with base open circuit and I_{cbo} *i.e.* collector-base leakage current with emitter open circuit. The former is always greater since, due to basic transistor action $I_{ceo} = (1 + \beta)I_{cbo}$. The leakage current value, even that for silicon transistors, can give rise for concern when the operating temperature exceeds $100°C$ and its effect must be taken into consideration when designing circuits.

THE PRACTICAL AMPLIFIER

The diagram of Fig. 249 shows a practical circuit for the commonest type of transistor amplifier. It uses the common-emitter configuration, one battery source and components values which are obtained with the aid of transistor static characteristics and a knowledge of the current-flow directions in the circuits. R_1 and R_2 provide the base-emitter bias by acting as a potential divider. The function of R_3 is covered under "stabilisation".

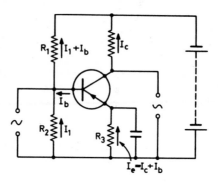

Fig. 249

Stabilisation. One of the main difficulties encountered in transistor circuit operation is a tendency for the operating conditions to alter or "drift", even under quiescent conditions. Such drift is caused by temperature changes and appropriate action must be taken to reduce this tendency. It is a fact that the resistance of germanium falls when the temperature is raised. This means a larger current flow which, in turn, tends to increase the heating of the transistor — eventually destroying it in extreme cases. Such destruction is said to be caused by "thermal runaway" and can be minimised, as can drift, by circuit stabilisation methods.

An efficient and common method of stabilising the base bias is by using a potential divider made up from R_1 and R_2. Since the current I_1 in these series resistors is much greater than the base current, it follows that the base potential is held fairly constant in spite of base current variation when an a.c. input signal is applied. The use of resistor R_3, inserted between the emitter and the battery +ve terminal has been explained in relation to Example 115 where this resistor was referred to as R_E. Its value is made very much larger than the transistor internal emitter resistance and the current is unaffected to any extent when the transistor temperature changes. Thermal runaway is thus prevented. Note that the use of silicon, in place of germanium for semi-conductor devices, is mainly to raise the permissible operating temperature and stabilising circuit techniques are still necessary for silicon transistors.

TRANSISTOR AS A SWITCH

This device can be made to operate as a highly efficient switch and is capable of working at a very high speed *viz* in micro-seconds. Furthermore, it can be energised with only a fraction of the main current to be controlled. It is thus highly sensitive and can operate with transducers sensing minute physical changes, temperature, pressure, strain, light etc.

For the switching function, the device is usually used in the common-emitter mode because of the high current gain. Since the resistance between the collector and emitter terminals is controlled by the base biasing, it follows that if the latter is made +ve with respect to the emitter, the base-emitter junction is reverse biased and no emitter and consequently no collector current flows. With the base made —ve with respect to the emitter, base current will flow, as will collector current. As the base is made progressively negative, the collector current continues to increase until it is limited by the collector supply voltage and the load resistance. Under this condition only a fraction of a volt is dropped across the collector-emitter junction and thus full supply voltage can be assured to be available for the load. The transistor can thus be seen as a switch. With a +ve bias on the base the switch is open and with a —ve potential, of suitable magnitude, the switch is closed.

PRACTICAL CIRCUIT. A basic circuit is shown in Fig. 250. Current will only pass between emitter and collector when the base is made at least 0·7 V to 1 V negative with respect to the emitter. The easiest way to achieve this is shown by using the related sub-circuits (a) or (b).

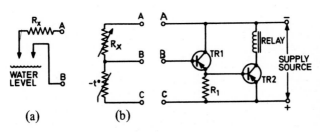

Fig. 250

Arrangement (a) is a simple remote control switch *viz* circuit closed by a rising water level. The appropriate bias, decided by the value of series resistor R_x is applied and the transistor TR1 is turned on. Current will then flow via R_1 through the emitter-collector junction.

The potential at the top of R_1 will adjust to about -1 V and TR2 is turned hard "on" causing the relay to operate. With the remote switch open, no bias, is applied to the base of TR1 and the transistor is "off" thus ensuring that TR2 is "off".

For arrangement (b) the bias on the base of TR1 is decided by a potential divider. A thermistor is shown connected in series with a variable resistor. It is evident that the potential of point B will vary according to the resistance value of the thermistor. This in turn varies with its temperature *i.e.* resistance decreases as temperature increases. The potential of B and hence the base of TR1 will thus alter with temperature and control can be effected. The series variable resistor R_x allows adjustment of the operating point.

THE THYRISTOR

This four-layer device, also termed a silicon-controlled rectifier (S.C.R.), is mainly used as a semiconductor switch, capable of controlling the application of a voltage to a load. It is provided with three terminals, the *anode, cathode* and *gate* or *trigger*. Fig. 251 illustrates the general construction and the appropriate symbols are also shown. It will be seen that two types are available, namely the *n*-gate and the *p*-gate. The latter is almost universally used with the anode made $+$ve to the cathode. Theory will show how the thyristor conducts in one direction only although it should be noted that bi-directional thyristors are now available. These are called "triacs".

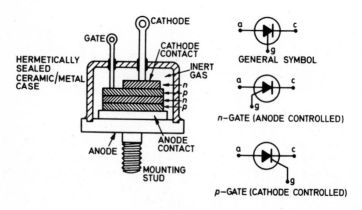

Fig. 251

If the anode of a *p*-gate thyristor is made −ve to the cathode, the two outer diodes of the four-layer sandwich—see Fig. 252, are reverse biased and conduction will not occur. With the anode made +ve to cathode, the centre diode only has a reverse bias which prevents conduction if no input is applied to the gate. The function of the gate is to cause the device to conduct or "fire" and when the appropriate potential is applied, current carriers from the outer layers are allowed to flow to the centre junction. When this occurs, the anode and cathode are considered to be connected *viz* a short-circuit occurs across the device. Once the gate has initiated conduction, the current carriers from the outer layers maintain the passage of main current even when the gate input is removed. Thus to "switch off" the current must be interrupted by using some external means or circuit condition. This is readily achieved if an alternating voltage is being controlled since the current value at the end of a half-cycle falls to zero. The applied voltage potential then reverses and a condition of non-conduction is obtained. The thyristor thus switches itself off at the end of a +ve half-cycle and will not conduct again until another +ve gate impulse is applied during a corresponding +ve half-cycle of mains voltage across the anode-cathode terminals. When operating to control a d.c. circuit, additional circuitry is required. Usually a second thyristor is connected in parallel with the main device with the anodes capacitively coupled. Triggering of the second thyristor will switch the load off at the period of capacitor charging.

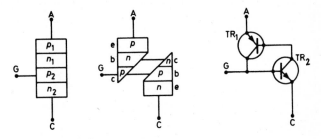

Fig. 252

THEORY OF OPERATION. The thyristor is represented by a four-layer sandwich (Fig. 252). If the centre *p-n* section is split to give the arrangement shown, we can develop the two complementary transistor analogy. TR1 is a *p-n-p* device and TR2 a *n-p-n* device. Assume a short positive current pulse to be applied to the base of TR2. With this correct bias on the base of TR2, this transistor is "turned on". The collector current of TR2 is drawn from the base of TR1. Since the emitter-base diode of TR1 is forward biased this unit is also "turned on". Collector current of TR1 can now supply the base of TR2 and gate current is no longer required. Both transistors now continue to conduct as long as the polarities of the supply to anode and cathode remain correct and the anode current does not fall below a value, called the *holding current*.

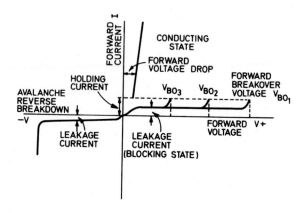

Fig. 253

If the current/voltage characteristic of a thyristor (Fig. 253) is considered, further explanation of the holding current can be made. With no gate connection, if an increasing voltage is applied between anode and cathode, a small but particularly constant current flows through the device between these terminals. At a certain value of voltage (V_{BO}), termed the *Breakover Voltage,* the current suddenly increases in value, being limited only by the load conditions. The device has switched itself into a state of low impedance by internal electron-hole rearrangement and will continue in this state until the current is reduced below the holding value. The thyristor will thus revert again to its "blocking state";

With the gate connected and an electrical bias of the correct relation to the anode applied, it is found that the value of Breakover Voltage can be controlled but, as already stated, once the thyristor has "fired" the gate is found to have no further effect.

CONTROL METHODS. The load-current value can be controlled by the length of time for which the thyristor is conducting and this can be varied by one of two ways: (a) Phase-shift Control (b) Burst-fire or Burst-triggering Control.

(a) Phase-shift Control. The diagram (Fig. 254) illustrates the
method. The device is triggered by a short-duration current pulse
applied to the gate during each +ve half-cycle of the supply voltage
to the anode, cathode and load circuit. The instant of pulse application or triggering can be altered in relation to the start of a supply
voltage half-cycle and thus the length of conduction time for the half-
wave can be controlled and the average current value through the
load, is varied. The triggering circuit usually involves a variable potential divider and capacitance circuit whose time constant can be adjusted.
The value of voltage built up across the capacitor is used to initiate
conductance through a transistor which then discharges through a
resistor, thus applying a pulse to the gate of the thyristors. The diagram shows how this method of control can trigger the thyristor at
an instant when the supply voltage is high and a large current will be
switched suddenly. This condition can give rise to radio interference
and suitable suppression arrangements may be necessary.

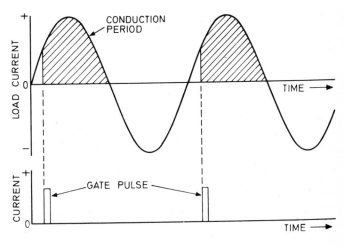

Fig. 254

(b) Burst-fire Control. The diagram (Fig. 255) shows the method.
Here the thyristor is switched on for a number of whole consecutive
positive half-cycles. Conduction is therefore for one or more half-waves
of load current with periods of non-conduction interposed. The

method is suitable for large loads where thermal or mechanical inertia ensures that the effect can be smoothed out to equate to that of a smaller constant current. The gate pulse is applied by a timing circuit at the instant just before the beginning of a +ve half-cycle of the main circuit voltage and as the latter is rising from zero, load current rises in a sinusoidal manner and falls similarly. This occurs for each conducting half-cycle and radio interference is thus minimised. It is apparent that the number of conduction cycles can be varied by the duration of pulsing the gate but in this context the heating which occurs at the gate electrode must be taken into account by the circuit designer. Since the gate supplies current during the pulsing period, this cause of additional temperature rise can be reduced by using a high-frequency supply for the trigger circuit.

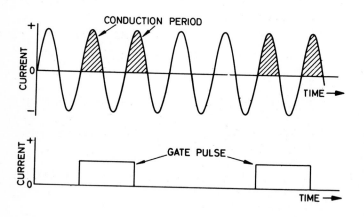

Fig. 255

CHAPTER 12

PRACTICE EXAMPLES

1.   A transistor connected in a common-emitter circuit shows changes in emitter and collector currents of 1.0 mA and 0.98 mA, respectively. What change in base current produces these changes and what is the current gain of the transistor?

2.   In a certain transistor, a change in emitter current of 1 mA produced a change in base current of 0.1 mA. Calculate the common-emitter and common-base short-circuit current gains.

3.   A certain transistor has a common-emitter short-circuit current gain α' or $\beta = 50$. Calculate the common-base short-circuit current gain.

4.   For a common-emitter transistor, the transfer characteristic is given by the values

I_b (μ microamperes)	20	40	60	80	100	120
I_c (mA)		1·4	2·5	3·6	4·7	5·8	6·9

Find the current gain.

5.   From the output characteristic shown in Fig. 246, deduce the transfer characteristic, and hence find the current gain for the transistor when connected in the common or grounded-emitter mode.

6.   The data given in the table refers to a *p-n-p* transistor in the common-base configuration.

Collector Voltage V_c (volts)	Collector Current I_c (milliamperes)				
	Emitter Current $I_e =$ 0 mA	Emitter Current $I_e =$ 2 mA	Emitter Current $I_e =$ 4 mA	Emitter Current $I_e =$ 6 mA	Emitter Current $I_e =$ 8 mA
− 5	0	− 1·9	− 3·7	− 5·7	− 7·6
− 30	− 0·1	− 2·0	− 3·8	− 5·8	− 7·7
− 55	− 0·2	− 2·1	− 3·9	− 5·9	− 7·8

Draw the collector-current/collector-voltage characteristic for the various values of emitter current and calculate the resistance of the transistor.

7.   From the data given for Example 6, plot the collector-current/emitter-current or transfer characteristic for $V_c = -30$ V and hence calculate the current gain (α). From this value deduce the current gain (β) for the common-emitter mode.

8.   The data given in the table refers to a *p-n-p* transistor in the common-emitter configuration.

Collector Voltage V_c (volts)	Collector Current I_c (milliamperes)			
	Base Current $I_b = -20\,\mu A$	Base Current $I_b = -40\,\mu A$	Base Current $I_b = -60\,\mu A$	Base Current $I_b = -80\,\mu A$
-3	-0.91	-1.6	$--2.3$	-3.0
-5	-0.93	-1.7	-2.5	-3.25
-7	-0.97	-1.85	-2.7	-3.55
-9	-1.0	-2.05	-3.0	-4.05

Plot the collector current/collector voltage characteristic for base currents of -20, -40, -60 and $-80\,\mu A$ and using these determine (a) the output resistance of the transistor for the $I_b = -60\,\mu A$ condition, and (b) the current gain when the collector voltage is -6 V.

9.   In the circuit shown, calculate the value of the bias resistor to give a standing base current of 40 μA. The supply voltage is 9 V and for the transistor $V_{BE} = -0.2$ and $V_{CE} = -5$ V.

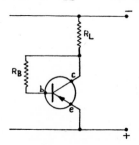

10.   The circuit shown below is for a typical common-emitter amplifier. If the current through the emitter resistor is 0·5 mA, determine the battery voltage. Assume a base-emitter voltage drop of 0·1 V.

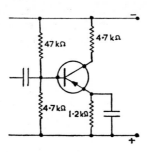

A.C. MACHINES–OPERATION

This chapter is devoted to the more common methods of control and operation of alternating current power equipment. In considering the different forms of a.c. machine in earlier chapters, little opportunity was presented for discussing operational procedures. Thus, for example, little has been said about the parallel operation of alternators or about the problems of load sharing. Since, for shipboard work, it is normally usual to run more than one machine for "electrical power supplies", it is necessary to consider the relevant theory, especially when it is appreciated that there are differences in the operating procedures for d.c. and a.c. generators.

In considering the overall aspects of operating a.c. machinery, it will also be recalled that little has been said about the starting and speed-control methods for a.c. motors. Shipboard practice requires such attention and this is discussed in the latter section of this chapter.

A.C. GENERATORS

Alternating current generators can be connected in parallel because, as will be seen, they control each other electrically by a circulating current which keeps them in step. The series connection of a.c. generators is not practicable unless the machines are coupled together mechanically. Without such mechanical coupling, and since provision for a circulating current does not exist, it is not possible for the machines to control each other electrically. Series running is not a feasible arrangement and will not be considered further.

A.C. GENERATORS IN PARALLEL

The reader is reminded that, like its d.c. counterpart, the a.c. machine can be made to function as either a generator or as a motor when it is connected to a supply source. Such a supply source would be "live busbars" which are already energised by a source of electrical power, *viz* other alternators.

The d.c. conditions for parallel running have already been discussed in Chapter 1. It was seen that, if the generated voltage of a machine under consideration was made greater than the busbar voltage then it would feed electrical power out, and would require to be driven by a prime-mover. It would contribute power into the "common pool" constituted by the busbar system, and would thus operate satisfactorily in parallel with others already "on the bars". If the generated voltage of the machine was lowered below that of the busbars then it would take electrical power to function as a motor. In this condition it can be used to provide mechanical power for driving associated machinery.

Somewhat similar conditions arise for the a.c. generator, and thus we can have it operating in the modes of an alternator or a synchronous motor. The control of the generated voltage of the machine will, as is to be seen, alter the operating mode but if parallel operation is required then, unlike d.c. operating conditions, a further adjustment is needed. Such an adjustment is associated with the prime-mover throttle valve and this must be carried out on each machine of the parallel arrangement, before satisfactory load sharing can be achieved.

SYNCHRONISING.   As seen in Chapter 9, an alternating-current generator or synchronous motor can only be connected to a live busbar system by ensuring correct synchronisation prior to closing the paralleling switch or circuit-breaker. The requirements for correct synchronising have already been mentioned, but these are now considered in detail.

To ensure correct synchronisation the conditions are:

1. The voltage of the incoming machine must equal that of the busbars.

2. The frequency of the incoming machine must be the same as that of the busbars.

3. The e.m.f. of the incoming machine must be in phase with the busbar voltage.

4. The phase sequence of the incoming machine must be similar to that of the busbars, and the switching arrangements must be such that, like phases of the machine and busbars are connected when the paralleling procedure is completed.

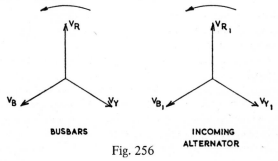

BUSBARS   INCOMING
                          ALTERNATOR

Fig. 256

If the conditions for the busbars and incoming machine are represented by the phasor diagrams of Fig. 256, then the synchronising requirements can be seen, in that:— (a) the respective voltage phasors must be equal. (b) the phasors must be rotating at the same speed—same frequency. (c) They must be in phase with each other *i.e.* corresponding line to line voltages must reach their positive maximum values at exactly the same instant. (d) The phasors must rotate in the same direction—same phase sequence.

The diagram of Fig. 257, illustrates the usual arrangement of instrumentation to assist the correct synchronising procedure. Permanent switchboard installations are provided with a synchronising panel comprising the necessary voltmeters, frequency meters and synchroscope. For marine work it is also customary to provide "synchronising lamps" as an additional facility and these will be considered separately.

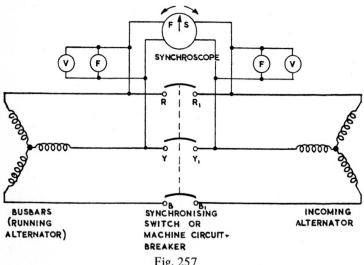

Fig. 257

THE SYNCHROSCOPE. Different forms of instrument, covered by various patents, are available but basically these comprise a fixed coil system of two windings and a freely rotatable pointer assembly. The simplest arrangement can be imagined as being made of a polyphase coil system, so connected to the busbars that a rotating magnetic field is set up, and a separate independent coil, connected between two lines of the "incoming" machine. This coil is fixed coaxially with the spindle so as to magnetise two soft-iron rotor pole pieces, carried by the spindle and constituting an armature. The pointer attached to the spindle is arranged to move over a scale marked "Fast" and "Slow" with appropriate arrows relating to the direction of pointer rotation. The instrument has no moving coils, control springs or ligaments and is usually "not continuously" rated. A connecting plug or switch is invariably provided to allow connection of the synchroscope to the busbars and the machine which is to be paralleled, and once synchronising has been completed, the instrument is isolated.

When energised, the rotor armature system of the synchroscope is magnetised in an alternating manner by a single-phase supply from the "incoming" machine. Thus alternate N-S poles are produced and the armature tends to align itself with the rotating magnetic field produced by the "busbar" supply. Assuming the correct synchronising condition to be shown by the 12 o'clock position, then no torque will be produced since the alternating rotor field of the instrument will complement the rotating field pole arrangement and, as the latter moves forward, the strength of the former falls off to reverse and build up in time to coincide with the next complementary pole arrangement, as arising from the next half cycle. If synchronising conditions are not correct and the poles of the rotating field are slightly ahead of the armature poles when they are fully energised, then there will be a torque exerted on the latter, tending to turn the pointer system in the ahead direction. This torque action, though pulsating, will produce rotation of the pointer in the "Slow" direction, showing that the incoming machine must be speeded up. The converse action will occur, if the incoming machine is running too fast. It should be noted that, the instrument will only respond to the frequency difference when this becomes small. It will not respond if the operating action, described above, is occuring too fast to be followed due to inertia of the rotor system. This is not a disadvantage since slight differences in frequency are shown clearly and the instant for synchronising can be determined with precision.

In practice, the engineer may find it difficult to adjust the speed of the incoming machine so that the pointer is stationary at 12 o'clock. Such a condition is not essential and a more practical proposition is to have the pointer rotating slowly in the "Fast" direction and to close the paralleling switch at about 11 o'clock. Due to the time-lag of the operating mechanism and human response, actual synchronising will thus take place nearer 12 o'clock, and the machine, running fast, will be slowed slightly and take a small proportion of load. If synchronised when running slow, the machine would take a motoring current which may well operate the reverse-power relay and "trip" the circuit-breaker of a machine already on the "bars".

SYNCHRONISING LAMPS. The lamp method of synchronising makes use of filament lamps, so connected across the contacts of the parallel-ing switch that the intensity of the illumination varies continuously. The error in the frequency of the incoming machine as compared with busbar frequency, is shown by the rate at which the lamps "darken" or "brighten". The diagram of Fig. 258, shows the usual method of lamp connection. This augments the instruments already shown on Fig. 257.

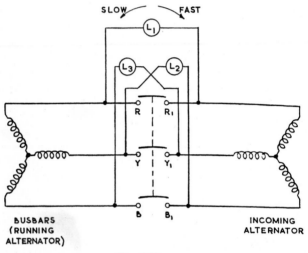

Fig. 258

In order to understand the operation of the lamps, it is better to first consider the synchronising of a single-phase a.c. generator with lamps connected as shown by Fig. 259a. Here, it is apparent that, equality of voltage and phase on either side of the switch, results in zero voltage across the lamps which are then completely dark. A dif-ference of frequency will cause the lamps to brighten and darken, the difference in frequency being shown by the rate of pulsation. This "dark lamp" method of synchronising makes difficult the judgement of the mid-point dark period—the instant of closing the paralleling switch, and is overcome by cross-connection of the lamps as shown by Fig. 259b.

The "bright lamp" connection, as shown by Fig. 259b, ensures that the lamps are at maximum brightness when the two supplies are in phase. It should be noted that, the voltage rating of each lamp should equal the working voltage of the system since this voltage is applied across each lamp when synchronism occurs. For the 180°

out-of-phase condition, the voltages of the busbars and incoming machine oppose each other, as far as the two-lamp circuit is concerned. The voltage acting around the lamp circuit is zero—hence the lamps are dark, and are only at maximum brightness when synchronism exists. The moment for synchronising can thus be determined with exactness. Frequently only one lamp is used, the other being replaced by a resistor, mounted within the switchboard.

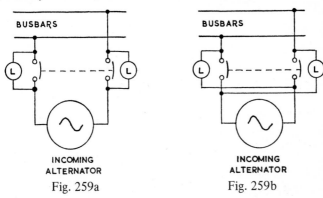

Fig. 259a   Fig. 259b

For three-phase systems, although direct connection of three lamps across the contacts of each line or cross-connecting of the lamps are methods which can be used, the Siemens-Halske arrangement, as shown in Fig. 258 and explained by the diagrams of Fig. 260, is favoured because it not only gives the correct instant for synchronising but also indicates when the incoming alternator is running fast or slow relative to the busbar voltage. From the superimposed phasor diagrams of Fig. 260, it will be seen that when running "slow", the lamps will light in the order $L_2 L_1 L_3 L_2 L_1 L_3$ *i.e.* in the order $L_1 L_3 L_2$. If the incoming machine is running "fast" the lamps will light $L_3 L_1 L_2 L_3 L_1 L_2$ *i.e.* in the order $L_1 L_2 L_3$. If the lamps are arranged in a triangular pattern they would tend to brighten in a clockwise sequence when running fast, and anti-clockwise when running slow. The moment for synchronising is with lamp L_1 "dark" and the other two L_2 and L_3 at equal but not full brilliance.

PARALLEL OPERATION

As stated earlier, there is a difference between the parallel operation of d.c. and a.c. generators and this dissimilarity is of particular importance to the marine engineer since he can be required to operate either a.c. or d.c. electrical generating and distribution systems. The procedure for putting an a.c. machine on the "bars" has already

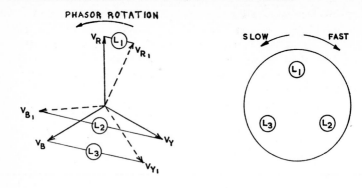

Fig. 260

been discussed in this chapter and in Chapter 9, the production of a torque which ensures that a machine runs in synchronism was also considered. Attention is now given to the action necessary to make a paralleled alternator take its share of the load and to study, in turn, the effects of the two adjustments possible, namely:– (1) operation of the field regulator *i.e.* excitation control and (2) operation of the throttle or steam valve *i.e.* speed control.

With correctly designed d.c. generators, parallel operation and load sharing can be achieved by adjustment of excitation, and reference to the governor characteristics of the prime-movers is not necessary. Basic consideration shows that if the field excitation of a machine is increased, then the generated e.m.f. of the machine is raised and it will take a larger portion of the load. This increase of load slows the d.c. generator down and allows the prime-mover to draw additional fuel or steam, since the governor will be actuated. The additional loading condition is thus carried.

With two alternators in parallel, an increase in excitation of one machine raises the generated e.m.f. and should thus tend to make it take more load. However, the machine cannot slow down since it is tied synchronously to the system, and thus the governor of the prime-mover is unaffected and no action results in causing the machine to take more load. As will be seen below, the operation of the excitation control merely causes a wattless current which circulates in the paralleled machines and busbar system. This current lags the generated e.m.f. by an angle ϕ such that the load can be equated to $E I \cos \phi$. The kW load thus remains constant to maintain an unvaried governor setting. To change the distribution of load between a.c. generators in parallel, the throttle valves must be manipulated. We thus see that, for two alternators in parallel, since the speeds (frequencies) must be identical, the *kW* loading on each machine must be related to the prime-mover input power *i.e.* to the amount of operation of the throttle valve and cannot be controlled by the excitation. The effect of excitation and throttle control is now considered in detail, but two problems are first introduced to revise the basics of a.c. load summation and power-factor control.

Example 123. A factory takes 200 kW at 660 V from a three-phase supply at a lagging power factor of 0·707. A synchronous motor is installed which takes an additional 100 kW. What must be the *kVA* rating of this motor to raise the power factor of the system to 0·866 (lagging).

Total active power $= 200 + 100 = 300$ kW

Total apparent power at 0·866 (lagging) power factor $= \dfrac{300}{0·866}$

$$= 346·4 \text{ kVA}$$

Here $\cos \phi = 0·866 \qquad \sin \phi = 0·5$

Total reactive power $= -346·4 \times 0·5 = -173·2$ kVAr

Original apparent power of load $= \dfrac{200}{0·707} = 282·9$ kVA

Here $\cos \phi = 0·707 \qquad \sin \phi = 0·707$

Original reactive power $= -\dfrac{200}{0·707} \times 0·707 = -200$ kVAr

Thus with the motor load, the reactive power is reduced by

$$-200 - (-173·2) = -26·8 \text{ kVAr}$$

Thus reactive power rating of motor must be 26·8 kVAr (leading). *i.e.* $+ 26·8$ kVAr

$\therefore$ Apparent rating of motor $= \sqrt{100^2 + 26·8^2} = 10\sqrt{10^2 + 2·68^2}$

$$= 10\sqrt{100 + 7·18} = 10\sqrt{107·18}$$

$$= 10 \times 10·35 = 103·5 \text{ kVA.}$$

Example 124. Two three-phase, 660 V, star-connected alternators in parallel supply the following loads: 40 kW at unity power factor, 40 kW at a lagging power factor of 0·85, 30 kW at a lagging power factor of 0·8, 80 kW at a lagging power factor of 0·7. The armature current of one machine is 100 A at a lagging power factor of 0·9. Determine the armature current, the power output and the power factor of the other machine.

Load	kVA	$kW = kVA \cos \phi$	$kVAr = kVA \sin \phi$	$\cos \phi$	$\sin \phi$
1	40	40	0	1	0
2	47·1	40	− 24·8	0·85	0·527
3	37·5	30	− 22·5	0·8	0·6
4	114·3	80	− 81·5	0·7	0·714
		190	− 128·8		

Machine 1.

Apparent power $= 100 \times 660 \times \sqrt{3} = 114$ kVA

Active power $= 114 \times 0·9 = 103$ kW

$$\phi_1 = 25°51' \quad \sin \phi_1 = 0·436$$

Reactive power $= 114 \times 0·436 = -49·8$ kVAr

Machine 2

Active power $= 190 - 103 = 87$ kW

Reactive power $= -128·8 - (-49·8) = -79$ kVAr

Apparent power $= \sqrt{87^2 + 79^2} = 10\sqrt{8·7^2 + 7·9^2}$

$= 10\sqrt{75·5 + 62·5} = 10\sqrt{138}$

$= 117·5$ kVA

Current of Machine 2 $= \dfrac{117·5 \times 10^3}{\sqrt{3} \times 660} = \dfrac{1175}{1·732 \times 6·6}$

$= 103$ A

Active power output of Machine 2 $= 87$ kW

Power factor of Machine 2 $= \dfrac{87}{117·5} = 0·742$ (lagging).

PARALLEL OPERATION (Continued).

The parallel operation of a.c. generators may be studied under two distinct considerations. The first would be, parallel working with "infinite busbars", as constituted by shore-based power stations tied together through a national transmission grid system. An ideal case of infinite busbars is one where the system is so large, in comparison with a single alternator, that its voltage and frequency are unaffected by the behaviour of the alternator. The second consideration is of major importance for the marine engineer, since it relates to shipboard working. Here busbar voltage and frequency can be altered by local conditions and the more common case, of two or more alternators running in parallel, is therefore given detailed attention in the study which follows.

(1). EXCITATION CONTROL. Assume two alternators to have been paralleled correctly. The voltage, frequency and phase of each are the same and this condition is represented by the phasor diagrams shown by Fig. 261. V is the busbar voltage *i.e.* that produced by machine No. 1 which is generating an e.m.f. of E_1 volts and is supplying the busbar load. E_2 is the e.m.f. of machine No. 2.

Note that the e.m.f.'s of the two machines in parallel are in phase relative to the external circuit, but in opposition when considered with respect to each other. This is shown here *i.e.* the local circuit is considered.

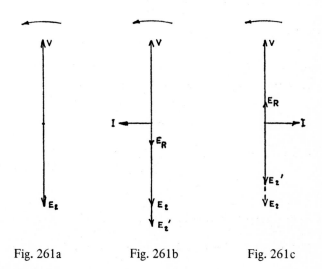

Fig. 261a   Fig. 261b   Fig. 261c

For diagram (a), the voltages are equal and in opposition. As a result, no current flows in the local circuit between the two alternators since the resultant voltage is zero. For diagram (b), the excitation of machine No. 2 is seen to have been increased. The generated e.m.f. E_2 is increased to E_2' and gives rise to a resultant voltage E_R acting round the local circuit. This circuit is mainly reactive and the resultant current I lags E_R by nearly 90°. This current represents no power flow either "from" or "to" the busbars and the prime-mover of alternator No. 2 is unaffected and the governor will be unaltered. The machine is considered as supplying a lagging current to, or taking a leading current from the busbars. This current will tend to neutralise the effect of any lagging load current taken from the busbars, and thus over-excitation of an a.c. generator improves the power factor of the paralleled system. Diagram (c), shows the effect of decreasing the excitation of machine No. 2. It is pointed out that, altering excitation does not appreciably alter the busbar voltage. This is explained by the reactive current I, for condition (b) having a lowering effect on the e.m.f. E_2', – lagging current has a demagnetising effect on the alternator field strength. In a similar manner, E_1, the generated e.m.f. of machine No. 1 is increased and the circulating current is such as to make E_2 and E_1 equal to V, – the busbar voltage.

Example 125. Two three-phase alternators in parallel, supply 125 kVA at a lagging power factor of 0·8. Each machine supplies the same power at the same current. If the excitation of each machine is changed so that the power factor of one falls to 0·5 (lagging), find the power factor of the other.

Note that as this is excitation control, the power output is unaltered.

Original apparent power $=$ 125 kVA $\cos \phi = 0·8$ $\sin \phi = 0·6$

$\therefore$ Total active power $=$ 125 × 0·8 $=$ 100 kW

The active power supplied by each machine $=$ 50 kW
Under the new condition.
For Machine 1

$\cos \phi_1 = 0·5$ and $\sin \phi_1 = 0·866$

Apparent power $= \dfrac{50}{0·5} = 100$ kVA

Reactive power $= -100 × 0·866 = -86·6$ kVA

Original reactive power $= -125 \sin \phi = -125 × 0·6 = -75$ kVAr

For Machine 2

Reactive power $= -75 - (-86 \cdot 6) = 11 \cdot 6$ kVAr

Active power $= 50$ kW

Apparent power $= \sqrt{50^2 + 11 \cdot 6^2} = 10\sqrt{5^2 + 1 \cdot 16^2}$

$\qquad\qquad\quad = 10\sqrt{26 \cdot 35} = 10 \times 5 \cdot 15 = 51 \cdot 5$ kVA

$\cos\phi_2 = \dfrac{50}{51 \cdot 5} = 0 \cdot 97$ (leading).

Since the lagging $kVAr$ of one machine has been made larger than the total lagging $kVAr$, it follows that the $kVAr$ of the other machine must be made anti-phase to produce the final result. Thus since No. 1 alternator operates at a lagging power factor, then No. 2 alternator must be operated at a leading power factor.

(2). THROTTLE CONTROL. Assume the governor control to be manipulated so that the fuel or steam valve of machine No. 2 is opened. Alternator No. 2 tends to speed up and phasor E_2 tries to overtake V, as is shown by Fig. 262. In connection with the local circuit, a resultant voltage E_R immediately becomes apparent to produce a current I lagging, as before, by almost 90°. This current is nearly in phase with E_2 which means that alternator No. 2 is now developing power as given by $E_2 I \cos\phi_2$. When this power output equals the increase of input power, as is brought about by the actuation of the throttle

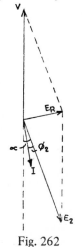

Fig. 262

valve through the governor, the tendency for prime-mover No. 2 to speed up increases, and this alternator set delivers power to the load. Alternator No. 1, thus relieved of load, speeds up slightly until its prime-mover governor operates to reduce the input power and bring about stable speed conditions.

The final distribution of load on each alternator is achieved by alternate operation of both machine throttle controls until the required loading is as shown by each alternator wattmeter, and the voltage and frequency of the system settle down to the desired condition.

If the driving power of alternator No. 2 is removed, because of some mechanical fault, such as a fuel stoppage, then the conditions would be as shown by Fig. 263. Voltage E_2' drops back behind the true synchronism position by an angle α'. There is now a resultant voltage E_R' acting round the local circuit, to produce current I', almost at $90°$ behind E_R' —the circuit, comprising the machine armature, being mainly reactive. The busbars are now supplying power VI' cos ϕ' to the machine and this will keep it running as a synchronous motor. The drop back of E_2' from the synchronous position, is only momentary and the machine is accelerated back into synchronism. The magnitude of the synchronising current and torque have already been discussed in Chapter 9.

Fig. 263

Note that, if an increased mechanical load was added to alternator, No. 2 when under the motoring condition, the machine e.m.f. E_2'

would drop back still further. E_R' and I' would increase so that the total power supplied from the busbars increases. This is the basis of operation of the synchronous motor, and although little work will be done on the a.c. machine when operating in this manner nevertheless, since it is usually used as the propeller motor for marine a.c. electric propulsion systems, it warrants a mention later in this chapter.

In the above treatment, reference was made to the synchronising current and this is now taken a stage further. Consider parallel operation as depicted by the diagram of Fig. 264. The alternators generate E_1 and E_2 volts to maintain a busbar voltage of V volts. Although these voltages are in phase with respect to the load, they are in direct opposition to each other. Suppose the excitations and powers developed by the prime-movers are set to cause currents I_1 and I_2 at power factors of $\cos \phi_1$ and $\cos \phi_2$. The total load is the phasor sum of I_1 and I_2. This could be shown on the phasor diagram but has been omitted in the interest of clarity. Assume now that the power input to No. 2 is increased and the set tries to accelerate. It advances by a small angle α. New load conditions are set up. E_2' and E_1 produce E_R acting round the local circuit. This causes the circulating current, which under no-load conditions was designated as the synchronising current I_S, lagging E_R by $90°$. This current I_S can be added by phasors to the original currents. Thus it combines with I_2 to give the new machine current I_2'. I_S is received by machine No. 1 and lessens the current output giving I_1' the resultant of I_1' and I_S.

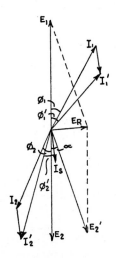

Fig. 264

The increased input to No. 2 makes it take more load so that its speed settles to that decided by the governor-actuated throttle-valve opening. Meanwhile machine No. 1, having been relieved of load, accelerates to a new speed (frequency), determined in the final stage by the overall loading of the system. Note therefore that I_S is a short-time circulating current, brought about by the transition conditions resulting from adjustment of the controls. Once the overall paralleled system settles down, we have operating conditions similar to those existing originally, except that I_1, I_2 and cos ϕ_1, cos ϕ_2 would have new values.

LOAD SHARING

Summarising the foregoing, we see that, increasing the excitation of a machine produces a wattless circulating current. This means that a change of generated voltage, relative to the busbars, changes the amount of reactive kVA which the machine supplies. Varying the power input tends to speed up the machine and power $E_2 I_2$ cos ϕ_2 is given out. Load sharing can therefore be considered from two viewpoints: (1) Sharing of kW and (2) Sharing of reactive kVA.

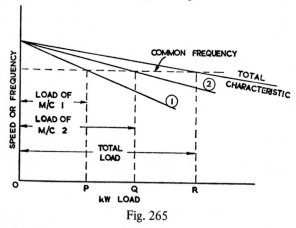

Fig. 265

kW LOAD SHARING. The speed regulation *i.e.* governor characteristics of the complete prime-mover and alternator set, determine the proportions of load taken by machines in parallel. The frequency/load characteristics of two sets are shown by Fig. 265

The characteristics of sets Nos. 1 and 2 are assumed unequal but, since the two frequencies are tied together, the total load delivered, at any given frequency, is obtained by adding together the individual loads at this frequency. Thus $OR = OP + OQ$. By repeating this addition for various frequencies, the common frequency curve is obtained. This curve indicates the fall of frequency with increase of total load and also the division of any total load, such as OR, into OP (m/c No. 1) and OQ (m/c No. 2). It is evident that, the greater proportion of the load is borne by the machine having the flatter governor characteristic.

It can also be pointed out that, as for load-sharing problems with d.c. generators, the characteristics can be plotted back-to-back as shown in Fig. 266. The point of intersection X, shows both the sharing of load and the common frequency. Altering the governor settings of No. 1 machine, raises curve 1 to 1_1. It is also seen that, if curve 1 is raised to 1_2, the point of intersection lies to the right of the right-hand vertical and the condition is represented when set No. 2 is being motored by No. 1, which is supplying not only the total load but this motoring power in addition.

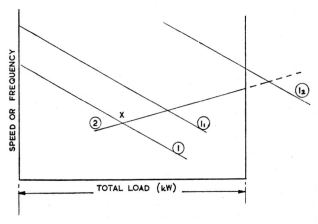

Fig. 266

Assuming straight line characteristics, a mathematical solution of problems is readily possible and procedure can be as follows. Assume two alternators Nos. 1 and 2,–of similar ratings, have prime-movers with straight line characteristics, such that one drops "a" per cent and the other "b" per cent between no load and full load (in kilowatts). Assume their speeds equal N at some total loading P(kW). Then, if load is increased (or decreased) by Q(kW), the increase (or decrease) in load will be shared so that both alternators fall (or rise) the same amount of speed. Conditions are shown in Fig. 267.

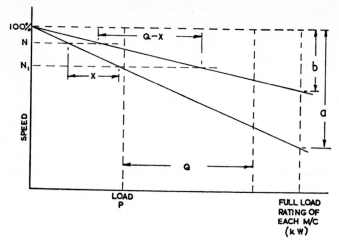

Fig. 267

Let the share of increased (or decreased) load taken by machine No. 1 be X (kW), then No. 2 will take $(Q-X)$ kilowatts.

Hence $\quad \dfrac{a\,N}{100} \times X \;=\; \dfrac{b\,N}{100}(Q-X) \quad \text{or} \quad aX \;=\; b(Q-X)$

$\therefore$ Load gained or lost by No. 1 is $\quad X \;=\; \dfrac{bQ}{a+b}$

and Load gained or lost by No. 2 is $\quad Q-X \;=\; \dfrac{aQ}{a+b}$

Example 126. Two identical alternators each rated 1·25 MVA at a power factor of 0·8 (lagging) supply a load in parallel. The governor setting on the prime-mover of No. 1 machine is such that it drops from 50 Hz on no load to 48 Hz on full load. The second machine (No. 2) drops from 50 Hz to 47 Hz. How will they share a load of 1·5 MW.

Machine No. 1 drops 2 Hz for 1000 kW (1250 kVA at 0·8 p.f.) *i.e.* 0·002 Hz per kW.

Machine No. 2 drops 3 Hz for 1000 kW (1250 kVA at 0·8 p.f.) *i.e.* 0·003 Hz per kW.

Let the load on No. 1 be X kilowatts.

Then load on No. 2 $= (1500 - X)$ kilowatts
The frequency must be equal, as they are paralleled

$\therefore$ $50 - 0{\cdot}002\,X = 50 - 0{\cdot}003\,(1500 - X)$
and $0{\cdot}002\,X = 4{\cdot}5 - 0{\cdot}003\,X$ or $0{\cdot}005\,X = 4{\cdot}5$

$\therefore$ $5X = 4500$ or Load on No. 1 $= 900$ kW
Load on No. 2 $= 1500 - 900 = 600$ kW.

The above is a direct solution of the problem which can also be answered by application of the expressions deduced earlier.
Thus for No. 1 machine, a speed change of 2 Hz in 50 Hz gives
$$a = 4 \text{ per cent}$$
and for No. 2 machine, a speed change of 3 Hz in 50 Hz gives
$$b = 6 \text{ per cent}$$
Here the load increase *i.e.* from no-load to full load $= Q$
$$= 1500 \text{ kW}$$

$$\therefore X = \frac{0{\cdot}06 \times 1500}{0{\cdot}04 + 0{\cdot}06} = \frac{0{\cdot}06 \times 1500}{0{\cdot}1} = 6 \times 150 = 900 \text{ kW}$$

and $Q - X = 1500 - 900 = 600$ kW.

kVA_rLOAD SHARING. The way in which machines running in parallel, share the reactive kVA is largely governed by their relative internal voltages. The voltage regulation characteristics of two machines are as shown by Fig. 268. Note that voltage is plotted against $kVAr$ load.

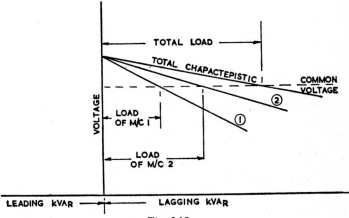

Fig. 268

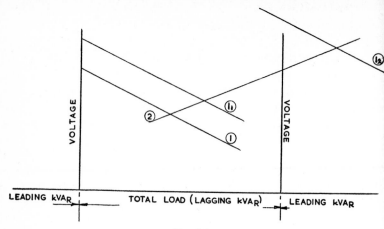

Fig. 269

As for *kW* load sharing, the characteristics can also be plotted back to back as shown by Fig. 269.

The positions of the characteristics are determined by the amount of excitation. An increase of this for one machine, such as No. 1, will raise the curve to 1_1. Machine No. 1 then takes a larger share of the *kVAr* load and the busbar voltage is raised. Condition 1_2 shows how No. 1 may be operated at a leading power factor even though the total load is lagging.

Fig. 268 clearly shows that the machine with the flatter characteristics takes the largest share of the load.

Example 127. Two alternators operate in parallel to supply a load of 1 MW at a power factor of 0·8 (lagging). If the steam supply of No. 1 machine is adjusted so that it is loaded to 600 kW and the excitation is adjusted so that its power factor is 0·707 (lagging) (a) find the power factor at which No. 2 operates. (b) If the load and steam supplies of both machines are unchanged and the excitations are altered so that the power factor of No. 2 is raised to become 0·8 (leading), find the new power factor of No. 1.

Total active load $= 1000 \text{ kW} \quad \cos \phi = 0\cdot8 \text{ (lagging)} \quad \sin \phi = 0\cdot6$

$$\therefore \text{ Total apparent power } = \frac{1000}{0\cdot8} = 1250 \text{ kVA}$$

$$\text{Total reactive power } = -1250 \times 0\cdot6 = -750 \text{ kVAr}$$

(a) Machine No. 1

$$\text{Active power} = 600 \text{ kW}$$

$$\text{Apparent power} = \frac{600}{0.707} = 848.7 \text{ kVA}$$

$$\therefore \quad \text{Reactive power} = -\frac{600}{0.707} \times 0.707 = -600 \text{ kVAr}$$

Machine No. 2

$$\text{Active power} = 1000 - 600 = 400 \text{ kW}$$

$$\text{Reactive power} = -750 - (-600) = -150 \text{ kVAr}$$

$$\therefore \quad \text{Apparent power} = \sqrt{400^2 + 150^2} = 10^2\sqrt{4^2 + 1.5^2}$$
$$= 10^2\sqrt{16 + 2.25} = 10^2\sqrt{18.25}$$
$$= 10^2 \times 4.275 = 427.5 \text{ kVA}$$

$$\cos \phi_2 = \frac{400}{427.5} = 0.94 \text{ (lagging)}.$$

(b) Even with increased excitation, the active power ouput of No. 2 machine is still 400 kW, although its power factor is now leading i.e. a positive *kVAr* condition.

$$\therefore \quad \text{Apparent power} = \frac{400}{0.8} = 500 \text{ kVA}$$

$$\text{Reactive power} = 500 \times 0.6 = 300 \text{ kVAr}$$

Thus for Machine No. 1

$$\text{Reactive power} = -750 + 300 = -450 \text{ kVAr}$$

$$\text{Apparent power} = 10^2\sqrt{36 + 20.25} = 10^2\sqrt{56.25}$$
$$= 10^2 \times 7.5$$
$$= 750 \text{ kVA}$$

$$\therefore \quad \cos \phi_1 = \frac{600}{750} = 0.8 \text{ (lagging)}.$$

THE SYNCHRONOUS MOTOR

The basic principles of the "motoring action" of an alternator, when connected to busbars, have already been covered. Here some attention is given to the machine when it is mechanically loaded *i.e.* when it takes electrical energy to operate as a motor. Such a motor is used to advantage in the larger sizes, because its power factor can be controlled by variation of the d.c. excitation. When operating in this fashion it can be used to improve the overall power factor of a load system, such as a factory, and when used as a ship's propulsion motor, unity power factor working will keep the current to a minimum and hence the copper losses of the arrangement and the size of the associated equipment will not be larger than that required for the appropriate kW rating.

OPERATING ACTION

Consider an alternator run up to speed and synchronised onto the busbars. If the machine is delivering no power it is said to "float on the bars". Suppose next the prime-mover is disconnected. The machine will draw power from the mains, sufficient to overcome the no-load losses and the rotor will slip back through an angle α_1. The phasor diagram is as shown by Fig. 270. The busbar voltage V and the generated e.m.f. E_1 now have a resultant E_{R_1} and the motor

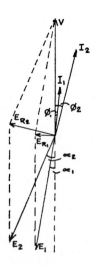

Fig. 270

draws a current I_1 from the supply. $I_1 = \dfrac{E_{R_1}}{Z_S}$ where Z_S is the synchronous impedance of the armature. I_1 will lag behind the applied voltage V by an angle ϕ_1. Power input $= VI \cos \phi_1$.

If the motor is loaded, the rotor will drop back further to an angle α_2. The resultant voltage is now E_{R_2} and the current will increase to a value I_2, lagging, as before, on V by an angle ϕ_2.

Power input is now $VI_2 \cos \phi_2$ and α_2 will adjust itself until sufficient power is drawn from the supply to deal with the load.

The phasor diagram can also be drawn, from the busbar point of view. In this case it would be like that shown by Fig. 271.

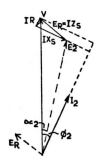

Fig. 271

To the basic voltage phasors have been added the voltage drops in the motor due to the resistance and synchronous reactance. Angle α is called the "angle of retard". The total mechanical power developed by the motor is $E_2 I_2 \cos(\phi_2 - \alpha_2)$ but, for a first approximation, it can be expressed as $V_2 I_2 \cos \phi_2$. The power available at the shaft is less than this by the amount of the iron, friction and windage losses.

An increased load torque causes the angle α to increase, this means a larger current. At the same time ϕ increases and the power factor $\cos \phi$ diminishes. For small values of ϕ, I increases at a greater rate than $\cos \phi$ decreases, so the power developed increases. A value of ϕ is obviously reached at which I increases at a lower rate than $\cos \phi$ decreases, and a condition of maximum power is obtained. Thus, once the increase in current is countered by a decrease in power factor, the power developed begins to fall and, if the load torque required is increased further, the motor will pull out of step and stall.

Under normal operating conditions if P is the total mechanical power developed in watts, V the supply voltage, I the motor current,

$\cos \phi$ the motor power factor and R_a the effective motor resistance then $P = VI \cos \phi - I^2 R_a$.

Example 128. A 500 V, synchronous motor develops 7·5 kW, the power factor being 0·9 (lagging). The effective armature resistance is 0·8 Ω. Iron and friction losses amount to 500 W and the excitation loss is 800 W. Find (a) the motor current, (b) the input power, (c) the overall effficiency.

Output power $= 7{\cdot}5$ kW $= 7500$ W

Mechanical power developed $= 7500 + 500 = 8000$ W

$\therefore \ 8000 = 500 \times I \times 0{\cdot}9 - I^2 \times 0{\cdot}8$

or $0{\cdot}8I^2 - 450I + 8000 = 0$

whence $I = \dfrac{450 \pm \sqrt{450^2 - 4 \times 0{\cdot}8 \times 8000}}{2 \times 0{\cdot}8}$

$= \dfrac{450 \pm \sqrt{450^2 - 25\,600}}{1{\cdot}6}$

Giving $I = 17{\cdot}2$ A

Power input to motor $= 500 \times 17{\cdot}2 \times 0{\cdot}9 = 7{\cdot}74$ kW

Total power input, inclusive of excitation loss $= 7{\cdot}74 + 0{\cdot}8$
$\hspace{11em} = 8{\cdot}54$ kW

Overall efficiency $= \dfrac{7{\cdot}5}{8{\cdot}84} \times 100 = 87{\cdot}6$ per cent.

Example 129. The input to a 11 kV, three-phase, star-connected synchronous motor is 50 A. The effective synchronous reactance and the resistance per phase are 29 Ω and 0·95 Ω respectively. Calculate the power supplied to the motor and the induced e.m.f. for a lagging power factor of 0·8.

Using the diagram of Fig. 271 it can easily be shown that E_2 is given by $E_2^2 = (V \cos \phi - IR)^2 + (V \sin \phi - IX_S)^2$

Resistance voltage drop per phase $= 50 \times 0{\cdot}95 = 47{\cdot}5$ V

Reactance voltage drop per phase $= 50 \times 29 = 1450$ V

Phase voltage $= \dfrac{11\,000}{\sqrt{3}} = 6352$ V

Then E_2^2 = $\{(6352 \times 0 \cdot 8) - 47 \cdot 5\}^2 + \{(6352 \times 0 \cdot 6) - 1450\}^2$

$\qquad$ = $5034^2 + 2361^2$ = $10^6(5 \cdot 034^2 + 2 \cdot 362^2)$

$\qquad$ = $10^6(25 \cdot 4 + 5 \cdot 57)$ = $10^6 \times 30 \cdot 97$

or E_2 = 5570 V

Thus induced e.m.f. = $\sqrt{3} \times 5 \cdot 57$ = $9 \cdot 635$ kV

$\qquad$ Input power = $\sqrt{3} \times 11\,000 \times 50 \times 0 \cdot 8 \times 10^{-3}$

$\qquad\qquad$ = 762 kW.

STARTING

The diagram of Fig. 272 depicts a polyphase, two-pole stator winding fed with polyphase alternating current. A synchronous rotating magnetic field, of constant strength, is produced which is assumed to rotate in a clockwise direction. The stator poles N_1 and S_1 are assumed to occupy the position shown at a particular instant, the two-pole rotor being stationary. Under these conditions, the rotor will be urged to move in an anti-clockwise direction but a half-cycle later, the stator poles will have reversed and S_1 will now be at the top and N_1 at the bottom. The rotor will now tend to move in a clockwise manner and, since it cannot respond to the alternating torque, it remains at rest and the motor is not self-starting.

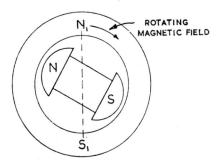

Fig. 272

If the rotor is run at synchronous speed, the opposite stator and rotor poles "lock-in" and the motor will continue to run as considered earlier under the heading of Operating Action. Thus, in order to operate, the motor must be brought up to speed, by some external means and synchronised onto the supply in the same way as an alternator. There are various practical methods of running the motor up to speed and the following are amongst the most usual.

1. An a.c. or d.c. pony motor, coupled to the main machine and of sufficient power to drive the latter up to slightly above synchronous speed, is used to "run up" the synchronous motor. An arrangement is also possible whereby, the d.c. exciter can be used as the pony motor to drive the main machine above synchronous speed. It is then switched off and reconnected, as a d.c. generator, to supply the field of the main motor. As the latter slows down and passes through synchronism, paralleling is affected and the load, to be driven, can then be taken up.

2. The synchronous motor can be provided with a cage winding on the rotor and this can be used to run the machine up to speed as an induction motor. The d.c. field is then applied and the rotor is pulled into synchronism. This is the method employed for the propeller motor of an a.c. ship-propulsion system. Briefly, since the propeller speed is controlled by the speed of the main alternator *i.e.* by the frequency of the supply, procedure is as follows.

Consider a 3 kV system.

The alternator is run up to about $\frac{1}{8}$ speed, when some $\dfrac{3000}{8}$ or 375 V are generated. Since the torque of an induction motor is proportional to the square of the supply voltage, it follows that a reduction of voltage to $\frac{1}{8}$ normal would produce a torque of $\frac{1}{64}$ normal. This reduced torque would be insufficient to provide any substantial power at the propeller, which may not even turn—hence the need for double excitation on the alternator. Returning to the operating sequence, at this stage, the motor would be connected to the alternator, the speed being meanwhile raised until at about $\frac{1}{5}$ speed, when only $\dfrac{3000}{5} = 600$ V are being generated, double excitation is applied to the alternator field. Thus if the alternator field is normally rated for 110 V d.c., at this stage in the operation, 220 V are applied. Since the field strength is doubled, $2 \times 600 = 1200$ V are generated, which are sufficient to cause the propeller motor to run up to speed as an induction motor. When synchronism is approached, the motor d.c. field is applied and the propulsion motor is pulled into step to operate as a synchronous motor. From now on, as the alternator speed is raised its excitation is reduced, so that by the time the normal operating speed value is reached, the alternator field will have been reduced to the rated value *i.e.* 110 V, for our example. Note that the induction motor cage winding is made up of copper bars, embedded in the rotor pole faces. These damper bars are connected together by short-circuiting rings carried round the rotor circumference.

INDUCTION MOTORS

These machines have been included here under the heading of "operation", since little has been said about their starting and control. Consideration is now given to these requirements. As the three-phase induction motor is the a.c. machine invariably used for marine applications, attention is confined to the usual operational practice with which it is associated.

STARTING

Cage-type motors are the most common and are considered first. Since the torque produced is proportional to the applied voltage squared, the direct application of line voltage has advantages. Unlike the d.c. motor which, under standstill conditions, offers a low resistance to the line voltage, a.c. motors have sufficient inherent impedance to limit the current to about 4–7 times the full-load current value. Thus the use of a rheostatic starter, such as is used for the d.c. motors, is unnecessary and a.c. motor-starting methods, for cage machines, are of three forms.

1. DIRECT-ON-STARTING.   Even though the starting current may be as high as seven times the full-load value, such a current will not necessarily burn out the machine, if it is limited to a short period. This method of starting is usual for small motors— up to about 5 kW, if shore practice is considered, but for marine work, when appropriately designed alternators are installed *i.e.* those which are arranged to maintain a fairly constant terminal voltage when a large starting current is being drawn, direct-on starting has been developed for motors rated up to 200 kW or more.

The value of starting current depends on the size and design of the motor. The motor is required to accelerate quickly and current falls rapidly. The main objection to this method of starting, is the large voltage drops in the cables which may affect other connected plant. *i.e.* lamps may dim or contactors may be de-energised. The high starting currents will produce large heating effects and the direct-on starting is not used for repeated operations, if a cooling interval is not provided. Starting is effected by means of a contactor, actuated by a push-button or control switch, which connects the motor to the supply. The contactor usually has a built-in isolator and H.R.C. fuses

are sometimes included for short-circuit protection. Over-current protection is provided by an overload unit, consisting of three series
current coils and dashpots but a thermal tripping arrangement can
be substituted which assists in ensuring a cooling period between
"starts". Full-voltage working for the closing coil, push-button circuits,
etc., is used, but for marine work it is also common practice to incorporate an auxiliary voltage transformer, giving 110 V for the control
and indicator lamp circuits.

2. STAR-DELTA STARTING. A two-step starter is used with a motor
whose phase windings are normally connected in delta. All six ends of
the phase windings are brought into the starter which is of the "two-
position" type. The first position connects the motor phases in star
and hence the voltage per phase at starting is $\dfrac{1}{\sqrt{3}}$ supply voltage and
the starting torque is $\frac{1}{3}$ that obtained by direct starting. The change-
over from star to delta is made by contact arms or segments in the
manually-operated starter or by contactors in the automatic starter.

The starting current is much reduced since it is $\dfrac{1}{\sqrt{3}}$ phase value or $\frac{1}{3}$
of the corresponding value obtained from direct switching. This
reduced starting current is shown by the following example and is the
main advantage of the method.

British practice uses the star-delta method of starting infrequently
for marine work because of the fact that, due to the interval between
the change-over, the motor is disconnected from the supply and any
induced e.m.f. is falling in value and frequency. When reconnection is
made to the supply, transient currents will flow which may have
peaks as large as those resulting from the direct-on switching method.
Thus high current conditions can occur and yet the torque is only $\frac{1}{3}$
of that obtained by direct starting. The method is also not considered
advantageous for marine work because: (a) if alternators can be
designed to cope with the current peaks of direct-on starting, then
the somewhat reduced conditions of star-delta starting offer little
advantage. (b) The reduced starting torque limits this method of
starting principally to centrifugal pump and fan motors.

Example 130. The stator winding of a three-phase motor is arranged
for star/delta starting. If the motor is rated at 7·5 kW, the voltage is
440 V, the full-load efficiency 85 per cent, the power factor 0·8

(lagging) and the short-circuit current equals four times the full-load current, calculate the current taken from the mains at the instant of starting.

Full-load output $= 7.5$ kW $= 7500$ W

Full-load input $= 7500 \times \dfrac{100}{85} = 8823$ W $= \sqrt{3}$ VI cos ϕ

$\therefore I = \dfrac{8823}{\sqrt{3} \times 440 \times 0.8} = \dfrac{25.06}{\sqrt{3}} = 14.5$ A

Short-circuit current $= 4 \times 14.5 = 58$ A

The short-circuit current/phase $= \dfrac{58}{\sqrt{3}} = 33.5$ A

$$Z_{ph} = \dfrac{440}{33.5} = 13.13 \ \Omega$$

Current in star $= \dfrac{\dfrac{440}{\sqrt{3}}}{13.13} = \dfrac{254}{13.13} = 19.35$ A

i.e. $\frac{1}{3}$ of direct-on starting.

3. AUTO-TRANSFORMER STARTING. By the use of a tapped three-phase auto-transformer, any fraction of the supply voltage can be applied to the motor without alteration to the motor connections. The reduced voltage results in a certain motor current and the mains current is further reduced by the transformer turns ratio. For example: for a 50 per cent tapping, the motor takes 0.5 times the current it would take if direct started and the mains current $= 0.5$ that of the motor current. Thus mains current $= 0.25$ times the motor current, if direct started. Torque is proportional to the square of the applied voltage *i.e.* it would be a quarter of that obtained by direct switching.

This type of starter is expensive and for marine work, it is only used for large motors. If switching, as described above, is used, the connections to the motor are broken when the tapping switches are moved and current surges or transients, as described for "star-delta" starting, can occur. To overcome this, methods based on the Korndorffer principle are commonly used, especially if the starters are of the automatic type.

The diagram of Fig. 273 shows the typical main circuit scheme for a marine auto-transformer type of starter. With operation of the "on" push-button—auxiliary circuits are not shown, contactors A and B close first. The motor receives a reduced voltage, the value of which depends on the tapping used on the transformer. Time-lagged switches in the auxiliary contactor closing-coil circuits are fitted and after a short period, contactor B opens. The motor now receives an increased voltage with a section of the transformer winding being used as a series reactor. After a further time interval, contactor C closes and contactor A opens. Thus the motor is connected to the supply direct and the transformer is disconnected. Circuit protection for over-current and under-voltage is built into contactor C since it is the main circuit or "running" contactor.

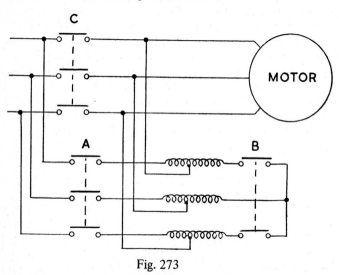

Fig. 273

Example 131. A cage-type induction motor which, if started directly on line, would take a current surge of 6·5 times normal full-load current and develop 70 per cent of normal full-load torque, is started by means of an auto-transformer on a 75 per cent tapping. Calculate the starting current in terms of normal full-load current and the torque.

With a 75 per cent tapping, since current is proportional to the voltage, motor current = 0·75 × 6·5 times normal full-load current.

The mains current is reduced because of the transformer ratio.

∴ Starting current, as drawn from the mains = $\frac{3}{4} \times \frac{3}{4} \times 6\cdot5$ times normal full-load current = 3·65 times normal full-load current.

On a 75 per cent tapping since torque is proportional to the voltage

squared then new torque $= \left(\dfrac{3}{4}\right)^2 \times \dfrac{70}{100}$ times full-load torque.

$$= \dfrac{9}{16} \times \dfrac{7}{10} = 0.394 \text{ times full-load torque.}$$

For the wound-rotor induction motor, theory has shown that this machine is capable of developing maximum torque at starting and during the acceleration period. Rotor resistance can be used for speed control but the resistor units are not usually rated for such a duty. The starter usually comprises a normal three-phase contactor, complete with the usual protection features, which controls the supply to the stator. The rotor resistor controller is interlocked with this contactor to ensure that starting can only be attempted with maximum resistance in the rotor circuit. For marine work, if speed control is required as for some types of a.c. winch or "barring" motor, the resistors are cut out by a drum controller or regulator similar to that used for d.c. machines. The rotor resistors are rated accordingly. Experience with this type of control for marine winch work has been disappointing and a cage-type, two-speed machine is usually favoured.

SPEED AND TORQUE CONTROL

The method by which such control can be achieved for the wound-rotor motor has already been considered. For the cage-type motor, some such control can be obtained by variations of the rotor design. The resistance and reactance of the cage can be adjusted by methods to be discussed but, as little has been said up to now about the constructional details of this machine, the chapter will be concluded with such considerations. The effect of different features of cage construction will be pointed out as the description proceeds.

Stator. Construction of the induction motor stator, for both the wound-rotor and cage types, is very similar to that of the a.c. generator in that it consists of low-loss electrical steel laminations which are pre-stamped with slots, compressed and secured into a fabricated steel or cast-iron frame which forms the body of the machine. Air circulation space is provided between the laminations and the frame, and radial air ducts can be arranged, if necessary, by fixing the stampings in stacks, separated by spacers. The stator slots are usually of the semi-enclosed type and, since the induction motor is a small air-gap machine, the bore of the stator is usually turned or ground true.

The windings are similar to those described for the alternator, usually of the single layer or concentric type. For relatively large output, low-voltage motors, it is possible to use a double layer, two bar per slot winding which has the following advantages. It is more rigid mechanically and due to the possibility of short-chording, the overall length of the machine can be reduced to save copper. The leakage reactance is also reduced by this technique.

For the medium size of motor, the windings can be either of wire or made up from rectangular bar. If wire coils are used they can be former wound and then roughly shaped, slot portions being untaped. The wires are introduced into the slots one at a time and the end sections pulled into shape by hand—this gives the common "Mush" winding. The coils are made up from copper wire covered with abrasion-resisting synthetic enamel and insulated from the core slots with troughs of micanite or mica supported by leatheroid or presspahn. The coils are made for a tight fit in the slots which are closed by fibre wedges. The stator is impregnated with a high grade thermo-setting resin varnish, which prevents movement of the coils and is resistant to moisture, salt spray and oily vapour. The end windings are taped and the insulation reinforced with suitable fibrous materials prior to the varnishing. For the large low-voltage motors already mentioned, when bar windings are used, the conductors of a one-turn coil can be of copper strip, insulated with tape—varnished and baked, or with micanite—hot-moulded under pressure direct on to the conductors. These can be pushed through the slots from one end only, a method which ensures a high degree of mechanical strength. All connections are thus on one side of the machine, being accessible for testing and inspection.

Rotor. The standard cage rotor is designed for optimum running performance and is used for applications where normal values of starting current and torques are acceptable. Such a rotor develops about 120–150 per cent full-load torque with 600-800 per cent full-load current when direct started. Where a smaller starting current and/or higher starting torque is required, alternative forms of rotor construction are available to give the desired characteristics. Because of the many variations in performance, it is difficult to distinguish between the "standard" and "high-torque" motor but appropriate examples are considered under these headings. Typical marine drives for the cage motor are centrifugal pumps, fans, refrigerating machinery, hydraulic steering-gear, *etc.* High starting torque features can be specified for the motor but in most cases starting is light and a standard rotor would be satisfactory. Even in the case of compressors, an "unloading valve" can be provided which is operated electrically to allow easy starting.

The core of a cage rotor is built up under pressure from similar materials and in a similar manner to the stator core. This laminated cylindrical core is keyed direct to the shaft for small motors, and to a cast-iron or cast-steel rotor hub or spider which is itself keyed to the shaft for larger machines. The core has slots which may be semi-enclosed or totally enclosed round its outer periphery. Ventilating ducts can be arranged to lie opposite corresponding ducts in the stator core along its length. The slots are usually skewed to ensure smooth starting and freedom from magnetic noise. The rotor periphery is accurately machined to ensure minimum air-gap between rotor and stator, consistent with good design. The shaft is of good quality tensile steel, proportional to withstand the torsional effects and mechanical shocks which may be transmitted from the driven machinery. The rotor is finally dynamically balanced after the cage winding is inserted.

STANDARD-CAGE ROTOR.   For small motors the winding consists of a die casting of special aluminium alloy. The bars, end-rings and cooling vanes are cast in one piece, this construction avoiding joints and producing a higher starting torque because of the slightly higher resistance of aluminium, as compared to copper.

For larger machines the rotor windings, as shown by Fig. 274, consists of solid copper or bronze bars brazed to substantial end-rings of the same material. The bars are a drive-fit in the slots and the profile

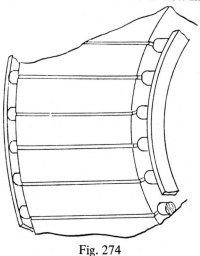

Fig. 274

and material, as for the end-rings, can be modified to give desired starting and running characteristics. Low resistance of the cage is a good feature for efficient running and small slip, but is a disadvantage for starting. Starting currents are large, power factor is low and the starting torque poor. The standard-cage rotor, with minor modifications is the most common since the majority of drives start light, and the current can be limited by various starting arrangements, such as the star-delta or auto-transformer starter.

HIGH-TORQUE CAGE ROTORS.   The advantages of the wound-rotor motor, with its external resistance to give improved starting, can be provided to some extent without forfeiting the robust construction and good running properties of the standard-cage motor. This is achieved by special constructions and two methods of giving this improved performance are considered.

(a) SHAPED AND EMBEDDED ROTOR BARS.   By appropriately dimensioned slots and conductors, a high starting characteristic is possible based on the fact that, if for large alternating currents, the conductors are of heavy cross section and made of solid copper, then large eddy currents are developed and an electromagnetic effect operates, tending to force the currents upwards in the conductors and towards the slot-openings. The current density rises over part of any one conductor, resulting in excessive copper loss. This in effect increases the resistance of the conductor, because a large part of its sectional area is free of current. The effect depends on frequency, the higher this is the more pronounced the localisation of the current. As the ship decreases and the rotor frequency falls, the effect disappears. The effect described, thus produces a high resistance value at starting and this can be further increased by shaping the conductors as shown by the examples of Fig. 275. The apparent increase in resistance of the cage bars during starting results in the desired increased starting torque.

Fig. 275a   Fig. 275b

(b) DOUBLE-CAGE ROTORS.   Here two circumferential rows of holes in the rotor iron constitute the conductor slots. The outer row is spaced from the inner by narrow radial air-gaps which link slot pairs and continue through to the rotor surface. The arrangement is shown by Fig. 276, from which it is seen that the radial air-gap ensures that the stator flux also links with the bottom conductors and does not take the shorter path which would be provided if no such radial air-gaps were introduced. The conductors of the outer cage are high resistance (bronze) and low reactance whilst the inner cage is of high conductivity copper with high reactance.

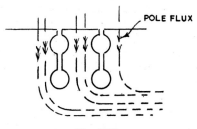

Fig. 276

At starting, the rotating magnetic field links mainly with the outer cage conductors since the rotor frequency is the supply frequency and the large reactance of the inner, more highly inductive, cage limits the current. The high-resistance outer cage is thus more operative and

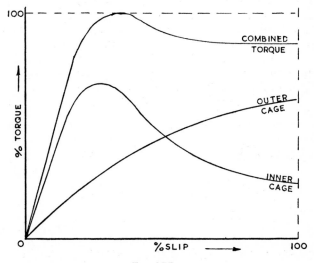

Fig. 277

the corresponding torque is high. As the motor accelerates, the frequency of the rotor currents fall, the reactance of the inner cage decreases and more current flows until it takes over the main part of the driving torque. The graphs of Fig. 277 show the torque/slip curve for a double cage rotor. It will be seen that by adjusting the separate torques by rotor slot design, the combined characteristic can be made to suit any condition.

SPEED ADJUSTMENT

The cage motor is a single-speed machine and is best suited to constant speed drives. It is possible to obtain a mechanical design of ventilating fan or pump, which will permit single-speed operation with no loss in efficiency. If a single speed is not acceptable, two or more speeds can be obtained by using a pole-change cage motor. This machine is a development of the standard motor and provides a convenient and efficient method of speed changing at low cost. It is similar to the standard motor but by an arrangement of multiple stator windings, two, three or four definite speeds are possible. Two-speed motors have been provided for marine work and only these are considered.

TWO-SPEED MOTORS.   These can be built as (a) dual wound or (b) consequent-pole machines.

(a)   The dual-wound arrangement has been used for marine duties—general service pumps and forced draught fans and consists of a standard motor which has two separate windings, one for each speed. The windings are pitched so as to divide the stator into two pole systems, one having double the number of the other. One winding only is connected into circuit at any one time and the usual forms of starting: direct-on, star-delta or auto-transformer can be used.

(b)   Consequent-pole machines have one winding only, which is tapped and connected to appropriate terminals. These motors can only be direct-started or used with auto-transformer starters. The arrangements for speed control are effected by suitable switching and the control gear, being more complicated than that for the dual-wound motor, is not favoured for marine duties. Further details will not therefore be considered.

Example 132.   On full-load, a four-pole, 50 Hz, wound-rotor induction motor operates with 5 per cent slip. Find the resistance to be inserted into each phase of the rotor circuit to reduce the speed to 1200 rev/min, if the torque is to remain constant. The ratio of stator standstill reactance to resistance is six to one.

Since $T = \dfrac{KV^2 s R_2}{R_2{}^2 + (sX_2)^2}$ then $T = \dfrac{CR_2}{\dfrac{R_2{}^2 + X_2{}^2 s}{s}}$ where C is a constant

thus $T = \dfrac{C}{\dfrac{R_2}{s} + \dfrac{X_2{}^2 s}{R_2}}$ and initially $T = \dfrac{C}{\dfrac{R_2}{0\cdot05} + \dfrac{36R_2{}^2 0\cdot05}{R_2}}$

Also $N_2 = N_1 - sN_1 = N_1(1-s)$ and $f = \dfrac{PN}{120}$

so $N_1 = \dfrac{50 \times 120}{4} = 1500$ rev/min.

Thus $1200 = 1500(1-s)$ and $\dfrac{4}{5} = 1 - s$ so $s = \dfrac{1}{5} = 0\cdot2$

Let mR_2 be the new value of rotor resistance per phase

Then $T = \dfrac{C}{\dfrac{mR_2}{0\cdot2} + \dfrac{36\,R_2^2\,0\cdot2}{mR_2}}$

and $\dfrac{R_2}{0\cdot05} + \dfrac{36R_2{}^2 0\cdot05}{R_2} = \dfrac{mR_2}{0\cdot2} + \dfrac{36R_2{}^2 0\cdot2}{mR_2}$

Thus $20 + 1\cdot8 = 5m + \dfrac{7\cdot2}{m}$ or $5m^2 - 21\cdot8m + 7\cdot2 = 0$

Whence $m = \dfrac{21\cdot8 \pm \sqrt{21\cdot8^2 - 4 \times 5 \times 7\cdot2}}{2 \times 5}$

$= 2\cdot18 \pm \dfrac{\sqrt{475\cdot24 - 144}}{10} = 2\cdot18 \pm \dfrac{\sqrt{331\cdot24}}{10}$

$= 2\cdot18 \pm \dfrac{10\sqrt{3\cdot3124}}{10} = 2\cdot18 \pm \sqrt{3\cdot3124}$

$$= 2 \cdot 18 \pm 1 \cdot 82 = 4$$

New resistance $\qquad = mR_2 = 4R_2$

$\therefore$ Additional resistance $= 3R_2$

CHAPTER 13

PRACTICE EXAMPLES

1. A synchronous motor having a power consumption of 50 kW is connected in parallel with a load of 200 kW having a lagging power factor of 0·8. If the combined load has a lagging power factor of 0·9, what is the value of the leading reactive kVA supplied by the motor and at what power factor is it working?

2. Two 3·3 kV, star-connected alternators when operating in parallel supply the following loads; (a) 800 kW at unity power factor, (b) 600 kW at a power factor of 0·707 (lagging). The current of one machine is 150 A at a power factor of 0·85 (lagging). Find the current, output and power factor of the other machine.

3. A ship's electrical system is supplied by two identical three-phase, star-connected alternators operating in parallel. The machines share a total load of 1000 kW at 440 V, 0·8 (lagging) power factor. If the kW loadings of the two machines are equal and one machine supplies a lagging current of 1000 A, find (a) the current supplied by the second machine, (b) the power factor of each machine, (c) the reactive current circulating between two machines.

4. Two alternators A and B operate in parallel. When tested individually, the frequency of machine A falls from 50 to 48·5 Hz when the load is 150 kW and that of machine B falls from 50 to 48·5 Hz when the load is 220 kW. If the total load is 200 kW, find the frequency of the paralleled system and the load on each machine.

5. A 15 kW, 440 V, three-phase star-connected synchronous motor has an armature of which the effective resistance per phase is 0·4 Ω. It is giving its full-load output at a leading power factor of 0·9. If the iron and friction losses amount to 500 W, find the armature current.

6. A 220 V, single-phase, synchronous motor has a synchronous impedance of 5 Ω and an effective armature resistance of 0·5 Ω. Calculate (a) the minimum armature current, (b) the generated e.m.f., (c) the angle of retard for a total load of 5 kW, which includes the iron and friction losses of the motor.

7.   A three-phase, 440 V, 37 kW induction motor has an efficiency of 82 per cent and operates at a lagging power factor of 0·85. When "direct-on" started the motor takes a current of six times full-load current and produces a torque of 1·5 times full-load torque. Calculate the current taken from the supply and the ratio of starting torque to full-load torque if the motor is started through an auto-transformer having a 75 per cent tapping.

8.   Find the ratio of starting to full-load current for a 15 kW, 415 V, three-phase, induction motor with a star-delta starter, given that the full-load efficiency is 85 per cent, the full-load power factor is 0·8 (lagging), the short-circuit current is 60 A at 220 V and the magnetising current is negligible.

9.   A 440 V, three-phase wound-rotor induction motor has a rotor resistance and standstill reactance of 0·02 ohm and 0·27 ohm per phase respectively. The stator to rotor phase turns ratio is 3:1 and the stator windings are connected in delta. If the motor is started by means of a resistance starter having a resistance of 0·25 ohm per phase, calculate the current taken by the motor from the supply, (a) at starting, (b) under full-load running conditions if the full-load slip is 4 per cent. What would be the current taken from the supply if the motor was accidently started with the starting resistance in the "run" position? Neglect the no-load current, the resistance and the reactance of the stator windings.

10.   A six-pole, three-phase induction motor on full load develops a useful torque of 162·4 Nm and the rotor e.m.f. makes 90 complete cycles per minute. Calculate the power. If the mechanical torque lost in friction is 13·6 Nm, find the rotor copper loss, the input to the motor and the efficiency. Assume the stator loss is 750 W.

CHAPTER 1

SOLUTIONS TO PRACTICE EXAMPLES

1.   Back e.m.f. $= 230 - (18 \times 0.9) = 230 - 16.2 = 213.8$ V

Power developed by armature $= 213.8 \times 18$ watts

$$= 3848.4 \text{ W} = 3.848 \text{ kW}$$

Power developed by shaft $= 3.848 - 0.112 = 3.736$ kW

Electrical input to motor $= (230 \times 18) + \left(230 \times \dfrac{230}{300}\right)$

$$= 4316.3 \text{ W} = 4.316 \text{ kW}$$

$\eta = \dfrac{3.736}{4.316} = 0.866$ or Efficiency $= 86.6$ per cent.

Alternative solution.

Field Copper Loss $= 230 \times \dfrac{230}{300} = 176.3$ W

Armature Copper Loss $= 18^2 \times 0.9 = 291.6$ W

Rotational Loss $= 112$ W

Total Losses $= 176.3 + 291.6 + 112 = 580$ W

Electrical input to motor $= (230 \times 18) + \left(230 \times \dfrac{230}{300}\right)$

$$= 4140 + 176.3 = 4316.3 \text{ W}$$

Output $= 4316.3 - 580 = 3736.3$ W

$$= 3.736 \text{ kW}$$

$$\eta = \frac{3736 \cdot 3}{4316 \cdot 2} = 0.866 \quad \text{or} \quad \text{Efficiency} = 86 \cdot 6 \text{ per cent.}$$

2. No load Condition. $I_{f_o} = \dfrac{110}{100} = 1 \cdot 1 \text{ A}$

$$I_{a_o} = 4 \cdot 5 - 1 \cdot 1 = 3 \cdot 4 \text{ A}$$

Input = Losses, since output of motor is zero

$\therefore$ 110 x 4·5 = Cu Losses + Rotational Loss

Note. Here P_R (the Rotational Loss) is assumed a constant to include Iron and Mechanical Losses for the same machine speed and flux density.

So 495 = (Arm.Cu Loss + Field Cu Loss) + P_R
and P_R = 495 − (3·4² x 0·12) − (110 x 1·1)
= 495 − 1·387 − 121 = 495 − 122·4 = 372·6 W

Load Condition. Output = Input − Losses
= (110 x 40) − (P_R + Cu Losses)
= 4400 − 372·6 − Arm. Cu Loss
− Field Cu Loss

Here $I_f = \dfrac{110}{100} = 1 \cdot 1 \text{ A}$ and $I_a = 40 - 1 \cdot 1 = 38 \cdot 9 \text{ A}$

Field Cu Loss = 1·1 x 110 = 121 W
Arm. Cu Loss = 38·9² x 0·12 = 181·66 W
So Output = 4400 − 372·6 − 181·66 − 121 = 3724·74 W

$$\eta = \frac{3724 \cdot 74}{4400} = \frac{9 \cdot 312}{11} = 0.8465$$

Efficiency = 84·65 per cent.

3.   No-load Condition. The reduced voltage on the armature is to ensure correct full-load speed otherwise this would be high with normal voltage. It must also be noted that, since data about brush drop is given, this must be taken into account. Such a voltage drop is responsible for a power loss which varies with current.

$$\text{Input to armature} = 3 \times 102 = 306 \text{ W}$$
$$= P_R + \text{Arm. Cu Loss} + \text{Brush Loss}$$

$$\therefore \ P_R = 306 - (3^2 \times 0 \cdot 1) - (3 \times 2) = 306 - 6 \cdot 9$$
$$= 299 \cdot 1 \text{ W}$$

Load Condition. $I_a = 40 - \dfrac{105}{95} = 40 - 1 \cdot 1 = 38 \cdot 9 \text{ A}$

$$\text{Input} = 105 \times 40 = 4200 \text{ W}$$

$$\text{Output} = \text{Input} - P_R - \text{Brush Loss} - \text{Arm. Cu Loss}$$
$$- \text{Field Cu Loss}$$
$$= 4200 - 299 \cdot 1 - (2 \times 38 \cdot 9) - (38 \cdot 9^2 \times 0 \cdot 1)$$
$$- (105 \times 1 \cdot 1)$$
$$= 4200 - (299 \cdot 1 - 77 \cdot 8 - 151 \cdot 3 - 115 \cdot 5)$$
$$= 4200 - 643 \cdot 7 = 3556 \text{ W} = 3 \cdot 56 \text{ kW}$$

$$\eta = \frac{3556}{4200} = 0 \cdot 846 \quad \text{or} \quad \text{Efficiency} = 84 \cdot 6 \text{ per cent}$$

The alternative solution is of interest.

$$\text{Back e.m.f. on load} = 105 - 2 - \text{Armature voltage drop}$$
$$= 105 - 2 - (0 \cdot 1 \times 38 \cdot 9) = 103 - 3 \cdot 89$$
$$= 99 \cdot 11 \text{ V}$$

$$\text{Therefore power developed by armature} = 99 \cdot 11 \times 38 \cdot 9$$
$$= 3855 \cdot 4 \text{ W}$$

$$\text{Power available at shaft} = 3855 - P_R$$
$$= 3855 - 299 = 3556 \text{ W}$$

Rest of solution as above.

4.   On Load. Output = 110 kW

$$\text{Input} \ = \ 110 \times \frac{100}{88} \ = \ 125 \text{ W}$$

$$\text{Input current} \ = \ \frac{125 \times 1000}{460} \ = \ 271{\cdot}7 \text{ A}$$

$$I_f \ = \ \frac{460}{46} \ = \ 10 \text{ A}$$

$$I_a \ = \ 271{\cdot}7 - 10 \ = \ 261{\cdot}7 \text{ A}$$

Then P_R = Losses on Load − Load Cu Losses

= (Input − Output) − Arm. Cu Loss
              − Field Cu Loss

$$\text{or } P_R \ = \ (125 - 110) - \frac{261{\cdot}7^2 \times 0{\cdot}03}{1000} - \frac{10 \times 460}{1000}$$

$$= \ 15 - 2{\cdot}055 - 4{\cdot}6 \ = \ 15 - 6{\cdot}655 \ = \ 8{\cdot}35 \text{ kW}$$

Solution is made easier if the Constant Loss P_C is used
= 8·35 + 4·6 = 12·95

On No Load.   Let I_o = the no-load motor current.

Then Input = No-Load Losses
VI_o = P_C + Arm. Cu Loss
and $460 \, I_o$ = $12\,950 + (I_o - 10)^2 \times 0{\cdot}03$
$460 \, I_o$ = $12\,950 + (I_o^2 - 20I_o + 100)\,0{\cdot}03$
or $46000 \, I_o$ = $1\,295\,000 + 3I_o^2 - 60I_o + 300$
Whence $3I_o^2 - 46\,060I_o + 1\,295\,300 \ = \ 0$
or $I_o^2 - 15\,353I_o + 431\,767 \ = \ 0$

$$\text{Solving for } I_o \ = \ \frac{15\,353 \pm \sqrt{15\,353^2 - (4 \times 431767)}}{2}$$

$$= \ \frac{15\,353 \pm 10^4 \, \sqrt{2{\cdot}357 - 0{\cdot}0173}}{2}$$

$$= \frac{15\,353 \pm 10^4\sqrt{2\cdot34}}{2} = \frac{15\,353 \pm 15\,297}{2}$$

Thus $I_o = \dfrac{56}{2} = 28$ A.

The following approximation would have made an easier solution.

Since the no-load armature current is small—some one-tenth of full-load current, then the armature copper loss on no load would be one hundredth of the full-load value $= \dfrac{2\cdot055}{100}$

$= 0\cdot02$ kW. It is sufficiently small to be neglected. If this was done then

On No Load $VI_o = P_C$ or $460\,I_o = 12\,950$

or $I_o = \dfrac{12\,950}{460} = 28\cdot15$ A.

5.   No Load.   At $16°C$ $I_{f_o} = 1\cdot2$ A
          and $I_{a_o} = 4\cdot2 - 1\cdot2 = 3$ A
Then $P_R = (500 \times 4\cdot2) - \{(3^2 \times 0\cdot2) + (1\cdot2 \times 500)\}$
          $= 2100 - (1\cdot8 + 600) = 2100 - 601\cdot8 = 1498\cdot2$ W

Since data on temperatures is given, it is obvious that this must be taken into account. However no temperature coefficient for copper is given, so we work to the approximation rule which gives results near enough for practical purposes. This rule amounts to allowing a $0\cdot4$ per cent increase in resistance for every $1°C$ rise of temperature.

On Load.

A $40°C$ temperature rise is allowed.

$\therefore$ The field and armature resistances can be increased by
     $0\cdot4 \times 40 = 16$ per cent.

Thus shunt-field resistance $= \dfrac{500}{1\cdot2} = 417\ \Omega$ (cold)

or $417 \times 1\cdot16 = 484\ \Omega$ (hot)

Armature resistance $= 0\cdot2\ \Omega$ (cold) or $0\cdot2 \times 1\cdot16$
$= 0\cdot232\ \Omega$ (hot)

Then $I_f = \dfrac{500}{484} = 1\cdot03$ A $\quad I_a = 80 - 1\cdot03 = 78\cdot97$ A

$$
\begin{aligned}
\text{Output of motor} &= \text{Input} - \text{Losses} \\
&= (500 \times 80) - (P_R + I_a^2\,R_a + I_f V) \\
&= 40\,000 - 1498\cdot2 + (78\cdot97^2 \times 0\cdot232) \\
&\qquad\qquad\qquad\qquad + (1\cdot03 \times 500) \\
&= 40\,000 - (1498\cdot2 + 1450 + 515) \\
&= 40\,000 - 3463 = 36\,537\ \text{W} \\
&= 36\cdot54\ \text{kW}
\end{aligned}
$$

Thus $\qquad \eta = \dfrac{36\,537}{40\,000} = 0\cdot913$

or Efficiency $= 91\cdot3$ per cent

Maximum efficiency occurs when

$I_a = \sqrt{\dfrac{P_C}{R_a}}\quad$ Here $P_C = P_R +$ shunt-field loss (hot)

$= 1498\cdot2 + (1\cdot03 \times 500) = 1498\cdot2 + 515$
$= 2013\cdot2$ W

$\therefore\ I_a = \sqrt{\dfrac{2013\cdot2}{0\cdot232}} = 10\sqrt{\dfrac{20\cdot132}{0\cdot232}} = 10\sqrt{86\cdot77}$

$= 10 \times 9\cdot315 \quad$ or $\quad I_a = 93\cdot15$ A

Then motor current would be $93\cdot15 + 1\cdot03 = 94\cdot18$ A.

6. No Load. $\quad$ Input $= (6\cdot5 + 2\cdot2)200 = 8\cdot7 \times 200$
$$= 1740 \text{ W}$$

Armature resistance $= \dfrac{3}{70} = 0\cdot043 \ \Omega$

Input $= P_R +$ Arm.Cu Loss $+$ Field Cu Loss
$$= P_R + (6\cdot5^2 \times 0\cdot043) + (2\cdot2 \times 200)$$
or $1740 = P_R + 1\cdot8 + 440$

Thus $\qquad P_R = 1740 - 441\cdot8 = 1298\cdot2 \text{ W}$

On Load. $\quad$ Let $I_a =$ the full load armature current

Then $\qquad$ Input $=$ Output $+$ Losses
or $200(I_a + 2\cdot2) = 15000 +$ (Arm. Cu Loss $+$ Field Cu
$$\text{Loss} + P_R)$$
$200 I_a + 440 = 15\,000 + (I_a^2 \times 0\cdot043) + 440 + 1298\cdot2$
$$= 15\,000 + 0\cdot043 I_a^2 + 1738\cdot2$$
or $\qquad 0\cdot043 I_a^2 - 200 I_a + 16\,298 = 0$

whence $\qquad I_a = \dfrac{200 \pm \sqrt{200^2 - (4 \times 0\cdot043 \times 16\,298)}}{2 \times 0\cdot043}$

$$= \dfrac{200 \pm \sqrt{200^2 - (0\cdot172 \times 16\,298)}}{2 \times 0\cdot043}$$

$$= \dfrac{200 \pm 100\sqrt{2^2 - (0\cdot172 \times 1\cdot6298)}}{0\cdot086}$$

$$= \dfrac{200 \pm 100\sqrt{4 - 0\cdot28}}{0\cdot086}$$

$$= \dfrac{200 \pm 100\sqrt{3\cdot72}}{0\cdot086}$$

$$= \dfrac{200 \pm 192\cdot9}{0\cdot086} = \dfrac{7\cdot1}{0\cdot086} = \dfrac{710}{8\cdot6}$$

Thus $I_a = 82\cdot56 \text{ A}$

Thus the motor armature current is $82\cdot56$ A and the field current $2\cdot2$ A. $\quad \therefore$ Total input current $= 84\cdot76$ A

The motor input power is 200×84.76 watts

$$\eta = \frac{15\ 000}{200 \times 84\cdot76} = \frac{750}{847\cdot6} = 0\cdot8848$$

or   Efficiency $= 88\cdot5$ per cent.

7. Substituting in the evolved expression for Iron Loss, we have

$$2400 = HN + EN^2 \qquad \dots\dots\dots\dots\dots (a)$$

$$\text{and } 800 = H\,\frac{N}{2} + E\left(\frac{N}{2}\right)^2 \qquad \dots\dots\dots\dots\dots (b)$$

$$\text{or } 2400 = HN + EN^2 \qquad \dots\dots\dots\dots\dots (c)$$

$$\text{and } 3200 = 2HN + EN^2 \qquad \dots\dots\dots\dots\dots (d)$$

Subtracting (c) from (d) $800 = HN$

Thus at 1500 rev/min the Hysteresis Loss is 800 W and the Eddy-Current Loss is $2400 - 800 = 1600$ W
Again Hysteresis Loss is proportional to speed so, at quarter speed, *i.e.* 375 rev/min,

$$\text{Hysteresis Loss} = \frac{800}{4} = 200\ \text{W}$$

Eddy-Current Loss is proportional to speed squared so, at quarter speed, this loss will be reduced by a sixteenth

$$\textit{i.e.}\quad \text{Eddy-Current Loss} = \frac{1600}{16} = 100\ \text{W}$$

Total Iron Loss at 375 rev/min is $200 + 100 = 300$ W.

8.   The solution is made by reference to the diagram of Fig. 7. A machine is rated at 150 kW, 220 V. Therefore the current would be

$$\frac{150\ 000}{220} = \frac{7500}{11} = 681\cdot8\ \text{A}$$

During testing, it can be assumed that, overloading is not desirable and therefore the machine which should be fully loaded would be the motor.

The motoring machine takes 681·8 A. Its field would be weaker than that of the generator, since the back e.m.f. would be below 220 V. Therefore motor field current $I_{f_M} = 5$ A and $I_{a_M} = 681·8 - 5 = 676·8$ A.

The generating machine supplies $681·8 - 140 = 541·8$ A

The field current $I_{f_G} = 7$A, so $I_{a_G} = 541·8 + 7 = 548·8$ A

$$\text{Input from the line} = 140 \times 220 \text{ W}$$
$$= 2P_R + \text{Arm. Cu Loss of 2 m/cs}$$
$$+ \text{Field Cu Loss of 2 m/cs}$$
$$\text{So } 30\,800 = 2P_R + 0·02(676·8^2 + 548·8^2) + 220(5 + 7)$$
$$= 2P_R + 0·02(457\,550 + 302\,000)$$
$$+ (220 \times 12)$$
$$= 2P_R + (0·02 \times 759\,550) + 2640$$
$$= 2P_R + 15\,191 + 2640 = 2P_R + 17\,831$$
$$\text{or } P_R = \frac{30\,800 - 17\,831}{2} = \frac{12\,969}{2} = 6485 \text{ W or}$$
$$6·485 \text{ kW}$$

Then estimated efficiency as a generator is deduced from

$$\eta_G = \frac{\text{Output (kW)}}{\text{Output (kW)} + \text{Losses (kW)}}$$

$$= \frac{150}{150 + \{P_R(\text{kW}) + \text{Arm.Cu Loss(kW)} + \text{Field Cu Loss(kW)}\}}$$

As a generator $I_{a_G} = 681·8 + 7 = 688·8$ A

$$\text{Armature Copper Loss} = \frac{688·8^2 \times 0·02}{1000} = 9·49 \text{ kW}$$

$$\text{Field Copper Loss} = \frac{7 \times 220}{1000} = 1·54 \text{ kW}$$

and $\eta_G = \dfrac{150}{150 + 6 \cdot 485 + 9 \cdot 49 + 1 \cdot 54} = \dfrac{150}{167 \cdot 5} = 0 \cdot 896$

$\therefore$ Generator efficiency = 89·6 per cent.

The estimated efficiency as a motor is given by

$$\eta_M = \frac{\text{Input (kW)} - \text{Losses (kW)}}{\text{Input (kW)}}$$

$$= \frac{150 - \left\{ P_R \text{(kW)} + \text{Arm. Cu Loss (kW)} + \text{Field Cu Loss (kW)} \right\}}{150}$$

Motor current = 681·8 A   Field Current = 5 A

and $I_{a_M} = 678 \cdot 8$ A

Armature Copper Loss $= \dfrac{676 \cdot 8^2 \times 0 \cdot 02}{1000} = 9 \cdot 151$ kW

Field Copper Loss $= \dfrac{5 \times 220}{1000} = 1 \cdot 1$ kW

$\eta_M = \dfrac{150 - (6 \cdot 485 + 9 \cdot 151 + 1 \cdot 1)}{150} = \dfrac{150 - 16 \cdot 736}{150}$

$= \dfrac{133 \cdot 264}{150} = 0 \cdot 889$

Thus motor efficiency = 88·9 per cent.

Efficiency of each machine, as estimated on the test bed, is

$$\eta^2 = \frac{\text{Generator output}}{\text{Motor input}} = \frac{220 \times 541 \cdot 8}{220 \times 681 \cdot 8} = 0 \cdot 794$$

$\eta = \sqrt{0 \cdot 794} = 0 \cdot 8911$

Assessed efficiency of each machine = 89·1 per cent.

9.   Let V be the voltage between the +ve busbar and the equalising connection and I_A the current in the armature of Generator A.

Let I_B be the current in the armature of Generator B. The problem is solved with the assistance of the accompanying diagram. Note that the shunt-field currents will be neglected.

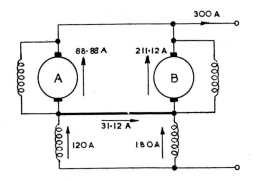

Using the equations. $E_A = V + I_A R_{a_A}$ and $I_A + I_B = 300$

$$E_B = V + I_B R_{a_B}$$

Then $235 = V + 0{\cdot}025 I_A$ ...   .   (a)

$237 = V + 0{\cdot}02 I_B$ or $237 = V + 0{\cdot}02(300 - I_A)$

$$= V + 6 - 0{\cdot}02 I_A$$

Whence $231 = V - 0{\cdot}02 I_A$ ...   (b)

Solving equations (a) and (b) by subtraction

$235 = V + 0{\cdot}025 I_A$
$231 = V - 0{\cdot}02 I_A$ giving $4 = 0{\cdot}045 I_A$

or $I_A = \dfrac{400}{4{\cdot}5} = 88{\cdot}88$ A

and $I_B = 211{\cdot}12$ A

Let $R =$ the equivalent resistance of the two series fields in parallel.

Then $\dfrac{1}{R} = \dfrac{1}{0{\cdot}015} + \dfrac{1}{0{\cdot}01} = \dfrac{1 + 1{\cdot}5}{0{\cdot}015} = \dfrac{2{\cdot}5}{0{\cdot}015}$

or $\quad R = \dfrac{0{\cdot}015}{2{\cdot}5} = 0{\cdot}006 \ \Omega$

Voltage drop across series fields $= 300 \times 0{\cdot}006 = 3 \times 0{\cdot}6$
$$= 1{\cdot}8 \ \text{V}$$

So Busbar Voltage $= V - 1{\cdot}8$

Solving for V from equation (a) $V = 235 - (0{\cdot}025 \times 88{\cdot}88)$

or $\quad V = 235 - \dfrac{8{\cdot}888}{4} = 235 - 2{\cdot}222 = 232{\cdot}778 \ \text{V}$

Busbar Voltage $= 232{\cdot}778 - 1{\cdot}8 = 230{\cdot}98 = 231 \ \text{V}$

Current in series field of Generator A $= \dfrac{1{\cdot}8}{0{\cdot}015} = \dfrac{1800}{15}$
$$= 120 \ \text{A}$$

Current in series field of Generator B $= \dfrac{1{\cdot}8}{0{\cdot}01} = 180 \ \text{A}$

Current in equalising connection $=$ current in series field of
Generator A–current in
armature of Generator A
$= 120 - 88{\cdot}88 = 31{\cdot}12$ A to make, with the field
current of Generator B (180 A), the armature current of
Generator B
$= 180 + 31{\cdot}12 = 211{\cdot}12$ A.

10.   The problem is solved in accordance with the Method 2,
described at the end of Chapter 1. The graphs of the load
characteristics are plotted as shown and current values added
for any assumed busbar voltage to give the total current
supplied. The combined characteristic is thus obtained and the
required answers read off.

Thus for a load current of 160 A, the busbar voltage is 228 V. Machine A supplies 83 A and Machine B 77 A.

Similarly for a load current of 200 A, the busbar voltage is 210 V. Each machine supplies 100 A.

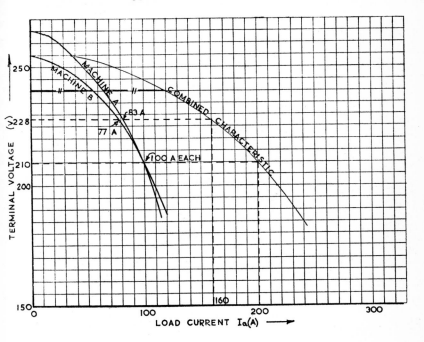

1. (a)   Turns ratio $\dfrac{3}{1} = \dfrac{N_1}{N_2} = \dfrac{V_1}{V_2}$ or $V_1 = 80 \times 3 = 240$ V

    (b)   For either winding 2 volts are induced by one turn

        $\therefore$ The primary voltage of 240 V is induced by

$$\frac{240}{2} = 120 \text{ turns}$$

        and the secondary voltage of 80 V is induced by

$$\frac{80}{2} = 40 \text{ turns}$$

    (c)   Since the kVA rating figure is the same for primary and secondary

        then primary current $= \dfrac{8 \times 1000}{240} = \dfrac{100}{3} = 33\cdot33$ A

        secondary current $= \dfrac{8 \times 1000}{80} = 100$ A.

2. (a)   With the required voltage/turn of 1·5 V for a 440 V supply

        $\dfrac{440}{1\cdot5} = \dfrac{880}{3} = 293\cdot33$ *i.e.* 293 turns are required

        The secondary turns would be $\dfrac{293\cdot3}{4} = 73\cdot32$

        i.e. 73 turns are required.

(b) From the transformer voltage equation

$E = 4.44 \, \Phi_m fN$ we have $440 = 4.44 \, \Phi_m \times 50 \times 293$

or $\Phi_m = \dfrac{440}{4.44 \times 50 \times 293} = \dfrac{440}{220 \times 293} = \dfrac{2 \times 10^{-2}}{2.93}$

$= 0.0068 \, \text{Wb}$

Using the relationship $\Phi_m = B_m A$

Then $A = \dfrac{\Phi_m}{B_m} = \dfrac{0.0068}{1.35} = 0.00506 \, \text{m}^2$

Thus area of core $= 5060 \, \text{mm}^2$

(c) Secondary current $= \dfrac{25\,000}{110} = 227.27 \, \text{A}.$

3. (a) This solution is made with the aid of the phasor diagram shown by Fig. 21.

Since $\dfrac{V_1}{V_2} = \dfrac{N_1}{N_2}$ $\therefore V_1 = V_2 \times \dfrac{N_1}{N_2} = 110 \times \dfrac{4}{1}$

or Primary voltage $= 440 \, \text{V}$

(b) Secondary Current $= \dfrac{100\,000}{110} = 910 \, \text{A}$

Again assuming

Primary ampere-turns $=$ Secondary ampere-turns

or $I_1' N_1 = I_2 N_2$ So $I_1' = 910 \times \dfrac{1}{4} = 227.5 \, \text{A}$

Given also $\cos \phi_2 = 0.8$ so $\sin \phi_2 = 0.6$

Similarly $\cos \phi_0 = 0.3$ and $\sin \phi_0 = 0.954$

(c)   Then from the diagram (Fig. 21)

$$I_1 \cos \phi_1 = I_0 \cos \phi_0 + I_1' \cos \phi_2$$
$$= (5 \times 0.3) + (227.5 \times 0.8) = 1.5 + 182 = 183.5$$

Similarly $I_1 \sin \phi_1 = I_0 \sin \phi_0 + I_1' \sin \phi_2$
$$= (5 \times 0.954) + (227.5 \times 0.6)$$
$$= 4.77 + 136.5 = 141.27 \text{ A}$$

$$\therefore \quad I_1 = \sqrt{(I_1 \cos \phi_1)^2 + (I_1 \sin \phi_1)^2}$$
$$= \sqrt{183.5^2 + 141.27^2} = \sqrt{33\,650 + 19\,900}$$
$$= \sqrt{53\,550} = 231.6 \text{ A}$$

(d)   Primary power factor $\cos \phi_1 = \dfrac{I_1 \cos \phi_1}{I_1} = \dfrac{183.5}{231.6}$
$$= 0.793 \text{ (lagging)}.$$

4.   Again, using the phasor diagram of Fig. 21 we have, since this is a step-down transformer in terms of voltage, the primary current reduced by the same ratio.

Thus $I_1' = I_2 \times 0.06 = 120 \times 0.06 = 7.2 \text{ A}$

Using the modified cosine rule, we can write

$$I_1^2 = I_0^2 + I_1'^2 + 2I_0 I_1' \cos \alpha$$
or $7.827^2 = 0.7^2 + 7.2^2 + (2 \times 0.7 \times 7.2 \cos \alpha)$
and $61.26 = 0.49 + 51.84 + 10.08 \cos \alpha$

$$\therefore \ 61.26 - 52.33 = 10.08 \cos \alpha \ \text{ or } \ \cos \alpha = \frac{8.93}{10.08} = 0.8859$$

giving $\alpha = 27°38'$

and since $\cos \phi_2 = 0.8$ $\therefore \ \phi_2 = 36° 52'$

and $\phi_0 = \alpha + \phi_2 = 27° 38' + 36° 52' = 64° 30'$
So $\cos \phi_0 = 0.43$.

5.   The substitution of values in the efficiency expression has been made in kW and kVA

Thus Efficiency $= \dfrac{\text{Output (kW)}}{\text{Output (kW)} + \text{Losses (kW)}}$

or $\eta = \dfrac{kVA \cos \phi}{kVA \cos \phi + P_{\text{Fe}} + P_{\text{Cu}}}$

$$= \frac{1 \times 0{\cdot}8}{(1 \times 0{\cdot}8) + 0{\cdot}015 + 0{\cdot}03} = \frac{0{\cdot}8}{0{\cdot}8 + 0{\cdot}045} = \frac{0{\cdot}8}{0{\cdot}845}$$

or $\eta = 0{\cdot}9467 \%$ Thus full-load efficiency $= 94{\cdot}67$ per cent.

6. (a)   Efficiency $= \dfrac{\text{Output (kW)}}{\text{Output (kW)} + \text{Losses (kW)}}$

or $\eta = \dfrac{kVA \cos \phi}{kVA \cos \phi + P_{\text{Fe}} + P_{\text{Cu}}}$

Note substitution of values for P_{Fe} and P_{Cu} is made in kW

$$\eta = \frac{20 \times 0{\cdot}81}{(20 \times 0{\cdot}81) + (0{\cdot}2 + 0{\cdot}18)} = \frac{16{\cdot}2}{16{\cdot}2 + 0{\cdot}38}$$

$$= \frac{16{\cdot}2}{16{\cdot}58} = 0{\cdot}976$$

Full-load Efficiency $= 97{\cdot}6$ per cent.

(b)   Since the kVA rating is now 30 kVA and it can be assumed that the voltage remains constant, therefore iron loss remains constant.

The new current is $\dfrac{3}{2}$ or $1{\cdot}5$ times the original current.

Copper Loss is proportional to current squared, thus new copper loss $= \left(\dfrac{3}{2}\right)^2$ times the original copper loss.

New Copper Loss $= \dfrac{9}{4} \times 180 = 405$ W

So $\eta = \dfrac{30 \times 0\cdot91}{(30 \times 0\cdot91) + 0\cdot405 + 0\cdot2} = \dfrac{27\cdot3}{27\cdot3 + 0\cdot605}$

$= \dfrac{27\cdot3}{27\cdot905} = 0\cdot976$

New Efficiency $= 97\cdot9$ per cent.

7.   For this solution, in accordance with theory, all secondary values are referred to the primary.

Then $\overline{R}_1 = R_1 + R_2 \left(\dfrac{N_1}{N_2}\right)^2 = 2\cdot5 + 0\cdot028 \left(\dfrac{2000}{220}\right)^2$

$= 2\cdot5 + 2\cdot315 = 4\cdot815\ \Omega$

Similarly $\overline{X}_1 = X_1 + X_2 \left(\dfrac{N_1}{N_2}\right)^2 = 2\cdot8 + 0\cdot032 \left(\dfrac{2000}{220}\right)^2$

$= 2\cdot8 + 2\cdot64 = 5\cdot44\ \Omega$

Full-load primary current $I_1 = \dfrac{20 \times 1000}{2000} = 10$ A

Equivalent primary impedance $\overline{Z}_1 = \sqrt{\overline{R}_1{}^2 + \overline{X}_1{}^2}$

or $\overline{Z}_1 = \sqrt{4\cdot815^2 + 5\cdot44^2} = 7\cdot265\ \Omega$

Primary short-circuit current would be $\dfrac{V_1}{Z_1} = \dfrac{2000}{7\cdot265} = 275$ A.

8. Full-load primary current $= \dfrac{27\cdot5 \times 1000}{450} = 61\cdot2$ A

Full-load secondary current $= \dfrac{27\cdot5 \times 1000}{121} = 227$ A

Full-load primary Cu Loss $= I_1^2 R_1 = 61\cdot2^2 \times 0\cdot055$
$= 206\cdot1$ W

Full-load secondary Cu Loss $= I_2^2 R_2 = 227^2 \times 0\cdot00325$
$= 167\cdot4$ W

Total full-load Cu Loss $= 206\cdot1 + 167\cdot4 = 373\cdot5$ W

Total full-load Losses $=$ Cu Losses $+$ Iron Loss
$= 373\cdot5 + 170$
$= 543\cdot5$ W

(a) η at full-load 0·8 (lagging) power factor
$$= \dfrac{27\cdot5 \times 0\cdot8}{(27\cdot5 \times 0\cdot8) + 0\cdot54} \times 100$$

or Efficiency $= 97\cdot59$ per cent

(b) For maximum efficiency, Copper Loss $=$ Iron Loss
Total Losses under this condition $= 2 \times 0\cdot17 = 0\cdot34$ kW

At Full load. Output $= 27\cdot5$ kVA when Copper Loss is
0·374 kW

Let $\dfrac{kVA \text{ at maximum efficiency}}{kVA \text{ at full load}} = x$

$\therefore$ At maximum efficiency, total Copper Loss $= 0\cdot374 \times x^2$ kilowatts

But at maximum efficiency, total Copper Loss $= 0\cdot17$ kW

So $0\cdot374x^2 = 0\cdot17$ or $x = \sqrt{\dfrac{0\cdot17}{0\cdot374}} = 0\cdot67$

$\therefore$ Output kVA at maximum efficiency $= 0\cdot67 \times 27\cdot5$
$= 18\cdot43$ kVA

(c)   Maximum efficiency at this *kVA* value and at a power

factor of 0·8 (lagging) $= \dfrac{18\cdot43 \times 0\cdot8}{(18\cdot43 \times 0\cdot8) + 0\cdot34}$

So η_{max} $= \dfrac{14\cdot74}{14\cdot74 + 0\cdot34} = \dfrac{14\cdot74}{15\cdot08} = 0\cdot9775$

Maximum Efficiency $= 97\cdot75$ per cent

9.   Although this and the next problem are not entirely concerned with transformer theory, they have been included to
remind the students of three-phase, line to phase voltage and
kW to *kVA* relationships,

The diagram for the problem is shown.

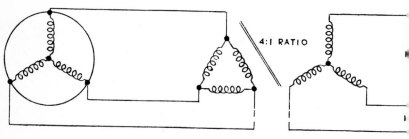

Secondary phase voltage of transformer $= \dfrac{120}{\sqrt{3}}$ volts

Primary phase voltage of transformer $= \dfrac{120}{\sqrt{3}} \times \dfrac{4}{1}$

$= \dfrac{480}{\sqrt{3}}$ volts

Line voltage of alternator $=$ primary phase voltage of transformer

$= \dfrac{480}{\sqrt{3}}$ volts

Phase voltage of star-connected alternator $= \dfrac{480}{\sqrt{3}} \times \dfrac{1}{\sqrt{3}}$

$= \dfrac{480}{3} = 160$ V.

10.

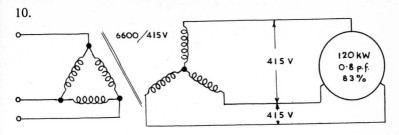

Motor output $= 120$ kW

Input $= \dfrac{120 \times 100}{83}$ kilowatts or $\dfrac{120 \times 100}{83 \times 0.8}$ kilovolt amperes

$= \dfrac{120 \times 1000}{664} = 180.7$ kVA

Line Current $= \dfrac{180\,700}{\sqrt{3} \times 415} = \dfrac{435.4}{\sqrt{3}}$ amperes

$= $ phase current of transformer secondary

Also transformer secondary phase voltage $= \dfrac{415}{\sqrt{3}} = 239.6$ V

Turns ratio $= \dfrac{\text{Primary phase voltage}}{\text{Secondary phase voltage}} = \dfrac{6600}{239.6} = \dfrac{27.5}{1}$

So primary phase current is reduced by this value

$= \dfrac{435.4}{\sqrt{3} \times 27.55} = \dfrac{15.8}{\sqrt{3}} = 9.13$ A

This could also be found by an alternative method

Transformer output $=$ transformer input $= 180.7$ kVA

$\therefore$ Transformer primary line current $= \dfrac{180\,700}{\sqrt{3} \times 6600}$ amperes

Transformer primary phase current $= \dfrac{180\,700}{\sqrt{3} \times 6600 \times \sqrt{3}}$

$= \dfrac{1807}{3 \times 66} = \dfrac{602.3}{66} = 9.13$ A

CHAPTER 3

1.
$$\text{Full-load current} = \frac{150\,000}{250} = \frac{600\,000}{1000} = 600 \text{ A}$$

$$\text{Current per conductor} = \frac{600}{4} = 150 \text{ A}$$

$$\text{Armature turns} = \frac{342}{2} = 171$$

$$\text{Armature ampere-turns} = 171 \times 150 = 25\,650$$

$$\text{Armature ampere-turns per pole} = \frac{25\,650}{4} = 6412 \cdot 5$$

$$\text{For the air-gap } B = \mu_0 H$$

$$\text{or } H = \frac{0 \cdot 32}{4\pi \times 10^{-7}} \text{ At/m}$$

$$\text{Ampere-turns for air-gap} = \frac{0 \cdot 32 \times 12 \times 10^{-3}}{4\pi \times 10^{-7}}$$

$$= \frac{32 \times 12 \times 10^2}{4\pi} = \frac{96}{3 \cdot 14} \times 10^2 = 3057 \text{ At/pole}$$

$$\text{Total ampere-turns per pole} = 6412 \cdot 5 + 3057 = 9470$$

$$\text{Number of turns per pole} = \frac{9470}{600} = \frac{94 \cdot 7}{6}$$

$$= 15 \cdot 78 \text{ (say 16 turns).}$$

2.   Number of armature turns $=$ 128 $\times$ 2 $=$ 256
      Number of armature conductors $=$ 256 $\times$ 2 $=$ 512
   Number of commutator segments $=$ number of coils $=$ 128

$$\text{Current per conductor} = \frac{600}{2} = 300 \text{ A}$$

Armature ampere-turns $=$ 256 $\times$ 300 $=$ 76 800 At
For this machine 128 segments $=$ 360 (mechanical degrees)

$$\text{or three segments} = \frac{360 \times 3}{128} = \frac{270^\circ}{32} \text{ (mechanical)}$$

If θ $=$ the brush shift in electrical degrees then, since this is a six-pole machine, one mechanical degree is equivalent to three electrical degrees or here $\theta = \dfrac{3 \times 270^\circ}{32}$ (electrical).

The ratio of demagnetising ampere-turns to the total number of armature ampere-turns is given by $\dfrac{2\theta}{180}$

$$\text{Demagnetising ampere-turns} = \frac{2\theta}{180} \times 76\ 800$$

$$= \frac{2 \times 3 \times 270}{180 \times 32} \times 76\ 800$$

$$= \frac{9 \times 76\ 800}{32} = 9 \times 2400$$

$$= 21\ 600$$

$$\text{Demagnetising ampere-turns per pole} = \frac{21\ 600}{6}$$

$$= 3600 \text{ At/pole.}$$

3.   Since $E \propto \Phi$ and N, then new voltage values can be deduced for each field-current value by multiplying by the speed ratio

i.e. $\dfrac{850}{700} = 1 \cdot 21$ Then

I_f values (amperes)
                0   0·1   0·24   0·5   0·77   1·2   1·92   3·43

E at 700 rev/min (volts)
                10   20   40   80   120   160   200   240

E at 850 rev/min (volts) $= 10 \times 1 \cdot 21 = 12 \cdot 1$ and so for all values

      $=$ 12·1   24·2   48·4   96·8   145·2   193·6   242   290·4

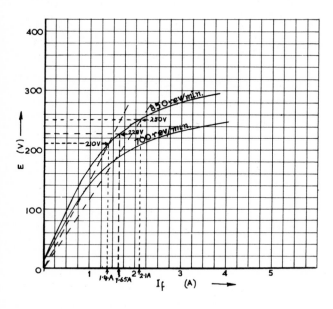

(a)   The results, when plotted, are shown by the above curves. Read 210 V on the 850 rev/min characteristic. Join this point to zero. This is the field resistance voltage-drop line for self-excitation to 210 V.

              The corresponding $I_f = 1 \cdot 4$ A

So $R_{f_1} = \dfrac{200}{1\cdot4} = 142\cdot9\ \Omega$

Similarly read 250 V on the 850 rev/min characteristic. Join this point to zero. This is the field resistance voltage-drop line for self-excitation to 250 V. Corresponding $I_f = 2\cdot1$ A.

So $R_{f_2} = \dfrac{250}{2\cdot1} = 119\ \Omega$

Series resistance for 210 V condition $= 142\cdot9 - 52 = 90\cdot9\ \Omega$

Series resistance for 250 V condition $= 119 - 52 = 67\ \Omega$

Range of field rheostat required is from 90·9 to 67 ohms

(b)   If armature current $= 200$ A, then voltage drop in armature must be $200 \times 0\cdot04 = 8$ V. If the terminal voltage is to be 220 V, then O.C. voltage must be $220 + 8 = 228$ V.

From O.C.C., field current $= 1\cdot65$A (shown on graph)

Field resistance $= \dfrac{228}{1\cdot65} = 138\cdot2\ \Omega$

Regulator $= 138\cdot2 - 52 = 86\cdot2\ \Omega.$

4.   For a shunt motor. $T\ \alpha\ I_a\ \Phi$ and here Φ can be assumed constant. Also Output Power $\alpha\ N\ T$

Originally, $E_{b_1} = V - I_{a_1}R_a = 220 - (60 \times 0\cdot15)$

$$= 220 - 9 = 211\ \text{V}$$

and since $E_b\ \alpha\ \Phi N$ and Φ is assumed constant then, if E_{b_2} is the back e.m.f., we have $\dfrac{E_{b_2}}{E_{b_1}} = \dfrac{N_2}{N_1}$

or $E_{b_2} = \dfrac{211\ \times\ N_2}{1200}$ ...   .   (a)

Using the output-power relationship then $\dfrac{P_1}{P_2} = \dfrac{N_1 T_1}{N_2 T_2}$

or since $P_2 = 1 \cdot 3\, P_1$ then $\dfrac{P_1}{1 \cdot 3\, P_1} = \dfrac{N_1\, I_{a_1}}{N_2\, I_{a_2}}$. Note that since

$T \propto I_a$, armature current has been substituted.

So $\dfrac{1}{1 \cdot 3} = \dfrac{1200 \times 60}{N_2\, I_{a_2}}$ or $I_{a_2} = \dfrac{1 \cdot 3 \times 1200 \times 60}{N_2}$ ... (b)

Also under the new speed condition. $E_{b_2} = 220 - 0 \cdot 15 \times I_{a_2}$
Using equations (a) and (b) then

$$\frac{211 \times N_2}{1200} = 220 - \frac{0 \cdot 15 \times 1 \cdot 3 \times 1200 \times 60}{N_2}$$

or $211\, N^2 = (22 \times 12 \times 10^3 \times N_2) - (15 \times 13 \times 12 \times 6 \times 12 \times 10^2)$

or $211\, N^2 - (264 \times 10^3 \times N_2) + (9 \times 13 \times 144 \times 10^3) = 0$

from which

$$N = \frac{(264 \times 10^3) \pm \sqrt{264^2 \times 10^6 - (4 \times 211 \times 117 \times 144 \times 10^3)}}{422}$$

or $N = \dfrac{(264 \times 10^3) \pm (2 \cdot 355 \times 10^5)}{422} = \dfrac{499\,500}{422} = 1183 \cdot 6$

New speed $= 1184$ rev/min

Thus speed reduction $= \dfrac{1200 - 1184}{1200} \times 100 = 1 \cdot 33$ per cen

5. Output of motor $= 10000$ W

Motor input $= 10000 \times \dfrac{100}{82}$ watts

Current I_1 at full load $= \dfrac{10000 \times 100}{82 \times 220} = 55 \cdot 4$ A

Back e.m.f. at 800 rev/min or E_{b_1} $= 220 - (55 \cdot 4 \times 0 \cdot 6)$
$= 220 - 33 \cdot 24$
$= 186 \cdot 76$ V

Half full-load current condition.

$$I_2 = \frac{55 \cdot 4}{2} = 27 \cdot 7 \text{ A}$$

$$E_{b_2} = 220 - (22 \cdot 7 \times 0 \cdot 6)$$

$$= 220 - 16 \cdot 62$$

$$= 203 \cdot 35 \text{ V}$$

For this series machine $\Phi \propto I$ and $N \propto \dfrac{E_b}{\Phi}$

$$\therefore \frac{N_2}{N_1} = \frac{E_{b_2}}{\Phi_2} \times \frac{\Phi_1}{E_{b_1}} \text{ or } N_2 = \frac{N_1 \times E_{b_2} \times I_1}{I_2 \times E_{b_1}}$$

Thus $N_2 = \dfrac{800 \times 203 \cdot 38 \times 55 \cdot 4}{27 \cdot 7 \times 186 \cdot 76}$ $= 1742$ rev/min (under new loading condition).

Also Torque $T \propto \Phi I$ so $T \propto I^2$

So $\dfrac{T_2}{T_1} = \left(\dfrac{I_2}{I_1}\right)^2 = \left(\dfrac{27 \cdot 7}{55 \cdot 4}\right)^2 = \left(\dfrac{1}{2}\right)^2 = \dfrac{1}{4} = 0 \cdot 25$

Thus $T_2 = 0 \cdot 25 T_1$

Percentage change of torque $= \left(\dfrac{T_1 - 0 \cdot 25 \, T_1}{T_1}\right) \times 100$

$$= \frac{0 \cdot 75 \, T_1}{T_1} \times 100$$

Thus change of torque $= 75$ per cent.

6.   The O.C.C. is plotted to give the curve shown below. The field voltage-drop line is drawn in by taking a current of 1 A to give the field voltage drop of $1 \times 480 = 480$ V. This point is joined to the origin and a line extended backwards, to cut the O.C.C. at 510 V. This would be the open-circuit terminal voltage resulting from self-excitation.

The load characteristic is deduced as follows:

For the chosen values of terminal voltage on the field voltage-drop line, the vertical ordinate is drawn. The O.C.C. voltage is noted and the armature voltage drop obtained. The armature current is found as shown and since the field current is neglected, values of load current are deduced and plotted giving V to an I_L base as required by the problem.

Terminal Voltage (volts)	Armature Voltage drop $= (E - V)$ (volts)	Load Current $= \dfrac{E - V}{R_a}$ (amperes)
510	0	0
450	$482 \cdot 5 - 450 = 32 \cdot 5$	$\dfrac{32 \cdot 5}{0 \cdot 1} = 325$
400	$455 - 400 = 55$	550
350	$422 \cdot 5 - 350 = 72 \cdot 5$	725
300	$385 - 300 = 85$	850
250	$335 - 250 = 85$	850
200	$260 - 200 = 60$	600
150	$180 - 150 = 30$	300

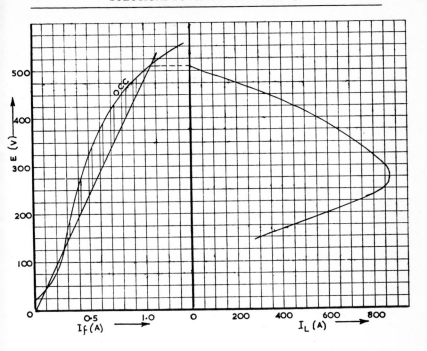

7. For a fan $T \propto N_1^2$ and for the two speed conditions

$$\frac{T_1}{T_2} = \frac{N_1^2}{N_2^2} = \frac{1000^2}{750^2} = \frac{16}{9}$$

Also for a series motor, $T \propto \Phi I$ but $\Phi \propto I$ $\therefore T = kI^2$

or $\dfrac{T_1}{T_2} = \dfrac{kI_1^2}{kI_2^2} = \dfrac{28^2}{I_2^2}$

Using the relation $\dfrac{T_1}{T_2} = \dfrac{16}{9}$ then $\dfrac{28^2}{I_2^2} = \dfrac{16}{9}$

or $I_2^2 = \dfrac{28^2 \times 3^2}{4^2}$ and $I_2 = \dfrac{28 \times 3}{4} = 21$ A

Again if E_{b_1} is the original back e.m.f. and E_{b_2} the final back e.m.f.

then $E_{b_1} = 110 - (28 \times 0.5) = 96$ V

Again, generated voltage $E_b \, \alpha \, \Phi N$ and since $\Phi \, \alpha \, I$ then $E_b \, \alpha \, IN$

So $\dfrac{E_{b_2}}{E_{b_1}} = \dfrac{kI_2N_2}{kI_1N_1} = \dfrac{21 \times 750}{28 \times 1000} = \dfrac{9}{16}$

and $E_{b_2} = \dfrac{96 \times 9}{16} = 54$ V

Let R = the required series resistance

Then $110 = I_2(0.5 + R) + 54$ or $21(0.5 + R) = 56$

and $0.5 + R = 2.67$ giving $R = 2.17 \, \Omega$

Required series resistance is 2·17 ohms.

8. As a motor. Input $= \dfrac{10 \times 1000}{0.85}$ watts

Full-load current $= \dfrac{10 \times 1000}{0.85 \times 230} = 51.15$ A

Shunt-field current $I_{f_1} = \dfrac{230}{115} = 2$ A

Armature current $= 51.15 - 2 = 49.15$A

Back e.m.f. $E_{b_1} = 230 - (51.15 \times 0.2) = 230 - 10.23$
$= 219.77$ V

As a generator. $E = V + I_aR_a$ so $E_2 = 220 + (51.15 \times 0.2)$

or $E_2 = 220 + 10.23 = 230.23$ V

Since E is proportional to Φ and N and Φ can be assumed proportional to I_f, then $E = kI_fN$

As a generator $V = 220$ volts $\quad I_{f_2} = \dfrac{220}{115} = 1.91$ A

and $\dfrac{E_2}{E_{b_1}} = \dfrac{kI_{f_2}N_2}{kI_{f_1}N_1}$ or $N_2 = \dfrac{E_2N_1I_{f_1}}{E_{b_1}I_{f_2}}$

$$= \frac{230 \cdot 23 \times 1000 \times 2}{219 \cdot 77 \times 1 \cdot 91}$$

$$= 1097 \text{ rev/min.}$$

Rating $= \dfrac{220 \times \text{Line current}}{1000} = \dfrac{220 \times (51 \cdot 15 - 1 \cdot 91)}{1000}$

$$= \frac{220 \times 49 \cdot 24}{1000} = 2 \cdot 2 \times 4 \cdot 924 = 10 \cdot 83 \text{ kW.}$$

9. The table is set out as a result of the assumption that, for any field-current value, generated e.m.f. is proportional to speed.

Field current I_f (amperes)

| 0·5 | 1·0 | 1·5 | 2·0 | 2·5 | 3·0 | 3·5 | 4·0 |

Generated e.m.f. E (volts)
at 900 rev/min

| 61 | 111 | 145·8 | 175·5 | 194·5 | 207 | 214 | 221 |

at 1000 rev/min

| 67·8 | 123·5 | 165 | 195 | 216 | 230 | 238 | 245 |

E.m.f. generated at 1000 rev/min $= \dfrac{1000}{900} \times$ e.m.f. generated

at 900 rev/min

When field current $= 3$ A, field voltage drop $= 3 \times 80 = 240$ V

This point is used to draw the field voltage-drop line. This intercepts the O.C.C. at 225 V *i.e.* the voltage to which the machine will self-excite.

Voltage ordinates between O.C.C. and field voltage-drop line

for field-current values of   0·5   1·0   1·5   2·0   2·5   Amperes
                          are   28   44   44   35   15   Volts

These give load-current values of
                          200   314   314   250   107   Amperes.

A load-current value is given by $\dfrac{\text{Voltage drop}}{\text{Armature Resistance}}$. Examples

are shown on the graph.

viz. $\dfrac{35}{0·14}$ = 250 A   and $\dfrac{28}{0·14}$ = 200 A

The deduced load characteristic can then be plotted, as is shown
by the following graph.

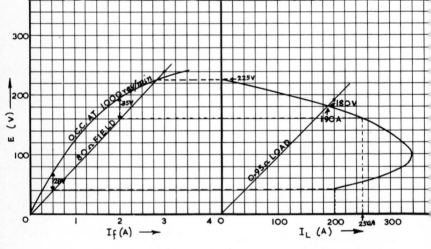

The load voltage-drop line is drawn by taking any current
value, say 200 A. ∴ Voltage-drop = 200 × 0·95 = 190 V.
Draw this voltage-drop line through the origin and 190 V. The
point of intersection with the load characteristic = 180 V and
gives the required terminal voltage at a current of 190 A.

10.   With $R = 0$ and a motor resistance of 0·3 Ω, the voltage-
drop when the current is 20 A is 20 × 0·3 = 6 V

Then back e.m.f. $E_{b_1} = 220 - 6 = 214$ V

With $R = 3\ \Omega$, a motor resistance of $0{\cdot}3\ \Omega$ and a current of
15 A the back e.m.f. $E_{b_2} = 220 - 15 \times 3{\cdot}3$
$$= 220 - 49{\cdot}5 = 170{\cdot}5\ \text{V}$$

Since the generated e.m.f. is proportional to speed and flux then

$$\frac{E_{b_1}}{E_{b_2}} = \frac{\Phi_1 N_1}{\Phi_2 N_2}$$

For the first speed condition $N_1 = 1200$ rev/min $E_{b_1} = 214$ V
and the flux is Φ_1. For the second speed condition $E_{b_2} =$
$170{\cdot}5$ V and $\Phi_2 = 0{\cdot}8\ \Phi_1$

Using the above deduction

$$N_2 = \frac{\Phi_1\ N_1\ E_{b_2}}{0{\cdot}8\ \Phi_1\ E_{b_1}} = \frac{1200\ \times\ 170{\cdot}5}{0{\cdot}8\ \times\ 214} = \frac{204\ 600}{171{\cdot}2}$$

or $\qquad\qquad\qquad N_2 = 1195$ rev/min

Torque is proportional to armature current and flux, whereas
power output is proportional to torque and speed. The
relationship can therefore be expressed as

$$\frac{P_1}{P_2} = \frac{N_1 T_1}{N_2 T_2} = \frac{N_1\ \Phi_1\ I_{a_1}}{N_2\ \Phi_2\ I_{a_2}}$$

or $\dfrac{P_1}{P_2} = \dfrac{1200 \times \Phi_1 \times 20}{1195 \times \Phi_2 \times 15} = \dfrac{1{\cdot}2 \times \Phi_1 \times 4}{1{\cdot}195 \times 0{\cdot}8\ \Phi_1 \times 3} = \dfrac{4{\cdot}8}{2{\cdot}868}$

So $\dfrac{P_1}{P_2} = \dfrac{1{\cdot}67}{1}$

Thus the power output at the original speed is $1{\cdot}67$ times the power
output at the second speed condition.

CHAPTER 4

1. Using the desired expression

$$\bar{R}_1 = R_1 + R_2 \left(\frac{N_1}{N_2}\right)^2 = 0{\cdot}36 + 0{\cdot}02 \left(\frac{460}{115}\right)^2$$

$$= 0{\cdot}36 + 0{\cdot}32 = 0{\cdot}68 \ \Omega$$

Similarly $\bar{X}_1 = X_1 + X_2 \left(\frac{N_1}{N_2}\right)^2 = 0{\cdot}83 + 0{\cdot}06 \times 4^2$

$$= 0{\cdot}83 + 0{\cdot}96 = 1{\cdot}79 \ \Omega$$

So impedance of transformer referred to primary side

is $\bar{Z}_1 = \sqrt{0{\cdot}68^2 + 1{\cdot}79^2} = \sqrt{0{\cdot}462 + 3{\cdot}2} = \sqrt{3{\cdot}662}$

$$= 1{\cdot}9 \ \Omega$$

∴ Current in the primary windings under short-circuit conditions

is $I = \dfrac{V}{\bar{Z}_1} = \dfrac{460}{1{\cdot}9} = 242$ A.

2. Given Iron Loss $\quad = \quad 165$ W or $0{\cdot}165$ kW

Full-load Copper Loss $= \quad 280$ W or $0{\cdot}28$ kW

Copper losses are proportional to current squared and current is proportional to the kVA loading.

Thus copper losses are proportional to kVA^2

or $\sqrt{\text{Copper Losses}} \ \alpha \ kVA$

Now transformer efficiency $= \dfrac{\text{Output (kW)}}{\text{Output (kW) + Losses (kW)}}$.

$$= \frac{x \times kVA \times \cos \phi}{(x \times kVA \times \cos \phi) + (x^2 \times \text{Full-load Cu Loss}) + \text{Fe Loss}}$$

where x is the fraction of full-load kVA

Then (a). At full load, (unity p.f.)

$$\eta = \frac{1 \times 25 \times 1}{(1 \times 25 \times 1) + (1^2 \times 0\cdot 28) + 0\cdot 165} = \frac{25}{25 + 0\cdot 28 + 0\cdot 165}$$

$$= \frac{25}{25\cdot 445} = 0\cdot 981 \quad \text{or} \quad \text{Efficiency} = 98\cdot 1 \text{ per cent}$$

At half full-load, (0·8 p.f.)

$$\eta = \frac{\frac{1}{2} \times 25 \times 0\cdot 8}{(\frac{1}{2} \times 25 \times 0\cdot 8) + (\frac{1}{4} \times 0\cdot 28) + 0\cdot 165}$$

$$= \frac{10}{10 + 0\cdot 07 + 0\cdot 165} = \frac{10}{10\cdot 235} = 0\cdot 9775$$

or Efficiency $= 97\cdot 75$ per cent.

Again (b). Transformer efficiency is maximum when the iron losses and the copper losses are equal.

If the Cu Loss is 0·28 kW at 25 kVA, it will be

$$0\cdot 165 \text{ kW at } \sqrt{\frac{0\cdot 165}{0\cdot 28}} \times 25 \text{ kilovolt amperes}$$

or at $\sqrt{0\cdot 589} \times 25 = 0\cdot 768 \times 25 = 19\cdot 2$ kVA
So Maximum efficiency occurs at 19·2 kVA.

3. Full-load current on primary $= \dfrac{17\,500}{450} = 38\cdot 9$ A

For efficiency.

From O.C. test.   Iron Loss   = 115 W
From S.C. test.   Copper Loss = 312 W

Then η on full-load (0·8 p.f.) $= \dfrac{17\cdot5 \times 0\cdot8}{(17\cdot5 \times 0\cdot8) + 0\cdot115 + 0\cdot312}$

$= \dfrac{14}{14 + 0\cdot115 + 0\cdot312} = \dfrac{14}{14\cdot427} = 0\cdot968$

or   Efficiency $= 96\cdot8$ per cent.

For voltage regulation

From S.C. test. Equivalent impedance referred to the primary

$$\overline{Z}_1 = \frac{15\cdot75}{38\cdot9} = 0\cdot404 \ \Omega$$

Equivalent Resistance referred to the primary $\overline{R}_1 = \dfrac{312}{38\cdot9^2}$

$$= 0\cdot206 \ \Omega$$

Equivalent Reactance referred to the primary $\overline{X}_1$

$$= \sqrt{0\cdot404^2 - 0\cdot206^2} \text{ or } \overline{X}_1 = \sqrt{0\cdot1635 - 0\cdot0424} = \sqrt{0\cdot1211}$$

$$= 0\cdot348 \ \Omega$$

Percentage voltage regulation at full load, 0·8 (lagging) power factor

$$= \frac{38\cdot9\left\{(0\cdot206 \times 0\cdot8) + (0\cdot348 \times 0\cdot6)\right\}}{450} = \frac{38\cdot9\,(0\cdot1648 + 0\cdot2088}{450}$$

$$= \frac{38\cdot9 \times 0\cdot3736}{450} \times 100 = 3\cdot22 \text{ or } 3\cdot22 \text{ per cent}$$

3·22 percentage voltage regulation would be a voltage drop of

$$\frac{121 \times 3·22}{100} = 3·89 \text{ V}$$

Thus secondary terminal voltage would be $121 - 3·89$

$$= 117·1 \text{ V}.$$

4. From the derived expression

Percentage voltage regulation $= \dfrac{I_1(\overline{R}_1 \cos \phi \pm \overline{X}_1 \sin \phi)}{V_1} \times 100$

Full-load current on H.V. side $= \dfrac{25 \times 1000}{450} = 55·6 \text{ A}$

So the transformer was tested up to full-load value.

Equivalent impedance referred to primary $\overline{Z}_1 = \dfrac{V}{I} = \dfrac{14·3}{55·6}$

$$= 0·257 \text{ }\Omega$$

Equivalent resistance referred to primary $\overline{R}_1 = \dfrac{W}{I^2} = \dfrac{316}{55·6^2}$

$$= 0·102 \text{ }\Omega$$

Equivalent reactance referred to primary $\overline{X}_1 = \sqrt{\overline{Z}_1^2 - \overline{R}_1^2}$

Thus $\overline{X}_1 = \sqrt{0·257 - 0·102^2} = \sqrt{0·066 - 0·0104}$

$$= \sqrt{0·0556} \qquad = 0·236 \text{ }\Omega$$

Percentage voltage regulation at full load, 0·8 (lagging) power factor

$$= \frac{55·6 \left\{(0·102 \times 0·8) + (0·236 \times 0·6)\right\}}{450} \times 100$$

$$= \frac{55 \cdot 6(0 \cdot 0816 + 0 \cdot 1416)}{450} \times 100 = \frac{55 \cdot 6 \times 0 \cdot 2232}{450} \times 100$$

$$= 2 \cdot 76 \text{ per cent (down)}$$

$$\therefore \text{ Secondary terminal voltage} = 121 - \left(\frac{2 \cdot 76}{100} \times 121\right)$$

$$= 121 - 3 \cdot 34 = 117 \cdot 66 \text{ V}$$

Percentage voltage regulation at full load 0·8 (leading) power factor

$$= \frac{55 \cdot 6 \left\{(0 \cdot 102 \times 0 \cdot 8) - (0 \cdot 236 \times 0 \cdot 6)\right\}}{450} \cdot \times 100$$

$$= \frac{55 \cdot 6(0 \cdot 0816 - 0 \cdot 1416)}{450} \times 100 = \frac{55 \cdot 6(- 0 \cdot 06)}{450} \times 100$$

$$= - 0 \cdot 74 \text{ per cent or } 0 \cdot 74 \text{ per cent (up).}$$

$$\text{Secondary terminal voltage} = 121 + \left(\frac{0 \cdot 74}{100} \times 121\right)$$

$$= 121 + 0 \cdot 895 = 121 \cdot 895 \text{ V.}$$

5. Full-load primary current $= \dfrac{27 \cdot 5 \times 1000}{450} = 61 \cdot 2 \text{ A}$

Full-load secondary current $= \dfrac{27 \cdot 5 \times 1000}{121} = 227 \text{ A}$

Full-load primary copper loss $= I_1^2 R_1 = 61 \cdot 2^2 \times 0 \cdot 055$
$$= 206 \cdot 1 \text{ W}$$

Full-load secondary copper loss $= I_2^2 R_2 = 227^2 \times 0 \cdot 00325$
$$= 167 \cdot 4 \text{ W}$$

Total full-load copper loss $= 206 \cdot 1 + 167 \cdot 4 = 373 \cdot 5$ W

$\qquad\qquad\qquad\qquad\qquad\quad = 0 \cdot 374$ kW

Total full-load Loss $\qquad =$ Copper Loss + Iron Loss

$\qquad\qquad\qquad\qquad\quad = 373 \cdot 5 + 170 = 543 \cdot 5$ W

$\qquad\qquad\qquad\qquad\quad = 0 \cdot 544$ kW

(a) At full load, 0·8 (lagging) power factor

$$\text{Efficiency} \quad = \quad \frac{kVA \cos \phi}{kVA \cos \phi + \text{Full-load Loss (kW)}} \times 100$$

$$= \frac{27 \cdot 5 \times 0 \cdot 8}{(27 \cdot 5 \times 0 \cdot 8) + 0 \cdot 54} \times 100 = \frac{22}{22 \cdot 54} \times 100$$

$$= \quad 97 \cdot 6 \text{ per cent.}$$

(b) At maximum efficiency, Copper Loss $=$ Iron Loss

$\therefore$ Total losses at maximum efficiency $= 2 \times 0 \cdot 17 = 0 \cdot 34$ kW

At full load, Output $= 27 \cdot 5$ kVA

and full-load Copper Loss $= 0 \cdot 374$

Then let $\quad \dfrac{kVA \text{ (maximum efficiency}}{kVA \text{ (full-load)}} \quad = \quad x$

So for maximum efficiency, total copper loss should equal $0 \cdot 374 \times x^2$ kilowatts. But it equals $0 \cdot 17$ kW

$$\therefore \quad 0 \cdot 374 x^2 = 0 \cdot 17 \text{ or } x = \sqrt{\frac{0 \cdot 17}{0 \cdot 374}} = 0 \cdot 67$$

Output kVA at maximum efficiency $= 0 \cdot 67 \times 27 \cdot 5$

$\qquad\qquad\qquad\qquad\qquad\qquad\qquad = 18 \cdot 43$ kVA

Thus $\eta_{\max} = \dfrac{18 \cdot 43 \times 0 \cdot 8}{(18 \cdot 43 \times 0 \cdot 8) + 0 \cdot 34} \times 100$

or Maximum Efficiency $= 97 \cdot 77$ per cent.

6. Percentage voltage regultation $= \dfrac{I_2(\bar{R}_2 \cos \phi + \bar{X}_2 \sin \phi) \times 100}{V_2}$

Now $\bar{R}_2 = \dfrac{\text{Copper loss}}{I_2{}^2}$ and $I_2 = \dfrac{50 \times 1000}{110} = 454 \cdot 5$ A

So $\bar{R}_2 = \dfrac{500}{454 \cdot 5^2} = 0 \cdot 00242 \ \Omega$

$\bar{Z}_1 = \dfrac{V_{1sc}}{I_1}$ and $I_1 = \dfrac{50 \times 1000}{440} = 113 \cdot 6$ A

So $\bar{Z}_1 = \dfrac{25}{113 \cdot 6 \text{ A}} = 0 \cdot 2201 \ \Omega$

Now $\bar{Z}_2 = \dfrac{\bar{Z}_1}{n^2}$ where $n = \dfrac{V_1}{V_2} = \dfrac{440}{110} = 4$

$\therefore$ So $\bar{Z}_2 = \dfrac{0 \cdot 2201}{16} = 0 \cdot 01375 \ \Omega$

giving $\bar{X}_2 = \sqrt{\bar{Z}_2{}^2 - \bar{R}_2{}^2}$

$= \sqrt{0 \cdot 01375^2 - 0 \cdot 00242^2} = \sqrt{0 \cdot 018314}$

$= 0 \cdot 01354 \ \Omega$

(a) Percentage voltage regulation

$= \dfrac{454 \cdot 5 \ \{(0 \cdot 00242 \times 0 \cdot 8) + (0 \cdot 01354 \times 0 \cdot 6)\}}{110} \times 100$

Thus voltage regulation $= 4 \cdot 15$ percent. (down).

(b) Copper loss $\propto$ current^2 or $\propto kVA^2$
Now for maximum efficiency. Cu Loss $=$ Fe Loss
$\therefore x^2$ copper loss $=$ iron loss, where x is the fraction of full load at which maximum efficiency occurs.

So $x = \sqrt{\dfrac{\text{Iron loss}}{\text{Copper loss}}} = \sqrt{\dfrac{250}{500}} = 0\cdot7071$

Thus maximum efficiency occurs at $0\cdot707$ of full-load.

7. Full-load primary current $= \dfrac{10\ 000}{440} = 22\cdot7$ A

From O.C. test.   Iron loss $= 75$ W

From S.C. test.   Copper loss $= 135$ W

Equivalent impedance referred to primary $\overline{Z}_1 = \dfrac{30}{22\cdot7}$

$= 1\cdot32\ \Omega$

Equivalent resistance referred to primary $\overline{R}_1 = \dfrac{135}{22\cdot7^2}$

$= 0\cdot262\ \Omega$

Equivalent reactance referred to primary $\overline{X}_1 = \sqrt{\overline{Z}_1{}^2 - \overline{R}_1{}^2}$

$= \sqrt{1\cdot32^2 - 0\cdot262^2} = 1\cdot29\ \Omega$

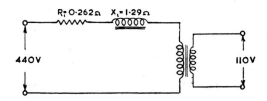

(a) Secondary terminal voltage on full load $0\cdot8$ (lagging) power factor

$=$ O.C. voltage $-$ voltage drop

$= 110 - \left[\left\{22\cdot7(0\cdot262 \times 0\cdot8) + (1\cdot29 \times 0\cdot6)\right\}\dfrac{110}{440}\right]$

$= 110 - \left[\left\{22\cdot7(0\cdot21 + 0\cdot774)\right\} \times \tfrac{1}{4}\right]$

$$= 110 - [\{22{\cdot}7 \times 0{\cdot}984\} \times \tfrac{1}{4}]$$

$$= 110 - 5{\cdot}59 = 104{\cdot}4 \text{ V}$$

(b) Percentage efficiency on full load, 0·8 (lagging) power factor

$$= \frac{10 \times 0{\cdot}8}{(10 \times 0{\cdot}8) + 0{\cdot}075 + 0{\cdot}135} \times 100$$

$$= \frac{8 \times 100}{8 + 0{\cdot}075 + 0{\cdot}135} = \frac{8}{8{\cdot}21} \times 100$$

Thus Efficiency $= 97{\cdot}5$ per cent.

8. Secondary voltage drop $= I_2 \, (\overline{R}_2 \cos \phi_2 \pm \overline{X}_2 \sin \phi_2)$

Where $I_2 =$ secondary full-load current $= \dfrac{50 \times 1000}{230}$

$$= 217{\cdot}5 \text{ A}$$

$\overline{R}_2 =$ equivalent resistance referred to the secondary

$$= R_2 + R_1 \left(\frac{N_2}{N_1}\right)^2 = 0{\cdot}015 + 0{\cdot}09 \left(\frac{230}{440}\right)^2$$

$$= 0{\cdot}04 \ \Omega$$

$\overline{X}_2 =$ equivalent reactance referred to secondary

$$= X_2 + X_1 \left(\frac{N_2}{N_1}\right)^2 = 0{\cdot}042 + 0{\cdot}19 \left(\frac{230}{440}\right)^2$$

$$= 0{\cdot}094 \ \Omega$$

$\therefore$ Secondary voltage drop $= 217{\cdot}5 \, \{(0{\cdot}04 \times 0{\cdot}8) + (0{\cdot}094 \times 0{\cdot}6)\}$

$$= 217{\cdot}5 \, (0{\cdot}032 + 0{\cdot}0564) = 19{\cdot}2 \text{ V}$$

and secondary terminal voltage $= 230 - 19\cdot2 = 211$ volts at full load, $0\cdot8$ (lagging) power factor

If the secondary terminals are short-circuited then the only circuit impedance is that of the transformer itself.

$$\overline{Z}_2 = \sqrt{\overline{R}_2{}^2 + \overline{X}_2{}^2} = \sqrt{0\cdot04^2 + 0\cdot094^2} = 0\cdot102\ \Omega$$

and $\overline{Z}_1$ is the total impedance referred to the primary

But $\overline{Z}_1 = \overline{Z}_2 \left(\dfrac{N_1}{N_2}\right)^2 = 0\cdot102 \times \left(\dfrac{440}{230}\right)^2 = 0\cdot372\ \Omega$

So primary short-circuit current $= \dfrac{V_1}{\overline{Z}_1} = \dfrac{440}{0\cdot372} = 1182$ A.

9. For the equivalent circuit refer to the text and diagrams of Chapter 4, Fig. 61.

  For this problem equivalent resistance referred to the secondary side is

$$\overline{R}_2 = 0\cdot008 + 0\cdot125 \left(\frac{115}{460}\right)^2 = 0\cdot008 + \frac{0\cdot125}{16}$$

$$= 0\cdot008 + 0\cdot0078 = 0\cdot0158\ \Omega$$

and equivalent reactance $\overline{X}_2 = 0\cdot025 + 0\cdot39 \left(\dfrac{115}{460}\right)^2$

$$= 0\cdot025 + \frac{0\cdot39}{16} = 0\cdot025 + 0\cdot0244 = 0\cdot0494\ \Omega$$

Full-load secondary current $I_2 = \dfrac{kVA \times 1000}{V_2} = \dfrac{17\cdot5 \times 1000}{115}$

$$= 152\ \text{A}$$

$$\text{Voltage drop in transformer} = I_2(\bar{R}_2 \cos \phi_2 + \bar{X}_2 \sin \phi_2)$$
$$= 152 \, [(0.0158 \times 0.8) + (0.0494 \times 0.6)]$$
$$= 152 \, [0.01264 + 0.02964]$$
$$= 152 \times 0.04228$$
$$= 6.43 \text{ V}$$

Full-load secondary terminal voltage $= 115 - 6.43 = 108.57$ V

$$\text{Efficiency} = \frac{kVA \cos \phi}{kVA \cos \phi + P_{Cu}(\text{kW}) + P_{Fe}(\text{kW})}$$

But $P_{Cu} = I_2^2 R_2$ and P_{Fe} is given

$$\therefore \eta = \frac{17.5 \times 0.8}{17.5 \times 0.8 + P_{Cu} + \dfrac{300}{1000}} \times 100 \text{ per cent}$$

$$P_{Cu} = 152^2 \times 0.0158 = 365 \text{ W} = 0.365 \text{ kW}$$

$$\text{so } \eta = \frac{14}{14 + 0.365 + 0.3} \times 100 = \frac{14}{14.665} \times 100$$

or Efficiency $= 95.4$ per cent.

10. From O.C. test. Iron loss P_{Fe} $= 250$ W

From S.C. test. Copper loss P_{Cu} $= 500$ W

Percentage efficiency at $\frac{1}{2}$ full-load, 0.7 (lagging) power factor

$$= \frac{0.5 \times 50\,000 \times 0.7}{(0.5 \times 50\,000 \times 0.7) + 250 + \dfrac{500}{4}} \times 100$$

or $\eta = \dfrac{17\ 500}{17\ 500 + 250 + 125} \times 100 = \dfrac{17\ 500}{17\ 875} \times 100$

Thus Efficiency $= 97 \cdot 9$ per cent.

The percentage voltage regulation is given by

$$\dfrac{I_1(\bar{R}_1 \cos \phi \pm \bar{X}_1 \sin \phi)}{V_1} \times 100$$

Here $\bar{R}_1 = \dfrac{P_{Cu}}{I_1^2}$ and $I_1 = \dfrac{50\ 000}{440} = 113 \cdot 6$ A

$\bar{R}_1 = \dfrac{500}{113 \cdot 6^2} = 0 \cdot 0388\ \Omega \quad \bar{Z}_1 = \dfrac{V_{sc}}{I_{sc}} = \dfrac{25}{113 \cdot 6}$

$= 0 \cdot 2201\ \Omega$

$\bar{X}_1 = \sqrt{0 \cdot 2201^2 - 0 \cdot 0388^2} = \sqrt{0 \cdot 0484 - 0 \cdot 0015}$

$= \sqrt{0 \cdot 0469} = 0 \cdot 2161\ \Omega$

Voltage regulation $= \dfrac{113 \cdot 6\{(0 \cdot 0388 \times 0 \cdot 8) + (0 \cdot 2161 \times 0 \cdot 6)\}}{440}$

$= \dfrac{113 \cdot 6\ (0 \cdot 031 + 0 \cdot 1297)}{440} = \dfrac{113 \cdot 6 \times 0 \cdot 16}{440} = 0 \cdot 0413$

or Voltage Regulation $= 4 \cdot 13$ per cent (down).

CHAPTER 5

1. (a) Since $i = I(1 - e^{-Rt/L})$ then $i = \dfrac{V}{R}(1 - e^{-Rt/L})$

or $i = \dfrac{100}{10}(1 - e^{-\frac{10 \times 0.1}{10}}) = 10(1 - e^{-0.1})$

$\therefore \quad i = 10 - 10e^{-0.1}$

Let $x = 10e^{-0.1}$ Then $\log x = \log 10 - 0.1 \log e$

or $\log x = 1 - (0.1 \times 0.4343) = 1 - 0.04343$

$\qquad = 0.95657$ giving $x = 9.048$

Thus $i = 10 - 9.048 = 0.952$ A

(b) $5 = 10(1 - e^{-t})$ $\therefore \ 1 - e^{-t} = 0.5$

or $e^{-t} = 0.5$ giving $e^{t} = 2$

So $t \log e = \log 2$ and $t \times 0.4343 = 0.3010$

$\therefore \quad t = 0.693$ second.

2. Here $v = V e^{-t/CR}$ so $300 = 500\, e^{-30/20 \times 10^{-6} \times R}$

or $0.6 = e^{-1.5 \times 10^{6}/R}$ giving $\dfrac{1}{0.6} = e^{1.5 \times 10^{6}/R}$

Thus $\log 1 - \log 0.6 = \dfrac{1.5 \times 10^{6}}{R} \log e$

$0 - \bar{1}.7782 = \dfrac{1.5 \times 10^{6} \times 0.4343}{R}$

$R = \dfrac{0.65145 \times 10^{6}}{0.2218}$

$= 2.93 \times 10^{6}$ ohms or 2.93 MΩ.

3.   It must be pointed out that, although the discharge resistor is in parallel with the field when the circuit is connected normally, it is actually in series with the field coil when the supply is switched off.

Since $i = Ie^{-Rt/L}$ and $R = 50 + 50 = 100\ \Omega$

Then $i = \dfrac{200}{50}\ e^{-100t/10} = 4e^{-10t}$

Thus $\log i = \log 4 - 10t \log e$

$= 0 \cdot 6021 - 10 \times 0 \cdot 04 \times 0 \cdot 4343$

$= 0 \cdot 6021 - 0 \cdot 1737 = 0 \cdot 4284$

giving $i = 2 \cdot 681$ A.

4.   Since $e^{-t/CR}$ is common to two parts of the problem, solution is made easier by evaluating this first. Thus:

Let $x = e^{-t/CR}$ Then $x = e^{-0 \cdot 2/50 \times 10^{-6} \times 100 \times 10^{3}}$

$= e^{-0 \cdot 2/5} = e^{-0 \cdot 04}$

so $\log x = -0 \cdot 04 \times \log e$

$= -0 \cdot 04 \times 0 \cdot 4343 = \overline{1} \cdot 9826$

or $x = 0 \cdot 9607$

Substituting back $i = Ix = \dfrac{V}{R}\,x = \dfrac{200}{100 \times 10^{3}} \times 0 \cdot 9607$

or $i = 2 \times 10^{-3} \times 0 \cdot 9607$

$= 1 \cdot 9214$ mA

The voltage across the capacitor is given by

$v = V(1 - e^{-t/CR}) = V - Ve^{-t/CR}$

$= 200 - 200x$ or $v = 200 - (200 \times 0 \cdot 9607)$

So $v = 200 - 192 \cdot 14 = 7 \cdot 86$ V

The voltage across the capacitor can also be obtained thus:

Voltage drop across the resistor $= 1·9214 \times 10^{-3} \times 100 \times 10$

$= 192·14$

So voltage across capacitor $= 200 - 192·14 = 7·86$ V

5.   For an inductive circuit, the time constant is given by $\dfrac{L}{R}$

Thus $\dfrac{15}{R} = 2$ or $R = \dfrac{15}{2} = 7·5\ \Omega$

Again $i = I(1 - e^{-Rt/L})$

or $i = \dfrac{300}{7·5}\ (1 - e^{-7·5 \times 0·2/15}) = 40\ (1 - e^{-0·1})$

$= 40 - 40\ e^{-0·1}$ Let $x = 40\ e^{-0·1}$

$\therefore\ \log x = \log 40 - 0·1 \log e$

$= 1·6021 - 0·1 \times 0·4343$

$= 1·6021 - 0·04343 = 1·5587$

$x = 36·2$ So $i = 40 - 36·2 = 3·8$ A

If $t =$ time taken to reach half value

Then $20 = 40\ (1 - e^{-7·5\ t/15})$ or $\dfrac{1}{2} = 1 - e^{-t/2}$

So $\dfrac{1}{2} = e^{-t/2}$ and $2 = e^{t/2}$

Thus $\log 2 = \dfrac{t}{2} \log e$

or $0.301 = \dfrac{t}{2} \times 0.4343$ and $t = \dfrac{0.602}{0.4343}$

or $t = 1.386$ seconds.

Note that from work done, it has been shown that the current reaches half value in $0.693 \times$ time constant of the circuit.

Here therefore. $t = 0.693 \times 2 = 1.386$ s (as deduced above).

6. Since on discharge $v = Ve^{-t/CR}$

then after 5 seconds $v = 50\,e^{-5/1 \times 10^{-6} \times 5 \times 10^{6}} = 50e^{-1}$

$$\therefore \quad \log v = \log 50 - \log e$$
$$= 1.6990 - 0.4343 = 1.2647$$

so $\quad v = 18.4$ V

With two resistors in parallel the discharge circuit has a resistance of R. Where $\dfrac{1}{R} = \dfrac{1}{5} + \dfrac{1}{5} = \dfrac{2}{5}$ or $R = 2.5$ MΩ

$\therefore$ At the end of a further 5 seconds

$$v = 18.4\,e^{-5/1 \times 10^{-6} \times 2.5 \times 10^{6}} = 18.4e^{-2}$$

So $\quad \log v = \log 18.4 - 2 \log e$
$$= 1.2647 - 2 \times 0.4343 = 1.2647 - 0.8686$$
$$= 0.3961$$

or $\quad v = 2.49$ V

The capacitor current would be $\dfrac{2.49}{2.5 \times 10^{6}}$

$$= 1 \times 10^{-6} \text{ amperes or } 1\,\mu\text{A (approx)}.$$

7. (a) The "operate time" of the relay is when i reaches a value of 10×10^{-3} amperes.

Substituting in $i = I(1 - e^{-Rt/L})$

then $10 \times 10^{-3} = \dfrac{50}{2 \times 10^3} (1 - e^{-2\,000\,t/100})$

or $\dfrac{10 \times 10^{-3} \times 2 \times 10^3}{50} = (1 - e^{-20\,t})$ giving $\dfrac{2}{5} = 1 - \dfrac{1}{e^{20\,t}}$

Then $\dfrac{2}{5} - 1 = -\dfrac{1}{e^{20\,t}}$

and $-\dfrac{3}{5} = -\dfrac{1}{e^{20\,t}}$

or $e^{20\,t} = \dfrac{5}{3} = 1\cdot67$

Thus $20t \log e = \log 1\cdot67$

or $20t \times 0\cdot4343 = 0\cdot2227$

and $t = \dfrac{0\cdot2227}{8\cdot686} = 0\cdot0257$ seconds or $25\cdot7$ ms.

(b) When the supply is removed, a back e.m.f. is generated by the fall of linked flux in the relay coil. This causes a current to pass through the coil and the resistor. The current is given by

$$i = I\,e^{-Rt/L}$$

The release time of the relay is when i reaches a value of 1×10^{-3} amperes

Then $\quad 1 \times 10^{-3} = \dfrac{50}{2 \times 10^3} \; e^{-3000\,t/100}$

or $\quad \dfrac{1 \times 10^{-3} \times 2 \times 10^3}{50} = e^{-30\,t}$

Then $\quad \dfrac{1}{25} = \dfrac{1}{e^{30t}} \quad$ and $\quad e^{30t} = 25$

or $\quad 30\,t \log e = \log 25 \;$ giving $\; 30\,t \times 0{\cdot}4343 = 1{\cdot}3979$

$$\text{or } t = \dfrac{1{\cdot}3979}{13{\cdot}029} = 0{\cdot}1073 \;\text{seconds}$$

$$\text{or } 107{\cdot}3 \text{ ms.}$$

8. (a) Assume the capacitor to be uncharged then $i = \dfrac{V}{R}$ where V is the applied voltage. So $i = \dfrac{100}{10^2 \times 10^3}$ amperes

or initial current $= \dfrac{10^2}{10^2} \times 10^{-3}$ amperes $= 1 \text{ mA}$

(b) Let i be the charging current when $v = 63{\cdot}2 \text{ V}$
Then $\quad 100 - 63{\cdot}2 = i\,100\,000$

$$\text{or } i = \dfrac{36{\cdot}8 \times 10^{-3}}{100} = 0{\cdot}368 \text{ mA}$$

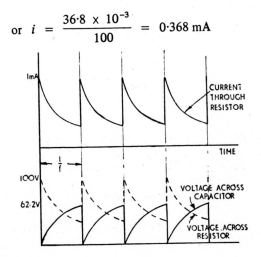

(c) After each instantaneous discharge of the capacitor, full voltage is applied to the circuit which is initially dropped wholly across R. As the charging current flows, it decreases because of the voltage rise across the capacitor terminals in accordance with the equation $v = V(1 - e^{-t/CR})$.

The time taken for the capacitor voltage to rise to 63·2 V is given by t

$$\therefore \ 63 \cdot 2 = 100 \, (1 - e^{-t/CR}) \ \text{ or } \ 0 \cdot 632 = 1 - e^{-t/2 \times 10^{-6} \times 10^{5}}$$

whence $0 \cdot 632 = 1 - e^{-t/0 \cdot 2}$

or $e^{-t/0 \cdot 2} = 0 \cdot 368$

$$\therefore \ e^{5t} = \frac{1}{0 \cdot 368} = 2 \cdot 715$$

or $5t \log e = \log 2 \cdot 715 = 0 \cdot 4338$

$$\therefore \ t = \frac{0 \cdot 4338}{5 \times 0 \cdot 4343} = \frac{1}{5} = 0 \cdot 2 \text{ second}$$

Here $\dfrac{1}{f} = 0 \cdot 2$ s or $f = \dfrac{1}{0 \cdot 2} = $ five sequences per second.

9. With d.c. and the rectifier conducting, the circuit current would be $\dfrac{2}{20} = 0 \cdot 1$ A or 100 mA

The moving-coil ammeter would read this and the thermo-couple ammeter, whose reading depends on the temperature difference between two junctions, which in turn is directly proportional to the heating effect of a current, will indicate a value dependent on the average heating power i.e. on its r.m.s. value. In this case it will read $\sqrt{100^2} = 100$ mA.

Thus as explained, the meter readings are identical. With a.c., only half-wave conduction occurs and the moving-coil instrument reads the average over a whole cycle even though current passes for a $\frac{1}{2}$ cycle only.

The peak value of the applied a.c. $= \dfrac{2}{0.707}$ volts

$$\text{and the current} = \frac{2}{20 \times 0.707}$$

$$= 141.4 \text{ mA}$$

The d.c. meter reads $\dfrac{1}{2}\left(\dfrac{2}{\pi} \times 141.4\right) = \dfrac{141.4}{3.14} \times 10^3 = 50 \text{ mA}$

The thermocouple meter would, as explained above, give a reading dependent on the average power. Thus for a full cycle, in this case, average power $= \frac{1}{2}$ the average power for a half-cycle.

So reading $= \dfrac{1}{2}$ r.m.s. value $= \dfrac{1}{2}\left(\dfrac{2}{20}\right) = \dfrac{1}{20} \times 10^3 = 50\text{mA}.$

10. Transformer.   A primary phase voltage $=$ 450 V
A secondary phase voltage $=$ 5 $\times$ 450
$=$ 2250 V

Rectifier. For a three-phase, full-wave rectifier $V_D = 1.35$ V

$\therefore V_D = 1.35 \times \sqrt{3}\ V_{ph}$ or $V_D = 2.34 \times 2250 = 5265$ V

Frequency of ripple $=$ six times that of the supply

$$= 6 \times 50 = 300 \text{ Hz.}$$

The voltage 30° on either side of the peak value is 0.866 times the maximum value. The voltage has thus dropped 13.4 per cent.

The value of the ripple voltage is thus $\dfrac{13.4}{2} = 6.7$ per cent.

or alternatively

Peak value of phase voltage $=$ $\sqrt{2} \times 2250$
Peak value of line voltage $=$ $\sqrt{2} \times \sqrt{3} \times 2250 = 5515$ V
Thus maximum value of rectified voltage $=$ 5515 V

and minimum value of rectified voltage $= 5515 \times \cos \dfrac{\pi}{6}$

$$= 5515 \cos 30$$
$$= 5515 \times 0\cdot866$$
$$= 4775 \text{ V}$$

or voltage dropped $= 5515 - 4775 = 740$ V peak to peak

$$= \frac{740}{5515} \times 100 = 13\cdot4 \text{ per cent.}$$

As before, ripple voltage $= \dfrac{13\cdot4}{2} = 6\cdot7$ per cent.

CHAPTER 6

1. For the voltage transformer, the appropriate ratio would be

   V.T. ratio $= \dfrac{6600}{110} = \dfrac{60}{1}$. The voltage will not vary appreciably

   and this ratio should be suitable.

   For the current transformer the appropriate ratio, to give

   full-scale deflection, would be C.T. ratio $= \dfrac{100}{5} = \dfrac{20}{1}$. Since

   current is a variable quantity depending on the load, in the opinion of the author, it is always safer to allow for overload conditions.

   Thus the full-load reading should not occur at full scale and this

   can be achieved by using a larger C.T. ratio. In this case $\dfrac{150}{5}$

   or $\dfrac{30}{1}$ would be suitable. Normal reading would then occur at

   about two-thirds full scale.

   By using a $\dfrac{30}{1}$ C.T. ratio, the wattmeter, being used on the

   5 A and 110 V ranges, must have its reading multiplied by the factors of 30 for the current and 60 for the voltage.

   The required constant is $30 \times 60 = 1800$.

2. Motor output $= 75$ kW

   Motor input $= \dfrac{75 \times 100}{80} = 93 \cdot 75$ kW

   or $\dfrac{93 \cdot 75}{0 \cdot 87} = 107 \cdot 76$ kVA

   The line current would be $\dfrac{107\,760}{\sqrt{3} \times 415} = \dfrac{107\,760}{718 \cdot 78}$

   $= 149 \cdot 92$ A

A 150/5 ratio current-transformer would be suitable but it is usual to allow a portion of the scale for noting starting and overload conditions and a 250/5 ratio C.T. would be the better proposition. If used, the meter reading would be $\dfrac{149 \cdot 92}{50}$, the $\dfrac{50}{1}$ C.T. ratio being taken into account.

Reading $= 3$ A for normal running conditions.

3. (a)　The three-phase apparent power $= \dfrac{12}{0 \cdot 8} = 15$ kVA

$$\text{The line current } = \frac{15\,000}{\sqrt{3} \times 415} = \frac{1000}{1 \cdot 732 \times 27 \cdot 67}$$

$$= \frac{1000}{47 \cdot 82} = 20 \cdot 91 \text{ A}$$

The wattmeter reading would be secondary C.T. current × phase voltage × load power factor.

Since the C.T. ratio $= \dfrac{25}{5}$ or $\dfrac{5}{1}$

$$\therefore \text{ reading } = \frac{20 \cdot 91}{5} \times \frac{415}{\sqrt{3}} \times 0 \cdot 8 = 800 \text{ W}$$

Note. The result could be arrived at, in a more direct fashion, since the power per phase $= \dfrac{12}{3} = 4$ kW or 4000 W

The C.T. ratio $= \dfrac{5}{1}$ $\quad \therefore$ reading $= \dfrac{4000}{5} = 800$ W

It is pointed out that, since the load is balanced and star-connected, the voltage-coil of the wattmeter is connected between a line and the star-point. The wattmeter is therefore subjected to $\dfrac{415}{\sqrt{3}}$ volts, *i.e.* a value within the rated range.

(b) The impedance per phase $= \dfrac{415}{\sqrt{3} \times 20 \cdot 91} = \dfrac{239 \cdot 6}{20 \cdot 91}$

$\qquad\qquad\qquad\qquad\qquad\quad = 11 \cdot 48 \ \Omega$

(c) The equivalent resistance $= 11 \cdot 48 \times 0 \cdot 8 = 9 \cdot 184 \ \Omega$
The equivalent reactance $= 11 \cdot 48 \times 0 \cdot 6 = 6 \cdot 888 \ \Omega$.

4.   Since the power is being measured by one wattmeter, it is evident that a neutral or star-point is available or has been artificially created and the total power will be three times the power measured — which is phase power.

Although phase voltage is applied to the instrument, reduced current in the ratio of $\dfrac{25}{5}$ or $\dfrac{5}{1}$ is being applied. $\therefore$ The wattmeter constant for the arrangement would be five and phase power would be $7 \times 5 = 35$ kW.

The total three-phase power $= 3 \times 35 = 105$ kW.

5.   Ratio of C.T.'s $= 25/5$ or $5/1$

Ratio of V.T.'s $= \dfrac{440}{110}$ or $4/1$

Ammeter voltage drop at 5 A $= 5 \times 0 \cdot 08 = 0 \cdot 4$ V
Ammeter burden $= 5 \times 0 \cdot 4 = 2$ VA

Wattmeter voltage drop at 5 A $= 5 \times 0 \cdot 1 = 0 \cdot 5$ V
Wattmeter burden per current coil $= 5 \times 0 \cdot 5 = 2 \cdot 5$ VA

$\therefore$ C.T. burden $= 2 + 2 \cdot 5 = 4 \cdot 5$ say 5 VA. This is a minimum value, a transformer capable of a larger burden say 10 VA would give greater accuracy on overload and would allow for extension of instrumentation should this be required.

Voltmeter current $= \dfrac{120}{3636} = \dfrac{10 \times 10^3}{303} = 33$ mA

Voltmeter burden $= 120 \times 33 \times 10^{-3} = 3 \cdot 3 \times 1 \cdot 2$
$\qquad\qquad\qquad = 3 \cdot 96 \, \text{VA}$

Wattmeter voltage-coil current $\qquad = \dfrac{120}{4000} = \dfrac{12 \times 10^3}{400}$

$\qquad\qquad\qquad\qquad\qquad\qquad = 30 \, \text{mA}$

Wattmeter burden per voltage coil $= 30 \times 10^{-3} \times 120$
$\qquad\qquad\qquad\qquad\qquad\qquad = 3 \times 1 \cdot 2 = 3 \cdot 6 \, \text{VA}$

$\therefore$ V.T. burden $= 3 \cdot 96 + 3 \cdot 6 = 7 \cdot 56$ say 10 VA. Here again, a transformer rating greater than the burden is usual.

6. Transformation ratio $\qquad = \dfrac{500}{400} = k = \dfrac{5}{4}$

Secondary current $\qquad = 100 \, \text{A}$

and since primary kVA $=$ secondary kVA

$\therefore \dfrac{500 \times I_1}{1000} = \dfrac{400 \times 100}{1000}$ or $I_1 = \dfrac{400 \times 100}{500} = \dfrac{400}{5}$

$\qquad\qquad\qquad\qquad\qquad\qquad\qquad\qquad\quad = 80 \, \text{A}$

Thus the single section of winding carries 80 A and the common section of winding carries $(100 - 80) = 20$ A

Now $\dfrac{\text{Wt of Cu in auto-transformer}}{\text{Wt of Cu in double-wound transformer}} = 1 - \dfrac{1}{k}$

$= 1 - \dfrac{1}{\dfrac{5}{4}} = 1 - \dfrac{4}{5} = \dfrac{1}{5}$

i.e. $\dfrac{4}{5}$ or 80 per cent of copper is saved.

7. Normal current rating of L.T. winding $= \dfrac{6000}{150} = $ 40 A

Normal current rating of H.T. winding $= \dfrac{6000}{250} = $ 24 A

With the L.T. winding connected in series with the H.T. winding so that the latter forms the common section, the output current would be $40 + 24 = $ 64 A. Thus the maximum load

kVA rating $= \dfrac{250 \times 64}{1000} = $ 16 kVA. A load of 12 kVA could

therefore be supplied, since the current rating of the windings would be in direct proportion to the kVA values.

Thus for 12 kVA, current in common section $= \dfrac{12}{16} \times 24$

$= $ 18 A

and for 12 kVA, current in series section $= \dfrac{12}{16} \times 40$

$= $ 30 A

Load between tapping point and common line

$= \dfrac{250 \times 48}{1000} = $ 12 kVA.

8. Transformer rating $= \dfrac{400}{0 \cdot 8} = $ 500 kVA

Primary current $= \dfrac{500\ 000}{\sqrt{3} \times 440} = $ 656 A

Secondary current $= \dfrac{500\ 000}{\sqrt{3} \times 550} = $ 525 A

Current in common section of a phase winding $= 656 - 525$

$= $ 131 A

Current in section of a phase winding supplying 550 V

$= $ 525 A.

9.   The most usual method of using a three-phase auto-transformer is to have the phase windings connected in star. The catch in this problem is the fact that, a delta connection is used and for such an arrangement, the secondary line voltage is not 70 per cent of the primary voltage, when the 70 per cent tapping is used. This is illustrated by the diagrams and solution

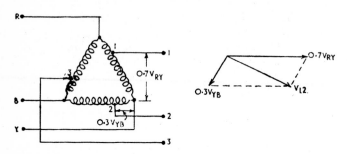

The phasor diagram shows the output voltage between lines as given by the solution

$$V_{1-2}^2 = \left\{(0{\cdot}7^2 + 0{\cdot}3^2) - (2 \times 0{\cdot}7 \times 0{\cdot}3 \times \cos 60°)\right\} V_L^2$$

Note here V_L is the line voltage $= V_{RY} = V_{YB}$

$$\therefore\ V_{1-2}^2 = \left\{(0.49 + 0.09) - (0.42 \times 0.5)\right\} V_L^2$$

$$= (0{\cdot}58 - 0{\cdot}21) V_L^2$$

or $V_{1-2} = \sqrt{0{\cdot}37}\ V_L = 0{\cdot}61\ V_L$ (approx.)

The voltage applied to the motor terminals would be

$$0{\cdot}61 \times 440 = 268{\cdot}4\ V.$$

10.   Motor power output $= 40\ 000\ W$

Motor power input $= \dfrac{40\ 000}{0{\cdot}82}$ watts

Motor three-phase rating $= \dfrac{40\ 000}{0{\cdot}82 \times 0{\cdot}85}$ volt amperes

$$\text{Full-load current} \ = \ \frac{40\ 000}{\sqrt{3}\ \times\ 0{\cdot}82\ \times\ 0{\cdot}85\ \times\ 440}\ =\ 75{\cdot}3\ \text{A}$$

Direct-on starting current $=\ 75{\cdot}3\ \times\ 6\ =\ 452$ A

As already noted for a star-connected arrangement, the output voltage is proportional to the transformer tapping. Thus the output voltage $=\ 75$ per cent of the input voltage, and in consequence the starting current will be only 75 per cent of the direct-on starting value because of the reduced impressed voltage on the motor.

Thus starting current $=\ 0{\cdot}75\ \times\ 452\ =\ 339$ A

Again due to transformer action and the fact that
Sec $kVA\ =\ $ Prim kVA, the primary or supply current is 75 per cent of the secondary current $=\ 0{\cdot}75 \times 339$ amperes.

Thus supply current $=\ 254$ A

Note. This latter part of the question is better answered with a knowledge of induction motor theory and is fully dealt with in the appropriate section of this book. Here we can say that, torque is proportional to the voltage squared and therefore, if the voltage is reduced to a $\dfrac{3}{4}$ value then, the torque will be

reduced to $\left(\dfrac{3}{4}\right)^{2}\ =\ \dfrac{9}{16}$ of the starting value $=\ \dfrac{9}{16}\ \times\ 1{\cdot}5$

of the full load value $=\ 0{\cdot}5625\ \times\ 1{\cdot}5\ \times\ $ full-load value
$$=\ 0{\cdot}844\ \times\ \text{full-load value.}$$

1.   A synchronous motor rotates at the speed at which the magnetic field rotates. The speed of this magnetic field, as for the synchronous speed of an alternator is decided by the frequency f of the supply and the number of poles P of the machine.

Thus for the motor $f = \dfrac{PN}{120}$ or $P = \dfrac{120f}{N} = \dfrac{120 \times 50}{120}$

$$= 50 \text{ poles}$$

For the alternator $P = \dfrac{120f}{N} = \dfrac{120 \times 50}{1000} = 6 \text{ poles.}$

2.   For 50 Hz working. $N = \dfrac{120f}{P} = \dfrac{120 \times 50}{8} =$

$$= 15 \times 50$$

$$= 750 \text{ rev/min}$$

For 60 Hz working, since speed is proportional to frequency

then new speed $= 750 \times \dfrac{60}{50} = 15 \times 60 = 900 \text{ rev/min}$

For constant excitation, generated e.m.f. is proportional to

speed. New voltage $= 440 \times \dfrac{900}{750} = 440 \times \dfrac{6}{5} = 44 \times 12$

$$= 528 \text{ V.}$$

3.   AC is the resultant of two e.m.f.'s displaced by $30°$ (as illustrated by the diagram).

Then $\dfrac{AX}{AB} = \cos 15°$

so $AX = AB \cos 15 = 20 \times 0{\cdot}9659$

or $AX = 19{\cdot}318$ V

$\therefore AC = 2 \times 19{\cdot}318 = 38{\cdot}64$ V.

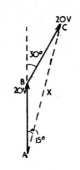

4. Breadth factor $K_D = \dfrac{\sin \dfrac{n\alpha}{2}}{n \sin \dfrac{\alpha}{2}}$

Where $n =$ no of slots/pole/phase.

Here $n = 72/6/3 = \dfrac{72}{18} = 4 \qquad \alpha = \dfrac{180}{12} = 15^{\circ}$

$\therefore K_D = \dfrac{\sin \dfrac{4 \times 15}{2}}{4 \sin \dfrac{15}{2}} = \dfrac{\sin 30}{4 \sin 7{\cdot}5}$

So $K_D = \dfrac{0{\cdot}5}{4 \times 0{\cdot}1305} = \dfrac{0{\cdot}5}{0{\cdot}522}$

or $K_D = 0{\cdot}958.$

5.   A pole spans $\dfrac{96}{8}$ = 12 slots   Full pitch being slot 1 to slot 13

A coil span covers 1 to 9 = 9 − 1 = 8 slots
If x = the coil pitch as some fraction of the full pitch, then

$$x = \frac{8}{12} = \frac{2}{3}$$

Coil pitch = $180 \times \dfrac{2}{3}$ = 120°

∴ θ = 180 − 120 = 60°

and K_S = $\cos \dfrac{60}{2}$ = $\cos 30°$

or K_S = 0·866.

6. $$P = \frac{120f}{N} = \frac{120 \times 50}{750} = \frac{120}{15} = \frac{40}{5}$$

= 8 poles

Voltage per phase = $\dfrac{3300}{\sqrt{3}}$ = 1905 V

Slots/pole/phase = $\dfrac{120}{8 \times 3}$ = 5 = n and $\alpha = \dfrac{180}{15}$ = 12°

Also $K_D = \dfrac{\sin \dfrac{n\alpha}{2}}{n \sin \dfrac{\alpha}{2}} = \dfrac{\sin \dfrac{5 \times 12}{2}}{5 \sin 6}$

$= \dfrac{\sin 30}{5 \sin 6} = \dfrac{0 \cdot 5}{5 \times 0 \cdot 1045}$

or $K_D = \dfrac{0 \cdot 5}{0 \cdot 5225} = 0 \cdot 956$

Thus $1905 = 2 \cdot 22 \times 0 \cdot 956 \times Z_{ph} \times 50 \times 0 \cdot 0448$

or $Z_{ph} = \dfrac{1905}{1 \cdot 11 \times 0 \cdot 956 \times 4 \cdot 48}$

$= 401 \cdot 2$ conductors per phase

The slots/phase $= \dfrac{120}{3} = 40$

$\therefore$ Conductors/slot $= \dfrac{401}{40} = 10$ (approx.)

7. Here $x = \dfrac{3}{4}$ and $\theta = 180 - \left(180 \times \dfrac{3}{4}\right)$

$= 180 - 135 = 45°$

$$\text{and} \quad K_S = \cos \frac{45}{2} = \cos 22 \cdot 5°$$

$$\text{or} \quad K_S = 0 \cdot 9239$$

Also since this is a mesh-connected machine then, e.m.f. per phase = 3300 V

As all other factors are constant, then

$$\frac{E_1}{E_2} = \frac{K \times 401 \cdot 2 \times 1}{K \times Z_{ph_2} \times 0 \cdot 924} \quad \text{or} \quad Z_{ph_2} = \frac{401 \cdot 2 \times 3300}{0 \cdot 924 \times 1905}$$

$$\text{Thus} \quad Z_{ph_2} = \frac{401 \cdot 2 \times 1 \cdot 732}{0 \cdot 924} = 751 \text{ conductors per phase}$$

$$\text{So conductors per slot} = \frac{751}{40} = 18 \cdot 8 \text{ say } 19.$$

8. Power output of motor   = 10 000 kW

$$\text{Input power to motor} = \frac{10\ 000 \times 100}{90} = 11\ 111 \text{ kW}$$

$$\text{Apparent power input} = \frac{kW}{\cos \phi} = \frac{11\ 111}{0 \cdot 95} = 11\ 696 \text{ kVA}$$

$$\text{Thus} \qquad 11\ 696 = \frac{\sqrt{3} \times 3300 \times I}{1000}$$

$$\text{or} \quad I = \frac{11\ 696}{\sqrt{3} \times 3 \cdot 3} = 2046 \cdot 3 \text{ A}$$

Therefore line current of system $= 2046 \cdot 3$ A

Also alternator phase current $= 2046 \cdot 3$ A

$$\text{The motor phase current} = \frac{2046 \cdot 3}{\sqrt{3}} = 1182 \text{ A}$$

Output kVA rating of alternator $=$ Input kVA rating of motor

$\dot{=}\ 11\ 696$ kVA

$$\text{or alternator output} = \frac{\sqrt{3} \times 3300 \times 2046 \cdot 3}{1000}$$

$$= \sqrt{3} \times 3 \cdot 3 \times 2046 \cdot 3$$

$$= 11\ 696 \text{ kVA}$$

Motor speed 110 rev/min.   Number of poles $= 50$

$$\therefore \text{ frequency of supply} = \frac{PN}{120} = \frac{50 \times 110}{120} = \frac{550}{12}$$

$$= 45 \cdot 8 \text{ Hz}$$

For the alternator, since $f = 45 \cdot 8$ Hz

$$\text{Then } 45 \cdot 8 = \frac{2 \times N}{120}$$

$$\text{or } N = 45 \cdot 8 \times 60 = 2748 \text{ rev/min.}$$

9. For a constant excitation, $V = 45$ volts and $I_{SC} = 30$ amperes

$$\text{Then } Z_S = \frac{45}{30} = 1 \cdot 5 \ \Omega$$

$$\therefore X_S = \sqrt{1 \cdot 5^2 - 0 \cdot 4^2} = \sqrt{2 \cdot 25 - 0 \cdot 16}$$
$$= \sqrt{2 \cdot 09} = 1 \cdot 44 \ \Omega.$$

10.   For this machine $I_L = \dfrac{kVA \text{ rating}}{\sqrt{3} \times \text{line voltage}} = \dfrac{750\ 000}{\sqrt{3} \times 3300}$

$$= \dfrac{2500}{1 \cdot 732 \times 11} = \dfrac{2500}{19 \cdot 05} = 131 \cdot 23 \text{ A}$$

Using the phasor diagram and working in phase values.

$$E_{ph} = \sqrt{(V_{ph} \cos \phi + IR)^2 + (V_{ph} \sin \phi + IX_S)^2}$$

V_{ph} here, is phase voltage $= \dfrac{3300}{\sqrt{3}} = 1905 \text{ V}$

$$\therefore E_{ph} = \sqrt{\{(1905 \times 0 \cdot 8) + (131 \cdot 23 \times 0 \cdot 5)\}^2}$$
$$\overline{\phantom{= \sqrt{}} + \{(1905 \times 0 \cdot 6) + (131 \cdot 23 \times 3)\}^2}$$

$$= \sqrt{(1524 + 65 \cdot 62)^2 + (1143 + 393 \cdot 69)^2}$$
$$= \sqrt{(1589 \cdot 6^2 + 1536 \cdot 7^2} = 10^3 \sqrt{1 \cdot 59^2 + 1 \cdot 537^2}$$
$$= 10^3 \sqrt{2 \cdot 53 + 2 \cdot 37} = 10^3 \sqrt{4 \cdot 9}$$
$$= 2 \cdot 214 \times 10^3 = 2214 \text{ V}$$

Voltage regulation $= \dfrac{2214 - 1905}{1905} = \dfrac{309}{1905}$

Thus voltage regulation $= 0 \cdot 162 \times 100$

$$= 16 \cdot 2 \text{ per cent.}$$

CHAPTER 8

1. Since $r_a = \dfrac{\delta V_a}{\delta I_a} = \dfrac{129 - 75}{(22 - 12) \times 10^{-3}} = \dfrac{54}{10} \times 10^3$

$= 5 \cdot 4 \times 10^3$ ohms

A.C. resistance of valve $= 5 \cdot 4 \text{ k}\Omega.$

2. For the 12 mA current condition

D.C. resistance $= \dfrac{75}{12 \times 10^{-3}} = 6 \cdot 25 \times 10^3 = 6 \cdot 25 \text{ k}\Omega.$

For the 22 mA current condition

D.C. resistance $= \dfrac{129}{22 \times 10^{-3}} = \dfrac{64 \cdot 5}{11} \times 10^3 = 5 \cdot 86 \text{ k}\Omega.$

3. The characteristic (shown overleaf) appears to be straight line between the 10·5 and 4·2 mA values and over this region the a.c. resistance would be

$r_a = \dfrac{\delta V_a}{\delta I_a} = \dfrac{25 - 15}{(10 \cdot 5 - 4 \cdot 2) 10^{-3}} = \dfrac{10 \times 10^3}{6 \cdot 3} = 1 \cdot 59 \text{ k}\Omega.$

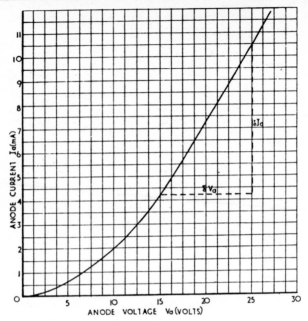

4.   The characteristic (shown overleaf) is plotted as shown and the load line is drawn in thus

Assume $I_a = 0$. Then voltage on anode would be 60 V to give point A.

Assume the valve resistance to be of minimum value. The anode current would then be $\dfrac{60}{300} = \dfrac{1}{5} = 0.2$ A or 200 mA.

Point B is thus obtained and AB gives the load line for a load-resistance value of 300 Ω. The point of intersection with the characteristic is at point P and the standing current is 100 mA.

The power dissipated in the load resistor is given by $I_a^2 R$

$$= (100 \times 10^{-3})^2 \times 300$$

$$= 10^{-2} \times 300 = 3 \text{ W}$$

or alternatively, from the load line for point P, $V_a = 30$ V

∴ Voltage dropped across load resistor $= V - V_a = 60 - 30$
$$= 30 \text{ V}$$

Thus power dissipated $=$ voltage across resistor × anode current
$$= 30 \times 100 \times 10^{-3} = 3 \text{ W}.$$

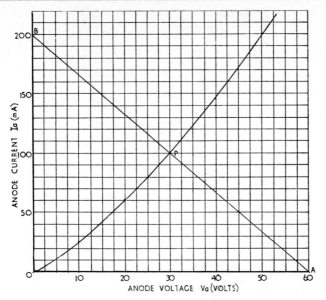

5. Since $I_a = V_a^{3/2}$ then a table of results can be compiled.

Anode voltage	0	1	2	3	4	5	(V)
Anode current	0	1	$\sqrt{8}$	$\sqrt{27}$	$\sqrt{64}$	$\sqrt{125}$	(mA)
Anode current	0	1	2·83	5·2	8	11·2	(mA)
Load voltage	0	1	2·83	5·2	8	11·2	(V)
Supply voltage	0	2	4·83	8·2	12	16·2	(V)

Specimen calculation.

$I_a = V_a^{3/2}$ So $I_a = 2^{3/2} = (2^3)^{1/2} = \sqrt{8} = 2\cdot83$ mA

The supply voltage V for this condition

$\qquad = $ load voltage drop + valve voltage drop

$\qquad = I_a R + V_a = 2\cdot83 \times 10^{-3} \times 10^3 + 2$

or $V = 2\cdot83 + 2 = 4\cdot83$ volts.

The deduced valve static and dynamic characteristics are plotted as shown overleaf and, from the latter, the value of anode current for an applied voltage of 8 V is seen to be 5·1 mA.

Note that this result could have been obtained by drawing a load line on the static characteristic for a resistor of 1 kΩ and a supply voltage of 8 V. This is shown dotted and the current is seen to be 5·1 mA. For this problem, since the dynamic characteristic has been deduced the load-line method is unnecessary.

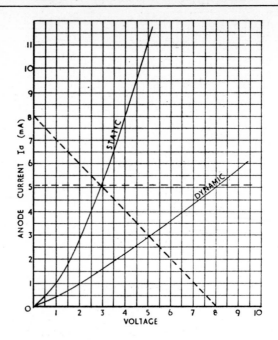

6. Since $g_m = \dfrac{\delta I_a}{\delta V_g}$ for a constant anode voltage, then

$$g_m = \frac{(17 \cdot 7 - 12)\, 10^{-3}}{-1 \cdot 5 - (-3)} = \frac{5 \cdot 7 \times 10^{-3}}{1 \cdot 5} = \frac{11 \cdot 4}{3} \times 10^{-3}$$

$$= 3 \cdot 8 \text{ mA/V.}$$

7. From the mutual characteristic (page 566), since

$$g_m = \frac{\delta I_a}{\delta V_g}, \text{ then for a constant anode voltage of 200 V}$$

$$g_m = \frac{(8 - 3) \times 10^{-3}}{-1 - (-2)} = \frac{5 \times 10^{-3}}{1} = 5 \text{ mA/V}$$

Also $r_a = \dfrac{\delta V_a}{\delta I_a}$ then, for a constant grid voltage of -1 V

$$r_a = \frac{200 - 140}{(8 - 3) \times 10^{-3}} = \frac{60}{5 \times 10^{-3}} = 12 \text{ k}\Omega$$

Also $\mu = \dfrac{\delta V_a}{\delta V_g}$ then, for a constant anode current of 3 mA

$$\mu = \frac{200 - 140}{-2 - (-1)} = 60$$

From the anode characteristic (shown overleaf), since

$$r_a = \frac{\delta V_a}{\delta I_a} \text{ then, for a constant grid voltage of } -1 \text{ V}$$

$$r_a = \frac{200 - 140}{(8 - 3) \times 10^{-3}} = \frac{60}{5 \times 10^{-3}} = 12 \text{ k}\Omega$$

Also $g_m = \dfrac{\delta I_a}{\delta V_g}$ then, for a constant anode voltage of 200 V

$$g_m = \frac{(8 - 3) \times 10^{-3}}{-1 - (-2)} = \frac{5 \times 10^{-3}}{1} = 5 \text{ mA/V}$$

Also $\mu = \dfrac{\delta V_a}{\delta V_g}$ then, for a constant anode current of 3 mA

$$\mu = \frac{200 - 140}{-1 - (-2)} = \frac{60}{1} = 60.$$

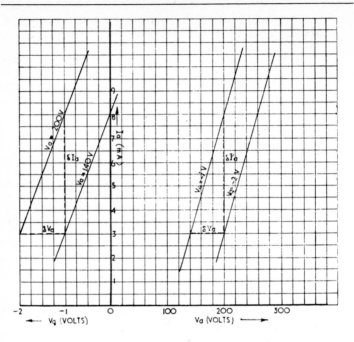

8.   Let A be a point on the $V_g = 0$ anode characteristic curve which must pass through the origin. If V_a is given an assumed value of — say 100 V and r_a is 20 kΩ then since

$$r_a = \frac{\delta V_a}{\delta I_a} \text{ , it follows that}$$

$$\delta I_a = \frac{\delta V_a}{r_a} = \frac{100}{20 \times 10^3} = 5 \times 10^{-3} \text{ or } \delta I_a = 5 \text{ mA}$$

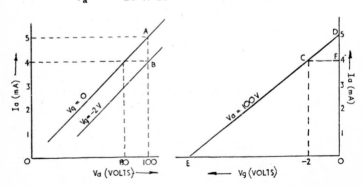

Point A can now be plotted and the $V_g = 0$ characteristic drawn, as is shown by the diagram on the left. A change of 100 V would result in a change of 5 mA.

The $V_g = -2$ V curve will be parallel to the $V_g = 0$ line and since $\mu = \dfrac{\delta V_a}{\delta V_g}$, its value can be found thus:

If a constant current value, say 4 mA is assumed, then on the $V_g = 0$ curve, for a constant current of 4 mA,

$$V_a = 100 \times \frac{4}{5} = 80 \text{ V}$$

From the data given in the question, point B can be shown and $\mu = \dfrac{100 - 80}{2 - 0} = \dfrac{20}{2} = 10$, for a constant current value of 4 mA.

"Cut-off" can be determined from the mutual-characteristic diagram on right, which is deduced thus:

When $V_g = -2$ volts then $I_a = 4$ milliamperes on the $V_a = 100$ volt line. Point C is shown. Point D is similarly obtained:

When $V_g = 0$ volts and $V_a = 100$ volts, then I_a is obtained from

$$r_a = \frac{\delta V_a}{\delta I_a} \text{ or } \delta I_a = \frac{\delta V_a}{r_a} = \frac{100}{20 \times 10^3} = 5 \times 10^{-3} \text{ A}$$

$$= 5 \text{ mA}$$

As deduced earlier, a change of 100 V would result in a change of 5 mA. Cut-off value OE is obtained from similar triangles

$$\frac{DF}{DO} = \frac{CF}{EO} \text{ or } EO = \frac{DO \times CF}{DF} = \frac{5 \times 2}{1} = 10 \text{ V}.$$

9.   After drawing in a suitable triangle such as ABC, we can deduce for a -1 V grid-voltage value

$$r_a = \frac{\delta V_a}{\delta I_a} = \frac{BC}{AB} = \frac{260 - 190}{(17 - 10)10^{-3}} = \frac{70 \times 10^3}{7} = 10 \text{ k}$$

For a 260 V anode-voltage value

$$g_m = \frac{\delta V_a}{\delta V_g} = \frac{AB}{\text{grid-voltage change}}$$

$$= \frac{(17 - 10)10^{-3}}{-1 - (-2)} = \frac{7 \times 10^{-3}}{1} = 7 \text{ mA/V}$$

For a 7 mA anode-current change

$$\mu = \frac{\delta V_a}{\delta V_g} = \frac{260 - 190}{-1 - (-2)} = \frac{70}{1} = 70$$

Check $\mu = r_a \times g_m = 10 \times 10^3 \times 7 \times 10^{-3} = 70$.

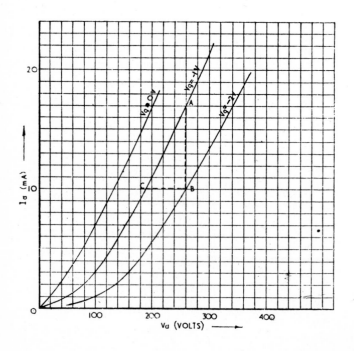

Ω

10.   Draw in a suitable triangle such as ABC
Then for the 200 V anode condition

$$g_m = \frac{\delta I_a}{\delta V_g} = \frac{(15\cdot5 - 10)\,10^{-3}}{-1-(-2\cdot6)} = \frac{5\cdot5 \times 10^{-3}}{1\cdot6} = 3\cdot4 \text{ mA/V}$$

Also for a -1 V grid condition

$$r_a = \frac{\delta V_a}{\delta I_a} = \frac{200 - 150}{(15\cdot5-10)10^{-3}} = \frac{50 \times 10^3}{5\cdot5} = 9\cdot1 \times 10^3$$

$$= 9\cdot1 \text{ k}\Omega$$

Again for a 5·5 mA anode-current change

$$\mu = \frac{\delta V_a}{\delta V_g} = \frac{200 - 150}{-1-(-2\cdot6)} = \frac{50}{1\cdot6} = 31$$

Check

$$\mu = r_a \times g_m = 9\cdot1 \times 10^3 \times 3\cdot4 \times 10^{-3} = 30\cdot94 = 31.$$

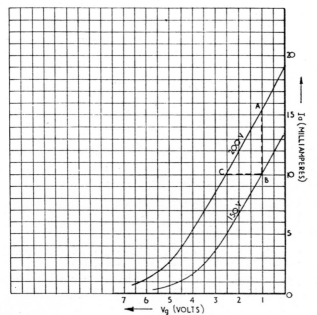

CHAPTER 9

1. The original speed of the machine is determined from $f = \dfrac{PN}{120}$

or $N_1 = \dfrac{f120}{P} = \dfrac{50 \times 120}{4} = 50 \times 30 = 1500$ rev/min

Also, for the generation of e.m.f., E is proportional to flux per pole and speed.

The flux per pole is in turn proportional to the field current

∴ $E \propto NI_f$ where I_f is the field current or $E = KNI_f$

Thus $\dfrac{E_1}{E_2} = \dfrac{KN_1 I_{f_1}}{KN_2 I_{f_2}}$ or $E_2 = \dfrac{E_1 N_2 I_{f_2}}{N_1 I_{f_1}} = \dfrac{200 \times 1200 \times 3}{1500 \times 4}$

∴ $E_2 = \dfrac{200 \times 4 \times 3}{5 \times 4} = 120$ V.

2. Since $E = \sqrt{(V \cos \phi + IR)^2 + (V \sin \phi + IX_S)^2}$

$= \sqrt{\{(500 \times 0.8) + (50 \times 0.2)\}^2 + \{(500 \times 0.6) + (50 \times 2.2)\}^2}$

$= \sqrt{(400 + 10)^2 + (300 + 110)^2} = \sqrt{410^2 + 410^2}$

$= 10^2 \sqrt{4.1^2 + 4.1^2} = 10^2 \sqrt{16.81 + 16.81}$

$= 10^2 \sqrt{33.62} = 10^2 \times 5.799$

$= 579.9$ or 580 V

Thus generated e.m.f. $= 580$ V.

3. Here $Z_S = \dfrac{1500}{250} = 6\Omega$ $X_S = \sqrt{6^2 - 2^2} = \sqrt{32} = 5.65\ \Omega$

Also $E = \sqrt{\{(6600 \times 0.8) + (250 \times 2)\}^2 + \{(6600 \times 0.6) + (250 \times 5.65)\}^2}$

$$= \sqrt{(5280 + 500)^2 + (3960 + 1412 \cdot 5)^2}$$
$$= \sqrt{5780^2 + 5372 \cdot 5^2} = 10^3\sqrt{5 \cdot 78^2 + 5 \cdot 37^2}$$
$$= 10^3\sqrt{33 \cdot 4 + 28 \cdot 8} = 10^3\sqrt{62 \cdot 2} = 7890 \text{ V}$$

Generated voltage = 7·89 kV.

4. Full-load line current of the alternator $= \dfrac{250 \times 1000}{\sqrt{3} \times 3000}$

$$\text{or } I = 48 \text{ A}$$

Since the machine is star connected, then $I_{\text{ph}} = 48$ A

The resistance per phase of the alternator $= \dfrac{2 \cdot 8}{2} = 1 \cdot 4 \ \Omega$

The synchronous reactance is 20 Ω per phase and the full-load

phase voltage is $\dfrac{3000}{\sqrt{3}} = 1731$ V

Open-circuit phase voltage E_{ph} when a full-load, lagging current at 0·8 p.f. is switched off, is given by

$$E_{\text{ph}} = \sqrt{\{(1731 \times 0 \cdot 8) + (48 \times 1 \cdot 4)\}^2 + \{(1731 \times 0 \cdot 6 \atop + (48 \times 20)\}^2}$$
$$= \sqrt{(1384 \cdot 8 + 67 \cdot 2)^2 + (1038 \cdot 6 + 960)^2}$$
$$= \sqrt{1452^2 + 1998 \cdot 6^2} = 10^3\sqrt{1 \cdot 452^2 + 1 \cdot 9986^2}$$
$$= 10^3\sqrt{2 \cdot 11 + 3 \cdot 99} = 10^3\sqrt{6 \cdot 1} = 2470 \text{ V}$$

Open-circuit line voltage when load is switched off

$$= \sqrt{3} \times 2470 = 4280 \text{ V}$$

For the 0·9 (leading) power factor condition

$$E_{\text{ph}} = \sqrt{\{(1731 \times 0 \cdot 9) + (48 \times 1 \cdot 4)\}^2 + \{(1731 \times 0 \cdot 4357) \atop - (48 \times 20)\}^2}$$

$$= \sqrt{(1557{\cdot}9 + 67{\cdot}2)^2 + (754{\cdot}2 - 960)^2}$$

$$= 10^3\sqrt{1{\cdot}625^2 + 0{\cdot}2058^2} \;=\; 10^3\sqrt{2{\cdot}64 + 0{\cdot}04}$$

$$= 10^3 \times 1{\cdot}637 \;=\; 1637\ \text{V}$$

Open-circuit line voltage $= \sqrt{3} \times 1637 = 2835\ \text{V}$.

5. From the short-circuit characteristic (S.C.C.) it is seen that a field current of 3·45 A is required to produce I viz a full-load current of 39 A. Note that $I = \dfrac{9000}{230} = 39{\cdot}1\ \text{A}$

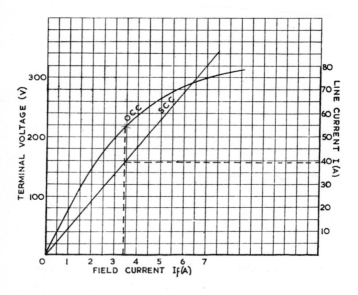

From the open-circuit characteristic (O.C.C.), a field current of 3·45 A produces a terminal voltage of 218 V.

From the above, the following can be deduced

$$\text{Synchronous Impedance } Z_s = \frac{218}{39} = 5{\cdot}6\ \Omega$$

$$\therefore X_s = \sqrt{5{\cdot}6^2 - 0{\cdot}25^2} = \sqrt{31{\cdot}36 - 0{\cdot}06} = \sqrt{31{\cdot}3}$$

$$= 5{\cdot}59\ \Omega$$

Thus $E = \sqrt{\{(230 \times 0\cdot8) + (39 \times 0\cdot25)\}^2 + \{(230 \times 0\cdot6) + 39 \times 5\cdot59)\}^2}$

$= \sqrt{(184 + 9\cdot75)^2 + (138 + 218)^2}$

$= \sqrt{193\cdot75^2 + 356^2} = 10^2\sqrt{1\cdot94^2 + 3\cdot56^2}$

$= 10^2\sqrt{3\cdot76 + 12\cdot67} = 10^2\sqrt{16\cdot43} = 10^2 \times 4\cdot055$

$= 405\cdot5$ V

Voltage regulation $= \dfrac{405\cdot5 - 230}{230} = \dfrac{175\cdot5}{230} = 0\cdot763$

Thus voltage regulation $= 76\cdot3$ per cent.

6.   If the resistance voltage drop is to be taken into account
then the voltage resistive component $= V + IR \cos \phi$

$= 230 + 39 \times 0\cdot25 \times 0\cdot8$

$= 230 + 7\cdot8 = 237\cdot8$ V

and the voltage reactive component $= IX \sin \phi$

$= 39 \times 0\cdot25 \times 0\cdot6 = 5\cdot85$ V

$\therefore E = \sqrt{237\cdot8^2 + 5\cdot85^2} = 10\sqrt{23\cdot78^2 + 0\cdot585^2}$

$= 10\sqrt{565\cdot5 + 0\cdot342} = 10\sqrt{565\cdot84} = 10^2\sqrt{5\cdot658}$

$= 10^2 \times 2\cdot38 = 238$ V

From the graph of the previous question it is seen that, the
field current required to produce this voltage on open circuit
$= 4\cdot05$ A. This is F of the m.m.f. diagram (Fig. 175a and b).

The field current required to produce the full-load current of
39 A on short circuit $= 3\cdot45$ A. This is $(F_a + F_x)$ of the m.m.f.
diagram.

Then from $F_o{}^2 = F^2 + (F_a + F_x)^2 + 2F(F_a + F_x)$
$\cos (90 - \phi)$

we have $F_o{}^2 = 4\cdot05^2 + 3\cdot5^2 + (2 \times 4\cdot05 \times 3\cdot5 \sin \phi)$

$= 16\cdot4 + 12\cdot25 + (8\cdot1 \times 3\cdot5 \times 0\cdot6)$

$= 16\cdot4 + 12\cdot25 + 17\cdot01 = 45\cdot66$

Thus $F_o = \sqrt{45\cdot66} = 6\cdot75$ A

Using the graph the generated voltage for 6·75 A would be 293 V.

Thus voltage regulation $= \dfrac{293 - 230}{230} \times 100 = \dfrac{63}{2·3}$

$= 27·4$ per cent.

Note the wide difference between the regulation figures as given by Examples 5 and 6. The usual procedure is to take the average value as given by both methods of prediction. This result would be nearer the true value of voltage regulation. Thus, for the examples being considered

Voltage regulation $= \dfrac{76·3 + 27·4}{2} = \dfrac{103·7}{2} = 51·8$ per cent

7.   The voltage per phase from the given line values are

(V)   0   75   146   194   227   254   276   302   315

Full-load current of the machine $= \dfrac{500 \times 10^3}{\sqrt{3} \times 440} = 655$ A

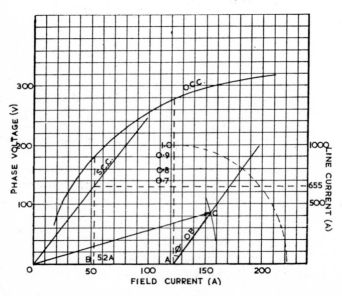

Using the Synchronous Impedance Method, we see that, from the S.C. characteristic curve, a field current of 52 A is required to produce 655 A line current. Also, from the O.C.C., a field current of 52 A produces an e.m.f. of 175 V/ph.

Then Z_S per phase $= \dfrac{175}{655} = 0\cdot267\ \Omega$

$\quad R$ per phase $= 0\cdot01\ \Omega$

$\therefore \quad X_S$ per phase $= \sqrt{0\cdot267^2 - 0\cdot01^2} = 0\cdot267\ \Omega$

Then open-circuit e.m.f. per phase is given by E where

$E = \sqrt{(V\cos\phi + IR)^2 + (V\sin\phi + IX_S)^2}$

or $E = \sqrt{\{(254\times0\cdot8) + (655\times0\cdot01)\}^2 \times \{(254\times0\cdot6) + 655\times0\cdot267)\}^2}$

$\quad = \sqrt{(203\cdot2 + 6\cdot55)^2 + (152\cdot4 + 175\cdot6)^2}$

$\quad = 10^2\sqrt{4\cdot393 + 10\cdot72} = 388\cdot7\ \text{V}$

O.C. terminal voltage $= \sqrt{3} \times 388\cdot7 = 673\ \text{V}$

Using the Ampere-Turn Method, since R per phase $= 0\cdot01\ \Omega$, the open-circuit e.m.f. required, if the resistance voltage drop to be accounted for, is $V + IR\cos\phi$. Thus: $254 + (655\times0\cdot01\times0\cdot8) = 254 + 5\cdot24 = 259\cdot24\ \text{V}$.

Note. The reactance voltage drop would be

$$IX\sin\phi = 655 \times 0\cdot267 \times 0\cdot6 = 105\ \text{V}$$

$\therefore\ E = \sqrt{259^2 + 105^2} = 10^2\sqrt{2\cdot59^2 + 1\cdot05^2}$

$\quad = 10^2\sqrt{6\cdot7 + 1\cdot1} = 10^2\sqrt{7\cdot8} = 2\cdot793 \times 10^2$

$\qquad\qquad\qquad\qquad\qquad\text{or } E = 279\cdot3\ \text{V}$

From the O.C.C., the field current required to produce this voltage (on open circuit) $= 123\ \text{A}$. If the graphical method of solution is applied, this field current is shown as OA.

The field current required to produce full-load current on short circuit $= 52\ \text{A}$, shown by OB.

The field current required to produce full-load current, at normal voltage, is the vector sum of these values displaced by

$(90 + \phi)$ degrees *i.e.* O.C. From the graph, OC = 160A and this current produces an O.C. phase voltage of 302 V. Thus when load is switched off, the terminal voltage will rise to $\sqrt{3} \times 302 = 523$ V

A reasonable value of voltage would be the average given by

$$\frac{673 + 523}{2} = 598 \text{ V}.$$

8. The voltage per phase from the given line values are
    (V)   90   166   229   254   274   306   328   342

Full-load current $= \dfrac{30 \times 10^3}{\sqrt{3} \times 440} = 39.4$ A

Full-load phase voltage $= \dfrac{440}{\sqrt{3}} = 254$ V

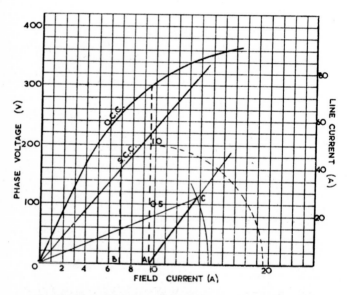

Synchronous Impedance Method.
     From the S.C. characteristic, a field current of 6·95 A is required to produce 39·4 A.

From the O.C.C., a field current of 6·95 A produces an e.m.f. of 252 V/ph.

From the above can be deduced

$$Z_S \text{ per phase} = \frac{252}{39\cdot4} = 6\cdot396\ \Omega \quad R \text{ per phase} = 0\cdot15\ \Omega$$

$$\therefore \quad X_S = \sqrt{6\cdot396^2 - 0\cdot15^2} = 6\cdot395\ \Omega$$

Open-circuit e.m.f./phase is given by E, where

$$
\begin{aligned}
E &= \sqrt{(V\cos\phi + IR)^2 + (V\sin\phi + IX_S)^2} \\
&= \sqrt{\{(254 \times 0\cdot8) + (39\cdot4 \times 0\cdot15)\}^2 + \{(2\cdot54 \times 0\cdot6) + (39\cdot4 \times 6\cdot395)\}^2} \\
&= \sqrt{(203\cdot2 + 5\cdot91)^2 + (152\cdot4 + 252)^2} = \sqrt{209\cdot1^2 + 404\cdot4^2} \\
&= \sqrt{43710 + 163\,500} = \sqrt{207\,210} = 455\cdot2\ \text{V}
\end{aligned}
$$

$$\text{Voltage regulation} = \frac{455\cdot2 - 254}{254} \times 100 = 79\cdot21 \text{ per cent.}$$

Ampere-turn Method.

The open-circuit e.m.f., if the resistance voltage drop is considered $= V + IR\cos\phi = 254 + (39\cdot4 \times 0\cdot15 \times 0\cdot8)$
$$= 258\cdot8\ \text{V}$$

The reactance voltage drop would be $IX\sin\phi$
$$= 39\cdot4 \times 6\cdot395 \times 0\cdot6 = 151\cdot2\ \text{V}$$

$$
\begin{aligned}
\text{Thus } E &= \sqrt{258\cdot8^2 + 151\cdot2^2} = 10^2\sqrt{2\cdot59^2 + 1\cdot51^2} \\
&= 10^2\sqrt{6\cdot7 + 2\cdot28} = 10^2\sqrt{8\cdot98} = 10^2 \times 2\cdot997 \\
&= 299\cdot7\ \text{V.}
\end{aligned}
$$

Using the Graphical Method of solution, it is seen from the diagram that, the field current required to produce a voltage of 299·7 V is OA = 9·6 A.

The field current required to produce full-load current on short-circuit = 6·95 A. This is OB.

By constructing a power-factor quadrant and marking off, at point A, a length equal to OB, point C is obtained.

From the graph OC = 15·8 A and this field current produces a phase voltage of 350 V.

$$\therefore \text{ Voltage regulation } = \frac{350 - 254}{254} \times 100 = \frac{96}{2 \cdot 54}$$

$$= 37 \cdot 8 \text{ per cent}$$

$$\text{Average voltage regulation } = \frac{79 \cdot 2 + 37 \cdot 8}{2} = \frac{117}{2}$$

$$= 58 \cdot 5 \text{ per cent.}$$

9.   Since $P = \dfrac{120f}{N} = \dfrac{120 \times 50}{1500} = 4$ poles, the total number

of electrical degrees around the circumference $= \dfrac{4}{2} \times 360$

$= 720°$, and the total number of mechanical degrees round the circumference $= 360°$. Thus for this machine,

$$1 \text{ mechanical degree } = \frac{720}{360} = 2 \text{ electrical degrees}$$

The angle of displacement $\alpha = 2$ electrical degrees

$$\text{Synchronising current } \quad I_S = \frac{E_r}{2X_{S/ph}}$$

$$\text{Phase voltage } = \frac{3300}{\sqrt{3}} = 1905 \text{ V}$$

$$E_r = 2E_2 \sin \frac{\alpha}{2} = 2 \times 1905 \sin \frac{\alpha}{2}$$

$$= 2 \times 1905 \times 0 \cdot 0175 = 66 \cdot 7 \text{ V}$$

$$\text{Synchronising current } I_S = \frac{66 \cdot 7}{2 \times 5} = 6 \cdot 67 \text{ A}$$

$$\text{Synchronising power (per phase) } = E_2 I_S \cos \frac{\alpha}{2} \text{ watts}$$

$$= \frac{1905 \times 6 \cdot 67}{1000} \times \cos \frac{2}{2} \text{ kilowatts}$$

$$= \frac{1905 \times 6 \cdot 67}{1000} \times 0 \cdot 9998 = 12 \cdot 7 \text{ kW}$$

Total synchronising power = 3 × 12·7 = 38·1 kW

Also since $\dfrac{2\pi NT}{60}$ = output power, then

$$T = \frac{38\ 100 \times 60}{2\pi \times N}\ \text{newton metres}$$

or $T = \dfrac{381 \times 60 \times 100}{2 \times 3·14 \times 1500} = 242·7\ \text{Nm.}$

10.   1 pole-pitch = 45 mechanical degrees = 180 electrical degrees.

∴ 1 mechanical degree $= \dfrac{180}{45}$ = 4 electrical degrees and

the angle of displacement α = 4 electrical degrees.

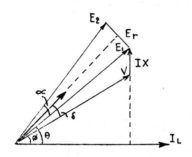

V = Alternator terminal voltage. E_1 and E_2 = Alternator generated voltage before and after displacement. E_r = Resultant voltage between alternator and busbar after phase displacement. I = Load current. I_S = Synchronising current. $\cos \phi$ = Power factor of alternator. α = Angle of displacement.

Full-load current $I = \dfrac{2000 \times 10^3}{\sqrt{3} \times 6000} = 192·4\ \text{A}$

$$\cos \phi = 0{\cdot}8 \quad \phi = 36°51' \quad \sin \phi = 0{\cdot}6$$

Here

$$
\begin{aligned}
E_1 \text{ or } E_2 &= \sqrt{(V \cos \phi)^2 + (V \sin \phi + IX_S)^2} \\
&= \sqrt{(6000 \times 0{\cdot}8)^2 + \{(6000 \times 0{\cdot}6) + (192{\cdot}4 \times 6)\}^2} \\
&= \sqrt{(4800)^2 + (4754)^2} = 10^3\sqrt{23{\cdot}03 + 22{\cdot}6} \\
&= 10^3\sqrt{45{\cdot}63} = 10^3 \times 6{\cdot}754 = 6754 \text{ V}
\end{aligned}
$$

$$\sin \theta = \frac{V \sin \phi + IX}{E_1} = \frac{4754}{6754} = 0{\cdot}7039. \therefore \theta = 44°45'$$

After angular displacement and using phase values

$$\text{Resultant voltage } E_r = 2E_2 \sin \frac{\alpha}{2} = 2 \times \frac{6754}{\sqrt{3}} \times \sin \frac{4}{2}$$

$$= \frac{13\,508}{\sqrt{3}} \times 0{\cdot}0349 = \frac{471{\cdot}4 \text{ V}}{\sqrt{3}} = 272{\cdot}2 \text{ V}$$

$$\text{Synchronising current } I_S = \frac{E_r}{X_S} = \frac{272{\cdot}2}{6} = 45{\cdot}36 \text{ A}$$

$$
\begin{aligned}
\text{Synchronising power per phase} &= E_2 I_S \cos \frac{\alpha}{2} \text{ watts} \\
&= \frac{6754}{\sqrt{3}} \times 45{\cdot}36 \times \cos 2° \\
&= 3900 \times 45{\cdot}36 \times 0{\cdot}999 \\
&= 3{\cdot}9 \times 45{\cdot}36 \times 0{\cdot}999 \\
&= 176{\cdot}7 \text{ kW}
\end{aligned}
$$

Total three-phase synchronising power $= 3 \times 176{\cdot}7 = 530 \text{ kW}$.

CHAPTER 10

1. Given $A = \dfrac{\mu R}{r_a + R}$ and $\mu = r_a g_m$

$\therefore \quad A = 16 \times 10^3 \times 2.5 \times 10^{-3}$

$\qquad = 16 \times 2.5 = 40$

Whence $A = \dfrac{40 \times 80 \times 10^3}{(16 \times 10^3) + (80 \times 10^3)} = \dfrac{40 \times 80}{16 + 80} = \dfrac{3200}{96}$

$\qquad = \dfrac{100}{3}$ Thus $A = 33.3$.

2. Using the equivalent circuit method of solution, as shown by Fig. 194 and neglecting C and C_S we have R_e as the effective load resistance.

$R_e = \dfrac{R R_g}{R + R_g} = \dfrac{40 \times 100}{40 + 100} = \dfrac{4000}{140} = \dfrac{200}{7} = 28.6 \text{ k}\Omega$

whence Stage Gain $A = \dfrac{\mu R_e}{r_a + R_e} = \dfrac{25 \times 28.6 \times 10^3}{(10 \times 10^3) + (28.6 \times 10^3)}$

$\qquad = \dfrac{25 \times 28.6}{10 + 28.6} = \dfrac{7.15 \times 10^2}{38.6} = \dfrac{715}{38.6}$

or $A = 18.5$.

3. Since $A = \dfrac{E_a}{E_g}$ or more correctly $\dfrac{\delta E_a}{\delta E_g}$

also $A = \dfrac{\mu R}{r_a + R}$

Then $\dfrac{\delta E_a}{\delta E_g} = \dfrac{\mu R}{r_a + R} = \dfrac{\mu r_a R}{r_a(r_a + R)}$

or $\delta E_g = \dfrac{\mu}{r_a} \times \dfrac{r_a R}{r_a + R} \times \delta E_g$

giving $3 = \dfrac{12}{10\,000} \times \dfrac{10\,000 \times 5000}{10\,000 + 5000} \times \delta E_g$

$= \dfrac{12 \times 5000}{15\,000} \times \delta E_g$ or $3 = 4\delta E_g$

$\therefore \quad \delta E_g = \dfrac{3}{4} = 0.75 \text{ V}$

Thus input voltage $= 0.75 \text{ V (r.m.s.)}$

4. Since $A = \dfrac{\mu R}{r_a + R}$

We have the two equations

$$10 = \dfrac{\mu\,10}{r_a + 10} \quad \text{and} \quad \dfrac{40}{3} = \dfrac{\mu\,20}{r_a + 20}$$

or $10\,\mu = 10(r_a + 10)$ and $60\,\mu = 40(r_a + 20)$

$\therefore \quad 10\,\mu = 10 r_a + 100$ or $60\,\mu = 60 r_a + 600$ ... ... (a)
and $60\,\mu = 40\,(r_a + 20)$ or $60\,\mu = 40 r_a + 800$ ... ... (b)

Subtracting (b) from (a) $0 = 20 r_a - 200$

or $20\,r_a = 200$ and $r_a = 10 \text{ k}\Omega$

Substituting back $10\,\mu = (10 \times 10) + 100$

$\text{or } 10\,\mu = 200 \text{ giving } \mu = 20.$

5. Given $r_a = 2 \text{ k}\Omega$ and $\mu = 7$ Also $\mu = g_m r_a$

$$\therefore g_m = \frac{\mu}{r_a} = \frac{7}{2 \times 10^3} = 3 \cdot 5 \times 10^{-3} \text{ or } 3 \cdot 5 \text{ mA/V}$$

Power dissipated in anode-load resistor is given by $I^2 R$
Thus one watt $= I^2 \times 3000$

Note. Here I is the r.m.s. value of the output current.

$$\therefore I^2 = \frac{1}{3000} \text{ or } I = \frac{1}{\sqrt{30}} \times \frac{1}{10} \text{ amperes}$$

giving $I = \dfrac{1}{5 \cdot 47} \times \dfrac{1}{10} \times 10^3 = \dfrac{100}{5 \cdot 47} = 18 \cdot 3 \text{ mA}$

Output voltage developed across the load resistor, by this
current, is $E_a = 18 \cdot 3 \times 10^{-3} \times 3 \times 10^3 = 18 \cdot 3 \times 3$
$= 54 \cdot 9 \text{ V (r.m.s.)}$

Now $A = \dfrac{\mu R}{r_a + R} = \dfrac{7 \times 3000}{2000 + 3000} = \dfrac{21\ 000}{5000} = 4 \cdot 2$

Also $A = \dfrac{E_a}{E_g}$ $\therefore$ $E_g = \dfrac{E_a}{A} = \dfrac{54 \cdot 9}{4 \cdot 2} = 13 \text{ V (r.m.s.)}$

Thus input voltage must be 13 V.

6. Anode slope resistance $r_a = \dfrac{\delta V_a}{\delta I_a}$, for a constant V_g

From the graph (overleaf), using given values for convenience, then
for a constant V_g, a δV_a of $(175 - 150) = 25 \text{ V}$

The corresponding δI_a is $(6 \cdot 2 - 4 \cdot 3) = 1 \cdot 9 \text{ mA}$

$$\therefore r_a = \frac{25}{1 \cdot 9 \times 10^{-3}} = 13\ 150 \text{ ohms} = 13 \cdot 15 \text{ k}\Omega$$

Mutual conductance $(g_m) = \dfrac{\delta I_a}{\delta V_g}$, for constant V_a

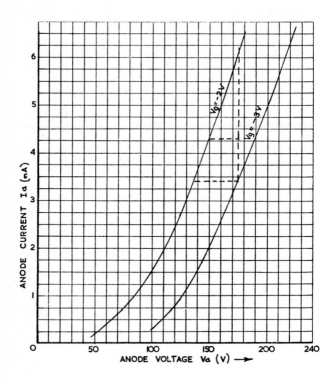

From the graph and using given values, for an anode voltage of 175 V, a change of grid voltage (δV_g) from -2 V to -3 V causes a change in anode current (δI_a) from 6·2 mA to 3·4 mA. This is $6·2 - 3·4 = 2·8$ mA.

$$\therefore g_m = \frac{2·8}{1} = 2·8 \text{ mA/V}$$

Thus amplification factor $\mu = \dfrac{\delta V_a}{\delta V_g}$, for a constant I_a

From the graph, a change in anode voltage (δV_a) from 138 V to 175 V changes the anode current from 3·4 mA to 6·2 mA when $V_g = -2$ volts. A change in V_g to -3 V then reduces the current back to 3·4 mA when $V_a = 175$ volts.

$\therefore \; \delta V_a \; = \; 175 \; - \; 138 \; = \; 37 \text{ volts and } \delta V_g \; = \; -2 - (-3)$
$$= 1 \text{ volt}$$

$$\therefore \; \mu \; = \; \frac{37}{1} \; = \; 37$$

Check. Since $\mu \; = \; r_a \times g_m \; = \; 13\cdot15 \; \times \; 10^3 \; \times \; 2\cdot4 \; \times \; 10^{-3}$
$$= \; 13\cdot15 \; \times \; 2\cdot8 \; = \; 36\cdot8, \text{ say } 37$$

Note. Since the h.t. voltage is not given, the load-line technique cannot be used and stage gain deduced. The graphical method is thus not readily adaptable and the equivalent circuit method is more straight-forward.

From the derived circuit. $I \; = \; \dfrac{\mu E_g}{r_a + R} \;$ and $\; E_a \; = \; IR$

$$\therefore \; I \; = \; \frac{37 \times E_g}{13\ 150 \; + \; 75\ 000} \; \text{ and } \; E_a \; = \; \frac{37 \times E_g \times 75\ 000}{88\ 150}$$

$$\text{or } \; E_a \; = \; \frac{37 \times 75 \times E_g}{88\cdot15}$$

also $\; A \; = \; \dfrac{E_a}{E_g} \; = \; \dfrac{37 \times 75}{88\cdot15} \; = \; 31\cdot5$

Thus Stage Gain $= \; 31\cdot5$

7. The load line is obtained as follows (characteristic overleaf).

When $I_a \; = \; 0 \quad V_a \; = \; 150 \; - $ a point on the line.

If the valve is assumed to have zero resistance, the anode current, $I_a \; = \; \dfrac{150}{20 \times 10^3} \; = \; 7\cdot5 \times 10^{-3} \; = \; 7\cdot5 \text{ mA}$

Hence, when $I_a \; = \; 7\cdot5 \text{ mA}, \; V_a \; = \; 0$

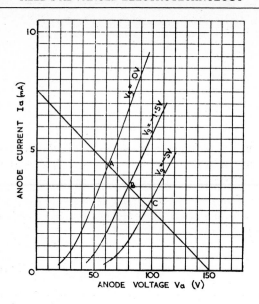

From the intersections of the load line with the three characteristic curves, the values of I_a are as follows

When $V_g = 0$, $I_a = 4\cdot4$ mA; when $V_g = -1\cdot5$, $I_a = 3\cdot5$ mA when $V_g = -3\cdot1$, $I_a = 2\cdot6$ mA.

Selecting B as the operating point, then

$$r_a = \frac{\delta V_a}{\delta I_a} = \frac{100 - 74}{5\cdot7 - 2\cdot9} = \frac{26}{2\cdot8} = \frac{13}{1\cdot4} = 9\cdot28 \text{ k}\Omega$$

$$g_m = \frac{\delta I_a}{\delta V_g} = \frac{5\cdot7 - 2\cdot9}{-1\cdot5 - (-3\cdot1)} = \frac{2\cdot8}{1\cdot6} = 1\cdot75 \text{ mA/V}$$

$$\mu = \frac{106 - 56}{3\cdot1} = \frac{50}{3\cdot1} = 16\cdot13$$

Check: $\mu = r_a g_m = 9\cdot28 \times 10^3 \times 1\cdot75 \times 10^{-3}$
    or $\mu = 16\cdot24$

Also $A = \dfrac{16 \cdot 2 \times 20\ 000}{(9 \cdot 28 + 20) \times 10^3} = \dfrac{16 \cdot 2 \times 20}{29 \cdot 28} = \dfrac{324}{29 \cdot 28}$

or $A = 11$ Thus Stage Gain $= 11$.

8.   This is a catch question and the hint for starting is to use the information that Class A working is desired. Straight line characteristics are assumed. Starting with the mutual characteristic and a 3 V (peak) swing on the grid, then point A is known.

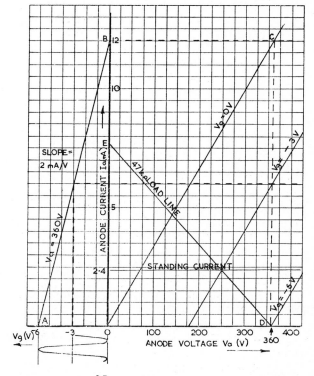

Since $g_m = \dfrac{\delta I_a}{\delta V_g}$ $\therefore$ the slope of the characteristic is given

and $\delta I_a = 2 \times \delta V_g = 2 \times 6 = 12$ mA – Point B.

Turning attention to the anode characteristic, we can assume that the $V_g = 0$ characteristic passes through the origin. The slope is known from $r_a = \dfrac{\delta V_a}{\delta I_a}$

Since $\mu = r_a g_m$ then $r_a = \dfrac{\mu}{g_m} = \dfrac{60}{2 \times 10^{-3}} = 30 \times 10^3$ or

$$= 30\ k\Omega$$

Also $r_a = \dfrac{\delta V_a}{\delta I_a}$. $\quad \therefore\ \delta V_a = 30 \times 10^3 \times 12 \times 10^{-3} = 360\ V$

Thus point C on the $V_g = 0$ characteristic is known and can be drawn in.

Next draw in the load line *i.e.* point D $= 360$ V and point E

$$= \dfrac{360}{47 \times 10^3} = 7\cdot66\ mA$$

Consideration of the anode characteristic shows that the other curves would be parallel, straight line and readily deducted. A line for $V_g = -6$ V is needed and must start at point D, since when $V_a = 360$ V, $I_a = 0$ for $V_g = -6$ V. The $V_g = -3$ V is parallel and mid-way between the others. Since, for Class A operation, the bias must be -3 V, then the standing current is 2·4 mA.

9. Point P is shown on the $V_g = -4$ V characteristic (overleaf). For $V_a = 200$ V, $I_a = 2\cdot5$ mA.

To determine the required parameters, draw triangle ABC about P so that B lies on the $V_g = -5$ V curve.

Then, for an anode voltage of 200 V

$$g_m = \dfrac{\delta I_a}{\delta V_g} = \dfrac{(3\cdot75 - 1\cdot4)\,10^{-3}}{-4 - (-5)} = 2\cdot35\ mA/V$$

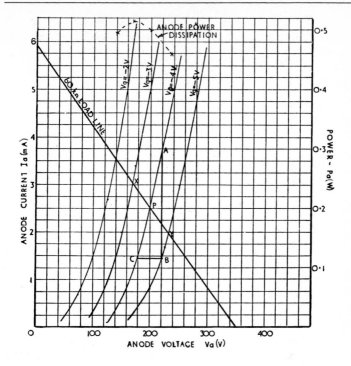

Also, for a current change of 2·35 mA

$$\mu = \frac{\delta V_a}{\delta V_g} = \frac{220 - 180}{-4 - (-5)} = 40$$

For the quiescent point P, anode current = 2·5 mA

∴ Voltage drop in load resistor = 2·5 × 10⁻³ × 60 × 10³
= 25 × 6 = 150 V

Voltage at anode = 200 V

∴ H.T. voltage = 200 + 150 = 350 V

The load line for the 60 kΩ resistor can be drawn in through points P and Q.

Using the intersections of the load line with the characteristics we see

(a) Anode voltage swings from 140 V to 230 V
$$= 90 \text{ V swing}$$

Anode current swings from 3·55 mA to 1·95 mA
$$= 1·6 \text{ mA swing}$$

(b) Voltage gain $= \dfrac{90}{-2 - (-5)} = \dfrac{90}{3} = 30$

(c) When $V_g = -2 \text{ V}$ $P_a = 140 \times 3·55 \times 10^{-3} = 0·497 \text{ W}$
$\phantom{\text{When }}$ $V_g = -3 \text{ V}$ $P_a = 170 \times 3·05 \times 10^{-3} = 0·519 \text{ W}$
$\phantom{\text{When }}$ $V_g = -4 \text{ V}$ $P_a = 200 \times 2·5 \times 10^{-3} = 0·5 \text{ W}$
$\phantom{\text{When }}$ $V_g = -5 \text{ V}$ $P_a = 230 \times 1·95 \times 10^{-3} = 0·45 \text{ W}$

These values are shown plotted to a base of anode voltage.
Assuming quiescent point P, then the input signal of 1 V (peak) will swing the working conditions along the load line between points X and Y.

Solution can be

A.C. power output $= \dfrac{(V_{a_{max}} - V_{a_{min}})\ (I_{a_{max}} - I_{a_{min}})}{8}$ milli-watts

where I_a is in milliamperes

$\therefore$ A.C. power ouput $= \dfrac{(230 - 170)\ (3·05 - 1·95)}{8}$

$$= \dfrac{60 \times 1·15}{8} = 11·5 \times \dfrac{3}{4} = 8·6 \text{ mW}$$

or alternatively

$$E_g = 0·707 \times 1 = 0·707 \text{ V (r.m.s.)}$$

Also $E_a = \mu E_g = \mu \times 0·707 = 40 \times 0·707 = 28·28 \text{ V}$

and $r_a = \dfrac{40}{2·35} = 17 \text{ k}\Omega$

$$\therefore \ I \ = \ \frac{28 \cdot 28}{(17 + 60)10^3} \ = \ \frac{28 \cdot 28}{77} \ = \ \frac{4 \cdot 04}{11} \ = \ 0 \cdot 367 \, \text{mA}$$

and $P \ = \ 0 \cdot 367^2 \times 10^{-6} \times 60 \times 10^3 \ = \ 0 \cdot 367^2 \times 6 \times 10^{-2}$
$\ = \ 0 \cdot 1347 \times 6 \times 10^{-2} \ = \ 8 \cdot 08 \times 10^{-3} \, \text{W} \ = \ 8 \cdot 1 \, \text{mW}.$

10. (a) Cathode current $\dfrac{3}{500} \ = \ \dfrac{6}{1000} \ = \ 6 \times 10^{-3}$ or $6 \, \text{mA}$

(b) Anode current $\dfrac{250 - 140}{22 \times 10^3} \ = \ \dfrac{110 \times 10^{-3}}{22} \ = \ \dfrac{10^{-2}}{2}$

$= \ \dfrac{10 \times 10^{-3}}{2} \ = \ 5 \, \text{mA}$

(c) Screen-grid current $= \ 6 - 5 \ = \ 1 \, \text{mA}$

(d) Screen-grid voltage $= \ 250 - (150 \times 10^3 \times 1 \times 10^{-3})$
$= \ 250 - 150 \ = \ 100 \, \text{V}$

(e) Control-grid voltage $= \ - 3 \, \text{V}$

Using the equation $A \ = \ \dfrac{\mu R}{(r_a + R)}$ we can modify it thus:

$$A \ = \ g_m \, r_a \, \frac{R}{(r_a + R)} \ = \ g_m \, \frac{R}{\dfrac{(r_a + R)}{r_a}} \ = \ g_m \, \frac{R}{1 + \dfrac{R}{r_a}}$$

Now for a pentode, the value of R is a fraction rather than a multiple of r_a. In Example 100, it was seen that r_a was some twenty times R and it follows that, for the above expression, if r_a is of this order, the value of the fraction $\dfrac{R}{r_a}$ reduces to become negligible (as a first approximation).

Thus we can write A $=$ g_m $\times$ R (approximately)

$$= 6 \times 10^{-3} \times 22 \times 10^3$$

$$= 132.$$

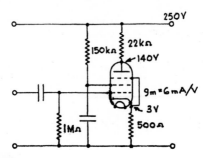

CHAPTER 11

1.   Rotor copper loss = slip × rotor input

Here slip $= \dfrac{3}{100}$ or $s = 0{\cdot}03$ Also,

        Rotor input = input to stator − stator losses
                    = 50 − 0·8 = 49·2 kW

∴ Rotor copper loss = 0·03 × 49·2 = 1·476 kW

        Rotor copper loss/phase $= \dfrac{1476}{3} = 492$ W

            Rotor output = rotor input − rotor copper loss
                       = 49·2 − 1·476 = 47·724 kW

Mechanical power developed = 47·7 kW

Note. Mechanical losses have been neglected.

2.   (a) Synchronous speed $N_1 = \dfrac{120f}{P} = \dfrac{120 \times 50}{6}$

                       = 1000 rev/min.

      Actual speed $N_2 = 990$ rev/min.

∴   Slip speed = 1000 − 990 = 10 rev/min.

or   Slip $= \dfrac{10}{1000} \times 100 = 1$ per cent.

   (b) Frequency of rotor e.m.f. = slip × supply frequency
                   = $s \times f$ = 0·01 × 50 = 0·5 Hz.

(c) Rotor copper loss = slip × rotor input power

But rotor input = rotor output + rotor losses

= (rotor output + rotor mechanical loss) + rotor copper loss

= 20 000 + x

where x = rotor copper loss

Note. 25 kW includes the mechanical loss.

Thus rotor copper loss = 0·01 (20 + x)

So x = 0·2 + 0·01 x

and 0·99 x = 0·2

∴ x = 0·202 kW or 202 W

(d) Input to motor = 20 + 0·202 + 2·5 = 22·7 kW

(e) Also $22·7 = \dfrac{\sqrt{3}\,VI\cos\phi}{1000}$

and $I = \dfrac{22·7 \times 1000}{\sqrt{3} \times 500 \times 0·85} = 30·84$ A.

3. $\dfrac{\text{Rotor copper loss}}{\text{Rotor output}} = \dfrac{s}{1-s} = \dfrac{0·04}{1-0·04} = \dfrac{0·04}{0·96} = \dfrac{1}{24}$

Rotor output (mechanical) = 15 kW

Rotor output (electrical) = 15 + 0·83 = 15·83 kW

Rotor copper loss = $\dfrac{15·83}{24}$ = 0·66 kW

Input to rotor = 15·83 + 0·66 = 16·49 kW

Output of stator $= 16.49$

Input to stator $= 16.49 + 0.95 = 17.44$ kW

Overall Efficiency $= \dfrac{15}{17.44} \times 100 = 86$ per cent.

4. (a) Synchronous speed $N_1 = \dfrac{120f}{P} = \dfrac{120 \times 50}{6}$

$\qquad = 1000$ rev/min.

Rotor frequency $f_2' = sf = \dfrac{90}{60} = 1.5$ Hz

Thus $s = \dfrac{f_2'}{f} = \dfrac{1.5}{50} = 0.03$

Rotor speed $N_2 = N_1 - sN_1 = 1000 - 0.03 \times 1000$

$\qquad = 1000 - 30 = 970$ rev/min

(b) Power output $= \dfrac{2\pi NT}{60} = \dfrac{2 \times 3.14 \times 970 \times 162.7}{60}$ watts

$\qquad = 16.518$ W

(c) $\dfrac{\text{Rotor copper loss}}{\text{Rotor output}} = \dfrac{s}{1-s} = \dfrac{0.03}{1-0.03} = \dfrac{0.03}{0.97}$

$\therefore$ Rotor output $= \dfrac{2 \times 3.14 \times 970 \times (162.7 + 13.56)}{60}$ watts

$\qquad = \dfrac{6.28 \times 97 \times 176.26}{6}$ watts

$$= 609 \times 29 \cdot 38 = 17\ 892 \cdot 4\ \text{W}$$

$$\text{Rotor copper loss} = \frac{17\ 892 \cdot 4 \times 3}{97} = 184 \cdot 46 \times 3$$

$$= 553 \cdot 4\ \text{W}$$

(d)   Motor input = rotor output (electrical)
              + rotor copper loss + stator loss

$$= (17\ 892 \cdot 4) + 553 \cdot 4 + 750$$

$$= 19\ 195 \cdot 8\ \text{W} = 19 \cdot 20\ \text{kW}$$

(e) Motor Efficiency $= \dfrac{16 \cdot 52}{19 \cdot 20} = 0 \cdot 86$

$$= 86\ \text{per cent.}$$

5. (a) Stator phase voltage $= 440\ \text{V}$

   Due to transformer action, the rotor-phase voltage

$$= \frac{440}{3} = 146 \cdot 7\ \text{V}$$

   At standstill rotor resistance per phase $= 0 \cdot 02\ \Omega$ and the starter resistance per phase $= 0 \cdot 25\ \Omega$

   Total rotor standstill resistance per phase $= 0 \cdot 02 + 0 \cdot 25$
   $$= 0 \cdot 27\ \Omega$$

   Rotor standstill reactance per phase $= 0 \cdot 27\ \Omega$

   $\therefore$ Total rotor circuit standstill impedance $= \sqrt{0 \cdot 27^2 + 0 \cdot 27^2}$
   $$= \sqrt{0 \cdot 1458} = 0 \cdot 382\ \Omega$$

   Rotor current per phase at standstill $= \dfrac{146 \cdot 7}{0 \cdot 382} = 384\ \text{A}$

By transformer action, stator current per phase $= \dfrac{384}{3} = 128$ A

∴ Line current taken from the supply at starting

$= \sqrt{3} \times 128 = 222$ A

(b) Under running conditions, the starting resistance is cut out and the slip is 4 per cent.

∴   Rotor-phase voltage   =   slip × standstill phase voltage
                                 = $0{\cdot}04 \times 146{\cdot}7 = 5{\cdot}87$ V

Rotor reactance per phase   =   slip × standstill reactance
                                   = $0{\cdot}04 \times 0{\cdot}27 = 0{\cdot}0108$ Ω

Rotor impedance per phase

$$= \sqrt{0{\cdot}02^2 + 0{\cdot}0108^2} = \sqrt{0{\cdot}0004 + 0{\cdot}000117}$$

$$= \sqrt{0{\cdot}000517} = 0{\cdot}0227 \ \Omega$$

Rotor current per phase $= \dfrac{5{\cdot}87}{0{\cdot}0227} = 259$ A

By transformer action, stator currents per phase $= \dfrac{259}{3}$

$$= 86{\cdot}3 \text{ A}$$

Line current taken from the supply $= \sqrt{3} \times 86{\cdot}3 = 149{\cdot}5$ A

At starting, with resistance cut out, rotor impedance per phase $= \sqrt{0{\cdot}02^2 + 0{\cdot}27^2} = \sqrt{0{\cdot}0004 + 0{\cdot}0729}$

$$= \sqrt{0{\cdot}0733} = 0{\cdot}271 \ \Omega$$

Rotor current per phase $= \dfrac{146{\cdot}7}{0{\cdot}271} = 541$ A

By transformer action, the stator currents per phase $= \dfrac{541}{3}$

$= 180{\cdot}3$ A and the line current taken from the supply
$= \sqrt{3} \times 180{\cdot}3 = 1{\cdot}732 \times 180{\cdot}3 = 312$ A

Note. The power rating given in the question is superfluous information.

6. Since $N_1 = \dfrac{120 f}{P}$ then $N_1 = \dfrac{120 \times 60}{12} = 600$ rev/min.

Slip $s = \dfrac{600 - 576}{600} = \dfrac{24}{600} = 0.04$

From torque expression as deduced

If T is the full-load torque, then $T = \dfrac{K\,440^2\,s\,R_2}{R_2^2 + (sX_2)^2}$

$$= \frac{K \times 440^2 \times 0.04 \times 0.02}{(0.02)^2 + (0.04 \times 0.27)^2} = \frac{K \times 4.4^2 \times 8}{0.0004 + 0.0108^2}$$

$$= \frac{K \times 4.4^2 \times 8}{0.0004 + 0.000117} = \frac{K \times 4.4^2 \times 8}{0.000517}$$

Again maximum torque T_m occurs when $R_2 = s_m \times X_2$

or $0.02 = s_m \times 0.27$ $\therefore$ $s_m = \dfrac{0.02}{0.27} = 0.074$

And $T_m = \dfrac{K \times 440^2 \times 0.074 \times 0.02}{0.02^2 + (0.074 \times 0.27)^2} = \dfrac{K \times 4.4^2 \times 14.8}{0.0004 + 0.0004}$

$$= \frac{K \times 4.4^2 \times 14.8}{0.0008}$$

Thus $\dfrac{T_m}{T} = \dfrac{K \times 4.4^2 \times 14.8}{0.0008} \times \dfrac{0.000517}{K \times 4.4^2 \times 8}$

$$= \frac{14.8 \times 5.17}{8 \times 8} = \frac{76.4}{64} = \frac{1.195}{1} \quad \text{or} \quad 1.2 \text{ to } 1$$

At maximum torque $s_m = 0 \cdot 074$ $\therefore$ $0 \cdot 074 = \dfrac{600 - N_2}{600}$

and $0 \cdot 074 \times 600 = 600 - N_2$

$\therefore$ Rotor speed $N_2 = 600 - 44 \cdot 4 = 555 \cdot 6$ rev/min.

7.   Stator phase voltage $= 440$ V

By transformer action, rotor-phase voltage $= \dfrac{440}{4} = 110$ V

Synchronous speed $N_1 = \dfrac{120 f}{P} = \dfrac{120 \times 50}{10} = 600$ rev/min

Slip at maximum torque $s_m = \dfrac{600 - 540}{600} = \dfrac{60}{600} \times 100$

$= 10$ per cent

Standstill reactance per phase $X_2 = \dfrac{R_2}{s_m} = \dfrac{0 \cdot 018}{0 \cdot 1} = 0 \cdot 18 \ \Omega$

For a full-load slip of 4 per cent, rotor current I_2 is given by

$$I_2 = \frac{s E_2}{\sqrt{R_2{}^2 + (s X_2)^2}} = \frac{0 \cdot 04 \times 110}{\sqrt{0 \cdot 018^2 + (0 \cdot 04 \times 0 \cdot 18)^2}}$$

$$= \frac{4 \cdot 4}{0 \cdot 0194} = \frac{440}{1 \cdot 94} = 227 \text{ A}$$

Rotor power factor $= \dfrac{R_2}{Z_2} = \dfrac{0 \cdot 018}{0 \cdot 0194} = 0 \cdot 925$ (lagging)

$$\frac{\text{Rotor copper loss}}{\text{Rotor output}} = \frac{s}{1 - s}$$

And rotor output $= \dfrac{\text{rotor copper loss} \times (1 - 0.04)}{0.04}$

Note. Phase values have been worked.

Rotor copper loss per phase $= 227^2 \times 0.018$ watts

$\therefore$ rotor output per phase $= \dfrac{227^2 \times 0.018 \times 0.96}{0.04}$

$\qquad\qquad = 227^2 \times 0.018 \times 24 = 22\,260$ W

Total rotor output $= 3 \times 22\,260 = 66\,780$ W

Hence shaft power $= 66.78$ kW

Note. Friction loss has been neglected.

Value of starting resistance/phase to give maximum torque at start $=$ rotor reactance/phase

Thus $R_2 = X_2 = 0.18\ \Omega$

$\therefore$ Starting resistance required per phase $= 0.18 - 0.018$
$\qquad\qquad\qquad\qquad\qquad\qquad\quad = 0.162\ \Omega.$

8. Power factor of no-load current $= \dfrac{2170}{\sqrt{3} \times 440 \times 19} = 0.15$

With full voltage applied to results of locked rotor test,

Standstill current $= \dfrac{440}{100} \times 70 = 308$ A

Power factor of this S.C. current $= \dfrac{3880}{\sqrt{3} \times 100 \times 70}$
$\qquad\qquad\qquad\qquad\qquad\quad = 0.32$ (lagging)

Current scale. 10mm $= 20$ A. (This was for the original diagram before photographic reduction.)

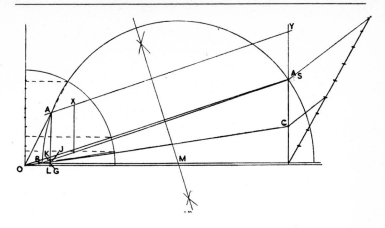

∴ Length of no-load current phasor = 9·5 mm
Length of standstill current phasor = 154 mm

Power component of output current = $\dfrac{50 \times 746}{\sqrt{3} \times 440}$ = 49 A

This is represented by a vertical phasor from the output line

= $\dfrac{49}{20}$ = 24·5 mm

On the circle diagram shown, line XY is drawn parallel to the output line BA$_S$, spaced 24·5 cm. The point of intersection with the circle is A. ∴ Input line current OA = 31 mm = 62 A. Power factor = 0·89 (from the power-factor quadrant).

Efficiency = $\dfrac{\text{Output}}{\text{Input}}$ = $\dfrac{\text{AK(mm)}}{\text{AL(mm)}}$ = $\dfrac{25}{28}$ = 0·8928

= 89·3 per cent.

Note. The copper losses have been divided in the ratio 4:5 by the simple geometrical construction shown *i.e.* DW is divided into nine parts and the appropriate parallel drawn to give point C.

$$\text{Slip} \ = \ \frac{KJ}{AJ} \ = \ \frac{10 \text{ mm}}{260 \text{ mm}} \ = \ 0.0385 \ = \ 3.85 \text{ per cent.}$$

9.   No-load test.

Current $= \ 10.5$ A

$$\text{Power factor} \ = \ \frac{1820}{\sqrt{3} \times 440 \times 10.5} \ = \ 0.228 \text{ (lagging)}$$

Standstill test.

If full voltage is applied, current $= \dfrac{440}{113} \times 50 = 195$ A

$$\text{Power factor} \ = \ \frac{3920}{\sqrt{3} \times 113 \times 50} \ = \ 0.4 \text{ (lagging)}$$

Current scale. 10 mm $=$ 10 A (on original diagram).

∴ Length of no-load current phasor $=$ 10.5 mm Length of standstill phasor $=$ 195 mm. The circle diagram is drawn in the accepted manner. Input line current is determined by finding the power component of the output current $= \dfrac{22\ 500}{\sqrt{3} \times 440} = 29.5$ A

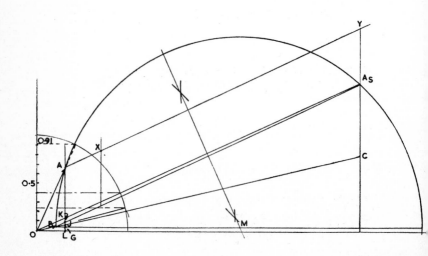

By drawing the line XY parallel to the output line and spaced 295 mm, we get point A and the output current, OA = 37·4 A.

Power factor = 0·91 (lagging), as read off the quadrant.

$$\text{Overall Efficiency} = \frac{AK}{AL} = \frac{28 \cdot 5 \text{ mm}}{33 \text{ mm}} = 0 \cdot 864$$

$$\text{or } 86 \cdot 4 \text{ per cent.}$$

$$\text{Slip} = \frac{KJ}{AJ} = \frac{1 \cdot 2 \text{ mm}}{30 \cdot 5 \text{ mm}} = 0 \cdot 0393 \times 100$$

$$= 3 \cdot 39 \text{ per cent.}$$

Developed torque is represented by AJ

Here AJ = 30·5 mm or 30·5 A

∴ Developed torque = $30 \cdot 5 \times \sqrt{3} \times 440$ synchronous watts

$$\text{or } T = \frac{30 \cdot 5 \times \sqrt{3} \times 440 \times 60}{2 \times 3 \cdot 14 \times N_1} \text{ newton metres}$$

Where N_1 is the synchronous speed of the motor

$$\text{Here } N_1 = \frac{120f}{P} = \frac{120 \times 60}{8} = 900 \text{ rev/min.}$$

$$\therefore \ T = \frac{30 \cdot 5 \times \sqrt{3} \times 440 \times 60}{2 \times 3 \cdot 14 \times 900} = 246 \cdot 75 \text{ Nm.}$$

10. No-load.

Current = 40 A

$$\text{Power factor} = \frac{3600}{\sqrt{3} \times 40 \times 415} = 0 \cdot 125 \text{ (lagging)}$$

Locked rotor.

$$\text{Current (full value)} = 80 \times \frac{415}{83} = 400 \text{ A}$$

$$\text{Power factor} = \frac{3450}{\sqrt{3} \times 80 \times 83} = 0.3 \text{ (lagging)}.$$

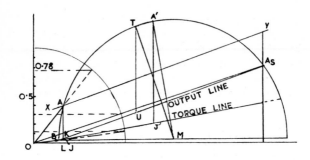

Current scale 10 mm = 30 A (on original diagram).

$$\text{Length of no-load current phasor} = \frac{40}{30} = 13.3 \text{ mm}$$

$$\text{Length of locked-rotor current phasor} = \frac{400}{30} = 133.3 \text{ mm}$$

The circle diagram is drawn in the accepted manner and the input line current is determined by drawing in the line XY parallel to the output line. The power component of the output current $= \dfrac{37\,000}{\sqrt{3} \times 415} = 52 \text{ A}$ or $\dfrac{52}{30} = 17.3 \text{ mm}$

i.e. the spacing of XY above the output line.

Line current OA = 25 mm = 75 A

Power factor from quadrant = 0.78 (lagging)

$$\text{Efficiency} = \frac{AK}{AL} = \frac{17.5 \text{ mm}}{19.8 \text{ mm}} = 0.88 \quad \text{or} \quad 88 \text{ per cent.}$$

Draw the perpendicular bisector of the output line to cut the circle. Drop the vertical TU = 46 mm = 138 A—the power component for maximum power.

$$\text{Maximum output} = \frac{\sqrt{3} \times 415 \times 138}{1000} = 99 \cdot 2 \text{ kW}$$

Full-load torque is proportional to vertical AJ = 17 mm.
Maximum torque is obtained from the vertical A′J′, drawn from the circle where it is intersected by the perpendicular bisector of the torque line extended to form a chord.

Maximum torque is proportional to A′J′ = 54 mm.

$$\therefore \quad \frac{\text{Full-load torque}}{\text{Maximum torque}} = \frac{1 \cdot 7}{5 \cdot 4} = \frac{0 \cdot 32}{1}.$$

CHAPTER 12

1. Using the values given, the change in base current can be deduced from $I_b = I_e - I_c$

   Here $\delta I_e = 1\,\text{mA}$ and $\delta I_c = 0.98\,\text{mA}$

   $\therefore\ \delta I_b = 1 - 0.98 = 0.02\,\text{mA}$

   The current gain $\beta = \dfrac{\delta I_c}{\delta I_b} = \dfrac{0.98}{0.02} = 49.$

2. $\delta I_b = 0.1\,\text{mA}$ $\quad \delta I_e = 1\,\text{mA}$ $\quad \therefore\ \delta I_c = 0.9\,\text{mA}$

   Thus $\beta = \dfrac{\delta I_c}{\delta I_b} = \dfrac{0.9}{0.1} = 9$

   $\alpha = \dfrac{\delta I_c}{\delta I_e} = \dfrac{0.9}{1.0} = 0.9$

   or $\beta = \dfrac{\alpha}{1-\alpha} = \dfrac{0.9}{1-0.9} = \dfrac{0.9}{0.1} = 9.$

3. α' or $\beta = \dfrac{\alpha}{1-\alpha}$ or $\alpha' - \alpha'\alpha = \alpha$ Thus $\alpha' = \alpha + \alpha'\alpha$

   giving $\alpha' = \alpha(1+\alpha')$ or $\alpha = \dfrac{\alpha'}{1+\alpha'}$ or $\dfrac{\beta}{1+\beta}$

   Here α' or $\beta = 50$ $\quad \therefore\ \alpha = \dfrac{50}{51} = 0.98.$

4.   The transfer characteristic is obtained by plotting the given values and its slope will give the current gain.

Thus β or $\alpha' = \dfrac{\delta I_c}{\delta I_b} = \dfrac{(5\cdot8 - 2\cdot5)\,10^{-3}}{(100 - 40)\,10^{-6}} = \dfrac{3\cdot3 \times 10^3}{60}$

$$= \dfrac{330}{6} \fallingdotseq 55.$$

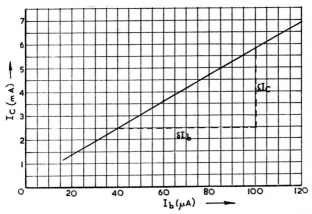

5.   For a constant V_c of $-4\cdot5$ volts, the following values can be read off from Fig. 246.

I_b (microamperes)	20	40	60	80	100	120	140
I_c (milliamperes)	1·2	2·3	3·4	4·5	5·6	6·7	8

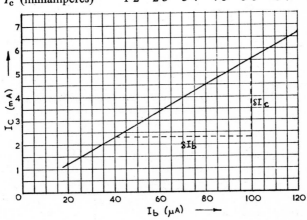

When the above results are plotted the following can be deduced from the transfer characteristic.

$$\beta = \frac{\delta I_c}{\delta I_b} = \frac{(5\cdot 6 - 2\cdot 3)\,10^{-3}}{(100 - 40)\,10^{-6}} = \frac{3\cdot 3 \times 10^3}{60} = \frac{330}{6}$$

Thus current gain $= 55$.

6. The collector-current/collector-voltage or "output" characteristic is shown by the graphs below and the output resistance r_0 can be obtained from $\dfrac{\delta V_c}{\delta I_c}$. Here, if the 6 mA graph is considered, then $r_0 = \dfrac{\delta V_c}{\delta I_c} = \dfrac{(55 - 5)}{(5\cdot 9 - 5\cdot 7) \times 10^{-3}}$

$$= \frac{50 \times 10^3}{0\cdot 2} = 250 \times 10^3 \text{ ohms}$$

Thus $r_0 = 250\,\text{k}\Omega$.

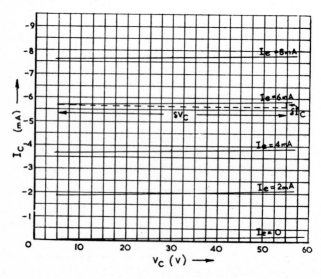

7.   The "transfer characteristic" is shown below for a collector voltage of − 30 V.

Since current gain α is given by $\dfrac{\text{change of output current}}{\text{change of input current}}$

then $\alpha = \dfrac{\delta I_c}{\delta I_e}$

Here $\alpha = \dfrac{(5\cdot8 - 2) \times 10^{-3}}{(6 - 2) \times 10^{-3}} = \dfrac{3\cdot8}{4} = 0\cdot95$

Also since β or $\alpha' = \dfrac{\alpha}{1 - \alpha}$

$\therefore \beta = \dfrac{0\cdot95}{1 - 0\cdot95} = \dfrac{0\cdot95}{0\cdot05} = 19.$

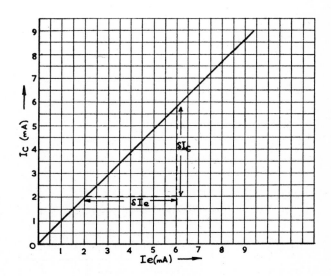

8.   Overleaf is shown the required output characteristic of a transistor connected in the common-emitter mode.

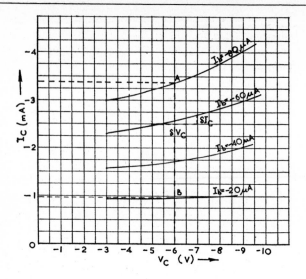

Then (a)   Output Resistance $r_0' = \dfrac{\delta V_c}{\delta I_c}$

Considering the 60 μA characteristic we have

$$\frac{\delta V_c}{\delta I_c} = \frac{(7-5)}{(2\cdot7-2\cdot5)\,10^{-3}} = \frac{2 \times 10^3}{0\cdot2}$$

Thus $r_0' = 10\,000$ ohms or $10\,k\Omega$

Also (b)   Current Gain $\beta = \dfrac{\delta I_c}{\delta I_b}$

For a collector voltage of -6 V

$$\delta I_c = AB = (3\cdot4 - 0\cdot95)10^{-3} = 2\cdot45 \times 10^{-3} \text{ amperes}$$

$$\delta I_b = 80 - 20 = 60 \times 10^{-6} = 60 \times 10^{-6} \text{ amperes}$$

So $\beta = \dfrac{\delta I_c}{\delta I_b} = \dfrac{2\cdot45 \times 10^{-3}}{60 \times 10^{-6}} = \dfrac{2\cdot45 \times 10^3}{60} = \dfrac{245}{6} = 40\cdot8$

Thus $\beta = 41$.

The current gain value can also be obtained from the transfer characteristic which can be deduced as shown below. The 3, 5 and 7 V graphs are drawn and the 6 V line deduced — shown dotted.

$$\text{Then } \beta = \frac{\delta I_c}{\delta I_b} = \frac{(3-1)\,10^{-3}}{(70-20)10^{-6}} = \frac{2}{50 \times 10^{-3}}$$

$$= \frac{2000}{50} = 40.$$

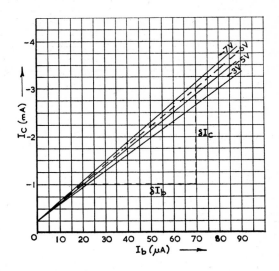

9.   Here the voltage across R_B is the base to collector voltage V_{B-C} which is the collector to emitter voltage less the base-emitter voltage.

$$\therefore V_{B-C} = V_{CE} - V_{BE} = 5 - 0.2 = 4.8 \text{ V}$$

$$\text{Then } R_B = \frac{V_{BC}}{I_b} = \frac{4.8}{40 \times 10^{-6}} = 0.12 \times 10^6 \text{ ohms}$$

$$= 120 \text{ k}\Omega.$$

10.   Voltage across 1·2 kΩ resistor $= 1·2 \times 10^3 \times 0·5 \times 10^{-3}$

$$= 0·6 \text{ V}$$

Voltage across base and + ve line $= V_{B-E} + 0·6$

$$= 0·1 + 0·6$$

$$= 0·7 \text{ V}$$

∴ Voltage across 4·7 kΩ resistor of potential divider $= 0·7$ V

Current through resistor $= \dfrac{0·7}{4·7 \times 10^3}$

$$= 0·149 \times 10^{-3} \text{ amperes}$$

$$= 0·149 \text{ mA}$$

Since no data is given for the base current, this is assumed to be negligible.

Voltage drop across 47 kΩ resistor

$= 0·149 \times 10^{-3} \times 47 \times 10^3 = 7 \text{ V}$

Supply voltage will need to be 7·7 V.

CHAPTER 13

1. For the 200 kW load, $\cos \phi_1 = 0 \cdot 8$ $\sin \phi_1 = 0 \cdot 6$

   For the 50 kW motor load, $\cos \phi_2$ is not known

   Thus ϕ_2 is unknown, also $\sin \phi_2$.

$$\text{For the combined load, } \cos \phi = 0 \cdot 9 \quad \phi = 25°50'$$

$$\sin \phi = 0 \cdot 4357$$

$$\text{Apparent power rating of 200 kW load} = \frac{200}{0 \cdot 8} = 250 \text{ kVA}$$

$$\text{Reactive power of 200 kW load} = 250 \times 0 \cdot 6 = 150 \text{ kVAr} \atop \text{(lagging)}$$

$$\text{Active power of total load} =. 50 + 200 = 250 \text{ kW}$$

$$\text{Apparent power of total load} = \frac{250}{0 \cdot 9} = 277 \cdot 77 \text{ kVA}$$

$$\text{Reactive power of total load} = 277 \cdot 77 \times 0 \cdot 4357$$

$$= 121 \cdot 1 \text{ kVAr (lagging)}.$$

The reactive power of the motor is leading in order to provide power-factor improvement and is obviously the difference between that of the 200 kW load and the total load. Thus:

$$\text{Reactive power of motor} = (150 - 121 \cdot 1)$$
$$= 28 \cdot 9 \text{ kVAr (leading)}$$

$$\text{Apparent power of motor} = \sqrt{50^2 + 28 \cdot 9^2} = 10\sqrt{25 + 8 \cdot 35}$$
$$= 10\sqrt{33 \cdot 35} = 57 \cdot 75 \text{ kVA}$$

$$\text{Power factor of motor} = \frac{50}{57 \cdot 75} = 0 \cdot 866 \text{ (leading)}.$$

2.

Load	kVA	kW	kVAr	$\cos \phi$	$\sin \phi$
(a)	800	800	0	1	0
(b)	750	600	-450	0·8	0·6
(c)	566	400	-400	0·707	0·707
		1800	-850		

$$\text{Apparent power of No. 1 machine} = \frac{\sqrt{3} \times 3300 \times 150}{1000}$$

$$= 1 \cdot 732 \times 33 \times 15 = 857 \cdot 34 \text{ kVA}$$

$$\text{Active power of No. 1 machine} = 857 \cdot 34 \times 0 \cdot 85 = 728 \cdot 74 \text{ kW}$$

$$\cos \phi_1 = 0 \cdot 85 \qquad \phi_1 = 31°47' \qquad \sin \phi_1 = 0 \cdot 5267$$

$$\text{Reactive power of No. 1 machine} = 857 \cdot 34 \times 0 \cdot 527$$
$$= 452 \text{ kVAr}$$

Thus active power output of machine No. 2
$$= 1800 - 729 = 1071 \text{ kW}$$

$$\text{Reactive power of machine No. 2} = -850 - (-452)$$
$$= -398 \text{ kVAr}$$

Apparent power of machine No. 2 =

$$= \sqrt{1071^2 + 398^2} = 10^2 \sqrt{10 \cdot 71^2 + 3 \cdot 98^2}$$
$$= 10^2 \sqrt{114 \cdot 7 + 15 \cdot 84} = 10^2 \sqrt{130 \cdot 54}$$
$$= 10^3 \sqrt{1 \cdot 305} = 10^3 \times 1 \cdot 142$$
$$= 1142 \text{ kVA}$$

$$\text{Current of No. 2 machine} = \frac{1142 \times 1000}{\sqrt{3} \times 3300}$$

$$= \frac{1142}{1 \cdot 732 \times 3 \cdot 3}$$

$$= \frac{1142}{5 \cdot 716} = 200 \text{ A}$$

$$\text{Power factor} = \frac{1071}{1142} = 0 \cdot 94 \text{ (lagging)}$$

3. Total load current $I = \dfrac{1000 \times 10^3}{\sqrt{3} \times 440 \times 0.8} = 1640$ A

Load per machine, or $P_1 = P_2 = \dfrac{1000}{2} = 500$ kW

Current of No. 2 machine is given as 1000 A

$\therefore$ Power factor of No. 2 or $\cos \phi_1 = \dfrac{500 \times 10^3}{\sqrt{3} \times 440 \times 1000}$

$$= 0.658 \text{ (lagging)}$$

Power factor of total load or $\cos \phi = 0.8$. $\therefore \sin \phi = 0.6$

Also $\cos \phi_2 = 0.658$. $\therefore \sin \phi_2 = 0.753$

Thus the active current component of No. 1 machine

$$= 1640 \times 0.8 - 1000 \times 0.658$$
$$= 1312 - 658 = 654 \text{ A}$$

Also the reactive current component of No. 1 machine

$$= 1640 \times 0.6 - 1000 \times 0.753$$
$$= 984 - 753 = 231 \text{ A}$$

Thus current supplied by No. 1 machine
$$\text{or } I_1 = \sqrt{654^2 + 231^2} = 10^2 \sqrt{6.54^2 + 2.31^2}$$
$$= 694 \text{ A}$$

Power factor of No. 1 machine,

$$\cos \phi_1 = \dfrac{654}{694} = 0.942 \text{ (lagging)}$$

The circulating current is found thus:

Since both machines supply equal power, then for No. 2

$$\text{Active current} = 1000 \times 0.658$$
$$= 658 \text{ A and for No. 1}$$

Active current $= 694 \times 0.942 = 654$ A or both currents are "in phase" and very nearly equal.

The circulating current is a wattless or reactive current obtained thus:

The load reactive current $= 1640 \times 0.6$
$$= 984 \text{ A}$$

This should be supplied by two machines or $\dfrac{984}{2} = 492$ A

from each machine

But No. 2 machine supplies a reactive current of $1000 \times 0.753 = 753$ A
$\therefore 753 - 492 = 261$ A must be passed to No. 1
This goes to make up the 492 A for the load $i.e.$ $231 + 261$
    Circulating current $= 261$ A.

4. Machine A drops 1·5 Hz for 150 kW $i.e.$ 0·01 Hz per kW
Machine B drops 1·5 Hz for 220 kW $i.e.$ 0·0068 Hz per kW
Let the load on machine A be x kilowatts. Then load on Machine B is $(200 - x)$ kilowatts

When the machines are in parallel and of common frequency
then $50 - 0.01 x = 50 - 0.0068 (200 - x)$
    or $0.01 x = (0.0068 \times 200) - 0.0068 x$

then $0.0168 x = 1.36$ or $x = \dfrac{136}{1.68} = 80.9$ or 81 kW

Thus machine A supplies 81 kW and machine B supplies $(200 - 81) = 119$ kW
The drop in frequency for machine A $= 0.01 \times 81 = 0.81$ Hz

$\therefore$ Frequency of the paralleled system $= 50 - 0.81$
$$= 49.19 \text{ Hz.}$$

5. Electrical output of motor $= (20 \times 746) + 500$
$$= 14\,920 + 500$$
$$= 15\,420 \text{ W}$$

Also since output $P = \sqrt{3}\, VI\cos\phi - 3\,I^2 R_{ph}$

we have $3\,I^2 R_{ph} - \sqrt{3}\, VI\cos\phi + P = 0$

or $I = \dfrac{\sqrt{3}\, V\cos\phi \pm \sqrt{(\sqrt{3}\, V\cos\phi)^2 - (4\times 3\,R_{ph}P)}}{2\times 3\,R_{ph}}$

$= \dfrac{\sqrt{3} \times 440 \times 0.9 \pm \sqrt{3(440\times 0.9)^2 - (12\times 0.4\times 15\,420)}}{6\times 0.4}$

$= \dfrac{\sqrt{3} \times 396 \pm \sqrt{3\times 396^2 - (12\times 6168)}}{2.4}$

$= \dfrac{686 \pm \sqrt{470\,448 - 74\,016}}{2.4} = \dfrac{686 \pm \sqrt{396\,432}}{2.4}$

$= \dfrac{686 \pm 10^2\sqrt{39.64}}{2.4} = \dfrac{686 \pm (10^2 \times 6.3)}{2.4}$

$= \dfrac{686 - 630}{2.4} = \dfrac{56}{2.4} = 23.3\ \text{A}.$

6.   Armature current is a minimum when the motor power factor is unity. Thus here $\cos\phi = 1$ and $P = VI - I^2 R$

Hence $5000 = 220\,I - 0.5\,I^2$

or $I^2 - 440\,I + 10\,000 = 0$

Solving $I^2 - 440\,I + 220^2 = -10\,000 + 220^2$

or $(I - 220)^2 = \sqrt{-10\,000 + 48\,400}$

$= \sqrt{38\,400}$

$\therefore\ I - 220 = \pm\,195.96 = 24.04\ \text{A}$

From Fig. 271.

E_2^2 or here $E^2 = (V\cos\phi - IR)^2 + (V\sin\phi - IX_S)^2$

Here $X_S = \sqrt{5^2 - 0.5^2} = \sqrt{24.75} = 4.975\ \Omega$

$\therefore\ E^2 = \{(220\times 1) - (24.04\times 0.5)\}^2$
$\qquad\qquad + \{(220\times 0) - (24.04\times 4.975)\}^2$

$$= (220 - 12 \cdot 02)^2 + (0 - 119 \cdot 6)^2$$

$$\text{or } E = \sqrt{207 \cdot 98^2 + 119 \cdot 6^2} = 10^2 \sqrt{2 \cdot 08^2 + 1 \cdot 2^2}$$

$$= 10^2 \sqrt{4 \cdot 32 + 1 \cdot 44} = 10^2 \sqrt{5 \cdot 76} = 240 \text{ V}$$

The angle of retard (α) can be obtained from the Sine Rule (see Fig. 271), but note that since the power factor is unity, IR is in phase with V and the angle between OV and E_R can be obtained from the impedance triangle.

$$\text{Thus cosine of this angle } = \frac{R}{Z_S} = \frac{0 \cdot 5}{5} = 0 \cdot 1$$

Angle between OV and E_R = $84 \cdot 3°$ and

$$\frac{E_R}{\sin \alpha} = \frac{E_2}{\sin 84 \cdot 3}$$

Here E_2 is $E = 240 \text{ V}$ and $E_R = 24 \cdot 04 \times 5$

$$\therefore \ \sin \alpha = \frac{E_R \times \sin 84 \cdot 3}{E_2} = \frac{120 \cdot 2 \times 0 \cdot 9951}{240}$$

$$= \frac{11 \cdot 96}{24} = 0 \cdot 498 \text{ and } \alpha = 29°52'.$$

7. Full-load current $= \dfrac{37\ 000}{\sqrt{3} \times 440 \times 0 \cdot 85 \times 0 \cdot 82} = 70 \text{ A}$

When direct-on started

Starting current $= 6 \times 70 = 420 \text{ A}$

When auto-transformer started

Line current to motor is 75 per cent of line current when direct started $= 0 \cdot 75 \times 420 = 315 \text{ A}$

Due to transformer action, primary line current is 75 per cent of secondary current.

Thus supply current $= 0 \cdot 75 \times 315 = 236 \cdot 25 \text{ A}$

Also torque ∝ voltage squared

∴ Torque = $0.75^2 \times 1.5$ times full-load torque
= $0.844 \times$ full-load torque   or
84.4 per cent of full-load torque.

8. Full-load current of motor = $\dfrac{15\ 000}{\sqrt{3} \times 415 \times 0.8 \times 0.85}$

= 30.5 A

With the starter in "star" position, the voltage per phase = $\dfrac{415}{\sqrt{3}}$

= 240 V

With the motor connected in "delta" and an applied voltage of 220 V, the current per phase = $\dfrac{60}{\sqrt{3}}$ amperes

The current under starting conditions, with $\dfrac{415}{\sqrt{3}}$ volts applied across each phase, would be $\dfrac{60}{\sqrt{3}} \times \dfrac{415}{\sqrt{3}} \times \dfrac{1}{220}$

$= \dfrac{6 \times 415}{3 \times 22} = \dfrac{415}{11} = 37.7$ A

Ratio of starting current to full-load current is given by

$\dfrac{\text{starting current}}{\text{full-load current}} = \dfrac{37.7}{30.5} = \dfrac{1.24}{1}$.

9. Stator phase voltage = 440 V Due to transformer action, the rotor phase voltage = $\dfrac{440}{3} = 146.7$ V

At standstill, rotor resistance per phase $= 0.02\ \Omega$ and the starter resistance per phase $= 0.25\ \Omega$

Total rotor resistance per phase $= 0.02 + 0.25 = 0.27\ \Omega$

Rotor standstill impedance per phase $= 0.27\ \Omega$

Motor standstill impedance per phase
$$= \sqrt{0.27^2 + 0.27^2} = \sqrt{0.1458} = 0.382\ \Omega$$

Rotor current per phase at standstill $= \dfrac{146.7}{0.382} = 384\ \text{A}$

By transformer action, stator current per phase $= \dfrac{384}{3}$
$$= 128\ \text{A}$$

Line current taken from the supply, at starting $= \sqrt{3} \times 128$
$$= 222\ \text{A}$$

Under running conditions the starting resistance is cut-out and the slip is 4 per cent.

$\therefore$ Rotor phase voltage $=$ slip $\times$ standstill phase voltage
$$= 0.04 \times 146.7 = 5.87\ \text{V}$$

Rotor reactance per phase $=$ slip $\times$ standstill reactance
$$= 0.04 \times 0.27 = 0.0108\ \Omega$$

Rotor impedance per phase $= \sqrt{0.02^2 + 0.0108^2}$

$= \sqrt{0.0004 + 0.000117} = \sqrt{0.000517} = 0.0227\ \Omega$

$\therefore$ Rotor current per phase $= \dfrac{5.87}{0.0227} = 259\ \text{A}$

By transformer action, stator current per phase $= \dfrac{259}{3}$
$$= 86.3\ \text{A}$$

$\therefore$ Line current taken from the supply $= \sqrt{3} \times 86.3 = 149\ \text{A}$

At starting, with starting resistance cut out, the rotor impedance

per phase $= \sqrt{0.02^2 + 0.27^2} = \sqrt{0.0004 + 0.0729}$

$\qquad = \sqrt{0.0733} = 0.271 \ \Omega$

Rotor current per phase $= \dfrac{146.7}{0.271} = 541$ A

By transformer action, the stator current per phase $= \dfrac{541}{3}$

$= 180.3$ A and line current taken from the supply
$= \sqrt{3} \times 180.3 = 312$ A.

10. Rotor e.m.f. makes 90 cycles per minute.

∴ rotor frequency $= 1.5$ Hz

Thus fractional slip $s = \dfrac{1.5}{50}$ or 3 per cent.

Synchronous speed $N_S = \dfrac{f \, 120}{P} = \dfrac{50 \times 120}{6} =$

$\qquad\qquad\qquad = 1000$ rev/min.

Output power $= \dfrac{2\pi \, NT}{60} = \dfrac{2 \times 3.14 \times 1000 \times 162.4}{60}$

$\qquad\qquad = \dfrac{628 \times 162.4}{6} = 17\,000$ W

Rotor output $= 17\,000$ W

Rotor input torque $= 162.4 + 13.6$

Rotor input power $= \dfrac{2\pi \, 1000 \times 176}{60} = 18420$ W

Rotor copper loss $= s \times$ rotor input

$\qquad\qquad = \dfrac{3 \times 18\,420}{100} = 552.6$ W

Total input $=$ rotor input $+$ stator loss

$$= (18\,420 + 750) = 19\,170 \text{ W}$$

$$\text{Efficiency} = \frac{17\,000}{19\,170} = 0\!\cdot\!887 \quad \text{or} \quad 88\!\cdot\!7 \text{ per cent.}$$

1.   In the diagram shown below, the device D, which takes
negligible current, is required to operate 5s after the potential
E is applied. If the minimum potential for operation is 63·2 V,
$E = 100$ V and $C = 2·5 \mu$F, what value of R is required?
If R is adjusted to $4 \times 10^6 \Omega$ (4 MΩ) what is then the new
delay period?
    Note. In a series circuit containing a capacitor of C farads
and a resistor of R ohms, the instantaneous voltage v across the
capacitor at time t seconds after applying a potential E across
the circuit is given by $v = E(1 - e^{-t/CR})$.

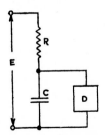

2.   Describe the star-delta method of starting a cage-type, three-
phase induction motor and state any disadvantage of this method
of starting.
    A three-phase, 440 V, 50 Hz, six-pole induction motor
develops 18 kW on full load with a speed of 970 rev/min. and
operating power factor of 0·88 (lagging). Calculate the full-load
(a) slip, (b) input in kW, (c) line current. The stator losses are
1·7 kW and the mechanical losses total 1·5 kW.

3.   As part of a control system, a d.c. generator is provided with
two sets of field windings, acting cumulatively, as shown in the
diagram and specified below.
(a) A control-field winding having 250 turns/pole and resistance
    25 Ω.
(b) A shunt-connected field winding having 500 turns/pole and
    resistance 50 Ω.

Tests show that the generated e.m.f. is directly proportional to excitation over the normal operating range and that with the control-field winding only excited and carrying 4 A, the open-circuit terminal p.d. of the machine is 50 V.

A constant load-current of 10 A is to be maintained while the load resistance, across the output terminals, varies from 2 Ω to 10 Ω. Determine the range of additional resistance, R_A, to be connected in series with the separately-excited, control-field winding across a 100 V d.c. source.

Neglect all voltage drops in the armature winding and the demagnetising effect of armature reaction. Assume speed of armature rotation to be constant.

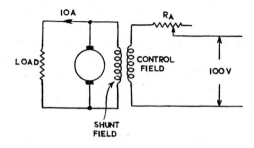

4.   The low-voltage release of an a.c. motor-starter consists of a solenoid into which an iron plunger is drawn against a spring. The resistance of the solenoid is 35 Ω. When connected to a 220 V, 50 Hz, a.c. supply the current taken is at first 2 A, and when the plunger is drawn into the "full-in" position the current falls to 0·7 A. Calculate the inductance of the solenoid for both positions of the plunger, and the maximum value of flux-linkages in weber-turns for the "full-in" position of the plunger.

5.   The circuit shown below is commonly used in control systems in order to combine voltages V_1, V_2, ... V_n prior to amplification.

By applying Kirchhoff's first law to point A, show that, if R is much greater than $R_1, R_2, ... R_n$ and if $R_1 = R_2 = R_n$ then, V is approximately equal to $(V_1 + V_2 + V_3 + ... V_n)\dfrac{1}{N}$.

Calculate the voltage input to the amplifier if, in a given system, two voltages only are to be combined and,
$R_1 = 25$ kΩ, $R_2 = 25$ kΩ, $R = 1$ MΩ, $V_1 = 750$ mV
and $V_2 = 250$ mV.

Compare the result obtained using the above expression for V with that obtained from a more exact calculation.

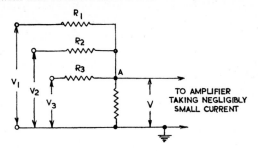

6. (a) An e.m.f. $E \sin 2\pi ft$, is applied across a non-resistive circuit consisting of a capacitor of C farads and an inductor of L henrys connected in parallel. Find the frequency at which zero current is taken from the supply (in terms of L and C) and give a brief description of the changes which occur within the parallel circuit.

(b) Describe briefly the effects a capacitor may produce when placed in a series circuit supplied by (1) an a.c. source, (2) a steady d.c. supply, (3) a fluctuating d.c. supply.

7. The potential of divider shown has a resistance r ohms connected across a fraction of the large resistance R ohms.

Prove that the current flowing in $r = \dfrac{VK}{r + RK - RK^2}$, where

K is the fraction of the tapping.

Find the minimum value of r to ensure that the potential divider does not exceed its rated current flow of 0·65 A when connected to a 220 V supply with the fraction of resistance

equal to $\dfrac{1}{4}$.

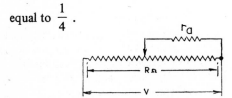

8.   A total load of 8000 kW at 0·8 power factor is supplied by two alternators in parallel. One alternator supplies 6000 kW at 0·9 power factor. Find the *kVA* rating of the other alternator and the power factor.

9.   A shunt motor runs at 900 rev/min when connected to a 440 V supply, the armature current being 60 A and armature resistance 0·4 Ω. At what speed will it run on a 220 V supply with an armature current of 40 A. Assume 60 per cent reduction in flux for the 220 V connection.

10.   Three equal resistors are connected to a three-phase system. If one resistor is removed, find the reduction in load if they are connected in (a) Star, (b) Delta.

11.   Three batteries A, B, and C have their negative terminals connected together. Between the positive terminals of A and B there is a resistor of 0·5 Ω and between B and C there is a resistor of 0·3 Ω.
Battery A   105 V. Internal resistance 0·25 Ω
Battery B   100 V.   ”   ”   0·2 Ω
Battery C   95 V.   ”   ”   0·25 Ω
    Determine the current values in the two resistors and the power dissipated by them.

12.   A 25 kVA transformer is connected to a supply of 3000 V, 50 Hz. There are 800 turns on the primary winding and 60 turns on the secondary. Calculate (a) the secondary terminal voltage, (b) the primary and secondary full-load current, and (c) the maximum flux in the core. State any assumptions made.

13.   A 220 V, d.c. shunt motor has an armature resistance of 0·5 Ω and an armature current of 40 A on full load. Determine the reduction in flux necessary for a 50 per cent reduction in speed. The torque for both conditions can be assumed to remain constant.

14.   A system, supplying a 100 kW lighting load at unity power factor and a 300 kW induction motor at 0·85 (lagging) power factor, is to have its power factor improved to 0·95 (lagging), by the use of a 100 kW synchronous motor. Determine the power factor and *kVA* of the synchronous motor.

15.   Two earth lamps rated at 120 V, 15 W are used on a 110 V distribution circuit. Due to damage to a motor, the resistance of the positive line is reduced to 10 Ω and that of the negative line to 5 Ω. Find the voltage across each lamp and the extra load on the generator due to the fault.

16.   A coil having a resistance of 10 Ω and an inductance of 0·15 H is connected in series with a capacitor across a 100 V, 50 Hz supply. If the current and voltage are in phase, what will be the value of the current in the circuit and the voltage drop across the coil?

17.   A load resistance of 2·1 Ω is fed through a twin-core cable 457 m long. The resistance of the cable is $4·38 \times 10^{-4}$ Ω/m. The d.c. supply voltage is 250 V. An earth fault somewhere along the positive line caused the supply current to rise to 120 A and the load current to fall to 98·7 A. At what distance along the cable did the fault occur?

18.   A 250 W discharge lamp takes a current of 2 A and is connected in series with a choke coil having negligible resistance. Calculate the inductance value of the coil so that the lamp can be used on a 250 V, 50 Hz a.c. supply. Find the value of the capacitor necessary to bring the power factor to unity.

19.   A balanced Wheatstone bridge consists of four resistors labelled clockwise P, Q, R and S, the supply being connected to the junctions of PS and QR. R is 450 Ω.

      If P and Q are interchanged, then a 5 Ω resistor has to be placed in parallel with R to maintain balance. Find S and the ratio of P to Q.

20.   Show that three sine-wave alternating currents, displaced from one another by 30°, can be represented (a) graphically by drawing their waveforms (b) by phasors. If the currents each have a maximum value of 100 A, calculate the resultant maximum value. What is its r.m.s. value?

21.   Find the current flowing in each branch of the following circuit. A 200 V d.c. supply charges, through a 10 Ω resistor, two batteries A and B in parallel.
A has an e.m.f. of 108 V and internal resistance 0·24 Ω
B has an e.m.f. of 105 V and internal resistance 0·26 Ω.

22.   An inductor has an inductance of 0·1 H and resistance of 10 Ω. When it is connected in series with a capacitor across a 220 V, 100 Hz supply the circuit resonates. Calculate the capacitance value of the capacitor and the voltage across the inductor.

23.   Ten 2 V batteries are connected (a) in series, (b) in parallel to an external resistance which remains the same for both conditions.

   If the power dissipated in the external resistance is 15 W for the parallel connection and 500 W for the series connection, find the internal resistance of each cell and the circuit current in each case.

24.   A 72 kVA transformer supplies (a) a heating and lighting load of 12 kW at unity power factor (b) a motor load of 70 kVA at 0·766 (lagging) power factor.

   Calculate the minimum rating of the power-factor improvement capacitors which must be connected in circuit to ensure that the transformer does not become overloaded.

25.   Two shunt generators X and Y work in parallel. Their external characteristics may be assumed to be linear over their normal working range. The terminal voltage of X falls from 265 V on no load to 230 V when delivering 350 A to the busbars, while the voltage of Y falls from 270 V on no load to 240 V when delivering 400 A to the busbars. Calculate the current which each machine delivers when they share a common load of 500 A. What is the busbar voltage under this condition and the power delivered by each machine?

26.   A 175 kVA, 6600/440 V, single-phase transformer has an iron loss of 2·75 kW. The primary and secondary windings have resistances of 0·4 Ω and 0·0015 Ω respectively. Calculate the efficiency on full load when the power factor is 0·9 (lagging).

27.   A non-linear resistor R_1 is in series with a linear resistor R_2 on a 125 V d.c. supply. $R_2 = 250$ Ω. R_1 has the following volt-ampere characteristic

Voltage (V)	0	25	50	75	100	125	150
Current (I)	0	0·038	0·082	0·132	0·2	0·3	0·485

   Resistor R_2 is bilateral, *i.e.* conducts equally well in either direction. Determine graphically or otherwise the current in the circuit.

28.   A 440 V load of 400 kW at 0·8 (lagging) power factor is jointly supplied by two alternators A and B. The kW load on A is 150 kW and the $kVAr$ load on B is 150 kVAr (lagging). Determine the kW load on B, the $kVAr$ load on A, the power factor of operation on each machine and the current loading of each machine.

29.   A diode valve, having the following characteristic

I_a (milliamperes)	0	5·5	13	22	32	42	52	59	63
V_a (volts)	0	25	50	75	100	125	150	175	200

is connected in series with a resistor of 10 000 Ω to a 240 V d.c. supply. If a resistor of 40 000 Ω is connected between the anode and cathode of the diode, determine the current through the diode.

30.   The following readings were obtained from a test on a vacuum diode.

V_a (volts)	3	5	10	15	20
I_a (milliamperes)	3·46	11·4	27·54	47·86	67·6

Show that the relationship between I_a and V_a is of the form $I_a = kV_a^n$ and determine the values of k and n.

The following specimen questions have been included to show the knowledge of electronic theory and circuitry now required for the D. of T. examinations.

31.   Explain the meaning of "p" and "n" type semi-conductor materials and give a brief description of the mechanism by which current passes through them.

32.   Explain what is meant by, and the significance of, four of the following terms   (i) voltage stabilisation, (ii) filter choke, (iii) impedance, (iv) rectification, (v) grid bias voltage.

33.   Define the term "valence electron" and explain, in terms of electronic structure, why some substances are better conductors of electricity than others.

34.   Describe the construction of a semi-conductor rectifier indicating the materials used. Sketch the output voltage wave of the bridge rectifier circuit shown below. If a capacitor was connected across the output terminals, what effect will it have on the waveform?

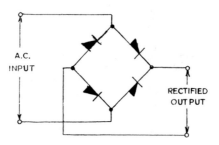

35.   Write notes on the following: (i) transistors, (ii) capacitors, (iii) inductors.

36.   Differentiate with the aid of simple sketches between two of the following types of electronic circuits (i) rectifier circuit, (ii) amplifier circuit, (iii) oscillator circuit.

37.   Shown below is a diagrammatic reproduction of an amplifier circuit. Explain the function of the various components and describe how a small electrical signal may be amplified.

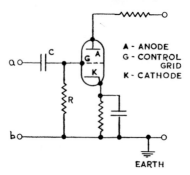

38.   With reference to "p" and "n" type germanium crystals explain the meaning of three of the following terms (i) donor ion, (ii) acceptor ion, (iii) valency electrons, (iv) co-valent bonds.

39.   Redraw the full-wave rectifier circuit shown below and
    (i) label the various components, (ii) for a single sinusoidal a.c.
    cycle, indicate the directions of the electron flow, (iii) explain
    the function of the resistor *R*.

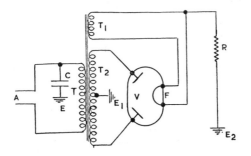

40.   Describe how liquid level and revolutions per minute may
    be converted to electrical signals for a data logger.

SOLUTIONS TO FIRST CLASS
EXAMINATION QUESTIONS

1.   A straight-forward substitution in the formula gives

$$63 \cdot 2 \;=\; 100 \, (1 - e^{-5/2 \cdot 5 \times 10^{-6} \times R})$$

$$\text{or } 0 \cdot 632 \;=\; 1 - e^{-2 \times 10^6 / R})$$

$$\text{Thus } e^{-2 \times 10^6} \;=\; 1 - 0 \cdot 632 \text{ or } \frac{1}{e^{2 \times 10^6 / R}} = 0 \cdot 368$$

$$\text{Hence } \quad e^{2 \times 10^6 / R} \;=\; \frac{1}{0 \cdot 368}$$

$$\text{or } \frac{2 \times 10^6}{R} \, \log e \;=\; \log 1 - \log 0 \cdot 368$$

$$\therefore \; \frac{2 \times 10^6}{R} \, \log 2 \cdot 718 \;=\; \log 1 - \log 0 \cdot 368$$

$$\frac{2 \times 10^6}{R} \times 0 \cdot 4343 \;=\; 0 - 1 \cdot 5658$$

$$=\; 1 - 0 \cdot 5658 \;=\; 0 \cdot 4342$$

$$\text{So} \frac{2 \times 10^6}{R} \times 0 \cdot 4343 \;=\; 0 \cdot 4342 \quad \text{or} \quad \frac{2 \times 10^6}{R} \;=\; 1$$

$$\text{Hence } R \;=\; 2 \times 10^6 \text{ ohms} \text{ or } 2 \text{ M}\Omega$$

If $R = 4 \times 10^6$ ohms

then $63 \cdot 2 = 100 \,(1 - e^{-t/2 \cdot 5 \times 10^{-6} \times 4 \times 10^6})$

$e^{-t/2 \cdot 5 \times 4} = 0 \cdot 368$

Hence $e^{t/10} = \dfrac{1}{0 \cdot 368}$

$\therefore \qquad \dfrac{t}{10} \log 2 \cdot 718 = \log 1 - \log 0 \cdot 368$

and $\dfrac{t}{10} \times 0 \cdot 4343 = 0 - \overline{1} \cdot 5658 = 1 - 0 \cdot 5658$

$\dfrac{t}{10} \times 0 \cdot 4343 = 0 \cdot 4342$

or $\dfrac{t}{10} = 1$ whence $t = 10$ seconds.

2.   Since $f = \dfrac{PN_S}{120}$ $\therefore N_S = \dfrac{120 f}{P} = \dfrac{120 \times 50}{6}$

$= 1000$ rev/min

Slip $= \dfrac{N_S - N}{N_S} \times 100 = \dfrac{1000 - 970}{1000} \times 100 = \dfrac{30}{10}$

or Slip $= 3$ per cent

Output power $= 18$ kW

Electrical output from rotor $= 18 + 1.5$

$$= 19.5 \text{ kW}$$

No mention of rotor copper losses are made, $\therefore$ these can be neglected.

So output from stator $= 18.5 \text{ kW}$

Input to stator $= 19.5 + 1.7 = 21.2 \text{ kW}$

Apparent power input $= \dfrac{21.2}{0.88} = 24.1 \text{ kVA}$

$\therefore$ Line current $= \dfrac{24\,100}{\sqrt{3} \times 440} = \dfrac{24\,100}{762}$

$$= 31.7 \text{A}.$$

3.   Tests with the control field show that with $4 \times 250 = 1000$ At/pole, an e.m.f. of 50 V is generated.
    Now "On load", neglecting all voltage drops and the armature reaction effects, for a load resistance of 10 Ω with a load current of 10 A, the e.m.f. to be generated $= 10 \times 10 = 100$ V. This 100 V would need an excitation of

$$1000 \times \frac{100}{50} = 2000 \text{ At/pole}$$

On 100 V, the shunt-field current $= \dfrac{100}{50} = 2 \text{ A}$

$\therefore$ Excitation due to shunt field $= 2 \times 500 = 1000 \text{ At/pole}$

    Now as 2000 At/pole are needed, the control field must produce $2000 - 1000 = 1000$ At/pole and the control field current must be $\dfrac{1000}{250} = 4 \text{ A}$

The control-field resistance must therefore be $\dfrac{100}{4} = 25\ \Omega$

Since this is the resistance value of the field alone, no series resistance is required.

Again "On Load", for a load resistance of 2 Ω with a load current of 10 A, the e.m.f. to be generated $= 2 \times 10 = 20$ V
This 20 V would need an excitation of

$$1000 \times \frac{20}{50} = 400 \text{ At/pole}$$

On 20 V the shunt-field current $= \dfrac{20}{50} = 0.4$ A

$\therefore$ Excitation due to shunt field $= 0.4 \times 500 = 200$ At/pole

Now as 400 At/pole are needed, the control field must produce $400 - 200 = 200$ At/pole and the control-field current

must be $\dfrac{200}{250} = 0.8$ A

The control-field resistance must therefore be $\dfrac{100}{0.8} = 125\ \Omega$

$\therefore 125 - 25 = 100\ \Omega$ must be added

The range of control-field resistance to give variation from 100 V to 20 V must be from 0 to 100 ohms.

4.   With plunger out.   Impedance $Z = \dfrac{220}{2} = 110\ \Omega$

$$R = 35\ \Omega$$

$$\therefore X = \sqrt{110^2 - 35^2} = 10\sqrt{121 - 12 \cdot 25}$$

$$= 10\sqrt{108 \cdot 75} = 100 \times \sqrt{1 \cdot 0875}$$

$$= 100 \times 1 \cdot 0425 = 104 \cdot 25\ \Omega$$

Thus $2\pi fL = 104 \cdot 25$ or $L = \dfrac{104 \cdot 25}{100 \times 3 \cdot 14} = \dfrac{1 \cdot 0425}{3 \cdot 14}$

$$= 0 \cdot 332 \text{ H}$$

With plunger in. Impedance $Z = \dfrac{220}{0 \cdot 7} = \dfrac{2200}{7}$

$$= 314 \cdot 28 \ \Omega$$

$$X = \sqrt{314 \cdot 28^2 - 35^2} = 10\sqrt{31 \cdot 43^2 - 3 \cdot 5^2}$$

$$= 10\sqrt{987 \cdot 84 - 12 \cdot 25} = 10\sqrt{975 \cdot 6} = 100\sqrt{9 \cdot 756}$$

$$= 100 \times 3 \cdot 123 = 312 \cdot 3 \ \Omega$$

Thus $L = \dfrac{312 \cdot 3}{100 \mp 3 \cdot 14} = 0 \cdot 995 \text{ H}$

Also $L = \dfrac{N\Phi}{I}$ or $L = $ Flux-linkages/ampere

Thus maximum flux-linkages $=$ inductance $\times$ peak value of current

$$= 0 \cdot 995 \times \sqrt{2} \times 0 \cdot 7$$

$$= 0 \cdot 995 \times 1 \cdot 414 \times 0 \cdot 7$$

$$= 0 \cdot 985 \text{ say 1 weber-turn.}$$

5. V is the voltage between A and the common line. Then
$V = V_1 - I_1 R_1$ where I_1 is the current in R_1
     Similarly $V = V_2 - I_2 R_2$ where I_2 is the assumed current and $V = V_n - I_n R_n$ where I_n is the assumed current

Adding: $NV = (V_1 - I_1 R_1) + (V_2 - I_2 R_2) + (V_3 - I_3 R_3)$

$$+ \ldots + (V_n - I_n R_n)$$

$$= (V_1 + V_2 + V_3 \ldots + V_n)$$

$$- (I_1 R_1 + I_2 R_2 + I_3 R_3 \ldots + I_n R_n)$$

$$\therefore \ NV \ = \ (V_1 + V_2 + V_3 \ \dots \ V_n) - r(I_1 + I_2 + I_3 \dots I_n)$$

Here $\ r \ = \ R_1 \ = \ R_2 \ = \ R_3 \ = \ R_n$

Now $I_1 + I_2 + I_3 + \ \dots I_n \ = \ I$ (the current in R)

Then $NV \ = \ (V_1 + V_2 + V_3 \ \dots \ V_n) - Ir$ But $\ V = IR$

$$\therefore \ \ \ \ NV \ = \ (V_1 + V_2 + V_3 \ \dots \ V_n) - r\frac{V}{R}$$

Now if r is very small compared to R then the factor $\dfrac{rV}{R}$ will be

small enough to be neglected. Thus $NV \ = \ V_1 + V_2 + V_3 \dots V_n$

or $\ V \ = \ \dfrac{V_1 + V_2 + V_3 \dots V_n}{N}$

For the values given

$$V \ = \ \frac{750 + 250}{2} \ = \ \frac{1000}{2} \ = \ 500 \text{ mV (the approximate} \atop \text{value)}$$

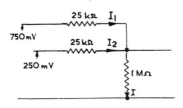

For a more precise value.

Let I_1 and I_2 be assumed currents and directions. Then the voltage V across the 1 MΩ resistor is given by

$$V \ = \ 0\cdot75 - 25 \times 10^3 I_1$$

and $\ V \ = \ 0\cdot25 - 25 \times 10^3 I_2$

The current in the 1 MΩ resistor $= I = I_1 + I_2$

Also $V = I \times 1 \times 10^6$

or $V = 10^6(I_1 + I_2) = 10^6 \left(\dfrac{0 \cdot 75 - V}{25 \times 10^3} + \dfrac{0 \cdot 25 - V}{25 \times 10^3} \right)$

$$= \frac{10^3(0 \cdot 75 + 0 \cdot 25 - 2\,V)}{25}$$

or $V = \dfrac{10^3(1 - 2\,V)}{25}$ giving $25\,V = 10^3 - 2 \times 10^3\,V$

or $2025\,V = 1000$

$\therefore\ V = \dfrac{1}{2 \cdot 025} = 0 \cdot 493$ volts or $V = 493$ mV.

6. (a) This is a condition of parallel resonance and, since the circuit has no resistance, then $I_{X_C} = I_{X_L}$ in magnitude but these are antiphase as shown by the phasor diagram.

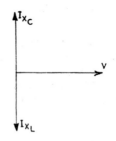

Now $I_{X_L} = \dfrac{V}{2\pi f L}$ and $I_{X_C} = \dfrac{V}{\dfrac{1}{2\pi f C}} = V 2\pi f C$

Under resonance conditions

$$\frac{V}{2\pi fL} = V 2\pi fC \quad \text{or} \quad f^2 = \frac{1}{(2\pi)^2 LC}$$

giving $f =. \dfrac{1}{2\pi\sqrt{LC}}$.

Since the capacitive and inductive circuit currents are anti-phase it is evident that while the capacitor is discharging *i.e.* current is falling, then current in the inductor is rising or the magnetic field is being established. Thus energy is being passed to and fro between these components and in theory the circuit would be in a state of continuous oscillation.

(b) (i). When an alternating potential is applied across a capacitor, a current will flow and its value is dependent on the size of the capacitor. The capacitor offers reactance which either increases or decreases that of the whole circuit, this result depending on whether the existing reactance of the circuit is capacitive or inductive. The capacitor thus permits the passage of an alternating current.

(ii). When introduced into a d.c. circuit the capacitor charges up, when the supply is first switched on, until the potential across the plates equals that of the applied voltage. The value of the circuit current is a maximum when the supply is applied and falls to zero after a time dependent on the resistance and capacitance values. This effect is minimised if inductance is present. The capacitor then continues to act as an insulator in the circuit.

(iii). When the d.c. voltage fluctuates, each change of supply voltage can be regarded as the application of a voltage equal to the change. This condition has been described under (b) (ii). Thus a charging current flows while the voltage is rising and a discharging current occurs when the voltage is falling.

7.   The equivalent resistance of resistors r and y in parallel, is given by the reciprocal of $\dfrac{1}{r} + \dfrac{1}{y} = \dfrac{y+r}{ry}$

or equivalent resistance $= \dfrac{ry}{r+y}$ ohms

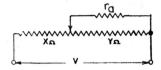

The current in section $\quad x = \dfrac{V}{x + \dfrac{ry}{r+y}} = \dfrac{V(r+y)}{rx + xy + ry}$ amperes

The voltage drop across $\ x = \dfrac{xV(r+y)}{rx + xy + ry}$ volts

The voltage drop across $\ r = V - \dfrac{xV(r+y)}{rx + xy + ry}$

$$= V\left(1 - \dfrac{x(r+y)}{rx + xy + ry}\right)$$

$$= V\left(\dfrac{rx + xy + ry - xr - xy}{rx + xy + ry}\right)$$

$$= \dfrac{Vry}{rx + xy + ry} \text{ volts}$$

The current in resistor r

$$= \dfrac{Vry}{r(rx + xy + ry)} = \dfrac{Vy}{rx + xy + ry} \text{ amperes}$$

If K is the fraction of the tapping point, then K

then $\quad K = \dfrac{y}{x+y} = \dfrac{y}{R}$

The current in the resistor r can now be written as

$$= \frac{Vy}{r(x+y) + xy} \text{ amperes}$$

and dividing numerator and denominator by $x + y$ we have

$$\frac{\dfrac{Vy}{x+y}}{\dfrac{r(x+y)+xy}{x+y}} = \frac{VK}{\dfrac{r(x+y)}{(x+y)} + \dfrac{xy}{(x+y)}} = \frac{VK}{r + xK}$$

Now $x + y = R$ so current is

$$= \frac{VK}{r + K(R-y)} = \frac{VK}{r + RK - yK} \text{ amperes}$$

Also since $K = \dfrac{y}{R} \qquad \therefore \ y = KR$

or current $= \dfrac{VK}{r + RK - RK^2} \text{ amperes}$

Using the above expression:

with $K = \frac{1}{4}$, current in $r = \dfrac{220 \times \frac{1}{4}}{r + \frac{1}{4}R - \dfrac{R}{16}}$

or $I_r = \dfrac{55 \times 16}{16r + 4R - R} = \dfrac{880}{16r + 3R} \text{ amperes}$

Voltage drop across $r = \dfrac{880r}{16r + 3R} \text{ volts}$

Current in resistor $y = \dfrac{880r}{16r + 3R} \bigg/ \dfrac{R}{4} \text{ amperes}$

Note. $K = \dfrac{y}{R} \qquad \therefore \ y = KR = \dfrac{R}{4}$

so current in resistor y $= \dfrac{880 \times 4}{R(16r + 3R)}$ amperes

$\therefore$ Total current $0 \cdot 65$ $= . \dfrac{880}{16r + 3R} + \dfrac{4(880r)}{R(16r + 3R)}$

$$= \dfrac{880}{16r + 3R} \left(1 + \dfrac{4r}{R}\right)$$

$$= \dfrac{880}{16r + 3R} \left(\dfrac{R + 4r}{R}\right) \text{amperes}$$

or $0 \cdot 65 (16rR + 3R^2) = 880(R + 4r)$

$10 \cdot 4rR + 1 \cdot 95R^2 = 880R + 3520r$

or $1 \cdot 95R^2 = 880R + 3520r - 10 \cdot 4rR$

$10 \cdot 4rR - 3520r = 880R - 1 \cdot 95R^2$

$$r = \dfrac{880R - 1 \cdot 95R^2}{10 \cdot 4R - 3520} \text{ ohms}$$

8. Alternator A supplies 6000 kW at 0·9 p.f. (lagging assumed)

Its apparent power rating $= \dfrac{6000}{0 \cdot 9} = 6667$ kVA

Also $\cos \phi_A = 0.9$ $\phi_A = 25° 50'$ $\sin \phi_A = 0 \cdot 4357$

Reactive power $= 6667 \times 0 \cdot 4357 = -2905$ kVAr

Alternator B supplies $8000 - 6000 = 2000$ kW

Apparent power rating $= \dfrac{8000}{0 \cdot 8} = 10\,000$ kVA

Since $\cos \phi = 0 \cdot 8$ (lagging assumed) $\sin \phi = 0 \cdot 6$

$\therefore$ Reactive power of load $= 10\,000 \times 0 \cdot 6 = -6000$ kVAr

Reactive power of alternator B $= -6000 - (-2905)$

$$= -3095 \text{ kVAr}$$

$\therefore$ Apparent power rating of alternator B $= \sqrt{2000^2 + 3095^2}$

$$= 10^3\sqrt{2^2 + 3\cdot095^2}$$

$$= 10^3\sqrt{4 + 9\cdot58} = 10^3\sqrt{13\cdot58}$$

$$= 3\cdot685 \times 10^3 = 3685 \text{ kVA}$$

Power factor of alternator B $= \dfrac{2000}{3685} = 0\cdot543$ (lagging)

A lagging power factor is assumed, if the power factors of both the load and alternator A are taken to be lagging.

9.   On 440 V $E_{b_1} = 440 - 60 \times 0\cdot4 = 440 - 24 = 416$ V
On 220 V $E_{b_2} = 220 - 40 \times 0\cdot4 = 220 - 16 = 204$ V

Now generated e.m.f. is proportional to flux and speed or $E_b \propto \Phi N$ and Φ is proprotional to the shunt field or here, from the data given $\Phi_2 = 0\cdot4\,\Phi_1$.

$$\therefore \frac{E_{b_1}}{E_{b_2}} = \frac{k\,\Phi_1 N_1}{k\,\Phi_2 N_2} \quad \text{or} \quad \frac{E_{b_1}}{E_{b_2}} = \frac{\Phi_1\,900}{0\cdot4\,\Phi_1 N_2}$$

giving $N_2 = \dfrac{900 \times E_{b_2}}{0\cdot4 \times E_{b_1}} = \dfrac{900 \times 204}{0\cdot4 \times 416} = \dfrac{900 \times 204}{166\cdot4}$

$$= \frac{183\,600}{166\cdot4} = 1104 \text{ rev/min.}$$

10.   Let $R =$ the resistance per phase and V the voltage between lines.

For the Star connection.

Voltage per phase $= \dfrac{V}{\sqrt{3}}$ and current per phase $= \dfrac{V}{\sqrt{3}\,R}$

$$\text{Power per phase } = I^2R = \left(\frac{V}{\sqrt{3}R}\right)^2 \times R = \frac{V^2}{3R}$$

$$\text{So three-phase power } = \frac{V^2}{3R} \times 3 = \frac{V^2}{R} \text{ watts}$$

Under the open-circuit condition *i.e.* say resistor No. 2 removed.

$$\text{Current per phase } = \frac{V}{2R} \quad \therefore \text{ power per phase } = \left(\frac{V}{2R}\right)^2 \times R$$

$$= \frac{V^2}{4R}$$

$$\text{Total power } = 2 \times \frac{V^2}{4R} = \frac{V^2}{2R} \text{ watts}$$

$$\therefore \quad \text{New power } = \tfrac{1}{2} \text{ or } 0{\cdot}5 \text{ times the original.}$$

For the Delta connection

$$\text{Current per phase } = \frac{V}{R} \quad \text{Power per phase } = \frac{V^2}{R^2}R = \frac{V^2}{R}$$

$$\text{and three-phase power } = \frac{3V^2}{R} \text{ watts}$$

Under the open-circuit condition *i.e.* say resistor No. 2 removed.

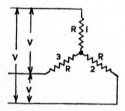
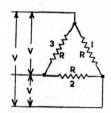

$$\text{Total power} = \frac{2V^2}{R}$$

i.e. New Power $= \dfrac{2}{3}$ or 0·666 times the original.

11.   Using Kirchhoff's laws.

$$105 - 100 = (I_1 \times 0.5) + (I_1 - I_2)\, 0.2 + (I_1 \times 0.25)$$
and $100 - 95 = (I_2 - I_1)\, 0.2 + (I_2 \times 0.3) + (I_2 \times 0.25)$
or $5 = 0.75\, I_1 + 0.2\, I_1 - 0.2\, I_2$
and $5 = 0.2\, I_2 - 0.2\, I_1 + 0.55\, I_2$
giving $5 = 0.95\, I_1$

To solve, multiply by 0·75.

Thus $3.75 = 0.7125\, I_1 - 0.15\, I_2$ ...   . (a)

Also for $5 = -0.2 I_1 + 0.75 I_2$

To solve, divide by 5

Thus $1 = -0.04 I_1 + 0.15\, I_2$ ...   . (b)

Adding (a) and (b) $4.75 = 0.6725\, I_1$

or $I_1 = \dfrac{475}{67.25} = 7.06\,\text{A}$ Also $5 + 0.2\, I_1 = 0.75\, I_2$

$\therefore I_2 = \dfrac{5 + (0.2 \times 7.06)}{0.75} = \dfrac{5 + 1.412}{0.75} = \dfrac{6.412}{0.75}$

Thus $I_2 = 8.55\,\text{A}$

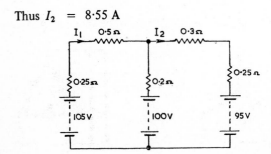

Power in 0·5 Ω resistor $= 7·06^2 \times 0·5 = 3·53 \times 7·06 = 24·9$ W

Power in 0·3 Ω resistor $= 8·55^2 \times 0·3 = 73·1 \times 0·3 = 21·9$ W.

12.   (a) Turns ratio for a transformer $\dfrac{N_1}{N_2} = \dfrac{V_1}{V_2} = \dfrac{800}{60} = \dfrac{40}{3}$

or   13·33:1

Thus secondary terminal voltage $= 3000 \times \dfrac{1}{13·33}$

$= 225$ V

(b) Primary current $= \dfrac{25\ 000}{3000} = 8·33$ A

Secondary current $= \dfrac{25\ 000}{225} = \dfrac{50\ 000}{450} = \dfrac{1000}{9}$

$= 111·1$ A

(c) For a transformer on no load, induced voltage $=$ the supply voltage

$\therefore V = E_1 = 4·44\ \Phi_m f N_1$ or $\Phi_m = \dfrac{3000}{4·44 \times 50 \times 800}$

or $\Phi_m = \dfrac{30}{2·22 \times 100 \times 8} = \dfrac{0·3}{17·76} = \dfrac{0·03}{1·776}$

$= 0·0169$ Wb.

13.   Here $E_{b_1} = 220 - 40 \times 0.5 = 220 - 20 = 200$ V

Also for a motor, torque is proportional to flux and armature current or $T \propto \Phi I_a$

Assuming a constant torque $T_2 = T_1$ amd

and $I_{a_1}\Phi_1 = I_{a_2}\Phi$ or $\dfrac{\Phi_2}{\Phi_1} = \dfrac{I_{a_1}}{I_{a_2}}$

If E_{b_2} is the second back e.m.f. condition

then $E_{b_2} = 220 - I_{a_2} \times 0.5$ $\therefore \dfrac{E_{b_2}}{E_{b_1}} = \dfrac{220 - 0.5\,I_{a_2}}{200}$

also $\dfrac{E_{b_2}}{E_{b_1}} = \dfrac{k\,\Phi_2 N_2}{k\,\Phi_1 N_1}$ If now $N_2 = 0.5\,N_1$

$\therefore \dfrac{E_{b_2}}{E_{b_1}} = \dfrac{k\,\Phi_2\,0.5\,N_1}{k\,\Phi_1 N_1}$

or $\dfrac{0.5\,\Phi_2}{\Phi_1} = \dfrac{220 - 0.5\,I_{a_2}}{200}$ giving $\dfrac{0.5\,I_{a_1}}{I_{a_2}} = \dfrac{220 - 0.5\,I_{a_2}}{200}$

or $200 \times 0.5 \times 40 = 220\,I_{a_2} - 0.5\,I_{a_2}^{\,2}$

and $0.5\,I_{a_2}^{\,2} - 220\,I_{a_2} + 4000 = 0$

thus $I_{a_2}^{\,2} - 440\,I_{a_2} + 8000 = 0$

$\therefore I_{a_2} = \dfrac{440 \pm \sqrt{440^2 - 4 \times 8000}}{2} = \dfrac{440 \pm 10^2\sqrt{4.4^2 - 3.2}}{2}$

or $I_{a_2} = \dfrac{440 \pm 10^2\sqrt{19.36 - 3.2}}{2} = \dfrac{440 \pm 10^2\sqrt{16.16}}{2}$

$$\therefore \; I_{a_2} \;=\; \frac{440 \pm 10^2 \times 4 \cdot 02}{2} \;=\; \frac{440 - 402}{2} \;=\; \frac{38}{2}$$

$$=\; 19 \text{ A}$$

Note. An alternative answer would be $I_{a_2} = \dfrac{440 + 402}{2}$

$= \dfrac{842}{2} = 421$ A. Under this condition the armature voltage

drop would be $421 \times 0 \cdot 5 = 210 \cdot 5$ V *i.e.* back e.m.f. would be $9 \cdot 5$ V. This would be explained by a reduced flux, the main flux being weakened by the armature reaction effect. Since $T \propto \Phi I_a$ then Φ will be reduced in proportion to the armature current increase so that here $T_3 = T_2 = T_1$.

This value of armature current can be discounted as an answer.

14. Initial condition. Lighting load 100 kW $\cos \phi = 1$ $\sin \phi = 0$

$$\therefore \; \text{Apparent power} \;=\; \frac{kW}{\cos \phi} \;=\; \frac{100}{1} \;=\; 100 \text{ kVA}$$

Reactive power $= kVA \sin \phi$

$$\doteqdot \; 100 \times 0 \;=\; 0 \text{ kVAr}$$

Induction motor load. 300 kW $\cos \phi = 0 \cdot 85$

Thus $\phi = 31°47'$ $\sin \phi = 0 \cdot 527$

Apparent power rating $= \dfrac{300}{0 \cdot 85} = 353$ kVA

Reactive power rating $= -353 \times 0 \cdot 527 = -186$ kVAr

Total true power of load $= 100 + 300 = 400$ kW

Total reactive power $= 0 + (-186) = -186$ kVAr

Total overall power *i.e* with synchronous motor
$= 400 + 100 = 500$ kW

$$\cos \phi \;=\; 0 \cdot 95 \qquad \phi \;=\; 18°12' \; \sin \phi = 0 \cdot 3123$$

Apparent power of overall load $= \dfrac{500}{0\cdot95} = 526\cdot3$ kVA

Reactive power of overall load $= 526\cdot3 \times 0\cdot3123$
$= -164\cdot36$ kVAr

The reduction of reactive power to be achieved by the synchronous motor $= -186 - (-164\cdot4) = -21\cdot6 = $ that of the motor $= 21\cdot6$ kVAr

Apparent power rating of synchronous motor

$= \sqrt{100^2 + 21\cdot6^2} = 10\sqrt{10^2 + 2\cdot16^2} = 10\sqrt{100 + 4\cdot66}$

$= 10\sqrt{104\cdot66} = 100 \times \sqrt{1\cdot05} = 100 \times 1\cdot025$

$= 102\cdot5$ kVA

Power factor of synchronous motor $= \dfrac{kW}{kVA} = \dfrac{100}{102\cdot5}$

$= 0\cdot98$ (leading).

15. Current from $+$ ve to $-$ ve line through fault to earth

$= \dfrac{110}{10 + 5} = \dfrac{110}{15} = \dfrac{220}{30} = 7\cdot33$ A

The voltage drop between the $+$ ve line and earth $=$ voltage across one lamp
  This voltage drop $= 7\cdot33 \times 10 = 73\cdot33$ V
The voltage drop between the $-$ ve line and earth $=$ voltage across the other lamp
  This voltage drop $= 7\cdot33 \times 5 = 36\cdot65$ V
Additional load on generator $= 7\cdot33 \times 110 = 806\cdot3$ W
$= 0\cdot81$ kW

16. This is a series circuit and since the current and voltage are in phase then the condition is one of resonance. The respective voltage phasors for the values of voltage across the inductor and capacitor are equal in magnitude, but in phase opposition.

At resonance the current $I = \dfrac{V}{R} = \dfrac{100}{10} = 10$ A

Reactance of the inductor at 50 Hz $= X_L = 2\pi fL$
$$= 314 \times 0.15 = 47 \, \Omega$$

Voltage drop across the inductor $= IX_L = 10 \times 47 = 470$ V
Because of resonance the voltage drop across the capacitor
$IX_C = 470$ V

$\therefore$ Impedance of coil $Z = \sqrt{R^2 + X_L{}^2}$

$\therefore Z = \sqrt{10^2 + 47^2} = 10\sqrt{1^2 + 4.7^2} = 10\sqrt{1 + 22.09}$

$\qquad = 10\sqrt{23.09} = 10 \times 4.805 = 48.05 \, \Omega$

Voltage drop across the coil $= IZ$
$$= 10 \times 48.05 = 480.5 \text{ V}.$$

17.   Each core of the cable will have a resistance of $457 \times 4.38 \times 10^4$
$$= 4.57 \times 10^2 \times 4.38 \times 10^{-4} = 20 \times 10^{-2} = 0.2 \, \Omega$$

Let the resistance between the supply and the fault condition
be x ohms. Then the circuit will be as shown. Applying
Kirchhoff's laws

$250 = 120x + 98.7(0.2 - x) + 98.7(2.1) + 98.7(0.2)$

$\qquad = 120x + 19.74 - 98.7x + 207.27 + 19.74$

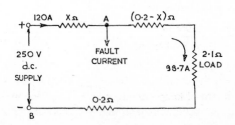

$$= 120x - 98 \cdot 7x + 246 \cdot 75$$

$$\text{or } 21 \cdot 3x = 250 - 246 \cdot 75 = 3 \cdot 25$$

$$x = \frac{3 \cdot 25}{21 \cdot 3} = 0 \cdot 1526 \ \Omega$$

The distance of the fault is therefore $\dfrac{0 \cdot 1526}{4 \cdot 38 \times 10^{-4}} = \dfrac{1526}{4 \cdot 38}$

$$= 348 \cdot 4 \ \text{m}$$

i.e. fault is 348·4 m from the supply point.

18. Impedance of circuit $Z = \dfrac{V}{I} = \dfrac{250}{2} = 125 \ \Omega$

Equivalent resistance of lamp from wattage rating is given by $I^2 R$

$$\therefore R = \frac{250}{2^2} = \frac{250}{4} = 62 \cdot 5 \ \Omega$$

The choke coil reactance X, since its resistance is negligible, is obtained from $X = \sqrt{Z^2 - R^2} = \sqrt{125^2 - 62 \cdot 5^2}$

Thus $X = 62 \cdot 5 \sqrt{2^2 - 1^2} = 62 \cdot 5 \sqrt{3}$

$$= 62 \cdot 5 \times 1 \cdot 732 = 108 \cdot 25 \ \Omega$$

Also $X = 2 \pi f L$

$$\therefore L = \frac{108 \cdot 25}{2 \times 3 \cdot 14 \times 50} = \frac{1 \cdot 0825}{3 \cdot 14} = 0 \cdot 344 \ \text{H}$$

Power factor of lamp circuit $= \dfrac{62 \cdot 5}{125} = \dfrac{1}{2} = 0 \cdot 5$ (lagging)

Here $\cos \phi = 0.5$ The phase angle would be $60°$

and $\sin \phi = \dfrac{\sqrt{3}}{2} = 0.866$

The reactive component of the current $= 2 \times 0.866 = 1.732$ A

Thus if the overall parallel circuit is to operate at unity power factor then the capacitor must take a leading reactive current sufficient to cancel out 1.732 A

$\therefore$ The capacitor current $I_C = 1.732$ A

and capacitor reactance $X_C = \dfrac{250}{1.732} = 144.2 \ \Omega$

$$\text{also} \quad X_C = \dfrac{10^6}{2\pi f C}$$

$$\therefore \ C = \dfrac{10^6}{2 \times 3.14 \times 50 \times 144.2} = \dfrac{10^2}{3.14 \times 1.442}$$

$$= \dfrac{100}{4.53} = 22 \ \mu\text{F}.$$

19. Initially at balance $\dfrac{Q}{P} = \dfrac{R}{S} = \dfrac{450}{S}$ ...   . (a)

With Q and P interchanged $\dfrac{P}{Q} = \dfrac{R_1}{S}$ where R_1 is the resistance of 450 Ω and 5 Ω resistors in parallel

Then $\dfrac{1}{R_1} = \dfrac{1}{450} + \dfrac{1}{5} = \dfrac{1 + 90}{450} = \dfrac{91}{450}$ whence $R_1 = \dfrac{450}{91}$

$$\therefore \quad \frac{P}{Q} \ = \ \frac{\dfrac{450}{91}}{S} \quad \dots \quad \dots \quad \dots \quad \dots \quad \dots \quad \dots \quad \dots \quad \text{(b)}$$

Multiplying (a) × (b) we have

$$\frac{Q}{P} \ \times \ \frac{P}{Q} \ = \ \frac{450}{S} \ \times \ \frac{\dfrac{450}{91}}{S} \quad \text{or} \quad 1 \ = \ \frac{\dfrac{450^2}{91}}{S^2}$$

$$\text{Thus} \ S^2 \ = \ \frac{450^2}{91} \quad \text{or} \quad S \ = \ \frac{450}{\sqrt{91}} \ = \ \frac{450}{9 \cdot 539} \ = \ 47 \cdot 3 \ \Omega$$

$$\text{From (b)} \ \frac{P}{Q} \ = \ \frac{450\sqrt{91}}{91 \times 450} \ = \ \frac{1}{\sqrt{91}} \ = \ \frac{1}{9 \cdot 539} \ = \ 0 \cdot 105$$

$$S \ = \ 47 \cdot 3 \ \Omega \qquad \text{Ratio P to Q} \ = \ 0 \cdot 105 : 1.$$

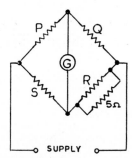

20.   Resultant of the three phasors of maximum value = 100 A
        Take horizontal and vertical components and obtain the
        resultant maximum value I_m

$$
\begin{aligned}
\text{Vertical components of} \ I_{1m} \ &= \ -0 \ \text{A} \\
\text{of} \ I_{2m} \ &= \ 100 \times 0 \cdot 5 \ = \ -50 \ \text{A} \\
\text{of} \ I_{3m} \ &= \ 100 \times 0 \cdot 866 \ = \ -86 \cdot 6 \ \text{A} \\
\text{Total} \ &= \ 0 - 50 - 86 \cdot 6 \\
&= \ -136 \cdot 6 \ \text{A}
\end{aligned}
$$

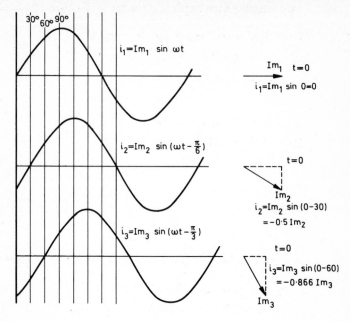

Horizontal component of I_{1m} = 100×1 = 100 A

of I_{2m} = $100 \times 0\cdot866$ = $86\cdot6$ A

of I_{3m} = $100 \times 0\cdot5$ = 50 A

Total = $100 + 86\cdot6 + 50 = 236\cdot6$ A

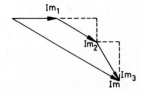

$$I_m = \sqrt{136\cdot6^2 + 236\cdot6^2} = 100\sqrt{1\cdot366^2 + 2\cdot366^2}$$

$$= 100\sqrt{1\cdot866 + 5\cdot598} = 100\sqrt{7\cdot464} = 100 \times 2\cdot732$$

$$= 273\cdot2 \text{ A}$$

R.M.S. value $= \dfrac{273\cdot2}{\sqrt{2}} = \dfrac{273\cdot2}{1\cdot414} = 193\cdot2$ A.

21.   By applying Kirchhoff's laws we get the following equations

Circuit from supply and through battery A

$$200 - 108 = 10(I_A + I_B) + 0.24 I_A$$
$$\text{or } 92 = 10.24 I_A + 10 I_B \quad \dddot{} \quad \dots \quad \dots \quad \dots \text{ (a)}$$

Circuit from supply and through battery B

$$200 - 105 = 10(I_A + I_B) + 0.26 I_B$$
$$95 = 10.26 I_B + 10 I_A \quad \dots \quad \dots \quad \dots \quad \dots \text{ (b)}$$

Solving

$$920 = 102.4 I_A + 100 I_B$$
$$\text{and } 95 \times 10.24 = (10 \times 10.24 I_A) + (10.26 \times 10.24 I_B)$$

giving $\quad 920 = 102.4 I_A + 100 I_B \quad \dots \quad \dots \quad$ (c)

and $\quad 972.8 = 102.4 I_A + 105.06 I_B \dots \quad \dots \quad$ (d)

Subtracting (c) from (d)

$$\therefore \qquad 52.8 = 5.06 I_B \text{ or } I_B = \frac{528}{50.6} = 10.43 \text{ A}$$

and $\quad 10 I_A = 95 - (10.26 \times 10.43) = 95 - 107.01$
$$= -12.01 \text{ or } I_A = -1.201 \text{ A}$$

Thus battery B is charged with 10.43 A; battery A discharges with 1.2 A and the supply current is $10.43 - 1.2 = 9.23$ A.

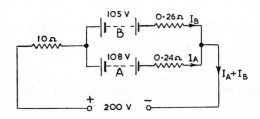

.22.   For series resonance the voltage drop due to the reactance of the inductor equals that across the capacitor of $IX_L = IX_C$ whence $X_L = X_C$

or $2\pi fL = \dfrac{10^6}{2\pi fC}$

At 100 Hz, $X_L = 6{\cdot}28 \times 100 \times 0{\cdot}1 = 62{\cdot}8\ \Omega$

$\therefore\ 62{\cdot}8 = \dfrac{10^6}{628{\cdot}4\ C}$ or $C = \dfrac{10^5}{62{\cdot}8^2} = \dfrac{10}{0{\cdot}394} = 25{\cdot}4\ \mu F$

At resonance, current is in phase with the supply voltage and is limited by the resistance only.

$\therefore\ I = \dfrac{220}{10} = 22\ A$

The impedance of the inductor

$= \sqrt{10^2 + 62{\cdot}8^2} = \sqrt{100 + 3944{\cdot}04} = 10\sqrt{1 + 39{\cdot}44}$

$= 10\sqrt{40{\cdot}44} = 10 \times 6{\cdot}359 = 63{\cdot}6\ \Omega$

$\therefore$ Voltage across inductor $= 22 \times 63{\cdot}6$

$\qquad\qquad\qquad\qquad\qquad = 1399{\cdot}2$ say 1400 V.

23.   Let r be the internal resistance of a cell. Then for the series condition:

Series current $I_S = \dfrac{20}{10r + R}$

Parallel current $I_P = \dfrac{2}{\dfrac{r}{10} + R} = \dfrac{20}{r + 10R}$

We are also given for the series condition $I_S^2 R = 500$ and for the parallel condition $I_P^2 R = 15$

$$\therefore \frac{I_P}{I_S} = \sqrt{\frac{15}{500}} = \frac{\dfrac{20}{r + 10R}}{\dfrac{20}{10r + R}} = \frac{10r + R}{r + 10R}$$

$$\therefore \sqrt{0 \cdot 03} = \frac{10r + R}{r + 10R} \quad \text{or} \quad 0 \cdot 1732 = \frac{10r + R}{r + 10R}$$

Whence $10r + R = 0 \cdot 1732r + 1 \cdot 732R$ or $9 \cdot 8268r = 0 \cdot 732\,R$

Thus $r = \dfrac{0 \cdot 732R}{9 \cdot 8268} = 0 \cdot 0749\,R$

$$I_S = \frac{20}{10r + R} = \frac{20}{0 \cdot 749R + R} = \frac{20}{1 \cdot 749R}$$

and power by $I_S{}^2R = 500 \quad \therefore\ 500 = \dfrac{400}{1 \cdot 749^2 R^2} \times R$

or $R = \dfrac{400}{1 \cdot 749^2 \times 500} = \dfrac{0 \cdot 8}{3 \cdot 06} = 0 \cdot 262\ \Omega$

and $r = \dfrac{0 \cdot 732R}{9 \cdot 827} = \dfrac{0 \cdot 732 \times 0 \cdot 262}{9 \cdot 827} = 0 \cdot 0196\ \Omega$

Also $I_P{}^2R = 15 \quad \therefore\ I_P{}^2 = \dfrac{15}{0 \cdot 262} = 57 \cdot 36$

or $\qquad I_P = 7 \cdot 57\ \text{A}$

Also $I_S{}^2R = 500 \quad \therefore\ I_S{}^2 = \dfrac{500}{0 \cdot 262} = 1908$

or $\qquad I_S = 43 \cdot 68\ \text{A.}$

24.   For the motor load

$\cos \phi = 0.766 \qquad \phi = 40° \quad \text{and} \quad \sin \phi = 0.6428$

$\therefore$ Power rating $= 70 \times 0.766 = 53.62$ kW

Reactive power $= 70 \times 0.6428 = -44.996$ kVAr

Total active power $= 12 + 53.62 = 65.62$ kW

The 72 kVA transformer is capable of supplying this power load and a reactive load, given by

$$\sqrt{72^2 - 65.62^2} = 10\sqrt{51.84 - 43.06} = 10\sqrt{8.78}$$
$$= -29.63 \text{ kVA}$$

The total reactive load must therefore be reduced by

$$-44.996 - (-29.63) = -15.366 = -15.4 \text{ kVAr}$$

This will then be the rating of the capacitors which will achieve the reduction by providing 15.4 kVAr at a zero (leading) power factor to oppose 15.4 kVAr of reactive load at zero (lagging) power factor.

25.   Since the generators have linear characteristics, a mathematical solution would be acceptable.

Since for a generator $\qquad V = E - I_a R_a$ then

then for machine X

$230 = 265 - 350 R_{aX}$

$\therefore R_{aX} = \dfrac{35}{350} = 0.1 \ \Omega$

For machine Y

$240 = 270 - 400 R_{AY}$

$\therefore R_{aY} = \dfrac{30}{400} = 0.075 \ \Omega$

Let I_x be the current given by machine X then $(500 - I_x)$ is the current given by machine Y

Thus $V = 265 - 0.1\,I_x$

and $V = 270 - 0.075\,(500 - I_x)$

Subtracting $0 = -5 - 0.1\,I_x + 0.075\,(500 - I_x)$

or $0 = -5 - 0.1\,I_x + 37.5 - 0.075\,I_x$

$\therefore 32.5 = 0.175\,I_x$ giving $175\,I_x = 32\,500$

or $35\,I_x = 6500$ and $I_x = \dfrac{1300}{7} = 185.7$ A

Thus machine X delivers 185·7 A and machine Y delivers 314·3 A

Also $V = 265 - 0.1 \times 185.7 = 265 - 18.57 = 246.43$ V

Machine X output $= \dfrac{246.43 \times 185.7}{1000} = 45.8$ kW

Machine Y output $= \dfrac{246.43 \times 314.3}{1000} = 77.4$ kW.

26.   Efficiency $= \dfrac{kVA \cos \phi}{kVA \cos \phi + \text{Copper Loss} + \text{Iron Loss}}$

Here Iron Loss or $P_{Fe} = 2.75$ kW

Also primary current $I_1' = \dfrac{175 \times 10^3}{6600} = \dfrac{175}{6.6} = 26.51$ A

Primary copper loss $= I_1^2 R_1 = \dfrac{26.51^2 \times 0.4}{1000} = 0.703 \times 0.4$

$= 0.2812$ kW

Similarly secondary current $I_2 = \dfrac{175 \times 10^3}{440} = \dfrac{175}{0.44}$

$$I_2 = 397.7 \text{ A}$$

$\therefore$ Secondary copper loss $= \dfrac{397.7^2 \times 0.0015}{1000} = 1.58 \times 0.15$

$$= 0.237 \text{ kW}$$

Total copper loss $P_{Cu} = 0.2812 + 0.237 = 0.5182 \text{ kW}$

$\therefore \eta \quad \dfrac{175 \times 0.9}{(175 \times 0.9) + P_{Cu} + P_{Fe}} = \dfrac{175 \times 0.9}{(175 \times 0.9) + 0.5182 + 2.75}$

$$= \dfrac{157.5}{157.5 + 0.5182 + 2.75} = \dfrac{157.5}{160.77} = 0.98$$

Thus Efficiency $= 98$ per cent.

27.　　The following graphs show the non-linear resistor character-istic and the voltage-drop line for R_2. This is obtained by assuming any current value and finding the corresponding volt-age drop.

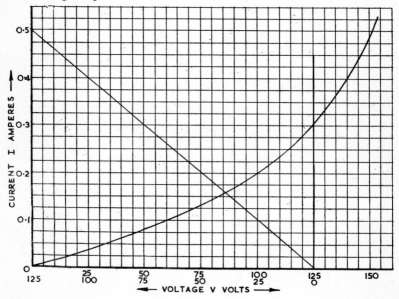

Thus with 125 V applied, the current would be

$$\frac{125}{250} = 0{\cdot}5 \text{ A}$$

If the characteristic of R_2 is plotted in reverse on the given curve for R_1, the point of intersection will give the circuit current and the voltage dropped across each component.

Note. That for R_2 when $V = 0$, $I = 0$ and when $V = 125$ volts, $I = 0{\cdot}5$ A *i.e.* two points on the characteristic.

The circuit current would be $0{\cdot}16$ A with 86 V dropped across R_1 and 34 V dropped across R_2.

28.   For the Load.

Active power $= 400 \text{ kW}$ $\cos \phi = 0{\cdot}8$ $\sin \phi = 0{\cdot}6$

Apparent power $= \dfrac{400}{0{\cdot}8} = 500 \text{ kVA}$

Reactive power $= 500 \times 0{\cdot}6 = -300 \text{ kVAr}$

For machine A, active power $= 150 \text{ kW}$ $\therefore$ active power supplied by machine B $= 400 - 150 = 250 \text{ kW}$

Similarly for machine B,
    Reactive power $= -150 \text{ kVAr}$

$\therefore$ Reactive power taken from machine A

$$= -300 - (-150) = -150 \text{ kVAr}$$

Thus apparent power loading of machine A

$$= \sqrt{150^2 + 150^2} = 100\sqrt{1{\cdot}5^2 + 1{\cdot}5^2}$$

$$= 100\sqrt{2{\cdot}25 + 2{\cdot}25} = 100\sqrt{4{\cdot}5} = 212{\cdot}1 \text{ kVA}$$

and loading of machine B

$$= \sqrt{250^2 + 150^2} = 100\sqrt{2{\cdot}5^2 + 1{\cdot}5^2}$$

$$= 100\sqrt{6{\cdot}25 + 2{\cdot}25} = 100\sqrt{8{\cdot}5} = 291{\cdot}5 \text{ kVA}$$

Power factor of machine A $= \dfrac{150}{212 \cdot 1} = 0 \cdot 707$ (lagging)

Power factor of machine B $= \dfrac{250}{291 \cdot 5} = 0 \cdot 857$ (lagging)

Current supplied by each machine (here three-phase alternators have been assumed)

For machine A $= \dfrac{212\,100}{\sqrt{3} \times 440} = \dfrac{212\,100}{762 \cdot 1} = 279$ A

For machine B $= \dfrac{291\,500}{762 \cdot 1} = 382 \cdot 5$ A.

29.   The problem is solved by plotting the valve characteristic and drawing the load line for the resistive circuit.

Thus consider the valve on open-circuit *i.e.* non-conducting. Then the current through the resistive circuit would be

$$\dfrac{240}{(10 + 50)\,10^3} = \dfrac{240}{60} \times 10^{-3} = 4 \text{ mA}$$

and the voltage drop across the anode resistor would be

$$4 \times 10^{-3} \times 10 \times 10^3 = 40 \text{ V}$$

$\therefore$ The p.d. between anode and cathode of the valve, for zero I_a would be $240 - 40 = 200$ V

This gives a value for point A on the load line.

Next consider the valve "short-circuiting" *i.e.* with no p.d. across it. The full 240 V would be applied across the anode resistor and the current would be $\dfrac{240}{10 \times 10^3} = 24$ mA.

This is for a zero V_a condition and gives point B on the load line. Join AB and the intersection point gives a value of 60 V applied to the value and an anode current of some 17 mA.

As a check we have

For 60 V across the valve, a current of $\dfrac{60}{50 \times 10^3} = 1\cdot2$ mA

through the 50 kΩ shunt circuit. With 60 V across the valve $240 - 60 = 180$ V would be dropped across the anode

resistor *i.e.* a current of $\dfrac{180}{10 \times 10^{-3}} = 18$ mA

∴ Valve current $= 18 - 1\cdot2 = 16\cdot8$ mA, as indicated by the graph.

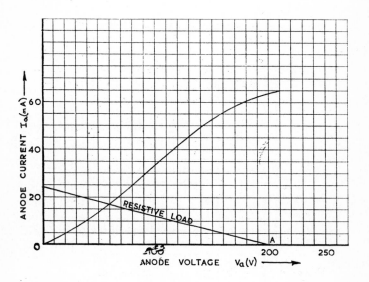

30.   A direct solution is effected by taking logarithms to the base e. Thus for $I_a = kV_a^n$

Then $\log_e I_a = \log_e k + n \log_e V_a$. This is of the form $y = mx + c$ *i.e.* for a straight line graph.

Then for the values of

I_a	3·36	11·4	27·54	47·86	67·6	(milliamperes)
$\log_e I_a$	1·24	2·43	3·31	3·86	4·214	
V_a	2	5	10	15	20	(volts)
$\log_e V_a$	0·693	1·61	2·303	2·71	3·0	

Then by plotting $\log_e I_a$ to a base of $\log_e V_a$ the slope and the value of k is obtained thus:

From the graph n is the slope $= \dfrac{2 - 0.35}{1.3 - 0} = \dfrac{1.65}{1.3} = 1.27$

and $\log_e k$ is the intercept $= 0.35$ or since $\log_e k = 0.35$

$$\therefore \quad k = 1.42$$

The equation becomes $I_a = 1.42 \, V_a^{1.27}$.

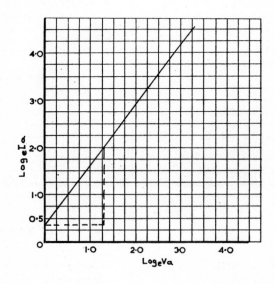

INDEX